Operations Research Proceedings 2001

Selected Papers
of the International Conference
on Operations Research (OR 2001)

Duisburg, September 3–5, 2001

Springer-Verlag Berlin Heidelberg GmbH

P. Chamoni · R. Leisten · A. Martin
J. Minnemann · H. Stadtler
Editors

Operations Research Proceedings 2001

Selected Papers
of the International Conference
on Operations Research (OR 2001)

Duisburg, September 3–5, 2001

With 88 Figures
and 38 Tables

Springer

Professor Dr. Peter Chamoni

Gerhard-Mercator-Universität Duisburg
Fakultät 3, Institut für Logistik und Informationsmanagement
Lehrstuhl für Wirtschaftsinformatik und Operations Research
Lotharstraße 63, 47057 Duisburg, Germany

Professor Dr. Rainer Leisten

Gerhard-Mercator-Universität Duisburg
Fakultät 3, Institut für Logistik und Informationsmanagement
Lehrstuhl für Produktionswirtschaft und Industriebetriebslehre
Lotharstraße 63, 47057 Duisburg, Germany

Professor Dr. Alexander Martin

Technische Universität Darmstadt
Fachbreich Mathematik · Fachgebiet Diskrete Optimierung
Schlossgartenstraße 7, 64289 Darmstadt, Germany

Dr. Joachim Minnemann

WestLB · Abteilung Mathematik/Operations Research
Herzogstraße 15, 40217 Düsseldorf, Germany

Professor Dr. Hartmut Stadtler

Technische Universität Darmstadt
Institut für Betriebswirtschaftslehre · Fachgebiet Fertigungs- und Materialwirtschaft
Hochschulstraße 1, 64289 Darmstadt, Germany

ISBN 978-3-540-43344-6 ISBN 978-3-642-50282-8 (eBook)
DOI 10.1007/978-3-642-50282-8

Library of Congress Cataloging-in-Publication Data applied for
Die Deutsche Bibliothek – CIP-Einheitsaufnahme
International Conference on Operations Research ‹2001, Duisburg›:
Selected Papers of the International Conference on Operations Research:
Duisburg, September 3–5, 2001; with 38 Tables / (OR 2001). P. Chamoni ... (ed.). –
Berlin; Heidelberg; New York; Barcelona; Hong Kong; London; Milan; Paris; Tokyo:
Springer, 2002
(Operations Research Proceedings; 2001)

http://www.springer.de

Originally published by Springer-Verlag Berlin Heidelberg New York in 2002.

Cover design: Erich Kirchner, Heidelberg

SPIN 10872140 42/2202-5 4 3 2 1 0 – Printed on acid-free paper

Preface

This volume contains the proceedings of the 2001 International Conference on Operations Research (OR 2001) held at the Gerhard-Mercator-University Duisburg, September 3-5, 2001.

OR 2001 was organized under the auspices of the German Society of Operations Research, Gesellschaft für Operations Research (GOR e. V.).

The conference and the annual general meeting were attended by 360 participants from 20 countries. The presentation of 220 papers was organized in 15 sections.

According to Duisburg as hosting city for this event OR 2001 emphasized on contributions of OR in the areas of energy, transport and traffic.

The program consisted of 2 plenary lectures (Reinhard Selten and Jörg Hennerkes) and 15 invited semiplenary lectures. 97 papers were submitted for publication. Following the advice of the section chairs the program committee decided to accept 59 papers for this volume.

The selected manuscripts will be published also in electronic form on the World Wide Web at

http://www.uni-duisburg.de/or2001.

We want to thank all referees and authors for delivering their final manuscript in due time. We are also grateful to the other members of the local organizing committee and especially to Stefan Krebs, Corinna Schu and David Betge for the perfect conference management. Roland Düsing, Ralph Gollmer and Steffen Stock supported us in editing the abstracts and the final version of this proceeding volume. Last but not least thanks to all the assistants and student assistants for their operations on OR 2001 in Duisburg.

Duisburg, December 2001 Peter Chamoni and Rainer Leisten (editors)

Section and Chairpersons

Section	Chairperson
Transport and Traffic	U. Zimmermann, Braunschweig A. Bachem, Köln
Energy and Environment	R. Schultz, Duisburg W. Pohl, Essen
Quantitative Methods in Marketing	R. Lasch, Dresden
Production, Logistics and Supply Chain Management	H. Kuhn, Eichstätt
Finance, Banking and Insurance	W. Kürsten, Jena
Scheduling and Project Planning	R. Kolisch, Darmstadt E. Pesch, Bonn
OR in the Health Service	S. Fleßa, Nürnberg
Continuous Optimization	F. Juhnke, Magdeburg F. Jarre, Düsseldorf
Discrete and Combinatorial Optimization	A. Martin, Darmstadt P. Brucker, Osnabrück
Stochastic Optimization and Simulation	W. Krug, Dresden K.-H. Waldmann, Karlsruhe
Econometrics, Statistics and Mathematical Economics	S. Huschens, Dresden W. Eichhorn, Karlsruhe
Decision Theory	W. Habenicht, Hohenheim
Artificial Intelligence and Fuzzy Logic	H. Rommelfanger, Frankfurt am Main B. Werners, Bochum
Information and Decision Support Systems. OR in Internet	S. Voß, Braunschweig W. Gaul, Karlsruhe
Didactics of OR, Public Understanding	L. Suhl, Paderborn C. Haehling von Lanzenauer, Berlin

List of Sponsors

BI-Con GmbH, Duisburg

cundus AG, Duisburg

DAF Trucks, Eindhoven (Netherlands)

DUALIS GmbH, Dresden

Etap-Hotel, Duisburg

Ferrotel Hotel, Duisburg

GAMS Software GmbH, Köln

Gerhard-Mercator-Universität Duisburg

Hotel Conti, Duisburg

Hotel Friederichs, Duisburg

Hotel IBIS, Duisburg

Hotel PLAZA, Duisburg

Hotel Steigenberger Duisburger Hof, Duisburg

IBM Deutschland GmbH, Heidelberg

ILOG Deutschland GmbH, Bad Homburg

Landschaftspark Duisburg-Nord GmbH, Duisburg

Maximal Software Inc.

Novotel Hotel, Duisburg

Springer-Verlag, Heidelberg

Stadt Duisburg, Büro der Oberbürgermeisterin, Duisburg

Stadt Duisburg, Museum der Deutschen Binnenschifffahrt, Duisburg

ThyssenKrupp Stahl AG, Duisburg

Verein der Freunde der Duisburger Wirtschaftsinformatik (VDWI) e. V., Duisburg

Table of Contents

Section 1

Transport and Traffic

Routing a Fleet of Vehicles for Dynamic Combined Pick-up and Deliveries Services *

Massimiliano Caramia[1], Giuseppe F. Italiano[1], Gianpaolo Oriolo[1], Andrea Pacifici[1], and Alessandro Perugia[2]

[1] DISP, Università di Roma "Tor Vergata",
via di Tor Vergata, 110, 00133 Roma, Italy
E-mail: caramia,italiano,oriolo,pacifici@disp.uniroma2.it
[2] Centro Ricerche ENEA Casaccia,
via Anguillarese, 301, 00060 S. Maria di Galeria, Roma
E-mail: perugia@casaccia.enea.it

Abstract. This work partially reports the results of a study aiming at the design and analysis of the performance of a multi-cab metropolitan transportation system. In our model we investigate a particular multi-vehicle many-to-many dynamic request dial-a-ride problem. We present a heuristic algorithm for this problem and some preliminary results. The algorithm is based on iteratively solving a single-vehicle subproblem at optimality: a pretty efficient dynamic programming routine has been devised for this purpose.

This work has been carried out by researchers from both University of Rome "Tor Vergata" and Italian Energy Research Center ENEA as a line of a reasearch program, regarding urban mobility optimization, funded by ENEA and the Italian Ministry of Environment.

1 Problem statement

The problem consists of dynamically routing each vehicle (from now on, *cab*) of a set C, for serving a set P of clients which demand for transportation in a urban territory. In our study, the road network is based on the actual map of the city of Rome, Italy. Nodes V in the network are called "centroids" and links represent physical links (roads) between them. Links' expected travel times are deterministic and known, and they may vary, depending on the particular day time. A minimum expected travel times d_{xy} between any two centroids $x, y \in V$ is thus computed, for any particular day time, with an off-line shortest path computation.

Clients' demands are basically of two types: booking request (service is required for "tomorrow") and immediate request services (one requires a ride in the next few hours). Demands pop up *dynamically*, that is, at any time

* This work is partially supported by the Italian Ministry of the Environment and ENEA Research Centre.

we do not have *any* information about future requests for services. Client j's demand is characterized by:

- a pick-up point u_j and a delivery point v_j from the set of centroids V;
- a time-window $[a_j, b_j]$ for pick-up;
- a stretch $\delta_j \geq 1$. With this value, the client specifies the maximum acceptable ratio, between actual and minimum expected travel time. As one cab may serve more than one client at the same time, actual travel time for client j may very likely exceed the minimum expected travel time between client's origin and destination points.

Each cab is characterized by:

- an origin (pick-up) point and a destination (delivery) point for the cab (from the set V);
- a time-window for duty;
- the capacity of the cab (in our experiments this is at most 6).

The problem consists of deciding:

- the demand allocation, i.e., the assignment of each client to a cab,
- the routing of each cab, i.e., a feasible ordering of pick-up and delivery points of the clients assigned to the cab

while minimizing a cost function related to the total distance travelled by the cabs.

One client's demand may be not accepted if it is not convenient. If this is not the case, we indicate with α_j (ω_j) actual pick-up and delivery times for client j induced by the routing decision.

2 Solution Algorithm

As our problem is dynamic, we devised an algorithm that tries to update the current solution any time a new event prompts (such as new demand, request cancellation, vehicles' breakdown,) or, if no new events arise, any time one would attempt to improve cost function. In both cases, this is done by local search techniques where the neighborhood search is based on the optimal solution of a sub-problem defined for a single cab.

In what follows we describe such a sub-problem and the corresponding solution algorithm. Given the current (global) solution, that is, feasible assignment for all the accepted clients and routing for the cabs, suppose a new demand pops up. Then we try to define a new feasible solution by assigning the new client to a specific cab h. The sub-problem consists in finding, if it exists, a new feasible routing for cab h including pick-up and delivery service for the new client. If a feasible routing is found, a new global solution (neighbor solution) simply modifies cab h's duty and leaves the others unchanged.

2.1 Single Vehicle Routing

Routine SCR (Single Cab Routing) is the basic step of the whole procedure. Basically, SCR tries to determine a new route for a specific cab given its current position and its duty, i.e., assigned clients and possible new clients. We indicate with $S_i(h,t)$, $i = 0,1,2$ respectively, the sets of clients j assigned to cab h that are on board ($\alpha_j \leq t < \omega_j$) if $i = 1$, still to be picked-up ($t < \alpha_j$) if $i = 2$, and already delivered ($t \geq \omega_j$) if $i = 0$. Recall that each client is characterized as above by its pick-up window $[a_j, b_j]$, and stretch δ_j.

The input to the problem is given by:

- current time t;
- a cab $h \in C$ along with its capacity, pick-up and delivery points, and (possibly estimated) position at time t;
- the sets $S_i(h,t)$, $i = 0,1,2$;

A feasible solution to this problem is a routing for cab h —that is, an *ordering* of the multiset whose elements are pick-up and delivery points of clients in $S_2(h,t)$, delivery points of clients in $S_1(h,t)$, the delivery point for cab j and, in case, the pick up point of cab j —satisfying the constraints on:

- stretch ($\omega_j - \alpha_j \leq \delta_j d_{u_j v_j}$);
- pick-up windows ($a_j \leq \alpha_j \leq b_j$) and;
- capacity of cab ($|S_1(h,t)| \leq 1, \forall h, t$).

In particular, an optimal solution to SCR is a feasible routing of minimum length (defined as the sum of expected travel times).

It is easy shown that the above problem is a generalization of the euclidean TSP (traveling salesperson problem) and therefore is $\mathcal{NP}$-hard. On the other hand, when the capacity of the cab is small, as it is in our case, an *exact* dynamic programming approach is suitable. A similar approach was originally proposed in [1,2]. Even better, in our experiments, this dynamic programming approach has been enhanced by a suitable use of algorithm A^*. (See for instance [3]). Hereafter we give a brief description of the basic dynamic programming approach used for solving the single cab sub-problem.

Dynamic programming approach As usual, this approach requires a convenient definition of "state" and the computation of a shortest path (if it exists) over a suitable "state graph".

In our system, SCR routine tries to allocate the demand of a specified set of clients on a given single cab. Recalling the definitions of $S_i(h,t)$, $i = 0,1,2$, a natural definition of "state" is a vector indicating the compositions of such sets. As routing is planned in a rolling horizon fashion, in our implementation we decided to keep fixed the maximum number of clients over which cab's routing is planned every time SCR is called. Let q indicate this number. The value of q is of course related to but may very likely exceed the capacity of the cab. We thus define "state" a vector $x \in \{0,1,2\}^q$. As an instance, consider a cab h which starts its duty. The first time SCR is called on this cab, the initial state is a vector $x = (2,2,\ldots,2)$ representing the fact that all the clients are still to picked up. Of course a state is feasible only if the number of clients on board does not exceed cab's capacity.

The state graph is built as follows. Vertices represent feasible states. A link from x_u to x_v exists if a transition between those two feasible states is possible and, in particular, if constraints on stretch, pick-up time window, and capacity are respected. The cost of such a link is the incremental cost (in terms of elapsed time or more complicated function of time and space) associated to this transition.

As an example consider a state $x_u = (1,2)$. This represents the fact that, say, client 1 is on board and another (client 2) is waiting to be picked-up. Possible transitions are therefore: (*i*) deliver client 1, or (*ii*) pick-up client 2. No other transitions are possible. Feasibility for such transitions has to be checked out. Possible links in the state graph are therefore (x_u, x_v) and (x_u, x_w) where $x_v = (0,2)$ and $x_w = (1,1)$. Associated costs are, respectively, the minimum expected time to perform delivery of client 1, and pick-up of client 2 from the current position.

A feasible (and optimal) routing is a (shortest) path from the initial to the final state. Note that simple Dijkstra's shortest path algorithm allows graph building and shortest path's computation to be executed simultaneously. Although the size of the graph increase exponentially with q (the number of vertices is $O(3^q)$), SCR is particularly efficient as q is a small number and, moreover, experiments showed that only a small fraction of states is generated before a shortest path is computed.

2.2 Main Routine

The main (dynamic) procedure MR is principally called upon the arrival of any new client's demand. Each time t a new client j pops up, the system must decide if his/her demand must be satisfied or rejected. To this purpose, for each currently available cab h, MR decide if it is possible to assign the new client j to cab h by calling SCR on a new instance where pick-up and delivery points of client j are added to the current $S_2(h,t)$ set for cab j. Eventually the most convenient cab, if any, is selected.

If no feasible routing can be produced the client is rejected. Even if feasibility is attained, MR rejects a demand if the ratio between the profit associated with its acceptance and its cost, in terms of increased travelled distance, does not exceed a given threshold.

Moreover, MR is called to on-line (re-)route cabs in case of exceptional events such as breakdown of cabs or cancelled requests.

Beside its use in a dynamic context, the main routine may also be used off-line, i.e., in a static fashion, to get improved routings for the cabs.

3 Computational Results and Future Research

Our test bed is based on the road network of the city of Rome, Italy. The network has about 300 vertices (centroids) and more than 1000 links (roads or stretches of roads).

In this preliminary phase of research, experiments have been organized as follows:

- The network scenario is fixed;
- Two types of objective functions are considered
  - average time for service, or
  - a linear combination of total distance travelled and total expected travel times which gives an estimate on the total cost of service in terms of fuel, vehicles consumption, drivers salary and clients satisfaction;

  without significant differences in the computational burden.
- Average expected distance traveled by clients is kept fixed and equal to 40 minutes in all the scenarios;
- Cabs capacity is fixed and equal to 4.

The algorithms were implemented in Ansi-C language and all the experiments have been performed on a Pentium III 350 MHz PC with 256 Mb RAM.

Table 1. Computational results for cabs' capacities equal to 4

Demand	Fleet	CPU time	Satisfied demands (%)	Cab utilization
	3	0.754	100	2.6
10	5	1.250	100	1.7
	10	1.250	100	0.8
	10	1.511	62%	1.9
30	15	2.401	90%	1.6
	25	3.538	100%	1.0
	15	2.611	58%	2.8
50	25	3.794	96%	2.0
	30	4.186	100%	1.1

Table 1 illustrates the results of more significant experiments. Here, 8-hours cab shifts are simulated. Links travel times are deterministic but they may vary: we identified six time bands per day in which the travel time of a link is keep fixed. The first two columns report the characteristics of different scenarios: in particular the demand in terms of number of clients per hour, and the fleet in terms of number of available cabs. CPU time (in seconds) measures the average time needed for computing a new solution when one new client demands for service. This is *the* critical factor as, in a real environment application, the response times to clients' demands are proportional to this time. The fourth column reports the fraction (percentage) of satisfied demands for each scenario. Finally, the last column reports cab's utilization in terms of average number of clients on board (with respect to the only cabs on duty.) Note that each table entry is an average value over 10 repetition of the same scenario. The objective function is, in this case, the average time for service. Satisfying results are obtained up to 20 cabs with demand frequency up to 50 clients per hour.

Table 2 illustrates results when the cabs' capacities are set equal to 6. As it is evident CPU time increases dramatically and performance deteriorates (for the latter scenario * indicates that computation was interrupted after about one minute.)

Table 2. Computational results for cabs' capacities equal to 6

Demand	Fleet	CPU time	Satisfied demands (%)	Cab utilization
	10	10.611	45%	1.9
50	25	26.839	87%	1.3
	30	*	*	*

Several different directions for future research may be foreseen, such as

- investigations of more sophisticated local search algorithms for the main routine and improved heuristics for the single cab sub-problem;
- experiments on more detailed (large sized) networks and with stochastic link travel times, and
- development of suitable graphical user interfaces allowing human decision makers to interact with the system.

References

1. Psaraftis, H. N., (1980) A Dynamic Programming Solution to the Single Vehicle Many-to-Many Immediate Request Dial-a-ride Problem. Transp. Sc. **14**, 130–154
2. Psaraftis, H. N., (1995) Dynamic Vehicle Routing: Status and Prospects. Ann. Op. Res. **61**, 143–164
3. Nicosia G., Oriolo G. (2000) Solving the Shortest Common Supersequence problem. Op. Res. Proc. 2000, eds. B. Fleischmann, R. Lasch, U. Derigs, W. Domschke, U. Rieder, Springer-Verlag, 2001

Simulating Delays for Realistic Time-Table Optimization

Ophelia Engelhardt-Funke, Michael Kolonko

Institut für Mathematik, TU-Clausthal, Erzstraße 1, 38678 Clausthal-Zellerfeld, engelhardt-funke@math.tu-clausthal.de

Abstract. We present a method for the macroscopic simulation of the accumulated delay along a sequence of tracks ("supertrack") in a railway network. In a naive approach perturbations in tracks or stations would be simulated separately to calculate their accumulation in the presence of slack times. Instead we analyze the delay process using results from queuing theory to approximate the distribution of the net delay on such a "supertrack". This allows to simulate on a reduced network. Simulating the performance of time-tables becomes faster and can be included in optimization schemes.

We compared scheduled and "real" transfer waiting times including delays during the optimization of time-tables. As was to be expected, there is a strong correlation between scheduled and "real" waiting time for time-tables with high waiting times. For time-tables with better scheduled waiting times that appear during the iterative optimization however, the variance of the "real" waiting time increases and time-tables with good scheduled waiting times behave increasingly poor with respect to "real" waiting time.

1 Transfer Optimization and Delays

When optimizing time-tables of railway networks it is a common goal to decrease the waiting times of the passengers changing lines (see [5]). If we denote a possible change at a station S from line L to line L' by (L, L', S) and let $w_F(L, L', S)$ be the passenger waiting time of this connection given a certain time-table F, then

$$W_{F,\text{sched.}} := \sum_{(L,L',S)} w_F(L, L', S) g(L, L', S)$$

can be an objective function measuring the (scheduled) synchronization of this time-table. Here, $g(L, L', S)$ is a weight indicating the importance of the connection (L, L', S) e.g. the number of passengers using it.

Considering delays, the real waiting time of passengers may differ significantly from the scheduled one due to missed connections even if we restrict ourselves to typical small perturbations as they are unavoidable in real operation. Especially time-tables with very small scheduled waiting time may actually have a poor performance in terms of real waiting time because these waiting times serve as buffers that allow passengers from slightly delayed feeder trains to still catch their connection. A more realistic approach to time-table optimization should therefore also try to include the real (mean) waiting time as an objective function.

2 Real Waiting Times

The actual waiting time of passengers that arrive with the i-th vehicle of line L at station S and want to change to line L', is a random variable $W_{i,\text{real}}(L, L', S)$ which depends on the arrival delay $A_i(L, S)$ of this vehicle at station S and the departure delay $B_j(L', S)$ if the receiver vehicle of line L' has number j. If we allow receiver vehicles to wait up to a fixed amount of time for delayed feeder vehicles, then arrival and departure delays also depend on delays of other lines at other stations because of delay propagation. Even the random variables $(A_i(L, S))_i$ representing delays of different vehicles of the same line at the same station are not necessarily independent or identically distributed if the network contains circular structures. This is why analytical expressions e.g. for the mean actual waiting time in the whole network are only available for very simple networks.

If analytical expressions are not available, it is still possible to simulate the performance of a time-table under typical small perturbations and to estimate the mean waiting time from the simulation results. If an objective function of an optimization problem can only be simulated, classical optimization methods that use the structure of the objective function cannot be used. In our multi-criteria optimization tool HiTT (see [3]) we use genetic algorithms combined with local search heuristics. Due to many evaluations of the objective function necessary during the optimization process it is important for the simulation to be fast.

A naive simulation has to model all possible sources of perturbation along all tracks and stations and then calculate the resulting delay at the end of that sequence taking into account the slack times given by the time-table. In deterministic optimization it is quite usual to aggregate sequences of successive tracks with no changing stations in between into one "supertrack" and to use this reduced network only.This is possible because the cost function "waiting time " only depends on changing stations. However, regarding delays the stations of minor importance in between cannot be neglected completely because e.g. passenger getting on and off trains may be a source of perturbation.

3 Reducing the Network Complexity

We present a method to approximate the distribution of the delay that comes from accumulated perturbations along such "supertracks". So the simulation can be done on the same reduced network that is used for deterministic transfer optimization and is fast because not every arrival and departure event in stations of minor importance has to be considered.

We use techniques from queuing theory and dam processes to model the accumulation and reduction of delay during a journey from some station S_0 to S_1 with possible other stations and sources of external perturbations in between. We are able to derive an approximation for the resulting delay distribution given the distributions of the external perturbations.

Neglecting breaking and acceleration processes, perturbations are modeled as sudden stops of a random amount of time at random places. We assume that the train driver reacts to delays in the following way: he goes faster whenever the vehicle is delayed and slows down to normal speed when it is in time again. Then only three different velocities are possible: speed a for vehicles that are in time, speed $b > a$ for delayed vehicles and speed 0 during the perturbation. Any delay can then be reduced while driving at speed b at rate of c time units per unit of track length, where

$$c := \frac{1}{a} - \frac{1}{b} \tag{1}$$

The i-th perturbation during the journey from S_0 to S_1 is modeled by a random variable (O_i, Z_i), meaning that the vehicle stops at place O_i for Z_i time units. This also includes the case that there is a scheduled stop at place O_i which lasts for Z_i time units longer than scheduled. Then O_i has a degenerated distribution.

We assume that $(X_i)_{i\in N} := (O_i - O_{i-1})_{i\in N}$ are independent and that $(Z_i)_{i\in N}$ form an iid sequence of exponentially distributed random variables with parameter ν. The two sequences are supposed to be independent from each other, so that the place where a perturbation occurs has no influence on its duration. It is also assumed that both sequences are independent from the initial departure delay, B, at the starting station of the journey. We are looking for a simple approximation of the delay A at the arrival station at the end of the journey, given the departure delay B and the distribution of $(O_i, Z_i)_i$ for the perturbations on the "supertrack" from S_0 to S_1.

Assume first $B = 0$ and let D_i be the accumulated delay just before the vehicle reaches place O_i. Then we have the following recursion for D_i:

$$D_1 = B := 0 \tag{2}$$

$$D_{i+1} := [D_i + Z_i - c(O_{i+1} - O_i)]^+, \tag{3}$$

$D_i + Z_i$ is the delay just after place O_i which is then reduced at rate c, given by (1) during the journey from O_i to O_{i+1}. Let

$$M_r := \max\{n | O_n \leq r\}$$

be the total number of perturbations up to place r. Then the actual delay at place r, given the departure delay was equal to zero, is

$$V_{r,B=0} = [D_{M_r} + Z_{M_r} - c(r - O_{M_r})]^+. \tag{4}$$

Formula (4) is the same as for the virtual waiting time at time cr in a single server queue (G/M/1-model) with arrival process $(cO_i)_{i\in N}$ and service times $(Z_i)_{i\in N}$ or the content of a reservoir in a dam process with input process (O_i, Z_i) and constant release rate c. (cp. [1]).

4 Approximating the Delay Distribution

From elementary queuing theory we obtain conditions under which the delay process $(V_r)_{r\geq 0}$ is stable or positive recurrent, i.e. the delay can always be reduced to 0 in finite (mean) time. Let

$$1/\lambda := \text{mean distance between successive } O_i,$$

then the stability condition is

$$\frac{\lambda}{c\nu} < 1.$$

In such a stable situation the distribution of the delay at place r, for large r, can be approximated by the steady state distribution of (4). Again from queuing theory we get

$$P(V_\infty \leq t) = 1 - \frac{\lambda}{c\nu}\exp(-\eta t). \tag{5}$$

In case of a general (non exponential) distribution for $(O_i - O_{i-1})$ the parameter η is only implicitly given. It is the unique positive solution to

$$1 = E[e^{\eta(Z_i - c(O_i - O_{i-1}))}]$$

and therefore depends on the distributions of $O_i - O_{i-1}$. However, $1/\eta$ is the conditional mean delay given the delay is positive and can therefore easily be estimated from statistical data.

In equation (5) we see that the additional delay a vehicle gets between the starting station and the next changing station at distance r, can be approximated by a random variable that has positive mass on zero and whose conditional distribution, given it is positive, is exponential. This type of distribution is often assumed in literature and has also been observed in real data (see [2]).

If the departure delay, B is nonzero, it can be reduced at rate c whenever there is no delay from external perturbations. In terms of queuing theory the amount of slack up to place r that is not needed to reduce delays from external perturbations, is the idle time of the server up to time cr. This can be approximated by the deterministic function $(1 - \frac{\lambda}{c\nu}) \cdot c \cdot r$ if cr is large. In terms of queuing theory this is the idle time of the server upto So the remainder of B at place r is approximately

$$[B - (c - \frac{\lambda}{\nu}) \cdot r]^+.$$

Hence we can approximate the arrival delay, A, at the end of a journey of length r by

$$A \simeq V_\infty + [B - (c - \frac{\lambda}{\nu}) \cdot r]^+. \tag{6}$$

For a simulation of delays at the end of a journey with external perturbations this means that we need only one call to a random generator for distribution (5) in (6) as long as the perturbation are independent and fulfill the homogeneity conditions given above.

5 Defining "Real" Waiting Times

Even if arrival and departure delays are given, there are several possibilities to define what is meant by "real" waiting time of a connection (L, L', S). If we simply take the time difference from the actual arrival to the actual departure time of a passengers, then the actual waiting time may be less than the scheduled one, e.g. if passengers arrive late but can still catch their receiver train which departs on time.

As this represents an unwanted reward of delays we defined our cost function "real waiting time" in the following way: if the i-th vehicle of line L arrives in station S and is $A_i(L, S)$ time units late, then the "real" waiting time $W_{i,\text{real}}(L, L', S)$ of those passengers that want to change to line L' at station S is

$$W_{i,\text{real},F}(L, L', S) := \begin{cases} w(L, L', S) & \text{if} \quad A_i(L, S) \le w_F(L, L', S) + \kappa(L, L', S) \\ \tau & \text{otherwise} \end{cases}. \tag{7}$$

Here $\kappa(L, L', S)$ is the maximum amount of time that receiver trains of line L' wait for delayed feeder trains of line L at station S before they depart. τ is the period of the receiver line. So $A_i(L, S) \le w(L, L', S) + \kappa(L, L', S)$ means that passengers arriving with line L can catch their connection to line L'. In the situation of $A_i(L, S) > w(L, L', S) + \kappa(L, L', S)$ the connection is missed and we give a penalty τ. The penalty τ represents the waiting time in the worst case of a missed connection where the arrival occurs just after the departure of the receiver vehicle and passengers have to wait for the full period to get on the next train of line L'.

This definition makes the "real" waiting time W_{real} comparable to the scheduled waiting time W_{sched}. If no connections are missed, then W_{real} and W_{sched} are equal and W_{real} has its minimum value. It also reflects the goal to obtain a time-table that is stable under typical perturbations apart from minimizing the time that individual passengers have to wait.

6 Simulation Results

In our experiments we used the method of repeated simulation. In a public transportation system there is a natural reset of the stochastic processes every morning when the first trains depart.

During one simulation run of a given time table F we calculate the mean waiting time during the day: Every time a vehicle i of line L arrives at station S we calculate the real waiting time of passengers using the connection (L, L', S), $W_{i,\text{real},F}(L, L', S)$, as given by (7). We update the mean waiting time

$$\bar{W}_{\text{real},F}(L, L', S) = \frac{1}{n} \sum_{i=1}^{n} W_{i,\text{real},F}(L, L', S)$$

for this connection. At the end of the simulation run we put

$$W_{\text{real},F} := \sum_{(L,L',S)} \bar{W}_{\text{real},F}(L, L', S) g(L, L', S)$$

to obtain a measure for the netwide performance of this time table F under perturbations. Let $W^j_{\text{real},F}$ denote the performance in the j-th simulation run. We dynamically decide how many simulation runs are necessary for a time-table F and stop simulation when the classical estimator

$$\bar{W}_{\text{real},F} := \frac{1}{n} \sum_{j} W^j_{\text{real},F}$$

becomes stable. This estimator becomes the value of our cost function "real waiting time" for the time table F.

Figure 1 shows how the real waiting time behaves when time tables are iteratively improved. Every dot represents a time table with scheduled and real waiting time. The dots at the upper right corner are randomly produced starting solutions. Here scheduled and real waiting time are almost the same. As time tables are getting better, first deterministic and scheduled waiting time are strongly correlated and both decrease. But after some optimization effort the variation in real waiting is getting larger. The time tables at the left hand side have gone through a longer improvement process with genetic algorithms and local search heuristics. Here the real waiting time is about three times larger than the scheduled one and does not decrease anymore, for some time-tables it is even getting worse.

7 Conclusion

We present a new approach to evaluate the waiting time of time tables. We show how the accumulation of delays along sequences of tracks can be approximated analytically using models of queueing theory. This allows a fast 'one-step' macroscopic simulation in reduced networks as they are typically used in time table optimization.

Our first simulation results show, that simply optimizing the scheduled waiting time may result in a time table that is very unstable and has high waiting times in practice. It is therefore necessary to include this aspect

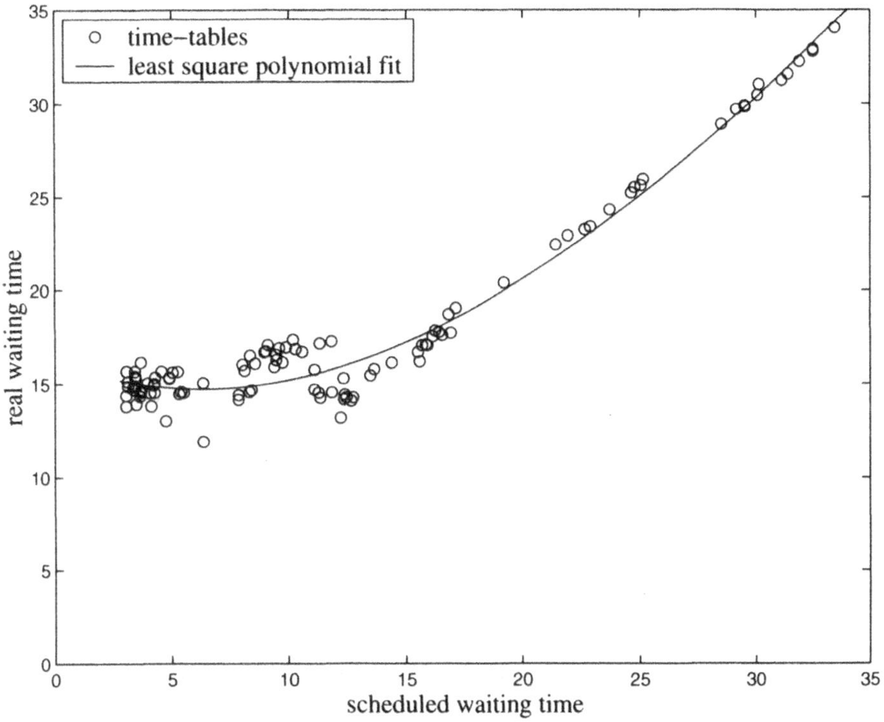

Fig. 1. simulated and deterministic waiting time (in 10 000 minutes · passengers) for several iteratively improved time-tables

into the optimization. The analytical results presented in this paper make it possible to use the real simulated waiting times as just another cost function into our multicriterial optimization system HiTT, see [3].

References

1. Asmussen, S. (1987) Applied Probability and Queues, Wiley
2. Herrmann, U. (1996) Untersuchungen zur Verspätungsentwicklung von Fernreisezügen auf der Datengrundlage der RZÜ Frankfurt am Main. Dissertation, TH-Darmstadt.
3. Kolonko, M., Engelhardt-Funke, O. (2001) Cost-benefit-analysis of Investments into Railway Networks with Periodically Timed Schedules. Lecture Notes in Economics and Mathematical Systems **505** Springer, Berlin Heidelberg
4. Muehlhans, E.(1990) Berechnung der Verspätungsentwicklung bei Zugfahrten. Eisenbahntechnische Rundschau **39**, 465-468
5. Nachtigall, K. (1998) Periodic Network Optimization and Fixed Interval Timetables. Habilitationsschrift, Universität Hildesheim

Was kann Operations Research für die Verkehrstelematik leisten?

Hans Hubschneider

PTV Planung Transport Verkehr AG, 76131 Karlsruhe

1. Was wollen wir unter „Verkehrstelematik" verstehen?

Der Begriff der Verkehrstelematik wird in unterschiedlichem Kontext mit deutlich verschiedenen Bedeutungen unterlegt. Im vorliegenden Beitrag soll unter Verkehrstelematik nicht nur ein Navigationssystem im Auto verstanden werden, und auch nicht nur die Übermittlung von Verkehrsinformationen über Mobilfunk. Das Spektrum für den Sammelbegriff Verkehrstelematik soll hier vielmehr eine ganze Prozesskette umfassen, in der verschiedene Teilbereiche des Verkehrsverhaltens modelliert und zunehmend integriert und durchgängig bearbeitet werden:

Ausgehend von der Verkehrserzeugung (wer will wann wohin) mit einer kollektiven Sicht auf den Verkehrsteilnehmer stellt sich in der Verkehrsverteilung die Frage „mit welchem Verkehrsmittel, wann und auf welchem Weg". Dynamik gewinnt der Prozess dann mit dem Verkehrsmanagement („wie kann der Verkehr optimal gelenkt werden?") aus der Sicht eines „Betreibers" der Verkehrsinfrastruktur und mit den Fragen „wie, wann und auf welchem Weg" aus der Sicht des einzelnen Verkehrsteilnehmers. Damit schließt sich der Kreis.

2. Die Grundlage: Verkehrsnetze

Verkehrsnetze sind einer der beiden notwendigen Teile von Verkehr: sie repräsentieren das Angebot an die Verkehrsteilnehmer. Auf Berechnungen in Verkehrsnetzen basiert auch die Mehrzahl der Anwendungen von OR in der Verkehrstelematik.

Abgebildet werden Verkehrsnetze als Knoten und Kanten – „ganz normal" also aus algorithmischer Sicht. Aber im Verkehr finden wir mehrere Fahrspuren, Ampeln, Gegenverkehr und Beschilderung. Deswegen sind Knoten räumlich fixiert, Kanten tragen Spuranzahl, Weglängen, Zeiten und Kosten. Zusätzliche Randbedingungen sind Abbiegeverbote und zeitliche Einschränkungen der Befahrbarkeit von Strecken. Spezielle Eigenschaften in Verkehrsnetzen sind die Eignung/Zulässigkeit für bestimmte Verkehrsmittel oder Verkehrsmodi (zu Fuß, mit Fahrrad, Pkw oder Lkw, oder für den Öffentlichen Verkehr). Letzterer prägt dem Verkehrsnetz bestimmte Linien als Folge mehrerer Kanten auf, sowie einzelne Fahrten als die zeitlich definierte Verfolgung einer Linie.

In Abhängigkeit von Art und Größe der Anwendungen haben wir es mit sehr unterschiedlichen Netzgrößen zu tun. Typische Netze einer Verkehrsplanungsumgebung umfassen rund 5000 Kanten und 2.000 Knoten; das aktuelle Modell für die Bundesverkehrswegeplanung enthält 630.000 Kanten; die Routensuche in einem detaillierten europäischen Straßennetz muss sich mit ca. 30 Mio. Kanten auseinandersetzen. Netzanalysen für den Öffentlichen Verkehr einer mittleren Großstadt basieren auf Netzen mit rund 1000 Haltestellen und 5.000 Linienfahrten. Das Netz der DB AG umfasst 8.000 Haltestellen und 40.000 Linienfahrten.

Die wesentlichste Operation in Verkehrsnetzen ist die Kurzwegsuche von zwischen zwei Punkten, von einem zu allen Punkten und die Suche nach Alternativrouten. Die besonderen Anforderungen kommen insbesondere aus der Vielzahl von zu berechnenden Routen – in vielen Anwendungen werden in kurzer Zeit Millionen von Routen gerechnet.

Häufig werden in den Verkehrstelematikanwendungen Dijkstra-Verfahren in unterschiedlichen Varianten eingesetzt. Großer Vorteil dieser Verfahren ist die gute Steuerbarkeit, gleichzeitige Berechnung von 1:n oder n:1 sowie die Verfügbarkeit effizienter Implementierungen.

Spezielle Verfahren finden insbesondere Anwendung, wenn 1:1-Routenberechnung in sehr performanten Umgebungen (z. B. Internet-Router) benötigt werden. Eine häufig genutzte Möglichkeit ist hier die hierarchische Strukturierung der Verkehrsnetze., die z. B. nach Fernstraßennetz, Hauptstrassennetz und Erschließungsnetz gegliedert werden. Für das Routing in solchen Multi-Level-Netzen wird jeweils innerhalb einer Netzmasche bis zum nächsthöheren Netzlevel gesucht, dann nur noch dieses Level betrachtet – und spiegelbildlich vom Zielpunkt rückwärts gesucht, bis sich die Wegebäume treffen. Diese Verfahren haben jedoch den grundsätzlichen Nachteil, dass sie keine Garantie für das Finden des wirklich optimalen Wegs ermöglichen – und außerdem sehr empfindlich gegenüber der Qualität des Levelling sind.

Alternativ zu dieser Vorgehensweise wurden sehr weitreichende Preprocessing-Verfahren entwickelt, die auf Basis statischer Bewertungen der einzelnen Kanten bestimmte Strukturen innerhalb der Graphen vorberechnen und auf dieser Grundlage extreme hohe Performance für das Routing ermöglichen. Damit werden Verbesserungen der Rechengeschwindigkeit um Größenordnungen erreicht.

Die Routensuche in Fahrplangraphen des Öffentlichen Verkehrs benutzt grundsätzlich ähnliche Algorithmen wie die Suche im normalen Straßennetz. Für spezielle Aufgabenstellungen des Öffentlichen Verkehrs, wie z.B. die Umlaufbildung für Fahrzeuge oder die Fahrerdienstplanung sind im Netzmodell zusätzliche Daten über die Lage der Betriebshöfe und mögliche Ablösepunkte für einen Fahrerwechsel zu definieren. Für die Umlaufbildung und die Dienstplanung gibt es Optimierungsverfahren. Aktuelle Forschungen betrachten hier die Umlaufplanung für mehrere Depots und eine integrierte Planung von Umlaufen und Diensten.

3. Wer will wann wohin?

Der zweite notwendige Bestandteil des Verkehrs sind die Verkehrsteilnehmer. Sie repräsentieren die Nachfrage, der das Verkehrsnetz als Angebot gegenüber steht.

Die Verkehrsnachfrage modelliert Personen, die zu einer bestimmten Zeit einen Reisewunsch von einem Startpunkt zu einem Zielpunkt haben. In vielen Fällen ist a priori die Verkehrsmittelwahl frei, manchmal auch die Abfahrts- oder Ankunftszeit. Nachgebildet wird Nachfrage in der Regel mittels Verkehrsbeziehungsmatrizen, die die Fahrtenanzahl pro Zeiteinheit zwischen Start- und Zielgebieten enthalten.

Für solche Matrizen gibt es zwei mögliche Quellen. In Haushaltsbefragungen werden eine Stichprobe von Verkehrsteilnehmern direkt zu ihren Mobilitätswünschen interviewt und die Ergebnisse hochgerechnet. Dabei wird die empirische Korrektheit mit einem hohen Aufwand für die personalintensive Durchführung bezahlt, häufig bevorzugt man deshalb stattdessen mathematische Modelle der Verkehrsnachfrageerzeugung. Diese gehen von meßbaren demographischen und strukturellen Parametern (Anzahl Einwohner, Arbeitsplätze, Einkaufsgelegenheiten) pro Gebiet und mittleren Fahrtenhäufigkeiten aus und prognostizieren zunächst die Menge von Fahrten für verschiedene Fahrtzwecke (z.B. Wohnen – Arbeit, Arbeit – Einkauf, Einkauf – Wohnen o.ä.). Die Verteilung auf die tatsächlich nachgefragten Verkehrsbeziehungen ergibt sich aus der Attraktivität der Gebiete für einen bestimmten Fahrtzweck und der Qualität der Verkehrsverbindung zwischen ihnen. Wegen der formalen mathematischen Ähnlichkeit nennt man solche Verteilungsmodelle auch Gravitationsmodelle.

Der Aufwand für die Datenversorgung und Kalibrierung eines Verkehrserzeugungsmodell lässt sich teilweise einsparen, wenn es aus früheren Untersuchungen bereits Nachfragematrizen gibt. Dann reichen für eine grobe Fortschreibung meist Zählungen von Fahrgästen an einigen wichtigen Haltestellen oder ausgewählten Strecken. Zur Fortschreibung vergleicht man die gezählten Fahrgastströme mit denen, die sich aus einer Umlegung (s.u.) der veralteten Matrix ergeben und justiert die Fahrtenmatrix so, dass die Abweichungen minimal werden. Dieses Problem ist in der Praxis stark unterbestimmt, weil nur eine kleine Menge von Zählwerten (z.B. bei Ein-/ Aussteigerzählungen die Randsummen der Matrix) zur Verfügung steht. In Ermangelung weiterer Information greift man zum Prinzip der Entropiemaximierung und bevorzugt unter den vielen mit den Messdaten kompatiblen Fahrtenmatrizen diejenigen, deren Fahrtbeziehungen in einem ähnlichen Zahlenverhältnis stehen wie bei der Originalmatrix. Die Minimierung der (nichtlinearen) Entropiefunktion unter linearen Nebenbedingungen für die Zählwerte erfolgt mit dem Standard-OR-Verfahren (z. B. Lagrange-Multiplier).

Die Attraktivität von derartigen Matrixschätzverfahren rührt aus der damit verbundenen Kostenersparnis bei der Datenbeschaffung. Deswegen ist auch die Auswahl der richtigen Zählstellen Gegenstand der Optimierung. Auf welche Haltestellen und Strecken sollte man sich konzentrieren, um den größten Informationsgewinn über die fortzuschreibende Fahrtenmatrix versprechen? Bei begrenztem Budget ähnelt die Frage einem Knapsack-Problem, jedoch mit einer subadditiven Wertfunktion, denn nahe beieinanderliegende Zählstellen sind zum Teil redundant, so dass sich ihre Informationsgewinne nicht einfach akkumulieren. Weil diese Eigenschaft bei exakten Knapsack-Algorithmen zu einer kombinatorischen Explosion führt, werden heute meist Abwandlungen der gängigen Heuristiken eingesetzt – hier herrscht mit Sicherheit noch Forschungsbedarf.

4. Wer fährt wo?

Wenn Tausende von Menschen morgens zur Arbeit fahren, wählen sie zuerst das Verkehrsmittel, dann die Startzeit und schließlich die Route oder den richtigen Umsteigepunkt. Modelle in der Verkehrsplanung zur „Umlegung" des Verkehrs bilden diese Entscheidungen ab. Für eine Modellierung mit mathematischen Verfahren müssen hierbei Zielfunktionen definiert werden. Man unterscheidet das **individuelle Optimum**, das sich einstellt, wenn jeder Verkehrsteilnehmer für sich selbst die optimale Entscheidung trifft – in der Regel, möglichst schnell am Ziel zu sein (Wardrop'sches Benutzergleichgewicht (1952): Reisezeiten auf unterschiedlichen benutzten Wegen zum gleichen Ziel sind gleich lang) und das **kollektive Optimum** – dieses entspricht der einfachen Zielfunktion, dass die Summe aller Reisezeiten minimal ist.

Für beide Zustände kann man einen mathematisch eindeutig beschreibbaren Gleichgewichtszustand berechnen. Dafür gibt es unterschiedliche heuristische Verfahren, die aufgrund der Problemgröße meist auf iterativen Ansätzen beruhen. Ausgehend von einer Startlösung, bei der alle Verkehrsteilnehmer die kürzeste Route wählen, wird die Nachfrage in weiteren Iterationsschritten auf alternative Routen verteilt.

Ergebnis einer solchen Umlegung ist die Information, wie viele Verkehrsteilnehmer welche Wege im Verkehrsnetz benutzen, wie einzelne Strecken belastet sind und welche Reisezeiten für die einzelnen Relationen gelten. Für die Nachbildung der Realität ist nur das Modell nach dem „individuellen Optimum" relevant; das kollektive Optimum ist zur Zeit nur eine Wunschvorstellung der Verkehrsplaner.

Für einfache Untersuchungen in der Verkehrsplanung genügen statische Modelle. Es werden z. B. vier unterschiedliche Nachfragematrizen für die Zeitbereiche Morgenspitze, Hauptverkehrszeit, Nachmittagsspitze sowie Nacht benutzt. Die entsprechenden Gleichgewichtszustände repräsentieren dann die jeweiligen Zeitbereiche; Tageswerte werden durch entsprechende Gewichtung hochgerechnet.

Eine spezielle, hochaktuelle Anwendungen der Verkehrsumlegung ist die Berücksichtigung von Mautgebühren für einzelne Strecken, Relationen oder Tageszeiten.

Die Nachbildung des individuellen Verhaltens unter diesen Randbedingungen ist wesentlich komplexer, weil die Optimierungszielgrößen der einzelnen Verkehrsteilnehmer unter den beiden gegenläufigen Teilzielen Zeit und Kosten relativ stark streuen. Daraus leiten sich insbesondere wesentlich andere Anforderungen an die Routensuche ab: es müssen effizient Bündel von Routen zwischen zwei Punkten gefunden werden, die für die Routenwahl in Frage kommen.

Eine andere Aufgabenstellung für die Umlegung ist die Fahrtenzuordnung für den Öffentlichen Verkehr, wo Fahrpläne, Zugfolgezeiten und Umsteigebedingungen eine entscheidende Rolle spielen. Anders als im Individualverkehr spielt hier allerdings die Abhängigkeit der Fahrzeit/Geschwindigkeit von der Verkehrsstärke keine Rolle: der Fahrplan ist vorgegeben. Dagegen kann die Abfahrtszeit hier nicht frei gewählt werden; der Fahrgast wählt zusammen mit der Abfahrtszeit in der Regel bereits eine bestimmte Verbindung. Gängige Anwendungen sind optimiert auf eine 1:1-Suche. Hier werden „normale" Kurzwegalgorithmen eingesetzt. Spannend aus der OR-Sicht sind jedoch die Suche von Alternativen mit unterschiedlichen Zielfunktionen - Zeit, Kosten und Bequemlichkeit können ähnlich divergent sein wie entsprechende Berechnungen bei der Straßenmaut. Sehr hohe Anforderungen werden gestellt, wenn für Umlegungsverfahren eine Vielzahl von Routen berechnet und möglichst alle relevanten Alternativrouten gefunden werden müssen (ansonsten ist ein Zug voll, und der Nachbarzug leer). Neuere Entwicklungen basieren hier auf Branch&Bound-Verfahren, die neben erheblicher Reduktion der Rechenzeiten (ca. Faktor 20 gegenüber normalen Kurzwegverfahren) eine erheblich bessere Ergebnisqualität unter komplexen Zielfunktionen bieten.

5. Von der Verkehrsplanung zum Verkehrsmanagement

Für die Verkehrssteuerung und das Verkehrsmanagement ist eine wesentlich detailliertere Modellierung der räumlichen und der zeitlichen Komponenten von Ortsveränderungen notwendig. Dynamik ergibt sich aus den kontinuierlichen Schwankungen der Verkehrsnachfrage, aber auch aus zeitlich variablem Angebot.

Statische Umlegungsverfahren betrachten Verkehr ähnlich wie eine kompressible Flüssigkeit, die sich im Netz bewegt. Es werden keine einzelnen Verkehrsteilnehmer unterschieden, die Bewegungen orientieren sich an pauschalen Zusammenhängen zwischen Verkehrsstärke (Fahrzeuge pro Stunde), Verkehrsdichte (Fahrzeuge pro km) und der Geschwindigkeit. Diesen „makroskopischen" Modellen stehen „mikroskopische" Modelle gegenüber, die mit unterschiedlicher Genauigkeit das Verhalten von einzelnen Verkehrsteilnehmern abbilden. Im Vordergrund stehen hier dynamische Effekte, also der Verkehr im zeitlichen Ablauf.

Die Wahl der Routen steht bei Verkehrsflussmodellen in der Regel nicht im Vordergrund, sondern wird als Teil der Modelleingabe betrachtet. Man hat hier die Möglichkeit, eine volle ausgerüstete Telematikumgebung zu simulieren: jedem Verkehrsteilnehmer kann zu jedem Zeitpunkt volle Information über die Zustände im Netz gegeben werden: jeder kann sein individuelles Optimum anstreben und seine Routenwahl entsprechend anpassen.

Leider kann damit aber höchstens die Reaktion von Verkehrsteilnehmern nachgebildet werden, die spontan auf eine gerade eintretende Störung reagieren. Ziel der dynamischen Umlegung ist jedoch die Ermittlung eines Gleichgewichtszustands, wie er sich im Falle einer regelmäßig wiederkehrenden Situation ergibt, bei der die einzelnen Verkehrsteilnehmer sukzessive aus ihren Erfahrungen lernen.

Die einzige bekannte Lösung für diese Aufgabenstellung ist derzeit die Iteration von individuellen Entscheidungsdurchläufen und die Verwendung von „gelernten“ Erfahrungen für den jeweils nächsten Iterationsschritt. Glücklicherweise konvergieren die verkehrlichen Aufgabenstellungen relativ gut; man kann mit 20 – 30 Durchläufen Stabilität erreichen. Aber in den meisten Fällen ist Konvergenz nicht beweisbar, und die Eindeutigkeit der Lösungen noch viel weniger. Die geschlossene Lösung von dynamischen Umlegungsaufgaben lässt noch auf sich warten.

Solche Simulationsmodelle mit Rückkopplung zur Realität (unter Nutzung der Informations- und Kommunikationskomponenten der Verkehrstelematik) ermöglichen heute auch den Einsatz der Simulation zur Verkehrsvorhersage. Hier wird im ersten Schritt Umfeldwissen (z. B. die Verkehrsnachfrage) kombiniert mit dem aktuell gemessenen Verkehrszustand; mit matrixorientierten Anpassungsverfahren (Maximierung der Entropie einer Abweichungsmatrix) kann man dann eine möglichst gute Ausgangssituation berechnen. Von dieser Startsituation rechnet die mikroskopische Simulation 30 Minuten in die Zukunft – und kann damit die Verkehrsentwicklung vorhersagen. Zum ersten Mal im Echteinsatz waren solche Verfahren im Verkehrsmanagement für die EXPO 2000 auf Basis einer Entwicklung der move GmbH, Hannover und der PTV AG, Karlsruhe.

6. Kann man Verkehr steuern?

Mit der Verkehrsvorhersage kann man erkennen, welche Situation sich unter Fortschreibung der aktuellen Bedingungen entwickeln würde. Optimal wäre, wenn man diese Entwicklung noch optimal gestalten könnte. Dies ist derzeit noch etwas Zukunftsmusik, aber die Entwicklungen laufen in großen Schritten in diese Richtung.

Kleinräumig finden wir Verkehrssteuerung an jeder lichtsignalgesteuerten Kreuzung. Hier wird ein konkrete Vorgabe für den Fahrer gemacht: Halt bei Rot, Fahrt bei Grün. Die Mehrheit dieser Anlagen wird über einfache zeitgesteuerte Schaltprogramme gesteuert, manche Anlagen können gezielt auf Anforderungen reagieren (Bevorzugung der Strassenbahn, Grünanforderung für Linksabbieger). Richtig anspruchsvoll werden die Aufgaben aber erst, wenn sehr komplexe Knotenpunkte oder eine Menge benachbarter Knoten gesteuert werden sollen. Zielfunktionen sind hier die Minimierung von Umlaufzeiten, des Auslastungsgrads eines Knotenpunkts oder eines Netzabschnitts und die Minimierung der Gesamtwartezeiten. Hier werden sehr komplexe Modelle für Prognose und Adaption an aktuelle Messwerte benutzt; die dahinter stehenden Verfahren reichen von linearer Optimierung über dynamische Optimierung bis zur Fuzzy-Regelung.

Das großräumige Verkehrsmanagement für Fernstrassen hat eine etwas andere Charakteristik, weil hier nicht einfach die Ampel auf „Rot" geschaltet werden kann. Der Verkehr muss hier im wesentlichen über Empfehlungen an das Kollektiv der Fahrer gesteuert werden (Wechselwegweisung – insbesondere als Angabe einer Umleitungsstrecke oder durch veränderte Zielangaben; alternativ Empfehlungen im Verkehrsfunk zur Benutzung von Umleitungsstrecken). Solche Steuerungsverfahren nutzen in vielen Fällen lineare oder nichtlineare Regler, überlagert mit Fuzzy-Verfahren, oder bilden Entscheidungen für mehrere Streckenabschnitte über dynamische Optimierung und Branch&Bound-Verfahren ab.

Trotz der Verfügbarkeit erster Optimierungsverfahren für die großräumige Verkehrssteuerung wird derzeit mehrheitlich noch die Erfahrung der Fachleute in Autobahnmeistereien und der Polizei genutzt. Die Optimierungsverfahren liefern primär Schaltvorschläge, und die menschliche Intelligenz plausibilisiert die OR. Insbesondere gibt auch der größte Teil der Verkehrsinformationen, die über Rundfunkanstalten verbreitet werden, nur aktuelle Ereignisse wieder – Unfall auf der Autobahn, Stau durch Überlastung -, aber sie werden nicht gezielt im Sinne des Verkehrsmanagements eingesetzt. Es gibt hier noch viel Spielraum für hochwertige Optimierungsverfahren.

7. Heute optimiert jeder Fahrer für sich selbst

In der Verkehrstelematik aus der Sicht eines Fahrers spielt OR noch keine große Rolle. Einzig im Navigationssystem finden wir Kurzwegalgorithmen – in der Regel sogar auf Basis heuristischer Verfahren. Die wesentliche Anforderung ist hier, in einer von Speicherplatz und Rechenkapazität sehr eingeschränkten Umgebung mit einem großen Straßennetz umzugehen (das in der Regel auf einer CD liegt).

Eine geschickte Mischung von Preprocessing und hierarchischer Aufteilung des Netzes ermöglicht es, in für den Benutzer akzeptablen Antwortzeiten eine Route zu berechnen. Die Qualität dieser Routen erfüllt oft nicht die Wunschvorstellungen – insbesondere aus Sicht eines OR-Fachmanns. Aber trotzdem ist die Routenwahl meist besser, als wenn man mit Karte und Kompass seinen Weg sucht.

Verkehrsinformationen werden relativ einfach berücksichtigt, indem die Widerstände einzelner Netzkanten modifiziert werden – damit ergeben in der Berechnung mögliche Alternativrouten, die dem Fahrer zur Auswahl angeboten werden.

Die lokalen, autonomen Routenberechnung in Verbindung mit Verkehrsinformationen hat in der Perspektive einen großen Nachteil: wenn ein Stau von 5% der Fahrer umfahren wird, weil das dynamische Navigationssystem vorbei leitet, hat der Nutzer einen Vorteil. Wenn aber 30% der Fahrer von ihrem System umgeleitet werden, wäre es vielleicht besser, man hätte kein Navigationssystem. Weil dann vielleicht die normale Route genügend entlastet wird, und der Stau sich auf die Umfahrung verlagert.

Die zunehmende Verfügbarkeit der Verkehrstelematik schafft sich hier also selbst neue Anforderungen. Eine Lösung für dieses Problem ist der Übergang zu sogenannter Offboard-Navigation: die eigentliche Routenberechnung wird auf ein zentrales System verlagert, das einen größeren Überblick hat und das individuelle Ziel mit der Gesamtverkehrslage verbinden kann. Aus Sicht der Optimierungsverfahren ist das aber die vollständige Integration mit dem Verkehrsmanagement, weil jetzt die Möglichkeit besteht, gezielt Routen so zu empfehlen, dass die Gesamtverkehrslage optimiert wird. Und trotzdem hoffentlich für den Einzelnen noch einen persönlichen Vorteil zu generieren. Dieses Feld wird derzeit in der Verkehrsforschung angegangen.

8. Optimierung im Verkehr der Zukunft

Die verkehrstelematische Vision für den Verkehr der Zukunft ist: Sie steigen in Ihr Auto ein, geben Ihr Ziel in das Navigationssystem ein, und dieses System erklärt Ihnen: „Sie können jetzt über den Mittleren Ring fahren, Fahrzeit 10 Minuten, Fahrpreis 0,50 Euro, oder Sie nehmen die Alternativroute, bezahlen nichts, aber brauchen 20 Minuten. Entscheiden Sie sich."

Dahinter steckt dann hochwertige Optimierung, weitreichende Abbildung von menschlichem Verhalten, vielleicht sogar im Sinne des volkswirtschaftlichen Gesamtwohls. Man könnte durchaus über gezielt eingesetzte Mautsysteme den Weg vom individuellen Optimum zum globalen Optimum des Verkehrs beschreiten. So ein System kann im Extremfall kostenneutral für den Benutzer sein: man bekommt Geld zurück, wenn man die belasteten, hoch attraktiven Strecken vermeidet. Die Optimierungsverfahren für ein solches Online-Verkehrsmanagement müssen noch entwickelt werden, und wir müssen auch erst noch alle Mitmenschen von der Sinnhaftigkeit überzeugen. Aber zumindest aus technologischer, aus mathematischer oder OR-Sicht sind diese Perspektiven interessant.

Und wenn diese neuen Telematikzentralen dann alle Ziele zurückmelden zur Verkehrsplanung, dann werden wir in vielleicht in die Lage kommen, ein geschlossenes Modell des Verkehrsverhaltens zu erstellen. Pech für das System, wenn Sie dann zuhause bleiben.

Some Practical Aspects of Periodic Timetabling

Christian Liebchen[1], Leon Peeters[2]

[1] TU Berlin, Combinatorial Optimization and Graph Algorithms, Straße des 17. Juni 136, D-10623 Berlin, Germany
[2] Rotterdam School of Management, Decision and Information Sciences, P.O. Box 1738, NL-3000 DR Rotterdam, The Netherlands

Abstract. We consider the design of periodic timetables for public transportation networks. The used model is a digraph in which temporal restrictions on the arcs relate periodically recurring events. Solution methods are commonly based on an integer programming approach; the number of integer variables equals the cyclomatic number of the digraph.

In timetabling, one major question is how to order lines that share a common track. In our paper we show, that in this context we can cut off $2^{\binom{n-1}{2}} - n!$ infeasible integer vectors, by adding only polynomially many well-known constraints.

Another point of interest is not to construct timetables that pay for short passengers' waiting times by an exorbitant amount of rolling stock required to perform a timetable. We will show that both routing the rolling stock and periodic timetabling can be modelled within one ILP. Although this leads to a quadratic objective function, we're able to derive a heuristic permitting good control of the considerable trade-off.

1 Introduction

In a fixed interval timetable for a public transportation network, every event, e.g. arrival of line L at station S, is repeated periodically. The restrictions stated most frequently are running times, stopping times, change activities, and separation constraints. But many other types are possible.

All these can be expressed in the Periodic Event Scheduling Problem (PESP) that has been introduced by Serafini and Ukovich[5]. An event that occurs once within the common period T is modelled by a node i, and a timetable has to assign a point of time $\pi_i \in [0, T)$ to every event i. Restrictions are modelled by requiring, periodically, a minimal and maximal gap between two events i and j, i.e.,

$$l_{ij} \leq \pi_j - \pi_i + p_{ij}T \leq u_{ij}, \tag{1}$$

where the periodic offset p_{ij} has to be integer.

From the theory of node potentials and tensions the following problem formulation can be obtained

$$\begin{aligned} &\min\ cx \\ &\text{s.t. } \Gamma x = pT \\ &\qquad l \leq x \leq u \\ &\qquad p \text{ integer,} \end{aligned} \tag{2}$$

where Γ denotes the cycle-arc incidence matrix for some fundamental cycle basis (cf. network matrix), c denotes a cost vector, and the p_{ij} variables that correspond to arcs of a spanning tree are set to zero. It is known that the PESP is $\mathcal{NP}$-hard [5].

Approaches to solve PESP instances cover backtracking strategies in a branch-and-bound context [5], genetic algorthims [3], and some classes of cutting planes [2], [4].

Odijk[4] discovered two classes of valid inequalities, in terms of the integer vector p, for every simple cycle of the digraph. A periodic timetable can be deduced from an integer vector p, if and only if p fulfills Odijk's inequalities on every simple cycle of the digraph. Hence, we are interested in finding some cycles for which Odijk's inequalities cut off many infeasible integer vectors p.

A common objective function is to minimize passenger waiting time. But results obtained by this approach risk to be useless in practice: The Berlin underground company (BVG) as well as the Berlin fast train company (S-Bahn Berlin GmbH) rejected several timetables that were computed at our institute, because they would have required additional vehicles. Thus, it is interesting to combine passenger and vehicle waiting times in one objective function, especially when allowing line changes of vehicles.

2 Passenger Waiting Time vs. Vehicle Waiting Time

Bus and tram networks usually allow many feasible periodic timetables, since only few static safety conditions have to be obeyed. Hence, we have to select a good timetable among a huge number of operable timetables. In practice, there are two major criteria for evaluating timetables: First, offer short waiting times to the passengers. Second, construct timetables that require not too much rolling stock. The question is how to combine these criteria in one objective function.

For the case that vehicles are not allowed to change lines, the vehicles' routing is trivially determined. We can thus restrict our considerations to one traffic line. Here, the timetable has only small effects on the costs for serving the line. These effects are even independent from the length and the period of the line.

Lemma 1. *Assuming fixed running and stopping times along a traffic line, for any two timetables, the number of vehicles required to serve both directions of this line differs by at most one.*

Proof. By our assumptions, the vehicles' only existing flexibility is their waiting time in the endpoints of the line. If the line is served every T time units, then no vehicle will wait T time units or even more, within one endpoint. Hence, on one hand, for any two timetables, the time elapsing before a vehicle executes the same action for the second time may vary by at most $2T - \varepsilon$ time units, for some positive ε.

But on the other hand, in periodic timetables an integer multiple of the period has to elapse before a vehicle carries out the same action for the second time. This integer multiple equals the number of vehicles required to serve both directions of the line. As there are at most two integer multiples of T contained in an interval of length $2T - \varepsilon$, for any two timetables the number of vehicles required to serve the traffic line differs by at most one, q.e.d.

Penalizing the need of one additional vehicle could be realized by introducing a positive coefficient for the corresponding integer variable. But as the vector p is of rather technical character, whereas the vector x directly models requirements from practice, we prefer a solution that omits coefficients for the p vector.

Consider the cycle C which a vehicle has to follow in order to get back to the same event, i.e. to the same station on the same line in the same direction. As is well known, $\sum_{a \in C} x_a = p_C T$ holds. Hence, for minimizing the number of vehicles p_C required to serve one public transportation line, we may raise the objective function's coefficients of all arcs $a \in C$ by a sufficiently large penalty M. Notice that we automatically maintain the piecewise constant behaviour of the cost function.

But forbidding line changes of vehicles may impose a very expensive restriction: Imagine N lines that could subsequently be served by one single vehicle within the period time T. Thus, we first give an exact, but intractable, model to support line changes, in order to derive a heuristic that behaves well in practice.

In any terminus station, a priori any arriving vehicle is allowed to continue on any departing line, so we have to add arcs from any arrival event to any departure event, within the considered station. But in every fixed timetable, for every arriving vehicle exactly one departing line will be chosen. Any feasible selection will translate into a bipartite perfect matching, and only the waiting time arising on the chosen arcs will be relevant for the total number of vehicles required. The problem can be stated by refining the original formulation (2) of the PESP:

$$\begin{aligned} &\min\ cx + \mathbf{M}\xi' x' \\ &\text{s.t. } \Gamma \begin{pmatrix} x \\ x' \end{pmatrix} = pT \\ &l \le \begin{pmatrix} x \\ x' \end{pmatrix} \le u \\ &p \text{ integer} \\ &\xi' \text{ matching in any terminus station} \end{aligned} \tag{3}$$

The variables x' correspond to the arcs within terminus stations and M is a sufficiently large penalty that has to apply to flexible running and stopping times as well. All this can be interpreted as covering all vehicle actions by some cycles. Just recall here, that we are only planning one abstract period, not one whole day of service. But one abstract period can be considered as the core of one whole day of service...

Since for instances of reasonable size, solution methods for the PESP (2) occasionally reach their limits, the quadratic term $\xi' x'$ in the objective function of (3) makes that problem intractable. We should fix ξ' to some value. Since we are not able to decide a priori which connections will not be effected in any optimal solution, in a good heuristic all coefficients should be non-zero. In order to save as much structure as possible, for a terminus station with n incoming – and thus n outgoing – lines, we just fix the matching variables ξ' heuristically to a fractional point inside the matching polytope, $\xi' \equiv \frac{1}{n}$, and hereby partition the penalty M symmetrically.

Notice that for the case that line changes are allowed, once a periodic timetable is fixed, the vehicles' routing is trivially determined, too. In every terminus station, the FIFO strategy leads to an optimal periodical assignment, as is shown in figure 1.

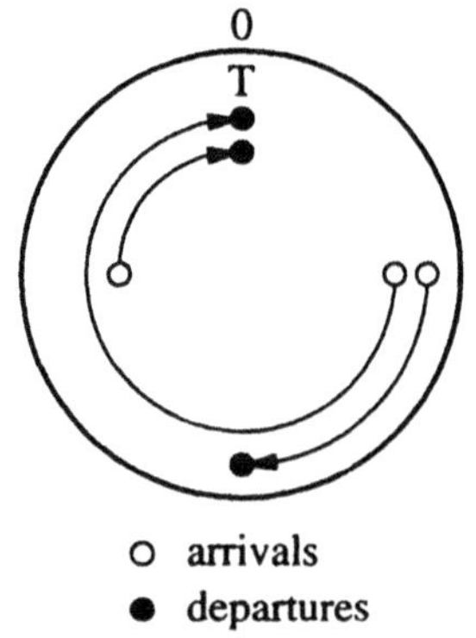

Fig. 1. The FIFO strategy leads to optimal periodic assignments

Computations on the tram network of the city of Halle have shown that there is indeed an important trade-off between passenger and vehicle waiting time. On one hand, the timetable that largely ignores vehicle waiting times, requires seven vehicles – which were about seven percent – more than the timetable that focusses on vehicle waiting time. On the other hand, passenger waiting time increases by more than 25% for the latter timetable. But our approach does allow even more: By varying the amount of the penalty M, we obtain several timetables that, in some way, approximate the Pareto-optima, as can be seen in figure 2.

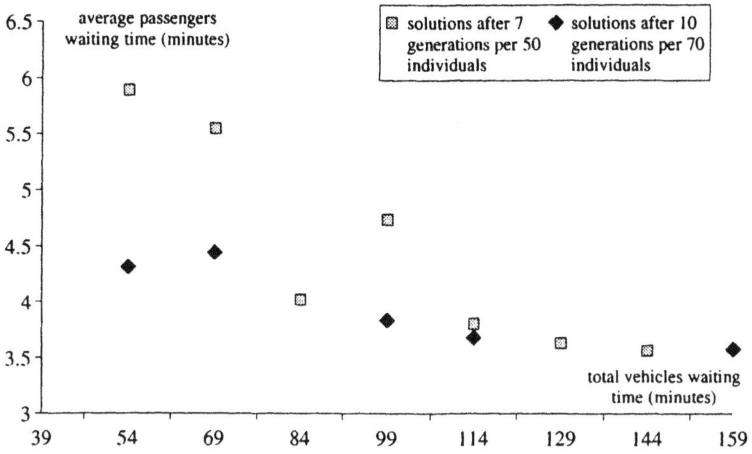

Fig. 2. Tradeoff between passenger and vehicle waiting time, observed in solutions generated by a genetic algorithm

Permitting line changes, two more vehicles can be saved. But it is even more interesting that for the same number of vehicles a timetable that allows vehicles' line changes may reduce passenger waiting time by about ten percent. The strategical question to be answered becomes:

> Which gain in passengers' waiting time justifies one supplementary vehicle, or in other words,
> what is the price of a passenger's waiting minute?

3 A Separation Heuristic for Sequencing

Urban railway networks often contain tracks that are used by several traffic lines. For example, during the rush-hour, the S-Bahn Berlin GmbH's backbone is frequented by seven trains every twenty minutes, and a safety distance of two minutes has to be ensured. Within such a context, a major question in timetable planning is how to order these lines.

As we will exploit the valid inequalities that were introduced by Odijk[4], we first cite Odijk's main result.

Theorem 1. *Let C be a cycle of the graph induced by a PESP instance. Denote by C^+ and C^- the forward resp. backward arcs of the cycle. Then*

$$\left\lceil \frac{\sum_{a \in C^+} l_a - \sum_{a \in C^-} u_a}{T} \right\rceil \leq \sum_{a \in C^+} p_a - \sum_{a \in C^-} p_a, \quad \textit{and} \tag{4}$$

$$\sum_{a \in C^+} p_a - \sum_{a \in C^-} p_a \leq \left\lfloor \frac{\sum_{a \in C^+} u_a - \sum_{a \in C^-} l_a}{T} \right\rfloor$$

are valid inequalities.

Note that here we explicitly mention an integer variable for every arc, as we did in the original equation (1). But of course, we can choose $|V|-1$ of them to be zero, as we did in the system (2).

In the PESP, sequencing n lines can be modelled by adding arcs with lower bound d and upper bound $T-d$, for some $d \in (0, \frac{T}{n}]$, such that the events of entering the common track form a clique. The tree necessary to define the fundamental cycle basis, will be chosen as a star tree, rooted at node number 0. Every arc $a=(i,j)$ is directed such that $i<j$, cf. figure 3. If we have $n \geq 3$ lines, the cycle space of this clique has dimension $\binom{n-1}{2}$,

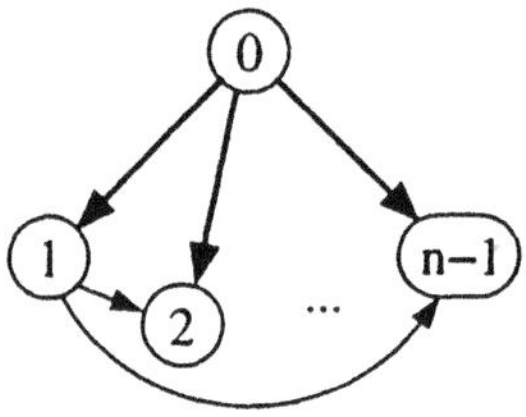

Fig. 3. Notation for the used clique, **bold** arcs are tree arcs

since every arc (i,j) that is not incident to node 0, leads to a fundamental cycle $(0,i,j,0)$. In this context, Odijk's inequalities from theorem 1 become

$$\lceil \frac{d+d-(T-d)}{T} \rceil \leq p_{ij} \leq \lfloor \frac{(T-d)+(T-d)-d}{T} \rfloor. \tag{5}$$

As $3d \leq 3 \cdot \frac{T}{n} \leq T$ for $n \geq 3$, every resulting integer variable p_{ij} is limited to be either zero or one. Hence, there are as many as $2^{\binom{n-1}{2}}$ vectors p fulfilling these box constraints – but only $n!$ solutions, or permutations, will make sense.

We are able to cut off any of the $2^{\binom{n-1}{2}} - n!$ useless integer vectors by adding only polynomially many constraints of Odijk's type. These inequalities can be derived from the $\binom{n-1}{3}$ non-fundamental triangles within the clique.

Theorem 2. *If an integer vector p does not correspond to a permutation of the n events, then there exists a triangle within the clique, for which one of Odijk's inequalities is violated.*

Proof. We have to start by understanding the variables p_{ij} in this context: p_{ij} equals zero if and only if event i occurs before event j, that is $\pi_i < \pi_j$.

Consider the fact that a vector p implies that three events $i<j<k$ cannot be ordered, i.e. $\pi_i < \pi_j < \pi_k < \pi_i$, or $\pi_i > \pi_j > \pi_k > \pi_i$ which behaves symmetrically. Hereby p does not correspond to a permutation of the n events. Then we know $p_{ij} = p_{jk} = 0$ and $p_{ik} = 1$. Exploiting Odijk's inequality (4) for the cycle (i,j,k,i), we get to

$$\lceil \frac{d+d-(T-d)}{T} \rceil \leq p_{ij} + p_{jk} - p_{ik}, \tag{6}$$

because the arc (i,k) will be traversed in its opposite direction. From $d \in (0, \frac{T}{n}]$, we deduce that the left hand side of inequality (6) is zero. We thus found a cutting plane separating this vector that does not correspond to a permutation.

In fact, these inequalities define the Linear Ordering Problem [1] on integer variables. Especially, if there would be a set of more than three events that is not linearly ordererd, there will be a subset of cardinality three, that is not ordered as well. Hence, each vector p that does not correspond to a permutation violates an inequality of Odijk's type on a triangle within the clique, q.e.d.

The Linear Ordering Problem is defined as follows:

$$\begin{aligned} \min\ & wz \\ \text{s.t.}\ & z_{ij} = 1 - z_{ji} \\ & z_{ij} + z_{jk} + z_{ki} \leq 2,\ \forall \{i,j,k\} \subseteq \{1,\ldots,n\} \\ & z \in \{0,1\}. \end{aligned} \tag{7}$$

Here, it is known that the above inequalities form the complete description of the convex hull of the feasible integer vectors, up to dimension five. However, in general the problem of finding a linear ordering of minimum weight is $\mathcal{NP}$-hard, as was shown by Grötschel, Jünger, and Reinelt[1].

From theorem 2 we get an alternative proof for the $\mathcal{NP}$-hardness of the PESP, by reduction from the Linear Ordering Problem.

Theorem 3. *The Linear Ordering Problem is polynomially reducible to the Periodic Event Scheduling Problem, with fixed interval time T.*

Proof. Consider an instance of the Linear Ordering Problem having $|V| = n-1$. We define the graph of the instance of the Periodic Event Scheduling Problem to be a clique, including one additional node 0. The lower bounds of the arcs are set to $l = \frac{T}{n}$, the upper bounds are set to $u = T - \frac{T}{n}$. From theorem 2 we know that the integer variables of the PESP instance correspond to the variables of the Linear Ordering Problem. Furthermore, the inequalities of the Linear Ordering Problem are implied by this instance of the Periodic Event Scheduling Problem.

Finally, we must ensure that a periodic timetable of minimal weight implies a linear ordering of minimal weight. We define $\tilde{w}_{ij} := w_{ij} - w_{ji}$ and $\tilde{w}_{ji} := 0$ in order to eliminate antiparallel arcs. System (2) allows to bind the rational variables x to the integer variables p, resp. z:

$$x_{0i} + x_{ij} - x_{0j} = Tp_{ij} \Leftrightarrow z_{ij} = \frac{x_{0i} + x_{ij} - x_{0j}}{T}.$$

Hence, the original objective function wz translates to

$$\sum_{i<j} \tilde{w}_{ij} \cdot \frac{x_{0i} + x_{ij} - x_{0j}}{T}.$$

Omitting the common divisor T and reordering, lets us define the coefficients c for the rational variables x of the PESP instance

$$c_{ij} := \begin{cases} \tilde{w}_{ij} & \text{if } 0 < i < j, \\ \sum_{k=j+1}^{n-1} \tilde{w}_{jk} - \sum_{k=1}^{j-1} \tilde{w}_{kj} & \text{if } i = 0. \end{cases}$$

The objective values in both problems are identical up to a constant summand and a constant factor.

In practice, it seems that this reduction of the search space speeds up running times. Indeed, we profit from the fact that complete descriptions of the Linear Ordering Polyhedron are known for small dimensions, and that in periodic timetabling, we rarely face ten or more traffic lines to be sequenced.

4 Conclusions

We are able to introduce central aspects of vehicle routing into the timetable optimization task. The exact model leads to a mixed integer program with quadratic objective function, but a nice relaxation has been derived. It permits us to get an idea of the notable trade-off between passenger and vehicle waiting time.

Investigating the special structure of subgraphs resulting from separation constraints, we discovered a source for promising cutting planes. This lead to an alternative proof for the $\mathcal{NP}$-completeness of the Periodic Event Scheduling Problem.

But there is still much work to be done. Solution methods, i.e. detecting an infeasible cycle, or selecting a very good timetable among a possibly huge number of feasible timetables, must be enhanced. Otherwise instances as the complete traffic of German Rail (DB Reise & Touristik) or Dutch Passenger Railways (NS Reizigers), as well as the complete public transport of the Ruhr area (Verkehrsverbund Rhein-Ruhr GmbH) can only be solved after decomposing the problem hierarchically.

References

1. Grötschel, M., Jünger, M., Reinelt, G. (1984) A cutting plane algorithm for the linear ordering problem. Operations Research **32**, 1195–1220
2. Nachtigall, K. (1996) Cutting Planes for a Polyhedron Associated with a Periodic Network. Deutsche Forschungsanstalt für Luft- und Raumfahrt e.V., IB 112-96/17
3. Nachtigall, L., Voget, S. (1996) A genetic algorithm approach to periodic railway synchronization. Computers and Operations Research **23**, 453–463
4. Odijk, M.A. (1994) Construction of periodic timetables, Part 1: A cutting plane algorithm. TU Delft Technical Report, 94-61
5. Serafini, P., Ukovich, W. (1989) A mathematical model for periodic scheduling problems. SIAM J. Disc. Math. **2**, 550–581

Pickup & Delivery-Probleme mit Umlademöglichkeit
Ein Tourenplanungsproblem aus der Automobilzulieferindustrie

Christopher Mues[1], Stefan Pickl[2]

[1] gedas telematics GmbH, Berlin, Christopher.Mues@gedas-telematics.de
[2] Mathematisches Institut, Zentrum für Angewandte Informatik Köln ZAIK, Universität zu Köln, D-50931 Köln, pickl@zpr.uni-koeln.de

Zusammenfassung Grundlage für die Modellierung ist das Pickup & Delivery-Problem. Dieses wird um die Möglichkeit erweitert, zusätzlich bei der Planung von Touren für jedes einzelne Gut die Transportart (Sammelgut, Milkrun) bestimmen zu können. Ein Gut wird als *Sammelgut* befördert, wenn es zunächst zum Zwischenlager des Spediteurs gebracht wird. Am Zwischenlager findet die Konsolidierung statt, d.h. am Zwischenlager werden einzelne Sendungen für den Weitertransport zu einem Werk zusammengefaßt. Daher muß bei der Beschreibung des Sammelgutverkehrs neben den üblichen Nebenbedingungen noch die zeitliche Reihenfolge von Vor- und Hauptlauf eingehalten werden. Als *Milkrun* wird ein Gut transportiert, wenn es ohne Umladung auf einer Tour vom Zulieferer zum Werk befördert wird.

Zur Lösung des Problems wird ein Spaltenerzeugungsverfahren und ein Lösungsalgorithmus vorgestellt. Die Ergebnisse beruhen auf realen Daten aus der Automobilzulieferindustrie. Die Lösung des Optimierungsproblems besteht somit aus einer Menge von Touren, durch die alle Güter transportiert werden, und aus der Information, ob Güter zwischen Touren umgeladen werden (Sammelgut) oder nicht (Milkrun).

Keywords: Pickup & Delivery, Vehicle Routing, Column Generation, Transshipment

1 Einleitung

Ein wesentlicher Kostenfaktor bei der Produktion von Fahrzeugen sind heute immer noch die Kosten für den Transport der für die Herstellung benötigten Güter von den Zulieferern zu den Automobilwerken. Um diese Kosten zu senken, wird auch versucht, Verfahren der Tourenoptimierung einzusetzen. Wir haben das Problem betrachtet, die Zulieferung innerhalb von Deutschland zu planen. In diesem Gebiet wird der Transport in der Regel mit LKW durch Spediteure abgewickelt. Hierzu werden zwei verschiedene Organisationsformen, der *Sammelgutverkehr* und der *Milkrun* (Abb. 1), angewendet.

Im *Sammelgutverkehr* werden Stückgüter von einem Spediteur bei den Lieferanten eingesammelt und zu dem *Umschlagsplatz* (Consolidation Center

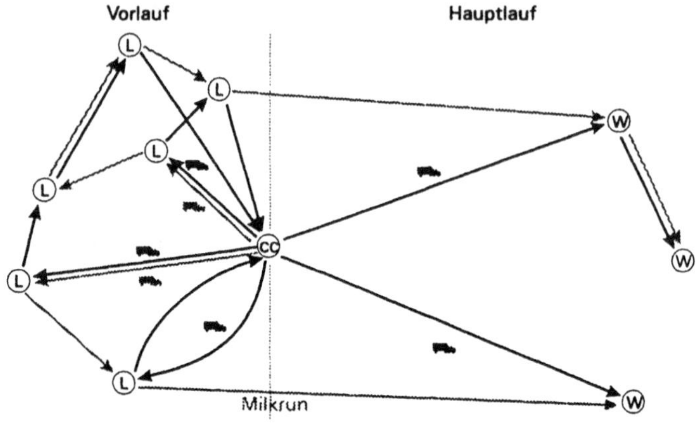

Abbildung1. Sammelgutverkehr vs. Milkrun

(CC)) des Spediteurs transportiert. Dieser Transport wird *Vorlauf* genannt. Am CC findet die *Konsolidierung* statt, d.h. Zusammenfassung einzelner Sendungen, damit die Güter gesammelt von dort zu den Werken transportiert werden können. Diese Tour heißt *Hauptlauf.* Hingegen ist der Milkrun eine Route von den Lieferanten zu den Werken, auf dem eine oder mehrere Sendungen ohne Umschlag durch den Spediteur transportiert werden.

2 Das Pickup & Delivery-Problem

Grundlage für die Modellierung ist das Pickup & Delivery-Problem, wie es in [Dumas91] beschrieben wird. Auf einer Tour soll es möglich sein, sowohl *verschiedene Lieferanten* als auch *unterschiedliche Werke* anzufahren. Im Gegensatz zu [Sol94] wird die Formulierung derart erweitert, daß bei der Planung von Touren für jedes einzelne Gut die jeweilige Transportart (Sammelgutverkehr bzw. Milkrun) bestimmt werden kann. Bei der Beschreibung des Sammelgutverkehrs muß neben den üblichen Nebenbedingungen noch die zeitliche Reihenfolge von Vor- und Hauptlauf eingehalten werden.

Sei A die Menge aller Aufträge und F die Menge aller Fahrzeuge. Start(i) für $i \in A \cup F$ bezeichnet den Beladeort eines Auftrags oder den Ausgangsort eines Fahrzeugs. Entsprechend ist Ziel(i) der Entlade- bzw. der Zielort. Ferner ist definiert $\text{Start}(A) := \{\text{Start}(a) \mid a \in A\}$ und $\text{Start}(F) := \{\text{Start}(f) \mid f \in F\}$.

Das *Pickup & Delivery-Tourenplanungsproblem mit Umladung (P)* wird wie folgt definiert. Gegeben ist der Graph $G = (V, E)$ mit

$$V_A := \text{Start}(A) \cup \text{Ziel}(A),$$
$$V_F := \text{Start}(F) \cup \text{Ziel}(F),$$

$$V := V_A \cup V_F \cup U,$$
$$E := E(V).$$

Gesucht sind Wege $P_i = v_1 \dots v_n$ für alle $i \in A \cup F$ mit

$$v_1 = \text{Start}(i),$$
$$v_k \in V_A \cup U,\ 1 < k < n,$$
$$v_n = \text{Ziel}(i)$$

und speziell für $f, f' \in F$, $f \neq f'$

$$V(P_f) \cap V(P_{f'}) \cap (V_A \cup V_F) = \emptyset,$$

so daß gilt

$$\bigcup_{a \in A} E(P_a) \subset \bigcup_{f \in F} E(P_f).$$

Anmerkungen:

- Eine Lösung von (P) ist in diesem Fall eine Menge von Wegen P_a und P_f.
- Die richtige Reihenfolge der Beladung, Umladung und Entladung eines Auftrags ist sichergestellt.
- Mehrfachumladungen sind möglich.

In Abbildung 2 wird der Fall dargestellt, daß ein Auftrag (Auftrag 2) umgeladen wird, während die übrigen Aufträge (Auftrag 1 und 3) ohne Umladung über den Umladeort transportiert werden. Hieraus ergibt sich auch die Notwendigkeit, daß Wege für Fahrzeuge bis auf die Umladeorte disjunkt sind, da sich ansonsten ungewollte Umladungen ergeben können.

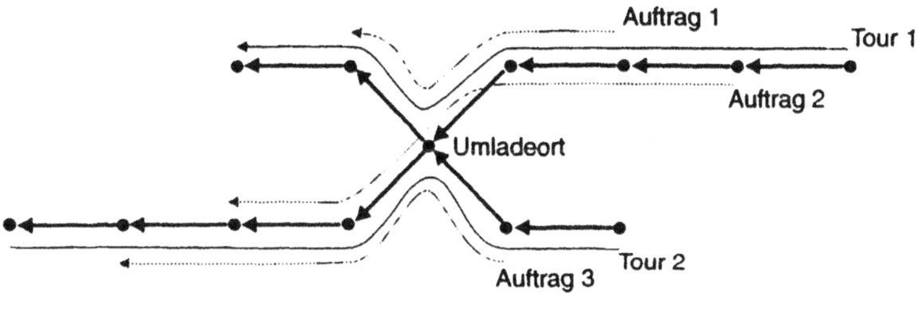

Abbildung2. Beispiel für eine Umladung zwischen Touren

2.1 Modellerweiterungen

In der Regel wird (P) noch um die Restriktionen erweitert, daß Zeitfenster und Fahrzeugkapazitäten eingehalten werden müssen.

Zeitfenster: Um Zeitfenster berücksichtigen und Umladungen zeitlich richtig beschreiben zu können, wird die Beschreibung des Umladeortes neu formuliert. Hierzu wird jeder Umladeort $u \in U$ durch Knoten $u(f)$ für alle $f \in F$ ersetzt. Entsprechend werden Kanten (v, u), (u, w) durch Kanten $(v, u(f))$, $(u(f), w)$ ersetzt und zusätzliche Kanten $(u(f), u(f'))$ eingefügt (Abbildung 3).

Eine Umladung läßt sich in diesem Modell leicht erkennen, da ein Auftrag genau dann umgeladen wird, wenn der Weg zu diesem Auftrag eine Kante aus $E(u(F))$ enthält[1].

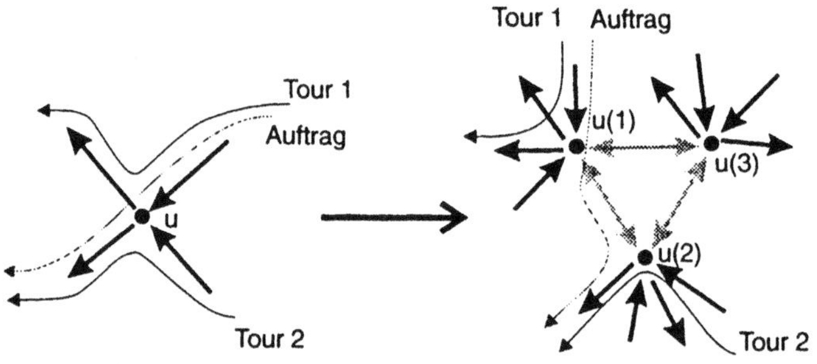

Abbildung3. Ersetzen eines Umladeortes

Hieraus ergibt sich eine erweiterte Formulierung des Problems.

Das *erweiterte Pickup & Delivery-Tourenplanungsproblem mit Umladung (P')* wird wie folgt definiert. Gegeben ist der Graph $G = (V, E)$ mit

$$\begin{aligned} V_A &:= \text{Start}(A) \cup \text{Ziel}(A), \\ V_F &:= \text{Start}(F) \cup \text{Ziel}(F), \\ U(f) &:= \bigcup_{u \in U} \{u(f)\}, \\ V &:= V_A \cup V_F \cup \left(\bigcup_{f \in F} U(f) \right), \\ E &:= E(V). \end{aligned}$$

Gesucht sind Wege $P_i = v_1 \dots v_n$ für alle $i \in A \cup F$ mit

$$\begin{aligned} v_1 &= \text{Start}(i), \\ v_k &\in \begin{cases} V_A \cup U(i) & \text{, falls } i \in F \\ V_A \cup \left(\bigcup_{f \in F} U(f) \right) & \text{, falls } i \in A \end{cases}, \; 1 < k < n, \\ v_n &= \text{Ziel}(i) \end{aligned}$$

[1] $u(F) := \{u(f) \mid f \in F\}$

und speziell für $f, f' \in F$, $f \neq f'$

$$V(P_f) \cap V(P_{f'}) = \emptyset,$$

so daß gilt

$$\left(\bigcup_{a \in A} E(P_a) \right) \setminus \left(\bigcup_{u \in U} E(u(F)) \right) \subset \bigcup_{f \in F} E(P_f).$$

Um die Einhaltung der Zeitfenster zu prüfen, müssen Zeiten t_v^a für einen Auftrag $a \in A$ für die Be- oder Entladung an dem Ort $v \in V$ berücksichtigt werden. Zudem fallen Fahrzeiten t_e für die Fahrt eines Fahrzeugs auf der Strecke $e \in E$ an. Es sei eine Kante $(v, w) \in P_i$ für ein $i \in F$ gegeben. Es muß gelten

$$t_w \geq \begin{cases} t_v + t_{vw} & , v \in V_F \\ t_v + t_{vw} + \sum\limits_{a \in A_{vw}} t_v^a & , v \in V_A \\ t_v + t_{vw} + \sum\limits_{a \in A_{vw}} t_v^{a,\text{Ent}} & , v = u(f'),\ w = u(f) \\ & \text{für ein } u \in U \\ t_v + t_{vw} + \sum\limits_{b \in u(F)} \left(\sum\limits_{a \in A_{bv}} t_v^{a,\text{Bel}} + \sum\limits_{a \in A_{vb}} t_v^{a,\text{Ent}} \right) & , v \in u(F), w \notin u(F) \\ & \text{für ein } u \in U \end{cases}$$

mit $A_{vw} := \{a \in A \,|\, (v, w) \in E(P_a)\}$ und den Be- und Entladezeiten $t_v^{a,\text{Bel}}$ und $t_v^{a,\text{Ent}}$ während der Umladung.

Fahrzeugkapazitäten: Fahrzeugen $f \in F$ wird eine maximale Kapazität q^f zugeordnet und Aufträgen $a \in A$ der Wert q^a, der die Ausnutzung einer Fahrzeugkapazität angibt. Für jeden Knoten $v \in V$ und jedes Fahrzeug $f \in F$ muß gelten

$$\sum_{a \in A_{vw}} q^a \leq \sum_{f \in F_{vw}} q^f.$$

mit $F_{vw} := \{f \in F \,|\, (v, w) \in E(P_f)\}$. Offensichtlich gilt $|F_{vw}| \in \{0, 1\}$.

3 Lösungsverfahren

Durch den gewählten Ansatz ergibt sich ein Mehrgüter-Fluß-Problem. Zu dessen Lösung haben wir den Ansatz der Spaltenerzeugung / Branch & Price wie in [Sol94] gewählt. Der Artikel [Desaulniers00] gibt einen Überblick über Verfahren, die heute zum Einsatz kommen. Ein neuer heuristischer Lösungsansatz wird in [Dörner01] beschrieben, der auf der Verwendung sogenannter Ameisen-Systeme beruht. [Noltemeier00] beschreibt einen Lösungsverfahren für das Problem der Generierung von Flugplänen. Schließlich entwickelt [Oertel00] eine Tabu Search-Verfahren für das Pickup & Delivery-Problem mit Umladung.

Für den von uns zuerst gewählten Ansatz wurde ein entsprechend geändertes Master- und Subproblem hergeleitet. Bedingt durch die Reihenfolgebeziehung von Vor- und Hauptlauf im Sammelgutverkehr gehen in der Zielfunktion des Subproblems die Ausführungszeitpunkte der Be- und Entladung von Gütern am CC ein. Daher kann das Subproblem nicht ohne Einschränkungen als Constrained-Shortest-Path Problem gelöst werden.

Hierfür wurde eine Heuristik zur Bestimmung einer Startlösung des Problems implementiert. Es werden enumerativ Touren erzeugt. Unter Ausnutzung des Prinzips der dynamischen Programmierung wird anhand verschiedener Kriterien der sich ergebende Enumerationsbaum eingeschränkt. Kriterien sind einerseits Bedingungen, anhand derer die Zulässigkeit einer Tour geprüft wird. Andererseits werden aus den 2-opt- und Or-opt-Kantenaustauschverfahren hergeleitete Kriterien verwendet wie auch eine Dominanzregel, um abzuschätzen, ob eine Tour in einer optimalen Lösung enthalten sein könnte.

Die Nachteile dieses Ansatzes liegen in der Schwierigkeit, gemischte Touren zu erzeugen, in denen Aufträge als Sammelgut und als Milkrun transportiert werden. Zudem beschränkt sich dieser Ansatz auf einen Umladepunkt. Daher wurde der in diesem Artikel beschriebene alternative Ansatz untersucht. Hierbei wird die Planung der Transportwege für Aufträge von der Planung der Wege für Fahrzeuge getrennt. Die Erzeugung der Wege erfolgt unabhängig von den Fahrzeugen und Aufträgen.

Es wird zunächst eine Menge von Wegen enumerativ erzeugt, die dann den Aufträgen und Fahrzeugen zugeordnet werden. Diese werden dabei auf ihre zeitliche und kapazitäre Zulässigkeit hin geprüft. Um bzgl. der auf einer Tour transportierten Aufträge vergleichbare Touren auszuschließen, wird zusätzlich eine Dominanzregel verwendet.

Bei der Zuordnung von Wegen zu Fahrzeugen soll nach Möglichkeit vermieden werden, daß sich Lösungen ergeben können, die sich bei gleichen Wegen nur durch die unterschiedliche Zuordnung dieser Wege zu Fahrzeugen unterscheiden. Daher werden Mengen von Touren mit der Eigenschaft gebildet, daß je zwei Touren aus einer dieser Mengen nicht in einer Lösung enthalten sind. Beispielsweise können zwei Touren dann nicht in einer Lösung vorhanden sein, wenn sie über den gleichen Beladeort führen und wenn dieser nur einmal angefahren werden darf.

Die Zuordnung von Wegen zu Aufträgen erfolgt anhand des Be- und Entladeortes. Diese Zuordnung ist eindeutig, da angenommen wird, daß keine zwei Aufträge mit gleichem Be- und Entladeort vorhanden sind. Die Menge der Wege, die Aufträgen zugeordnet sind, kann anhand der Zuordnung von Wegen zu Fahrzeugen weiter eingeschränkt werden. Durch die Zuordnung von Wegen zu Fahrzeugen ergibt sich eine Zuordnung von Kanten zu Fahrzeugen. Ein Weg für einen Auftrag ist dann unzulässig, wenn es für jeden Teil (vom Beladeort zum Umladepunkt, von einem Umladepunkt zu einem weiteren, ...) des Weges nicht ein Fahrzeug gibt, daß allen Kanten zugeordnet ist.

Das so gebildete kombinatorische Optimierungsproblem wird dann als ganzzahliges lineares Optimierungsproblem formuliert und gelöst.

4 Zusammenfassung

Die Laufzeiten des zuerst implementierten Verfahrens lagen für die verfügbaren realen Datensätze in einem Bereich von etwa 2 bis 6 Minuten, wobei jeweils etwa 70 Aufträge verplant werden mußten. Die so gefundenen Startlösungen bestanden bis auf Ausnahmen nur aus Sammelgutverkehr. Als Vergleich für die Güte der Lösung wurde eine Optimierungssoftware herangezogen, die unter weitaus günstigeren Modellannahmen (10-fache Laufzeit, Verletzung von Zeitfenstern) für das Problem optimale Lösungen bestimmte und die gegenüber dieser Heuristik lediglich 5% bessere Ergebnisse erzielte.

Eine potentielle Weiterentwicklung des vorgestellten Ansatzes wird innerhalb der Verfahren der dynamischen Spaltenerzeugung liegen.

Literatur

1. BLASUM, U. (2000): *Anwendung des Branch & Cut Verfahrens auf das kapazitierte Vehicle-Routing Problem*, PhD thesis, University of Cologne
2. DESAULNIERS, G., DESROISIERS, J. und SOLOMON, M.N. (1999): *Accelerating Strategies in Column Generation Methods for Vehicle Routing and Crew Scheduling Problems*, Les Cahiers du GERAD G-99-36
3. DESAULNIERS, G., DESROISIERS, J., ERDMANN, A., SOLOMON, M.M. und SOUMIS, F. (2000): *The VRP with pickup and delivery*, Les Cahiers du GERAD, G-2000-25
4. DESROCHERS, M., DESROSIERS, J. und SOLOMON, M. (1992): *A new Optimization Algorithm for the Vehicle Routing Problem with Time Windows*, Operations Research, 342-354:2
5. DÖRNER, K.F., HARTL, R.F. und REIMANN, M. (2001): *Ants Solve Time Constrained Pickup and Delivery Problems with Full Truck Loads* in: FLEISCHMANN, B. et al. (eds.) (2001), Operations Research Proceedings 2000, Springer, Berlin
6. DUMAS, Y., DESROSIERS, J. und SOUMIS, F. (1991): *The Pickup and Delivery Problem with Time Windows,* European Journal of Operational Research, 7-22:54
7. JOHNSON, E.L., NEMHAUSER, G.L., und SAVELSBERGH, W.P. (2000): *Progress in Linear Programming-Based Algorithms for Integer Programming: An Exposition*, INFORMS Journal of Computing, 2-23:12
8. KOHL, N., DESROISIERS, J., MADSEN, O.B.G., SOLOMON, M.M. und SOUMIS, F. (1999): *2-Path Cuts for the Vehicle Routing Problem with Time Windows*, Transportation Science, 101-116:33
9. LÜBBECKE, M.E. (2001): *Combinatorial Restrictions on Pickup and Delivery Paths*, Technical Report, Department of Mathematical Optimization, Braunschweig University of Technology

10. NOLTEMEIER, A. (2000): *Modelle und Lösungsverfahren zur Generierung von Flugplänen*, PhD thesis, University of Cologne
11. OERTEL, P. (2000): *Routing with Reloads*, PhD thesis, University of Cologne
12. SOL, M. (1994): *Column Generation Techniques for Pickup and Delivery Problems*, PhD thesis, Eindhoven University of Technology

Section 2

Energy and Environment

Zentrale Datenbasis im Energiehandel

Carsten Felden[1], Stefan Florek[2]

[1] Gerhard-Mercator-Universität Duisburg, Lotharstraße 65, 47048 Duisburg, C.Felden@uni-duisburg.de

[2] RWE Trading GmbH, Huyssenallee 2, 45128 Essen, stefan.florek@rwetrading.com

1 Ausgangslage

Im Zuge der Liberalisierung der Energiemärkte, insbesondere der Strommärkte, innerhalb der Europäischen Union sahen sich die Vereinigten Elektrizitätswerke Westfalen Energie Aktiengesellschaft (VEW Energie AG) zahlreichen neuen Herausforderungen gegenübergestellt. Die Sicherstellung der Wettbewerbsfähigkeit ist daher zu einem elementaren Unternehmensziel geworden. Die Beschaffungsoptimierung zur Kostenreduzierung umfasst nun nicht mehr nur die Beschaffung von Primärenergien, die in den eigenen Kraftwerken in elektrischen Strom umgewandelt werden, sondern auch die Beschaffung von Strom selbst. Die Möglichkeit des Handels mit Strom führte zu einer großen Vielfalt an neuen Produkten und potenziellen Handelspartnern. Dies alles hatte eine Steigerung der Komplexität des Strommarktes zur Folge. Die VEW Energie AG trug dieser Entwicklung Rechnung, indem sie einen sogenannten Trading Floor einrichtete, dessen Mitarbeiter sich vorwiegend der Aufgabe des Energiehandels widmen. Zu ihrer Unterstützung sollte ein Marktinformationssystem bereitgestellt werden, das allen Mitarbeitern des Handelsbereichs als Datenbank für Marktdaten zur Verfügung steht. Auf Grundlage der Marktdaten können Marktanalysen, statistische Auswertungen und Preisprognosen erstellt werden. Daraus resultiert, dass zu diesem Zweck heterogene Datenformate in der Datenbank zusammen geführt werden müssen, die strukturiert oder unstrukturiert[1] aus internen oder externen, z. B. dem World Wide Web (WWW), Quellen stammen.

Die Entwicklung des Marktinformationssystems erfolgte unter dem *Namen Systematic Analysis and Research Tool* (SMART) in mehreren Stufen, an deren Ende jeweils ein funktionsfähiges Informationssystem steht, dessen Leistungsvermögen von Stufe zu Stufe wächst. Auf diese Weise wurde sichergestellt, dass bereits nach relativ kurzer Zeit ein nutzbares Produkt vorliegt. Ferner wurde in Erwägung gezogen, das Marktinformationssystem als Ausgangsbasis für ein

[1] Zu strukturierten und unstrukturierten Daten vgl. Hansen (1998, S. 6 f.).

unternehmensweites Analytisches Informationssystem zu verwenden. Damit weist das Marktinformationssystem die Merkmale eines Data Mart[2] auf, das zu einem Data Warehouse[3] weiterentwickelt werden soll.

2 Der Energiemarkt in der Bundesrepublik Deutschland

Die alte monopolistische Energiemarktstruktur wurde 1998 schlagartig durch Wettbewerb ersetzt. Die Fortschritte der Liberalisierung sind in den einzelnen europäischen Ländern unterschiedlich. Eines kann aber durchgehend beobachtet werden. Wie andere Energieträger wird Elektrizität heute hoch volatil an Spot- und Terminbörsen gehandelt. Trading und Risikomanagement, Produktentwicklung, Marketing und Kundenbetreuung sind existenzielle Schlüsselfähigkeiten der Unternehmen geworden. Durch Wettbewerb wuchs der Kostendruck. Für ein erfolgreiches Zusammenspiel von eigener Erzeugung, Handel und Vertrieb ist eine zentrale Datenhaltung, die Daten aus internen und externen Quellen unter Sicherung der Datenqualität zusammenfasst, notwendig.

Die Bewertung längerfristiger Produkte und Lieferungen war auf Grund einer fehlenden Terminbörse problematisch. Bis zu einem gewissen Grad wurden Quotierungen unabhängiger Broker herangezogen. Soweit wie möglich musste dieses Manko durch Daten aus anderen Märkten (z. B. Brennstoffe), physikalisch / technische Informationen (z. B. Stromverbrauch), sowie sonstige Daten (Feiertags- und Ferienkalender) ausgeglichen werden.

3 Entwicklung des Marktinformationssystems bei der VEW Energie AG

Zur Durchführung des Smart-Projekts, auf die sich die weiteren Ausführungen beschränken, wurde ein Team gebildet und eine projektübliche phasenorientierte Vorgehensweise gewählt. Im Folgenden wird auf zentrale Projekttätigkeiten eingegangen.

---

[2] Zum Data Mart vgl. Schinzer/Bange (1999, S. 52).

[3] Zum Data Warehouse vgl. Chamoni/Gluchowski (1999, S. 13 ff.).

3.1 Informationsbedarfsanalyse

Anhand der ermittelten Kenntnisse[4] über die Tätigkeiten der Anwender wurden allgemein formulierte Szenarien erstellt. Die einzelnen Szenarien wurden zu Themengruppen, zusammen gefasst. Diese Gruppen bilden auch die späteren multidimensionalen Datenstrukturen, von denen insgesamt sechs abgebildet wurden.

Durch die Erstellung repräsentativer Aufgaben kann zunächst ein gegenseitiges Verständnis und bereits ein frühzeitiger Systemprototyp geschaffen werden.[5] Im SMART-Projekt wurden 29 verschiedene repräsentative Aufgaben identifiziert, die den Anwendern noch einmal zur Bewertung vorgelegt wurden. Daraus werden diejenigen mit der höchsten Wichtigkeit in der täglichen Arbeit ausgewählt. Aus den individuellen Bewertungen wurde das arithmetische Mittel gebildet und die Themengruppen, z. B. Product Price History, zusammen gefasst. Aus der Bewertung resultiert die Erkenntnis, dass zunächst zur Prototypentwicklung die multidimensionale Struktur *Product_Price_History* zu Grunde gelegt wird.

3.2 Entwicklung des Datenmodells

Für die Abbildung der Ergebnisse der Informationsbedarfsanalyse wurde das Application Design for Analytical Technologies (ADAPT)[6] als Modellierungsmethode gewählt. Diese Entscheidung resultiert aus den Unzulänglichkeiten des Entity Relationship Modelling (ERM) bei der Darstellung multidimensionaler Sachverhalte. Da eine relationale Datenbank zur Verfügung stand, wurde die multidimensionale Datenstruktur in Form eines Star Schemas[7] realisiert. Im Rahmen des Ermessensspielraums, den ADAPT bei der Verwendung der Beschreibungselemente bietet, wurde darauf geachtet, möglichst wenige unterschiedliche Symbole zu verwenden. Da das Datenmodell im Rahmen des Prototyping[8] intensiv zum Gedankenaustausch mit dem fachlichen Projektleiter genutzt werden sollte, war eine verständliche Darstellungsweise geboten. Die folgende Abbildung zeigt den multidimensionalen Datenraum *Product Price History* in der ADAPT-Notation.

---

[4] Zu Erhebungsmethoden der Informationsbedarfsanalyse vgl. Stahlknecht (1995, S. 259).

[5] Vgl. Nielsen (1993, S. 100).

[6] Zu ADAPT vgl. Bulos (1998, S. 253 ff.).

[7] Zum Star Schema vgl. Hahne (1999, S. 152 ff.).

[8] Zum Prototyping vgl. Balzert (2001, S. 100 ff.).

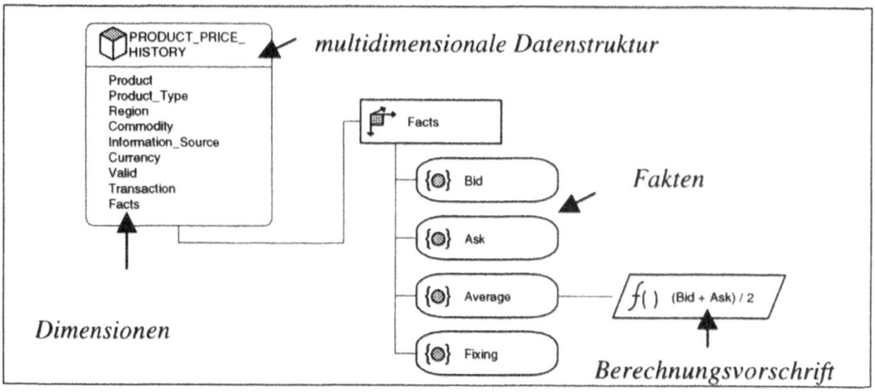

Ziel dieser Analyse ist, die jeweiligen Bid-, Ask- (inklusive deren Durchschnitt)[9] und Fixingkurse[10] der Stromhändler im Zeitverlauf darzustellen. Dabei können sich die Kurse auf verschiedene Regionen (z. B. CEPI, EIS)[11] oder auf verschiedene Produktarten (z. B. Base oder Peak) beziehen. Zu beachten ist, dass zum einen eine Gültigkeitszeit (Valid) und zum anderen eine Transaktionszeit (Transaction)[12] existiert. Diese sind eingeführt worden, damit bei Fehlerkorrekturen der jeweils gültige Wert anhand der aktuellen Transaktionszeit festgestellt werden kann. Parallel zum bisherigen Vorgehen wurde ein Glossar und ein Data Dictionary[13] aufgebaut, um sich zum einen auf die Begriffswelt des Energiehandels einzustimmen und zum anderen die Strukturen des Analytischen Informationssystems zu dokumentieren.

3.3 Implementierung

Die Implementierung von SMART erfolgte in einer Microsoft Access-Datenbank. Logische Datenmodellierung und Implementierung fallen hier zusammen, da bei der Festlegung der Beziehungen zwischen den Tabellen automatisch das entsprechende Star Schema erzeugt. Da sich das semantische Datenmodell sehr gut in eine relationale Datenstruktur überführen ließ, war eine vorausgehende logische Modellierung nicht erforderlich. Die Dimensionen des Datenmodells wurden in Dimensionstabellen transformiert. Dimensionsattribute und Hierarchiestufen bildeten die jeweiligen Attribute, Dimensionselemente beziehungsweise -ausschnitte sind Bestandteil der Datensätze. Aus den Datenwürfeln gingen die

[9] Bei Bid und Ask handelt es sich um Kauf- und Verkaufgebote.
[10] Das Fixing ist der jeweilige Schlusskurs eines Handelstages.
[11] CEPI bezeichnet Norddeutschland; EIS bezeichnet Süddeutschland.
[12] Zur Modellierung zeitbezogener Daten im Data Warehoue vgl. Stock (2001).
[13] Zum Data Dictionary vgl. Wertz (1986).

Faktentabellen hervor. Da mehrere Faktentabellen vorliegen, handelt es sich beim logischen Datenmodell der SMART-Datenbank um eine Galaxie[14] mit 1:n-Beziehungen zwischen den Dimensions- und Faktentabellen.

4 Entwicklungen an der Gerhard-Mercator-Universität Duisburg

An der GMU konzentrierte sich die Weiterentwicklung darauf, externe Daten aus bisher unerschlossenen Quellen zu finden, zu extrahieren und in der Datenbank zu erfassen. Des Weiteren wurde die Schwierigkeit erkannt, heterogene Daten in einer Benutzeroberfläche zu präsentieren. Aus diesem Grunde wurde das Konzept der Informationsvisualisierung als Lösungshilfe untersucht.

4.1 Integration externer Daten aus unbekannten Quellen

In einem zusätzlichen Ansatz wird angestrebt, die Datenbasis mit externen Daten zu erweitern. Ziel ist, diese Quellen im World Wide Web (WWW) zu finden und den Analysten zur Verfügung zu stellen. Die folgende Abbildung zeigt den Verlauf.

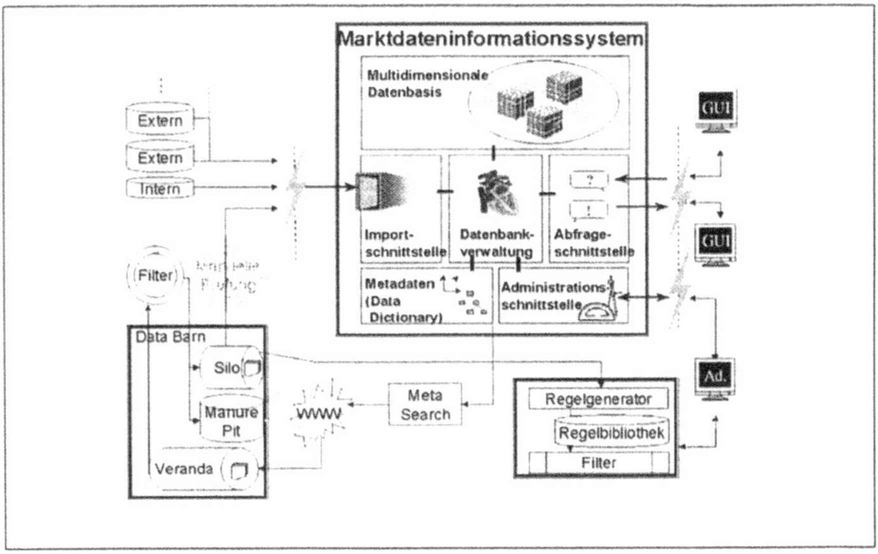

---

[14] Zur Galaxie vgl. Hahne (1999, S. 158 f.).

Ausgewählte Metadaten aus dem Data Dictionary bilden die Suchbegriffe für eine Suche im WWW (Meta Search). Die Suchergebnisse werden in die Data Barn übertragen. Die Data Barn stellt transiente (Veranda und Silo) und persistente (Manure Pit) Speicher für die Bearbeitung der Internetseiten zur Verfügung. Den zentralen Problembereich der Integration bildet in diesem Kontext die Regelgenerierung. Zur Klassifizierung wendet ein Filter die zuvor generierten Regeln an. Dazu wird ein Multilayer Perceptron (MLP)[15] mit Deskriptoren, Deskriptorenergänzungen sowie Datenqualitätsanforderungen angewendet.[16] Zum Abschluss des Integrationsprozesses müssen die interessanten Internetseiten mit den jeweiligen multidimensionalen Datenstrukturen verbunden werden, um bei den Analysen den Zugriff zu ermöglichen. Dies geschieht in Form einer weiteren Dimension für diese Datentypen im Star Schema.[17] Die uninteressanten Internetseiten verbleiben als Trainingsdatenbestand in der persistenten Manure Pit.

4.2 Nutzung des Visualisierungskonzeptes für eine einheitliche Benutzeroberfläche

Ein zentrales Problem besteht darin, die Daten aus heterogenen Quellen in einer Oberfläche dem Analysten zur Verfügung zu stellen, so dass dieser eine tatsächliche Entscheidungsunterstützung erfährt.

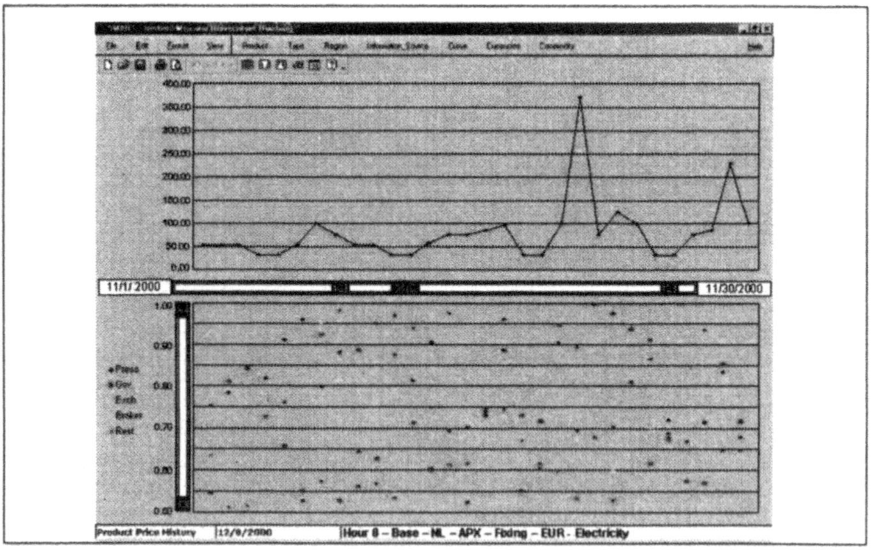

[15] Zum MLP vgl. Bishop (1995, S. 117 ff.).

[16] Zu beachten ist, dass zurzeit das Ergebnis der Filterung manuell zu überprüfen ist.

[17] Aus diesem Grunde ist eine objektorelationale Datenbank erforderlich.

Der geteilte Bildschirm wird optisch durch den mittig positionierten Timeslider erreicht. Das obere Diagramm stellt die Marktdaten aus bekannten Quellen in einem Kurvendiagramm dar, das untere Diagramm die in der Datenbank erfassten Internetdokumente in Form eines sogenannten Starfield Display[18]. In diesem Kontext ist zu beachten, dass ein kausaler Zusammenhang zwischen den Kursverläufen und den ergänzenden Internetdaten nicht zwingend besteht. Beide Diagramme orientieren sich jeweils an der in der Datenbank erfassten Gültigkeitszeit (der Timeslider bezieht sich auf beide Darstellungen), dabei resultiert diese für die Internetdokumente aus dem Datum, an dem das Dokument auf dem jeweiligen Internetserver platziert wurde. In der Statusleiste rechts unten ist zu erkennen, welche Selektion in den Dimensionsmenüs vorgenommen wurde. Die möglichen Vollbilddarstellungen werden für Analysen der Energiehändler verwendet. Die rechte und linke Seite des Timeslider zeigt die Einstellung zugunsten der unteren und oberen Grenze des abgebildeten Zeitintervalls. Die durch Slidebuttons veränderbaren Einstellungen werden angezeigt. Durch Verschiebung der Pfeilbutton wird das Datum herauf- oder entsprechend heruntergesetzt. Analog dazu werden auch Slidebuttons für die Bestimmung der Grenzen für die Internetdokumente verwendet. Diese befinden sich direkt neben dem Diagramm der Dokumentenanzeige.

5 Fazit

Durch die Entwicklung des Marktinformationssystems ist die Zusammenführung interner externer Daten gelungen. Diese Datenbasis ermöglicht Entscheidungsträgern verbesserte Analysemöglichkeiten, um Tendenzen im Energiemarkt frühzeitig zu erkennen. Dabei gewährt das Modell und das System eine hohe Flexibilität bei Anpassungen an veränderten Rahmenbedingungen im Energiemarkt. Es verbleibt für die Weiterentwicklung des Informationssystems die Aufgabe, die Bewertung externer Informationen zu optimieren sowie den schon jetzt weitestgehend automatisierten Integrationsprozess vollständig ohne menschliche Eingriffe ausführen zu lassen.

Literaturverzeichnis

Balzert (2001): Balzert, H.: Lehrbuch der Softwaretechnik I, 2. Auflage; Heidelberg, Berlin, 2001.

Bulos (1998): Bulos, D.: OLAP Database Design - A New Dimension, in: Chamoni/Gluchowski (1998), S. 251-261.

[18] Zur Visualisierung und insbesondere zum Starfield Display vgl. Card/MacKinlay/Shneiderman (1999).

Card/MacKinlay/Shneiderman (1999): Card, S. K.; Mackinlay, J. D.; Shneiderman, B.: Readings in Information Visualization – Using Vision to Think, San Francisco, 1999.

Chamoni/Gluchowski (1998): Chamoni, P.; Gluchowski, P.: Analytische Informationssysteme - Data Warehouse, On-Line Analytical Processing, Data Mining; Berlin, [u. a.], 1998.

Chamoni/Gluchowski (1999a): Chamoni, P.; Gluchowski, P.: Analytische Informationssysteme - Data Warehouse, On-Line Analytical Processing, Data Mining, 2. Auflage; Berlin [u. a.], 1999.

Chamoni/Gluchowski (1999b): Chamoni, P.; Gluchowski, P.: Analytische Informationssysteme - Einordnung und Überblick, in: Chamoni/Gluchowski (1999a), S. 3-25.

Hahne (1999): Hahne, M.: Logische Datenmodellierung für das Data Warehouse - Bestandteil und Varianten des Star Schemas, in: Chamoni/Gluchowski (1999a), S. 145-170.

Hansen (1998): Hansen, H. R.: Wirtschaftsinformatik I, 7. Auflage, Stuttgart, 1998.

Lewis/Rieman (1993): Lewis, C.; Rieman, J.: Task-centered User-Interface Design – A practical Introduction, URL: http://bmrc.berkeley.edu/courseware/cs160/fall99/Book/, 1993, 31.08.2001.

Nielsen (1993): Nielsen, J.: Usability Engineering; Boston [u. a.], 1993.

Schinzer/Bange (1999): Schinzer, H. D.; Bange, C.: Werkzeuge zum Aufbau analytischer Informationssysteme - Marktübersicht, in: Chamoni/Gluchowski (1999a), S. 45-74.

Stahlknecht (1995): Stahlknecht, P.: Einführung in die Wirtschaftsinformatik, 7. Auflage, Berlin, Heidelberg, New York, [u. a.], 1995.

Stock (2001): Stock, S.: Modellierung zeitbezogener Daten in einem Data Warehouse. Diss.; Wiesbaden, 2001.

Wertz (1986): Wertz, C. J.: The Data Dictionary - Concepts and Uses; Amsterdam, 1986.

Der Einsatz der Clusteranalyse zur Definition von Referenzanlagen

Sandrine Nunge, Jutta Geldermann, Otto Rentz

Deutsch-Französisches Institut für Umweltforschung (DFIU / IFARE)
Universität Karlsruhe (TH), Hertzstr. 16, D-76187 Karlsruhe

1 Problemstellung

Für die Ableitung und Formulierung von Luftreinhaltestrategien zur Bekämpfung der weitreichenden länderübergreifenden Luftverschmutzung ist eine konsistente Abbildung und Aggregation der technologischen Realität sowie deren techno-ökonomische Charakterisierung erforderlich. Dabei besteht ein erster Schritt in der Identifizierung der relevanten Emissionsquellen. Insbesondere bei der Schadstoffgruppe flüchtige organische Verbindungen (engl. Volatile Organic Compounds, VOC) ist die technologische Realität durch eine große Anzahl heterogener, teils kleiner Anlagen gekennzeichnet, die sich im einzelnen voneinander durch die eingesetzten Prozesse, die generierten Emissionen, die implementierten Minderungsmaßnahmen und die Emissionsminderungskosten unterscheiden. Die VOC-Problematik besteht darin, dass kleine Anlagen zwar im einzelnen einen geringen Beitrag zu den Gesamtemissionen leisten, jedoch in der Gesamtheit betrachtet keinesfalls vernachlässigbar sind. Im Gegensatz zu SO_2-relevanten Emissionsquellen, wie beispielsweise große Feuerungsanlagen, die genehmigungsbedürftig sind, unterliegt die Mehrheit der VOC-relevanten Emissionsquellen nicht der Genehmigungsbedürftigkeit. Daher ist es schwierig, innerhalb der vergleichsweise kurzen Zeit, die für die Erarbeitung von Luftreinhaltestrategien den politischen Entscheidern zur Verfügung steht, jede individuelle Anlage eines Sektors einzeln zu ermitteln. Einen wesentlichen Beitrag zur vereinfachten Darstellung der technologischen Realität und zur geeigneten Erfassung von technischen, umweltrelevanten und ökonomischen Eigenschaften von VOC-Emissionsminderungsmaßnahmen auf Anlagen- bzw. Prozessebene, liefert der *Referenzanlagenansatz*.

1.1 Der Referenzanlagenansatz

Der Referenzanlagenansatz besteht darin, jede reale Anlage einer homogenen Anlagenkategorie zuzuweisen, wobei jede Anlagenkategorie durch eine durchschnittliche bzw. repräsentative Anlage, eine sog. *Referenzanlage*, repräsentiert wird. Bislang wurde bei der Festlegung und Definition von Referenzanlagen eine pragmatische Vorgehensweise verfolgt, indem die Abstimmung in Expertenkreisen durchgeführt wurde (vgl. [RENTZ 1999]). In diesem Beitrag wird die Eignung des methodischen Verfahrens der Clusteranalyse als Basis für die Referenzanlagenbildung untersucht.

1.2 Clusteranalyse

Ziel einer Clusteranalyse ist die Zusammenfassung der zu klassifizierenden Objekte zu Klassen, so dass die Objekte innerhalb einer Klasse möglichst ähnlich und die Klassen untereinander möglichst unähnlich sind [BACKHAUS 1994, SPÄTH 1983]. Eine Klassifizierung von *m* Objekten (hier: technische Anlagen, die als M = {1, 2, ..., m} durchnumeriert sind) besteht darin, *n* Teilmengen (auch Klassen, Kategorien, Cluster genannt) $C_j \subset M$ zu identifizieren, die diese *m* Objekte beinhalten und den folgenden Anforderungen gerecht werden:

$$C_1 \cup C_2 \cup \ldots\ldots \cup C_n = M$$

$$C_i \cap C_k = \varnothing \quad (i \neq k)$$

$$m_j := |C_j| \geq 1 \quad (j=1, 2, \ldots., n)$$

$$1 \leq n \leq m$$

Für die vorliegende Problemstellung sollen die Anlagen eines gegebenen Sektors nach zwei Eigenschaften bzw. Gruppen von Eigenschaften in Klassen aufgeteilt werden: Eingesetzte Prozesse/Verfahren und Anlagengröße/-kapazität. Für ausgewählte VOC-Emissionsquellen ist eine Übersicht von Anlagen und deren zugrundeliegenden Parametern in *Tabelle 1* zusammengefasst.

***Tabelle 1:** Qualitative und quantitative Variablen für Anlagen ausgewählter VOC-relevanter Sektoren*

Sektor	Qualitative Variable	Quantitative Variable
Feuerungen im Haushaltsbereich (< 50 MWth)	Brennstofftyp, z.B. Steinkohle, Braunkohle, Heizöl, Gas, Holz, ...	thermische Kapazität
Verarbeitung von Mineralölerzeugnissen in Raffinerien	verschiedene Prozesse z.B. Erdölverarbeitung, physikalische Trennverfahren, Konversionsverfahren, ...	Erdölverarbeitungskapazität
Bandbeschichtung (Coil coating)	Anlagentyp, z.B. selbstständige oder betriebseigene Anlage Beschichtetes Material (Al, Stahl)	Produktionskapazität für jeweils Stahl und Aluminium
Chemische Reinigung	Reinigungsmaschinetyp, z.B. offene, geschlossene Maschine Verwendete Lösemittel	Kapazität der Reinigungsmaschine Kapazität gereinigter Textilien

In diesem Beitrag wird die Eignung der Clusteranalyse zur Referenzanlagenbildung am Beispiel der Verarbeitung von Mineralölerzeugnissen in Raffinerien untersucht. Dafür werden die 337 Raffinerien in 30 Ländern der UN/ECE-Region herangezogen (vgl. [RHODES 1996]. Zunächst soll eine Clusteranalyse nach Prozessen durchgeführt werden: Das Ergebnis wird eine Anzahl *K* an Clustern sein. Jeder dieser *K* Cluster soll anschließend in Größen- bzw. Kapazitätsklassen unterteilt werden. Dabei besteht der erste Schritt in der Auswahl eines geeigneten Verfahrens zur Klassenbildung.

2 Auswahl eines geeigneten Verfahrens zur Klassenbildung

Zunächst muss die Ähnlichkeit zwischen den zu klassifizierenden Raffinerien ermittelt werden. Dabei ist wesentlich, ob die Variablen über eine quantitative oder eine qualitative Ausprägung verfügen.

2.1 Bestimmung der statistischen Maßzahlen

Die Homogenität bzw. die Heterogenität kann durch einen paarweisen Vergleich der zu klassifizierenden Raffinerien mittels einer statistischen Maßzahl (*Ähnlichkeitsmaß* bzw. *Unähnlichkeitsmaß (Distanzmaß)*) quantifiziert werden. Ein Ähnlichkeitsmaß spiegelt die Ähnlichkeit zweier Objekten wieder, d.h. je größer der Wert eines Ähnlichkeitsmaßes, desto ähnlicher sind sich die zwei Raffinerien. Ein Distanzmaß dient ebenfalls dazu, die Ähnlichkeit zwischen zwei Objekten zu ermitteln, jedoch gilt: Je kleiner der Wert des Distanzmaß, desto ähnlicher sind sich die verglichene Objekte. Das Ermittlungsverfahren der Ähnlichkeits- bzw. Distanzmaße hängt davon ab, ob es sich um qualitative oder quantitative Variablen handelt. [BACKHAUS 1994] In der Literatur ist eine Reihe von statistischen Maßzahlen aufgeführt, wie z.B. der Tanimoto-Koeffizient für qualitative Variable und die quadrierte Euklidische Distanz für quantitative Variable (vgl. [BACHER 1996]). Mit Hilfe von statistischen Maßzahlen kann die Ähnlichkeitsmatrix gebildet werden, die als Ausgangaspunkt für die Durchführung der Clusteranalyse dient.

2.2 Auswahl eines geeigneten Verfahrens zur Klassenbildung

In der Literatur werden zahlreiche Verfahren zur Clusteranalyse beschrieben. Eine Einteilung ist in *Bild 1* aufgeführt (für Erläuterungen siehe [BACHER 1996, NUNGE 2001]). Zur Lösung der vorliegenden Problemstellung muss auf die Anwendung graphentheoretischer Verfahren verzichtet werden, da sie für die Klassifizierung einer großen Menge an Objekten (337 Raffinerien) nicht geeignet sind. Informationen über die zu klassifizierenden Objekte sind verfügbar in der Form einer Liste von realen Raffinerien mit ihren Eigenschaften in Bezug auf eingesetzte Prozesse und Kapazitäten. In dieser Liste sind die Raffinerien nicht nach Raffinerietypen geordnet, so dass keinesfalls von einer vorläufigen Klassifizierung nach Raffinerietyp ausgegangen werden kann. Die partitionierenden Verfahren, die ausgehend von einer gegebenen Ordnung, die einzelnen Objekte austauschen und neu ordnen, können daher nicht angewendet werden. Deshalb wird zur Problemlösung auf die hierarchischen Verfahren zurückgegriffen. Da die Raffinerien einzeln aufgelistet sind und somit jede Raffinerie einem Cluster entspricht, wird ein agglomeratives, hierarchisches Verfahren herangezogen.

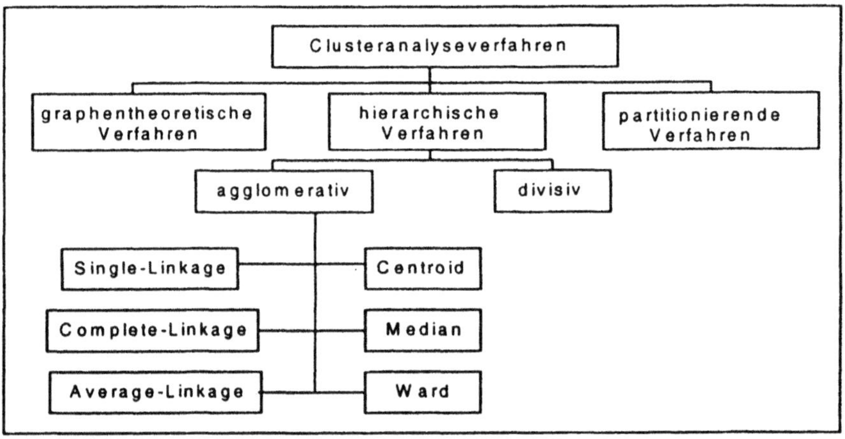

***Bild 1**: Übersicht über ausgewählte clusteranalytische Verfahren*

Der Algorithmus für die *hierarchisch agglomerative Clusterbildung* ist durch folgende Schritte gekennzeichnet [BACKHAUS 1994, BACHER 1996]:

1: Zu Beginn bildet jedes zu klassifizierende Objekt ein selbstständiges Cluster. Die Clusterzahl *K* entspricht der Anzahl von Objekten *n*.

2: Es wird nach dem Clusterpaar *({p},{q})* mit der größten Ähnlichkeit bzw. der geringsten Unähnlichkeit in der Ähnlichkeits- bzw. Distanzmatrix gesucht.
Das Clusterpaar wird zu einem neuen Cluster *{p,q}* verschmolzen.
Die Clusterzahl *K* wird um 1 reduziert.

3: Es wird geprüft, ob sämtliche Objekte untersucht wurden.
Bei "ja" wird der Algorithmus beendet. Bei "nein" wird zu Schritt 4 gegangen.

4: Die Ähnlichkeit bzw. Unähnlichkeit des neu gebildeten Clusters *{p,q}* zu den verbleibenden Clustern wird wie folgt berechnet [BACHER 1996]:

$$u_{\{p,q\},i} = \min\left(u_{p,i}, u_{q,i}\right) \quad \text{bzw.} \quad \ddot{a}_{\{p,q\},i} = \max\left(\ddot{a}_{p,i}, \ddot{a}_{q,i}\right)$$

mit:

$u_{p,i}$ bzw. $u_{q,i}$: Unähnlichkeitsmaß zwischen den Raffinerien p und i bzw. zwischen q und i

$ä_{p,i}$ bzw. $ä_{q,i}$: Ähnlichkeitsmaß zwischen den Objekten (z.B. Raffinerien) p und i bzw. zwischen q und i

5: Zurück zu Schritt 2.

2.3 Bestimmung der Clusteranzahl

Aufgrund des kurzen Zeitraums, der für die Erarbeitung von Luftreinhaltestrategien eingeräumt werden kann, besteht eine realistische Durchführung in der Charakterisierung einer sinnvollen Anzahl von Referenzanlagen. Da darüber hinaus eine Vielzahl von Sektoren zu berücksichtigen sind (etwa 40 für VOC), sollte eine Anzahl von etwa 20 Klassen, also 20 Referenzanlagen, pro Sektor nicht überschritten werden.

3 Durchführung der Clusteranalysen[1]

Zielsetzung ist es, die 337 Raffinerien in 30 Staaten der UN/ECE-Region in Clustern aufzuteilen.[2]

3.1 Clusteranalyse nach Prozessen (qualitative Variablen)

In der Prozessbeschreibung muss auf Unterschiede in der technischen Ausführung bei Prozessen (vgl. [ECKER 2000, CONCAWE 1999]) detailliert eingegangen werden, so dass nicht nur Informationen über das Vorhandensein oder Fehlen der verschiedenen Prozesse vorliegen, sondern auch detaillierte Informationen (wie z.B. über Betriebsbedingungen, Stoff- und Energieverbräuche) wesentlich sind. Diese detaillierten Informationen gehen allerdings bei der Umwandlung in Ähnlichkeitsmaße verloren. Deshalb muss eine problemadäquate, in diesem Fall empirische Klassifizierung durchgeführt werden. Dabei werden zunächst verschiedene Prozessgruppen unterschieden:

I. Raffinerien mit ausschließlicher Rohölverarbeitung (21 Anlagen)

---

1 Die praktische Durchführung der Clusteranalysen erfolgt mit Hilfe des statistischen Computerprogramms SPSS (Superior Performing Software System).

2 Die Liste der zu klassifizierenden Raffinerien sowie eine kurze Übersicht über die eingesetzten Prozesse in Raffinerien gibt [NUNGE 2001].

II. Raffinerien ohne Rohölverarbeitung (28),
III. Raffinerien mit physikalischen Trennverfahren (ohne Konversionsverfahren) (34)
IV. Raffinerien mit Konversionsverfahren (ohne physikalische Trennungsverfahren) (102),
V. Raffinerien mit thermischen Konversionsverfahren (140),
VI. Raffinerien mit katalytischen Konversionsverfahren (9),
VII. Raffinerien mit thermischen und katalytischen Konversionsverfahren (3).

Das Kriterium, nach dem diese empirische Klassifizierung durchgeführt wurde, ist die Einsetzbarkeit von möglichen Maßnahmen zur Minderung von VOC-Emissionen, damit die Abbildungsgenauigkeit gewährleistet werden kann. Insbesondere bei prozessintegrierten Maßnahmen ist die Unterscheidung der Prozesstypen von vorrangiger Bedeutung.

3.2 Clusteranalyse nach Kapazitäten (quantitative Variablen)

Damit die Ergebnisse der Clusteranalysen so wenig wie möglich verfälscht werden, sollen in einem ersten Schritt Ausreißer aus der gegebenen Menge an Raffinerien entfernt werden. Die Identifizierung von Ausreissern erfolgt mittels Boxplots. Im vorliegenden Fall entsprechen Ausreißer Raffinerien, für die die VOC-Emissionen und die Kosten nicht mehr denjenigen der entsprechenden Referenzanlage zugeordnet werden können.

Nach der Identifizierung von Ausreißern wird die Distanzmatrix beispielhaft für die 136 Raffinerien des Raffinerietyps V ermittelt. Zunächst wird zu diesem Zweck das Distanzmaß der quadrierten Euklidischen Distanz herangezogen. Die Durchführung der Clusteranalyse zur Bildung von drei Clustern führt zu Raffineriekategorien mit sehr unterschiedlich großen Objektmengen (Cluster 1: 8, Cluster 2: 46 und Cluster 3: 82 Objekte je Cluster). Die Festlegung einer Anzahl von vier Clustern ergibt relativ homogene Cluster, wobei die jeweiligen Objektmengen homogener sind (Cluster 1: 8, Cluster 2: 46, Cluster 3: 46 und Cluster 4: 36)[3].

Für jedes der vier Cluster wird nun eine Referenzanlage definiert, deren repräsentative Erdölverarbeitungskapazität dem arithmetischen Mittelwert der jeweiligen Bandbreite an Kapazitäten im Cluster entspricht. Zusätzlich zu dieser charakteristischen Kapazität wird jede Referenzanlage durch die untere und/oder obere Kapazität der Bandbreite spezifiziert. Die folgenden Referenzanlagen für den Raffinerietyp V können im Anschluss an diese erste Clusteranalyse definiert werden:

- Referenzanlage 1 (entspricht Cluster 1): 13,7 Mio. Tonnen verarbeitetes Rohöl pro Jahr, Bereich ab 12 Mio. t/a
- Referenzanlage 2 (entspricht Cluster 2): 7,9 Mio. Tonnen verarbeitetes Rohöl pro Jahr, Bereich von 6 Mio. bis 12 Mio. t/a

---

[3] Cluster 1 und Cluster 2 wurden beibehalten, Cluster 3 wurde in zwei Cluster aufgeteilt.

- Referenzanlage 3 (entspricht Cluster 3): 4,2 Mio. Tonnen verarbeitetes Rohöl pro Jahr, Bereich von 3 Mio. bis 6 Mio. t/a
- Referenzanlage 4 (entspricht Cluster 4): 2,2 Mio. Tonnen verarbeitetes Rohöl pro Jahr, Bereich bis 3 Mio. t/a.

Zur erneuten Überprüfung der Ergebnisse wird die Clusteranalyse nach einem hierarchisch agglomerativen Clusteranalyseverfahren anschließend für weitere Distanzmaße eingesetzt. Dabei ergeben sich keine gravierenden Abweichungen.

4 Diskussion

4.1 Validierung der Ergebnisse

Die Überprüfung bzw. Validierung der Ergebnisse einer Clusteranalyse kann durch Anwendung des formalen Verfahrens der Diskriminanzanalyse erfolgen. Die Diskriminanzanalyse ist ein strukturen-prüfendes Verfahren, das zur Analyse von Gruppenunterschieden herangezogen werden kann. (vgl. [BACKHAUS 1994]) Jedoch sollten bei der Entwicklung von Emissionsminderungsstrategien auf supranationaler Ebene die Ergebnisse nationalen (Branchen-)Experten[4] zur Prüfung und Diskussion vorgelegt werden, um die Ergebnisse der Clusteranalyse zu validieren.

4.2 Übertragbarkeit der Vorgehensweise auf ähnlich strukturierte Sektoren

Die für den Sektor der Verarbeitung von Mineralölerzeugnissen angewandte Vorgehensweise kann auf weitere gut dokumentierten Sektoren übertragen werden, z.B. die chemische Reinigung oder die Bandbeschichtung (Coil Coating). Für schwach dokumentierte Sektoren kann daher derzeit allein auf Expertenwissen oder auf die umweltrechtlichen Regelwerke zurückgegriffen werden. So kann sich die Bildung von Anlagenkategorien und die anschließende Definition von Referenzanlagen beispielsweise an den Kapazitätsklassen und Schwellenwerte für den jährlichen Lösemittelverbrauch orientieren, die in der EU-Lösemittel-Richtlinie (1999/13/EG) genannt werden. Dabei dienen die Schwellenwerte als Unter- bzw. Obergrenzen für die jeweiligen Referenzanlagen.

---

4 Beispielsweise im Rahmen von Arbeitsgruppen, wie bei Protokollverhandlungen unter der UN/ECE oder in den sog. Technical Working Groups bei der Bestimmung von besten verfügbaren Techniken auf europäischer Ebene.

4.3 Zusammenfassung

Zur Darstellung der technologischen Realität bietet die Bildung von Referenzanlagen für die Klassifizierung von Anlagen und Prozessen eine problemadäquate Lösung: Der Referenzanlagenansatz ermöglicht es, die Charakterisierung von Minderungsmaßnahmen auf eine sinnvolle Anzahl von Anlagenkategorien zu beschränken.

Anhand des Beispielsektors "Verarbeitung von Mineralölerzeugnissen" kann aufgezeigt werden, dass die sektorweise Klassifizierung von Anlagen und anschließende Definition von Referenzanlagen mit Hilfe der Clusteranalyse grundsätzlich durchgeführt werden können. Die Übertragbarkeit der vorgestellten Vorgehensweise auf weitere Sektoren ist somit grundsätzlich möglich. Jedoch liegt bei der Durchführung der Klassenbildung von Anlagen das wesentliche Problem in der Datenverfügbarkeit bzw. -qualität. Die Lösung besteht in einem iterativen Prozess, indem die Datengrundlage kontinuierlich verbessert wird.

5 Literatur

[BACHER 1996] BACHER, J.: *Clusteranalyse - Anwendungsorientierte Einführung*, 2. Auflage, München, Wien, Oldenburg, 1996

[BACKHAUS 1994] BACKHAUS, K.; ERICHSON, B.; PLINKE, W.; WEIBER, R.: *Multivariate Analysemethoden - Eine anwendungsorientierte Einführung*, 7. Auflage, Springer-Verlag, Berlin, Heidelberg, 1994

[CONCAWE 1999] CONCAWE: *Best available techniques to reduce emissions from refineries*, Prepared for the CONCAWE Air and Water Quality Management Groups by its Special Task Forces AQ/STF-55 and WQ/STF-28, Brüssel, Mai 1999

[ECKER 2000] ECKER, A.; WINTER, B.: *Stand der Technik bei Raffinerien im Hinblick auf die IPPC-Richtlinie*, Monographien, Band 119, Umweltbundesamt, Wien, 2000

[NUNGE 2001] NUNGE, S.: *Der Referenzanlagenansatz zur Ableitung von Luftreinhaltestrategien*, VDI Verlag, voraussichtlich 2001, zugl.: Universität Karlsruhe, Dissertation, 2001

[RENTZ 1999] RENTZ, O.; NUNGE, S.; LAFORSCH, M.; HOLTMANN, T.: *Technical Background Document for the Actualisation and Assessment of UN/ECE Protocols related to the Abatement of the transboundary Transport of Volatile Organic Compounds from Stationary Sources*, Report of the Task Force on the Assessment of Abatement Options/Techniques for Volatile Organic Compounds from Stationary Sources, on behalf of the German Federal Ministry for the Environment, Nature Conservation and Nuclear Safety and the Federal Environmental Agency, Karlsruhe, September 1999

[RHODES 1996] RHODES, A.K.: *Worldwide Refining - Distillation Capacity Exceeds 76 million b/d, Hydrotreating Surges*, in: Oil & Gas Journal Special, December 23, 1996

[SPÄTH 1983] SPÄTH, H.: *Cluster-Formation und -Analyse, Theorie, FORTRAN-Programme und Beispiele*, R. Oldenbourg Verlag, , München Wien, 1983

MESAP/TIMES - Advanced Decision Support for Energy and Environmental Planning

Uwe Remme[1], Gary A. Goldstein[2], Ulrich Schellmann[1], Christoph Schlenzig[1]

[1] Institute of Energy Economics and the Rational Use of Energy (IER), University of Stuttgart, Pfaffenwaldring 31, 70550 Stuttgart, Germany, ur@ier.uni-stuttgart.de
[2] International Resources Group (IRG), 1211 Connecticut Ave. NW, Suite 700, Washington DC 20036, ggoldstein@irgltd.com

Abstract. In view of the requirements for climate and environmental protection and the increased international competition within a deregulated energy market, authorities and energy providers want to improve their competence of strategic planning in order to identify robust decisions for the upcoming threats and opportunities. This paper presents the planning tool MESAP/TIMES which integrates the TIMES optimization model and the MESAP software environment for energy and environmental planning. Questions that may be studied with MESAP/TIMES range e.g. from the planning of a local energy system, over the evaluation of energy technologies on a national scale up to the analysis of carbon permit trading strategies in the international Kyoto protocol discussion.

1 Introduction

TIMES (**T**he **I**ntegrated **MARKAL-EFOM** **S**ystem) is a mathematical modelling scheme for representing, optimizing and analyzing energy systems on a flexible time and regional scale. The TIMES model has been developed under the auspices of the International Energy Agency (IEA) within the Energy Technology Systems Analysis Programme (ETSAP) [2]. The TIMES development pursues the goals of merging the advantages of existing energy system models like MARKAL [3] and EFOM [7], of eliminating some shortcomings in these models and of creating a modelling environment being readily adaptable to new ideas and methodologies. It has been implemented in the equation-based modelling language GAMS. The model has been designed for the long-term analysis of energy, environmental and economic (E3) issues over a time-horizon ranging from several years to decades. Typical questions analyzed with such a model are for example the assessment of greenhouse gas abatement strategies or capacity expansion planning in the electricity sector.

MESAP, the **M**odular **E**nergy **S**ystem **A**nalysis and **P**lanning toolbox developed at IER since 1984, was designed to analyze strengths and weaknesses of energy systems, to validate possible future strategies using computer based energy system models and to monitor the effectiveness of the implemented strategies [5]. MESAP uses a flexible process-engineering oriented technique for modelling energy systems and a user friendly relational database management system to integrate different operations research methodologies with one single database. The MESAP shell for TIMES supports the user in the process of building an energy system and facilitates a structured analysis of the results.

This paper is divided in two sections. The first section describes the TIMES methodology. In the second part the MESAP system and the integration of TIMES as new module within the MESAP environment are presented.

2 The Energy System Model TIMES

TIMES follows a so-called bottom-up systems engineering approach that allows a detailed technical description and economic evaluation of the energy technologies. The energy system modelling approach used by TIMES can be divided into the four parts model topology, numerical data, mathematical structure and scenarios. Due to this separation and the generic formulation of the model equations, the TIMES methodology can be easily applied to different case studies ranging from a local energy system of a municipality over the national energy system of a country to the analysis of multinational energy system. Therefore, TIMES can be considered as a model generator.

2.1 Model topology and time dimension

Reference energy system (RES) The mathematical description within TIMES is based on the concept of a reference energy system (RES). The RES models the energy system as a network of processes and commodities being connected by commodity flows. Depending on the analyzed energy system, the RES may cover the entire energy system of a country from the primary energy supply over conversion, transport and distribution to the end-use sectors as shown simplified in Fig. 1.

By doing this, competition and interaction between different parts of the system may be studied, e.g. competition between increased usage of energy efficient or conservation technologies in the end-use sectors and fuel-switching to less carbon intensive energy carriers in the conversion sector to reduce CO_2 emissions. One characteristic aspect of an RES is, that in most cases a process does not depict a single plant, but rather an entire technology type being available in the energy system. The commodities being produced or consumed by the processes may be energy carriers, energy services, materials, emissions or money. Another feature of an RES is, that besides the existing stock of technologies used today it also contains a portfolio of new technologies that are still under research or too expensive today, but might become competitive in the future.

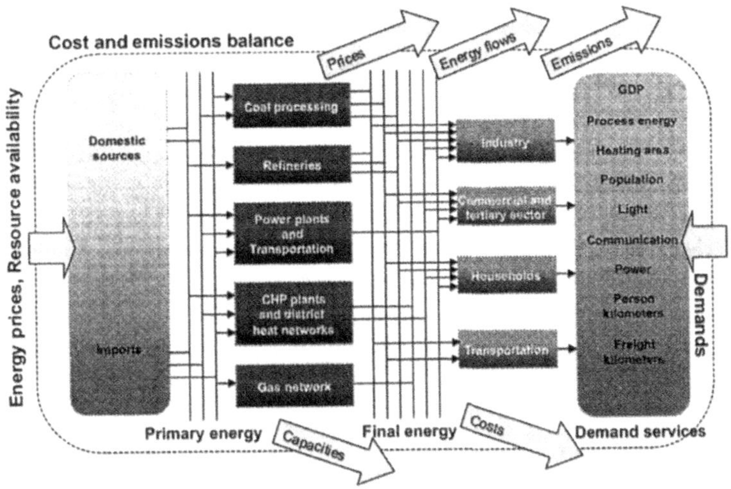

Fig. 1. Concept of a reference energy system (RES).

Regions A model in TIMES can consist of a variable number of regions (Fig. 2). The regions are connected by inter-regional exchange processes describing the trade activities between the internal model regions. The exchange processes might represent abstract trade flows but can also model physical transport processes like electricity grids or pipelines. A multi-regional model might for example be useful to study the benefits of international cooperation in greenhouse gas abatement by carbon permit trading or to analyze the trade of energy intensive products and the development of electrical or natural gas grid capacity in liberalized energy markets.

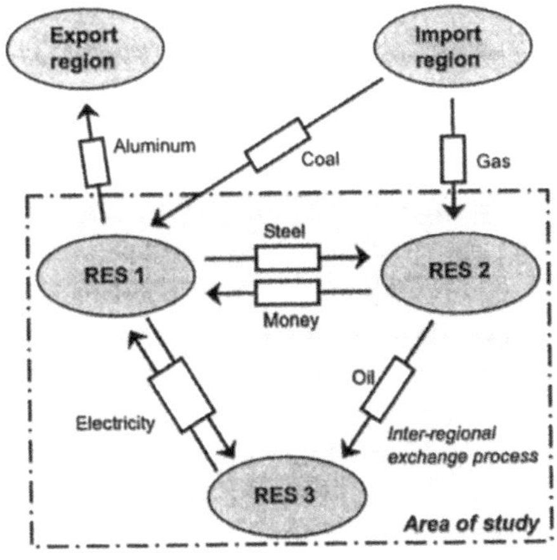

Fig. 2. Building multi-regional models by exchange processes.

Time horizon and time segments In order to describe dynamic aspects over time, the time horizon is divided into periods which may have durations from one to several years. The number of periods and their durations are completely under user-control. Assigning existing capacities to their installation year (in contrast to the usage of residual curves) makes changes in the starting year of the model horizon relatively easy. The average year of a period can again be divided in different subannual time-segments on up to four levels (annual, seasonal, weekly, daily) to model seasonal, weekly or daily variations of e.g. the electricity and heat demand. TIMES uses a load curve and not a load duration curve, so that for example the concurrent production of heat and electricity in combined heat and power plants can be captured in the model.

2.2 Numerical data

The numerical data required by the model as inputs are either global or are assigned to a process, a commodity or a commodity flow. Typical input data comprise technical and economic description of the processes (e.g. efficiencies, availabilities, investment costs, existing capacities), end-use demands, resource availability, import prices of energy carriers and policy measures (e.g. subsidies or taxes) or targets (e.g. greenhouse gas reduction goal).

2.3 Mathematical structure

Using the RES abstraction, the energy system is described by a generic system of linear equations including for example mass and energy balances, transformation equations for the processes relating the input to the output flows, capacity constraints limiting the activity of a process by its available capacity and capacity transfer constraints between successive time periods. In an addition to the standard set of equation types, TIMES offers the user a generic equation framework to add non-standard, user-specific constraints to the model without having to write an equation in GAMS code, for example to put a lower bound on the electricity produced from renewables. The decision variables of the model are the commodity flows between the processes (e.g. the choice of energy carriers), the production or activity of the processes and new installation of technologies. The objective function of TIMES is expressed as the discounted sum of annual costs minus revenues, so as to provide year-by-year reporting of net costs. The model accepts technology-specific discount rates as well as a general discount rate. It is also possible to distinguish between the technical and economic lifetime of a technology. The capacity whose technical life has not expired at the end of the modelling horizon is salvaged in the objective function. This means, that all investment related costs taking place after the last model year are compensated and thus do not effect optimization. Further features of the TIMES methodology include:

- option to use price elastic demands,
- storage processes between time-slices of the load curve and between periods,
- representation of the intensity of attributes, such as sulfur content of fuel,
- vintaging of a technology allowing to describe the technical characteristics of installed capacity (e.g. its efficiency) as a function of its construction year,
- possibility to alter technical parameters of an installed capacity over its lifetime, e.g. decline of the availability of a power plant at the end of its lifetime due to aging,
- flexible process description giving the option to let the model determine the optimal ratio of input/output flows for processes with several inputs/outputs.

2.4 Scenarios

Due to the longevity of most of the energy technologies as well as their high capital costs, investment decisions in the energy sector usually have to take into a account a time horizon of several decades. Several key assumptions made in energy system models, like future evolution of end-use demand or energy prices, cannot be predicted over such a long horizon. To address these uncertainties, usually different scenarios describing possible future developments are created to compare their possible outcomes.

3 MESAP

Building and analyzing an energy system model is a complex task. The modeler has not only to create an abstract simplified representation of the real-world system which still accurately depicts the technical characteristics of the system. Since the optimization is driven by the cost parameters, one also has to ensure that the price relationships, e.g. the shadow prices of different energy carriers induced by import, transportation and process related cost parameters, are realistic. This situation is complicated by the fact, that energy system models usually contain a large number of technologies (up to several hundreds). In addition, the model size and the dependencies within the model are increased by the introduction of time periods, load curves and multiple regions. Furthermore, the model is usually applied for several scenarios creating a huge amount of results to be analyzed. Because of these reasons a software tool is essential to support the modeler in the tasks of building a model and analyzing its results. Moreover, such a planning tool helps to make the model - often considered as a black box from outsiders - and its results more transparent.

3.1 Philosophy of MESAP

For these purposes the integrated energy and environmental planning tool MESAP has been developed at IER. The main characteristics of MESAP are:

- a set of integrated planning models sharing the same database,
- user friendly data entry and model handling,
- powerful and flexible analysis tools,
- common language for energy system data enabling the easy data sharing between case studies and models,
- modular architecture facilitating methodological enhancements.

Fig. 3 summarizes the architecture of MESAP which consists of three parts: the calculation modules, the MESAP information system and additional tools for specific modules or purposes.

The calculation modules include INCA an investment calculation module, PlaNet an energy system simulation model, PROFAKO an operational planning model for electricity and district heat, a general XTRACTOR creating sets and parameters for GAMS models, CalQlator a general equation editor and the optimization model TIMES. The MESAP information system comprises the central database, the Explorer for building and navigating through the database and basic tools for entering data (DataSheet) and analyzing them (Analyst, DataCube). Additional tools are available for special modules, e.g. the RES Editor for the energy system models PlaNet and TIMES, or for special purposes like accessing the database via the Internet.

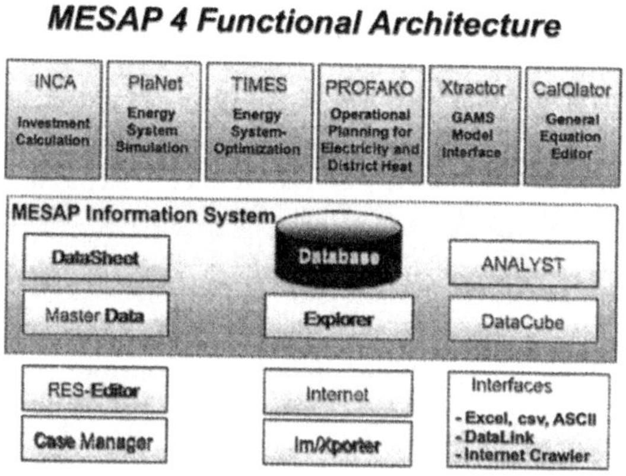

Fig. 3. The architecture of MESAP4.

3.2 Integration of TIMES in MESAP

Due to the modularity of the MESAP concept and its standardized data interface, a continuous integration of new methodologies and planning models has been possible. Currently the integration of TIMES in MESAP is underway. The different tasks in the course of building, running and analyzing a model are:

- defining time horizon and model topology,
- specifying processes and commodities,
- entering and editing data,
- defining scenarios,
- running the model,
- analyzing the results.

A great part of these tasks can be accomplished by MESAP components. For the definition of the model topology the RES editor (Fig. 4) can be used.

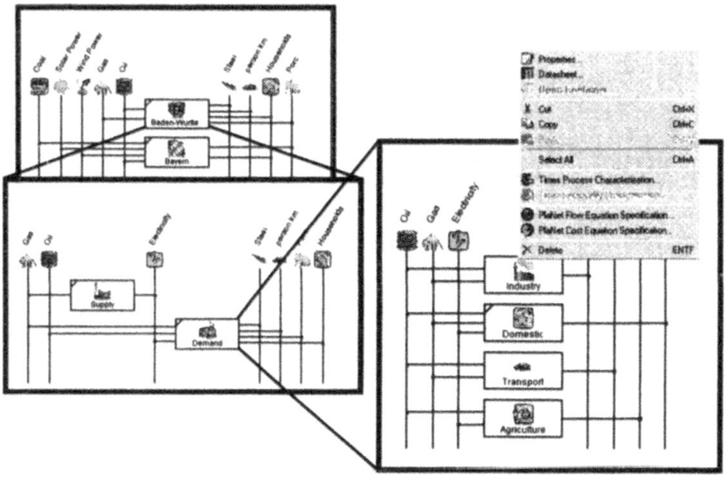

Fig. 4. RES Editor with containers for structuring the RES and direct access to further information.

Besides regions the RES editor also allows additional structuring of the RES in sector or technology containers. Within the RES editor the user can intuitively navigate through the model. The graphical process and commodity objects offer a direct link to additional forms for the specification of processes and commodities (e. g. determining the commodity type) and to the DataSheet for entering and editing data. In addition to a pure process or commodity oriented view, the user can create and save specific views on the data, e.g. investment costs of all coal power plants.

To run the model, the input information needed for TIMES is extracted out of the database in an ASCII file. In the opposite direction MESAP will allow the import of already existing TIMES models. The Analyst is developed as an COM-Add-In programme for Microsoft Excel. It offers within Excel the possibility to create tables, graphics and reports, which through a direct reference to the database can be automatically updated, if new results are present. Another analysis tool, the DataCube, opens the opportunity to pivot rows and columns in a table or to change the sort order of rows and columns in a drag-and-drop manner providing different perspectives on the results. In the DataCube it is also possible to drill-down to the underlying values of an aggregated number, e.g. to check the individual electricity production of each coal plant constituting the total electricity generated by coal.

4 Outlook

The TIMES model has been developed with the goal to tackle some drawbacks in existing energy system models and to create a modelling platform being open to new model paradigms and ideas. Currently endogenous technology learning (ETL) [6], which allows to describe the investment costs of a technology as a function of the cumulative installed capacity, is implemented in TIMES. Another possible extension is the introduction of an uncertainty index in the parameters and sets, which will support stochastic programming or fuzzy logic. Further future enhancements may include the linkage of TIMES to other models like a general equilibrium model similar to MARKAL-MACRO [4] or the connection with an environmental impact assessment model like EcoSense [1]. In this context, the ongoing integration of MESAP into TIMES will provide powerful and flexible tools, so that the modeler or planner can spend his time on the central issues of his work not being sidetracked by cumbersome data converting and processing tasks.

References

1. European Commission DG XII, Science Research and Development, JOULE (editor): Externalities of fuel cycles - ExternE Project, Vol.7 - Methodology 1998 Update, EUR 19083 EN, Brussels-Luxemburg, 1998.
2. ETSAP, The New TIMES: a Model for the Millenium, In: Hill, D. (Editor): ETSAP News. 7, No. 1, January, 2001, (available under `http://www.ecn.nl/unit_bs/etsap/newslet/main.html`)
3. Goldstein, G. A., Greening, L. A. (2001) Energy planning and the development of carbon mitigation strategies - Using the MARKAL family of models, (available under `http://www.ecn.nl/unit_bs/etsap/reports/markal.html`).
4. Manne, A., Wene, C.-O. (1982) MARKAL-MACRO: A linked model for energy-economy analysis, BNL-47161, Brookhaven National Laboratory.
5. Schlenzig, C., Reuter, A. (1996) An Information and Decision Support System for Energy and and Environmental Planning. In: Carraro, C., Haurie, A. (editors): Operations Research and Environmental Management, Kluwer Academic Publishers, Dordrecht, Netherlands.
6. Seebregts, A.J., Kram, T. et. al. (1998) Endogenous technological learning: Experiments with MARKAL, Report ECN-C-98-064, Petten, Netherlands.
7. Van der Voort, E., Doni, E. et. al. (1984) EFOM-12C Mark I, Energy Supply Modelling Package, Mathematical Description, Rapport8896.

Modellierung und Implementierung von Stammdaten für die Demontageplanung und -steuerung mit ERP-Systemen

Frank Schultmann, Magnus Fröhling, Otto Rentz

Institut für Industriebetriebslehre und Industrielle Produktion, Universität Karlsruhe (TH) Hertzstraße 16, 76187 Karlsruhe
E-Mail: frank.schultmann@wiwi.uni-karlsruhe.de

Abstract:
Vor dem Hintergrund gestiegener Anforderungen an die Produktverantwortung von Herstellern gewinnen Fragestellungen einer wirtschaftlichen Demontage zunehmend an Bedeutung. Neben der Entwicklung demontagespezifischer Planungssysteme bietet sich grundsätzlich auch die Möglichkeit an, Aufgaben der Demontageplanung und -steuerung (DPS) mit Enterprise Resource Planning-Systemen (ERP-Systemen) wahrzunehmen. Eine Schlüsselrolle in diesem Zusammenhang spielen Stammdaten als zentrale Daten- und Planungsgrundlage dieser Systeme. Im vorliegenden Beitrag werden daher die demontagespezifischen Anforderungen an Stammdaten für die DPS umrissen. Anhand des ERP-Systems SAP® R/3® wird exemplarisch untersucht, wie diese Anforderungen von ERP-Systemen erfüllt werden können.

1 Einleitung

Produzierende Unternehmen werden mit einer zunehmend strengeren Produktverantwortung für die von ihnen hergestellten Güter konfrontiert. Diese beschränkt sich nicht ausschließlich auf die Herstellung und Inverkehrbringung von Produkten, sondern erstreckt sich mittlerweile über deren gesamten Lebenszyklus [1]. Fragestellungen der Rücknahme und Entsorgung von Produkten am Ende ihrer Gebrauchsphase gewinnen daher zunehmend an Bedeutung. Eine Schlüsselrolle in diesem Zusammenhang spielt eine kosteneffiziente Demontage zurückgenommener Altprodukte. Unter Demontage soll dabei die zerstörungsfreie oder zerstörende Zerlegung von Produkten in Bauteile und -gruppen mit dem Ziel einer Verwendung oder Verwertung verstanden werden. Neben dem Einsatz spezieller

Planungswerkzeuge für die Demontage von Produkten[1] lassen sich Planungsaufgaben für Demontage und Recycling auch in die Produktionsplanung und -steuerung (PPS) integrieren [2]. Ein vielversprechender Ansatz zur Realisierung einer wirtschaftlichen Demontage ist daher die Integration von deren Planungs- und Steuerungsaufgaben in die von Enterprise Resource Planning-Systemen (ERP-Systemen)[2] wahrgenommene PPS. Den ERP-Stammdaten und damit auch deren Definition, Auswahl und Abstimmung kommt dabei zentrale Bedeutung zu, da sie die Basis sämtlicher Geschäftsprozesse in ERP-Systemen bilden [3]. Ziel des vorliegenden Beitrags ist daher die Untersuchung, wie nötige Produktdaten ermittelt, dargestellt und implementiert werden können, um eine Demontageplanung und -steuerung (DPS) mit ERP-Systemen zu ermöglichen. Hierbei wird nach einem kurzen Überblick über die Stammdatenermittlung und -modellierung für die DPS deren Implementierung exemplarisch am Beispiel von *SAP® R/3®* verdeutlicht[3].

2 Ermittlung und Modellierung von Stammdaten für die Demontageplanung und -steuerung

Eine zentrale Voraussetzung zum Einsatz von ERP-Systemen zur Demontageplanung und -steuerung ist die Gewinnung der hierfür relevanten Daten und deren Überführung in eine Form, die eine einfache Überrtragung in die Stammdatensätze des ERP-Systems, hier *SAP® R/3®*, ermöglicht. Die Anforderungen an die Ermittlung und Modellierung von Stammdaten zur DPS variieren dabei abhängig vom Ausgangspunkt der Betrachtungen. Grundsätzlich kann eine Stammdatenermittlung bei Unkenntnis von Stammdaten der Produktion, bei Kenntnis von Stammdaten der Produktion und zum Zeitpunkt der Konstruktion eines neuen Produktes erfolgen.

Zusätzlich sind hierbei zum Teil konkurrierende Kriterien, wie etwa eine möglichst automatische Datengenerierung [4], die Berücksichtigung der sich während des Produktgebrauchs u.U. ändernden Produktstruktur [5], die Einbeziehung von Teilen, deren Demontage nicht zwingend notwendig ist [6] sowie die Notwendigkeit evtl. andere Demontagewege als die Umkehrung der Montage zu nutzen [7] zu beachten und dabei abhängig von der jeweiligen Ausgangslage unterschiedlich zu gewichten.

---

[1] Vgl. z.B. [10], [1], [17].

[2] Zum Begriff des Enterprise Resource Planning-Systems vgl. [18].

[3] Dem vorliegenden Beitrag liegen Untersuchungen auf dem Releasestand 4.0B von *SAP® R/3®* zu Grunde. Zu Aufbau und Funktionsumfang von *SAP® R/3®* vgl. z.B. [3].

Stammdaten für die Demontage umfassen dabei hauptsächlich die Produktstruktur, die Mengenbeziehungen einzelner Komponenten sowie die durchzuführenden Demontagevorgänge. Zu deren Modellierung können zunächst *Strukturbilder*, *Recycling-*, *Verwertungs-* und *Demontagegraphen* sowie *Demontagestrukturen* herangezogen werden (vgl. [8], [2], [9], [10]). Das Ergebnis einer solchen Modellierung kann jeweils die Darstellung der Demontagedaten in Form eines Demontagegraphen bilden, da dessen Struktur der Aufteilung der Stammdaten in ERP-Systemen sehr ähnlich ist. Im Folgenden werden konkrete Maßnamen zur Implementierung der so vorliegenden Daten am Beispiel von *SAP® R/3®* verdeutlicht.

3 Implementierung von Stammdaten für die Demontageplanung und –steuerung in *SAP® R/3®*

Die z.B. in Demontagegraphen dargestellten Demontageinformationen lassen sich in den Stammdatenstrukturen *Material*, *Baukasten-Stückliste*, *Arbeitsplatz* und *Arbeitsplan* von *SAP® R/3®* speichern. Für jedes Produkt, jede Baugruppe und jedes Teil sind hierfür zunächst Materialstammsätze anzulegen. Produktstruktur und Mengenbeziehungen lassen sich über Baukasten-Stücklisten abbilden. Daten zu benötigten Ressourcen (z.B. Maschinen) werden in Arbeitsplätzen, Demontageaktivitäten bzw. -vorgänge in Arbeitsplänen gespeichert. Die prinzipielle Übertragbarkeit wird in Abbildung 1 veranschaulicht. Bei der Implementierung sind jedoch einige Besonderheiten der Demontage im Vergleich zur „herkömmlichen" Produktion zu berücksichtigen. Hierauf wird in den folgenden Abschnitten eingegangen.

3.1 Speicherung demontage- und recyclingrelevanter Informationen

Der Einsatz von ERP-Systemen zur Demontageplanung und -steuerung erfordert naturgemäß die Möglichkeit, demontage- und recyclingrelevante Informationen in diesem System ablegen zu können [5], [11]. Hierbei sind eine Vielzahl von Daten wie etwa Qualitätsmerkmale, Informationen zur Lagerfähigkeit, zu Aufarbeitungs-, Aufbereitungs- Verwendungs- und Verwertungsmöglichkeiten von Komponenten [12], [13] und Ausschuss, der möglichst in Abhängigkeit von Nutzungsdauer, Verschleiß und Beanspruchung abgebildet werden sollte [14], [9], zu berücksichtigen. Zudem sollten auch Informationen über Verbindungen einzelner Produktkomponenten sowie über Stoffe, insbesondere hinsichtlich deren Einordnung in Gefahrstoffklassen darstellbar sein [9], [15], [1].

Eine große Anzahl der geforderten demontage- und recyclingrelevanten Informationen lässt sich in den standardmäßig vom System *SAP® R/3®* bereitgestellten Datenfeldern der Materialstammsätze hinterlegen. Hierzu zählen u.a. Daten zur Lagerfähigkeit sowie Zuordnungen zu Gefahrstoffklassen. *SAP® R/3®* stellt zudem mit dem Modul *QM* (*Quality Management*) ein umfassendes Werkzeug zum Qualitätsmanagement bereit. Über prozentuale Ausschussmengen lässt sich ebenso der bei der Demontage vermehrt anfallende Ausschuss grob berücksichtigen, auch wenn es bislang nicht möglich ist, diese Ausschussmengen in Abhängigkeit von den o.g. Parametern anzugeben. Reichen die standardmäßig bereitgestellten Datenfelder zur Speicherung der benötigten Daten nicht aus, ist es im Rahmen der Systemanpassung (sog. Customizing) möglich, weitere Datenfelder und Bildschirmmasken hinzuzufügen.

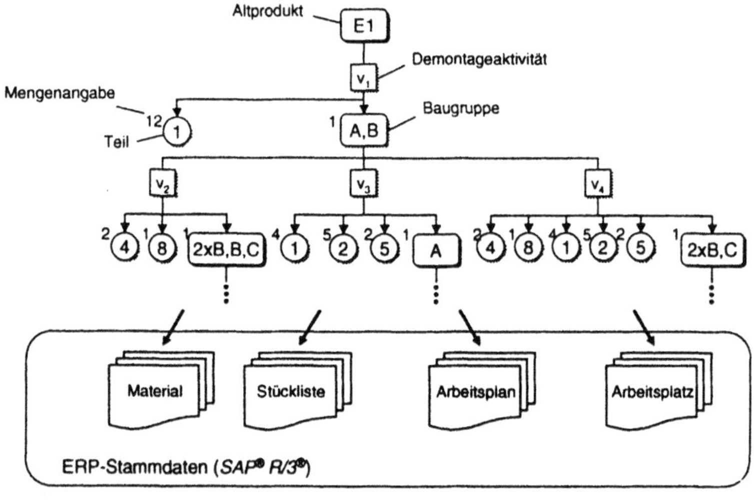

Abbildung 1: Übertragbarkeit von Demontagestammdaten in ERP-Systeme

3.2 Unterscheidung von Primär-, Alt- und Sekundärprodukten

Im Rahmen der Integration von Aufgaben der DPS in die PPS ist es sinnvoll, Produkte, Baugruppen und –teile unterschiedlicher Qualitäten und Eigenschaften in getrennten Beständen zu führen [14]. Hierfür bietet sich eine Gliederung in Primär-, Alt- und Sekundärprodukte an, da diese einer Gliederung des Güterkreislaufs in Produktion, Demontage und Aufarbeitung/Aufbereitung folgt. Zusätzlich lassen sich so die Qualitätsdifferenzen zwischen Primär-, Alt- und Sekundärprodukten herausstellen.

In *SAP® R/3®* lässt sich eine solche Unterscheidung auf zwei komplementäre Arten erreichen. Zum einen können in der Systemanpassung neue Materialarten definiert werden, zum anderen lässt sich diese, in erster Linie systeminterne,

Unterscheidung durch unterschiedliche Materialnummern, z.B. durch Einführen einer neuen Stelle, erreichen.

3.3 Abbildung von Kuppelproduktstrukturen

Die Demontage von Produkten lässt sich als mehrstufiger, linearer Kuppelproduktionsprozess mit starrer oder elastischer Kopplung modellieren [10], [1]. Daher muss ein ERP-System, welches zur Planung und Steuerung der Demontage eingesetzt werden soll, die Möglichkeit besitzen, solche Kuppelproduktionsprozesse adäquat abzubilden.

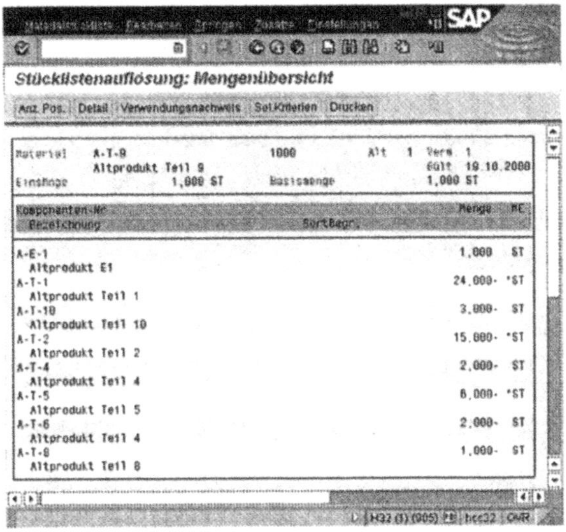

Abbildung 2: Demontagestückliste des Beispielproduktes *E1*

In *SAP® R/3®* werden Kuppelproduktionsprozesse mit Hilfe von *Haupt-*, *Kuppel-* und *Nebenprodukten* modelliert. Das Produkt, dessen Fertigung Hauptzweck eines Kuppelproduktionsprozesses ist, wird dabei als Hauptprodukt bezeichnet. Für Hauptprodukte werden Stücklisten angelegt, in denen die bei der Produktion dieses Hauptproduktes entstehenden Kuppelprodukte mit negativen Bedarfsmengen aufgeführt werden. Abhängig davon, ob diesen Kuppelprodukten Kosten zugeordnet werden sollen oder nicht, wird von Kuppel- bzw. Nebenprodukten gesprochen. Abbildung 2 zeigt exemplarisch eine Stücklistenauflösung für die Demontage des in Abbidlung 1 eingeführten Beispielproduktes E1. Es ist zu erkennen, dass eine Demontage eines Altproduktes als Produktion einer ihrer Komponenten modelliert wird. Das Altprodukt geht mit positiver Mengenangabe ein, entstehende Kuppelprodukte sind durch negative Bedarfsmengen gekennzeichnet. Das Hauptproblem bei einer auf diese Weise abgebildeten Demontage liegt darin, dass sich eine Demontage nicht durchgängig planen lässt, sobald auf einer Demontagestufe mehr als ein weiter zu demontierendes Altprodukt entsteht.

3.4 Variantenmanagement

Die durch die stärker gewordene Kundenorientierung ohnehin schon große Anzahl an Produkvarianten vervielfacht sich, wenn deren Demontage und Recycling betrachtet werden. Änderungen im Fertigungsablauf während der Zeitspanne, in der die Produkte hergestellt werden [9], vor allem aber Modifikationen während des Produktgebrauchs sowie unterschiedliche Qualitätszustände durch Gebrauch, Verschmutzung und Korrosion insbesondere bei langlebigen Gütern, sind die Ursachen hierfür [16], [1]. ERP-Systeme, die für die DPS eingesetzt werden müssen in der Lage sein, dieser Variantenvielfalt Rechnung zu tragen.

Die wesentlichen Anforderungen hinsichtlich des Variantenmanagement lassen sich mit der Funktion der Variantenkonfiguration von *SAP® R/3®* lösen, auch wenn diese sehr komplex und mit einem enormen Pflegeaufwand der Daten verbunden ist. Da sich diese Funktion im Rahmen dieses Beitrages nicht auch nur annähernd darstellen lässt, wird auf weitere Ausführungen verzichtet[4].

3.5 Implementierung demontagebezogener Ressourcen- und Prozessdaten

Ressourcen für die PPS (z.B. Fertigungshilfsmittel), in *SAP® R/3®* in sog. Arbeitsplätzen abgelegt, unterscheiden sich datentechnisch nicht signifikant von Ressourcen für die DPS. Lediglich Kosten- und Leistungsdaten für die in der Demontage eingesetzten Ressourcen sind u.U. zu aktualisieren [9]. Generelle Anforderungen an Ressourcendaten für die Demontage (vgl. [13]) werden i.d.R. von heutigen ERP-Systemen erfüllt, es kann daher an dieser Stelle auf weitere Betrachtungen verzichtet werden.

Analog zu den behandelten Ressourcendaten sind auch die in ERP-Systemen i.d.R. in Arbeitsplänen abgelegten Prozessdaten für Produktion und Demontage sehr ähnlich [6], [14]. Arbeitspläne können neben Arbeitsgangfolgen zur Erstellung von Gütern oder Dienstleistungen, u.a. auch Vorgabewerte sowie Zuordnungen von benötigten Materialien und Fertigungshilfsmitteln enthalten. Anforderungen in diesem Bereich betreffen im Wesentlichen die Zuordnung der Arbeitspläne zu Materialien und die Flexibilität der Auswahl von Arbeitsgangfolgen. Sollen ERP-Systeme auch für die DPS eingesetzt werden, so müssen mehrere Arbeitspläne, bspw. für Produktion und Demontage einem Material zugeordnet werden können [2]. Zudem müssen die Arbeitspläne flexibel ausgestaltet sein [6]: Alternative Arbeitsgangfolgen, die z.B. unterschiedliche Demontagewege beschreiben, sollten ebenso darstellbar sein wie kurzfristige Änderungen, die sich etwa ergeben wenn bei der Demontage unerwartet Abweichungen in der Produktstruktur vorgefunden werden.

---

[4] Vgl. hierzu [19].

Generell lassen sich die Prozessdaten für die DPS in Arbeitspläne in *SAP® R/3®* überführen. Probleme entstehen, wenn unterschiedliche Demontagefolgen zur Demontage eines Produktes gespeichert werden sollen. Arbeitspläne werden den entstehenden Produkten zugeordnet was in diesem Fall problematisch ist, da unterschiedliche Demontagefolgen häufig auch zu unterschiedlichen Produkten führen.

4 Zusammenfassung und Ausblick

Zusammenfassend lässt sich feststellen, dass die Anforderungen an die Modellierung und Implementierung von Stammdaten für eine Demontageplanung und -steuerung mit Hilfe von *SAP R/3-PP* weitgehend erfüllt werden können. Neben den angesprochenen Problemen bei der Abbildung von Kuppelproduktstrukturen, dem sehr komplexen Variantenmanagement und der Arbeitsplanzuordnung bereitet nur eine fehlende manuelle Auswahl der zu verwendenden Stücklisten gewisse Probleme. Werden diese Probleme umgangen oder die betreffenden Funktionen geändert, lassen sich die Planungs- und -steuerungsfunktionen von *SAP R/3-PP* für die grundlegenden Aufgaben der Demontageplanung und -steuerung einsetzen. Die Ergänzung bislang fehlender demontaggespezifischer Planungswerkzeuge wie z.B. zur Prognose des Altproduktaufkommens, zur Bestimmung der optimalen Demontagetiefe oder zur Verfahrenswahl[5] würden die Vorteile integrierter mit denen spezieller Systeme verbinden. Es stünden somit umfassende Werkzeuge zur Demontageplanung und -steuerung zur Verfügung.

Literatur

1. Schultmann, F. (1998) Kreislaufführung von Baustoffen: stoffflußbasiertes Projektmanagement für die operative Demontage- und Recyclingplanung von Gebäuden, Erich Schmidt, Berlin

2. Rautenstrauch, C. (1997) Fachkonzept für ein integriertes Produktions-, Recyclingplanungs- und -steuerungssystem (PRPS-System), DeGruyter, Berlin u.a.

3. Hildebrand, K.; Rebstock, M. (2000) Betriebswirtschaftliche Einführung in SAP® R/3®, Oldenbourg, München und Wien

4. Kurbel, K.; Schneider, B.; Etzrodt, A. (1995) Von Produktionsdaten zu Recyclinginformationen, in Haasis, H.-D.; Hilty, L. M.; Kürzl, H.; Rautenstrauch, C.: Umweltinformationssysteme in der Produktion, Metropolis, Marburg, 165-170

5. Huber, A. (2000) Demontage-Primärbedarfsplanung. Umwelt, 6, 56-59

[5] Vgl. z.B. [17].

6. Kurbel, K.; Rautenstrauch, C. (1996) Integrated Planning of Production and Recyling Processes – an MRP II-based Approach. Proceedings of the 2nd International Conference on Managing Integratied Manufacturing. Management Centre, University of Leicester, Großbritannien, 189-194

7. Bönker, T.; Möller, T. (1995) Wirtschaftliche Demontage von Elektrogeräten und Leiterplatten durch Teilautomatisierung. Maschinenmarkt, 33, 33-39

8. Gehrmann, F. (1986) Konstruktion und werterhaltendes Recycling niederwertiger technischer Gebrauchsgüter, dargestellt am Beispiel Hauhaltskleinmaschinen, VDI, Düsseldorf

9. Schneider, B. (1999) Recycling-Informationssysteme: Integration von Produktion und Recycling, Gabler, Wiesbaden

10. Spengler, T. (1994) Industrielle Demontage- und Recyclingkonzepte, Erich Schmidt, Berlin

11. Kraus, M.; Heimig, I.; Scheer, A.-W. (1995) Informationsmodell für die integrierte Entsorgungssicherung, in: Haasis, H.-D.; Hilty, L. M.; Kürzl, H. (Hrsg.): Betriebliche Umweltinformationssysteme (BUIS) – Projekte und Perspektiven, Metropolis, Marburg

12. Corsten, H.; Reiß, M. (1991) Recycling in PPS-Systemen. DBW, 51, Nr. 5, 615-627

13. Haasis, H.-D.; Rentz, O. (1992) Umwelt PPS – Ein weiterer Baustein einer CIM-Architektur?, in: Görke, W.; Rininsland, H.; Syrbe, M. (Hrsg.): Information als Produktionsfaktor, Springer, Berlin u.a., 235-241

14. Kurbel, K.; Rautenstrauch, C. (1997) Integration des Produktrecycling in die Produktionsplanung und -steuerung, in: Weber, H. (Hrsg.): Umweltmanagement: Aspekte einer umweltbezogenen Unternehmensführung, Schäffer-Poeschel, Stuttgart

15. Kaase, W.; Ebach, H.; Roth, N. (1996) DV-gestütztes Recycling Management-System. Industrie Management, 2,13-18

16. Feldmann, K.; Meedt, O. (1996) Demontage elektronischer Geräte unter Nutzung effizienter Werkzeuge und lebenszyklusübergreifender Produktdaten, Industrie Management, 2, 30-34

17. Rudolph, A. (1999) Altproduktentsorgung aus betriebswirtschaftlicher Sicht, Physica, Heidelberg

18. Kurbel, K. (1999) Produktionsplanung und -steuerung: Methodische Grundlagen von PPS-Systemen und Erweiterungen, 4. Auflage, Oldenbourg, München u.a.

19. Fröhling, M.; Schultmann, F.; Rentz. O (2001) Variantenmanagement mit SAP® R/3® im Rahmen der Produktdemontage. Working Paper. Institut für Industriebetriebslehre und Industrielle Produktion, Universität Karlsruhe (TH)

Energy/Environmental Modeling with the MARKAL Family of Models

Ad J. Seebregts[1], Gary A. Goldstein[2], Koen Smekens[1]

[1] Energy Research Centre of the Netherlands (ECN), Policy Studies unit, P.O. Box 1, 1755 ZG Petten, The Netherlands, e-mail: seebregts@ecn.nl
[2] International Resources Group Ltd., 1211 Connecticut Ave. NW, Suite 700, Washington, DC 20036, USA, e-mail: ggoldstein@irgltd.com

Abstract:
This article presents an overview and a flavour of almost two decades of MARKAL model developments and selected applications. The MARKAL family of models has been contributing to energy/environmental planning since the early 1980's. Under the auspices of the International Energy Agency's (IEA) Energy Technology Systems Analysis Programme (ETSAP) the model started as a linear programming (LP) application focused strictly on the integrated assessment of energy systems. It was followed by a non-linear programming (NLP) formulation which combines the 'bottom-up' technology model with a 'top-down' simplified macro-economic model. In recent years, the family was enlarged by members to model material flows, to employ stochastic programming (SP) to address future uncertainties, to model endogenous technology learning using mixed integer programming (MIP) techniques, and to model multiple regions (NLP/LP).

1 Introduction

MARKAL (acronym for MARKet ALlocation) is a widely applied bottom-up, dynamic, originally and mostly a linear programming (LP) model developed by the Energy Technology Systems Analysis Programme (ETSAP) of the International Energy Agency (IEA) (ETSAP, 2001). MARKAL depicts both the energy supply and demand side of the energy system. MARKAL provides policy makers and planners in the public and private sector with extensive detail on energy producing and consuming technologies, and it can provide an understanding of the interplay between the macro-economy and energy use. As a result, this modeling framework has contributed to national and local energy planning, and to the development of carbon mitigation strategies. The MARKAL family of models is unique, benefiting from application in wide variety of settings and global technical support from the international research community.

Implementation in more than 40 countries and by more than 80 institutions, including developed, transitional, and developing economies indicates wide acceptability.

This article provides a brief overview of the various model variants illustrated by a selection of results from the vast and still growing amount of MARKAL experience world-wide. Section 2 will give a short description of the main structure of the model and the various members of the family, followed by a few examples in Section 3. More examples and further references can be found at the ETSAP web site.

Along with its general introduction to MARKAL and its potential uses, this paper provides potential MARKAL users with a sense of the international commitment now driving innovation and evolution of the modeling framework. The framework constantly benefits from the contributions of talented researchers in institutions around the world, unlike those models that are the product of a single institution.

2 MARKAL: Overview

As with most energy system models, energy carriers in MARKAL interconnect the conversion and consumption of energy. This user-defined network includes all energy carriers involved with primary supplies (e.g., mining, petroleum extraction, etc.), conversion and processing (e.g., power plants, refineries, etc.), and end-use demand for energy services (e.g., boilers, automobiles, residential space conditioning, etc.). The demand for energy services may be disaggregated by sector (i.e., residential, manufacturing, transportation, and commercial) and by specific functions within a sector (e.g., residential air conditioning, heating, lighting, hot water, etc.). The building blocks depicted in Fig. 1 represent this network, referred to as a Reference Energy System (RES).

The optimization routine used in the model's solution selects from each of the sources, energy carriers, and transformation technologies to produce the least-cost solution subject to a variety of constraints. The user defines technology costs, technical characteristics (e.g., conversion efficiencies), and energy service demands. As a result of this integrated approach, supply-side technologies are matched to energy service demands. The specification of new technologies, which are less energy- or carbon-intensive, allows the user to explore the effects of these choices on total system costs, changes in fuel and technology mix, and the levels of greenhouse gases and other emissions. Therefore, MARKAL is highly useful for understanding the role of technology in carbon mitigation efforts and other energy system planning settings.

A variety of different constraints may be applied to the least-cost solution. These constraints include those related to a consistent representation of the energy system, such as balancing energy inputs and outputs, utilization of capacity, replacement of expended capacity by new investments and satisfaction of demand. In addition, environmental or policy issues, such as greenhouse gas emissions, may be examined in several ways, including sectoral or system-wide emissions limits on an annual basis or cumulatively over time. Alternatively, the imposition of a carbon tax or other fee structure could be modeled if desired. As a result, various costs for carbon may be generated for different levels of emission reductions. In this way, future technology configurations are generated and may be compared. If constraints are also placed on the types of technologies and rates of penetration, the configuration of the entire energy system will change. In all cases, MARKAL will produce the least-cost solution which meets the provided set of constraints.

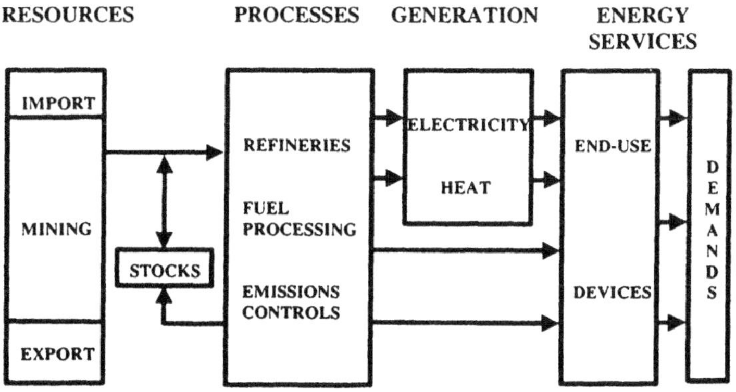

Fig. 1. MARKAL Building Blocks

Table 1 provides an overview of the current MARKAL family of models. With few exceptions, individual versions are additive, and they can be used in combination with each other where appropriate. In some instances however, features are mutually exclusive as they represent different modeling techniques that address the same needs. For example MACRO/MICRO/MED each address changes in demand levels that respond to changes in energy prices. However, because each of these versions has a different underlying theoretical development they may not be used together.

MARKAL's collaborative approach to model development is implemented through an open architecture provided by the General Algebraic Modeling System (GAMS, see (Brooke et al. 1992)). Although the theory and mathematics underlying the model are complex, MARKAL users can effectively work with the model without a complete command of the computational methods employed. This goal is achieved through the use of a modern data handling and analysis support shell: the Windows based ANSWER, developed by ABARE (Australian Bureau of Agricultural and Resource Economics). ANSWER is implemented in Visual Basic. The MARKAL input data and results are stored in MS-Access databases, with flexible import and output routines (e.g. based on Excel files).

Table 1. Overview of the MARKAL family of models

Member/version	Type of model	Short description (Reference or example)
MARKAL	LP	Standard model. Exogenous energy demand. (Fishbone et al. 1983)
MARKAL-MACRO	NLP	Coupling to macro-economic model, energy demand endogenous. (Hamilton et al. 1992)
MARKAL-MICRO	NLP	Coupling to micro-economic model, energy demand endogenous, responsive to price changes. (Regemorter and Goldstein 1998)
MARKAL-ED (MED)	LP	As MARKAL-MICRO but with step-wise linear representation of demand function. (Loulou and Lavigne 1996)
MARKAL With multiple regions	NLP	Linkage of multiple country specific MARKAL-MED and MARKAL-MACRO, including trade of emission permits. (Bahn et al. 1998)
MARKAL with material flows	LP	Besides energy flows (electricity, heat) material flows and recycling of materials can be modelled in the RES. (Gielen et al. 1998)
MARKAL with uncertainties	SP	Stochastic Programming. Only with standard model. (Ybema at al. 1998)
MARKAL-ETL	MIP	Endogenous technology learning based on learning-by-doing curve. Specific cost decreases as function of cumulative experience. (Barreto and Kypreos 1999)

3 Energy and Environmental Questions Answered by MARKAL

This section illustrates the use of MARKAL in answering typical energy/environmental policy and planning questions. These examples concern: carbon dioxide emission reduction, market-based instruments, technology dynamics and R&D. These topics are only a selection of the various questions that can be answered by the use of the MARKAL model. More examples and further references can be found at the ETSAP web site (ETSAP 2001).

How to reach carbon dioxide reduction? The widest current applications are for the analysis of policies designed to reduce carbon emissions from energy and materials consumption. Emissions of CO2 due to fossil fuel burning are virtually certain to be the dominant influence on the trends in atmospheric CO2 concentrations during the 21st century (IPCC, 2001), in ways that are expected to adversely affect the climate. Therefore, the development of the world's future energy system is critical in mitigating these adverse effects. Large reductions in CO2 emissions are required in order to stabilise CO2 concentrations. Models like MARKAL can aid in identifying least-cost solutions for meeting future emission reduction targets. Fig. 2 shows a recent example of the study (Gielen et al. 2000).

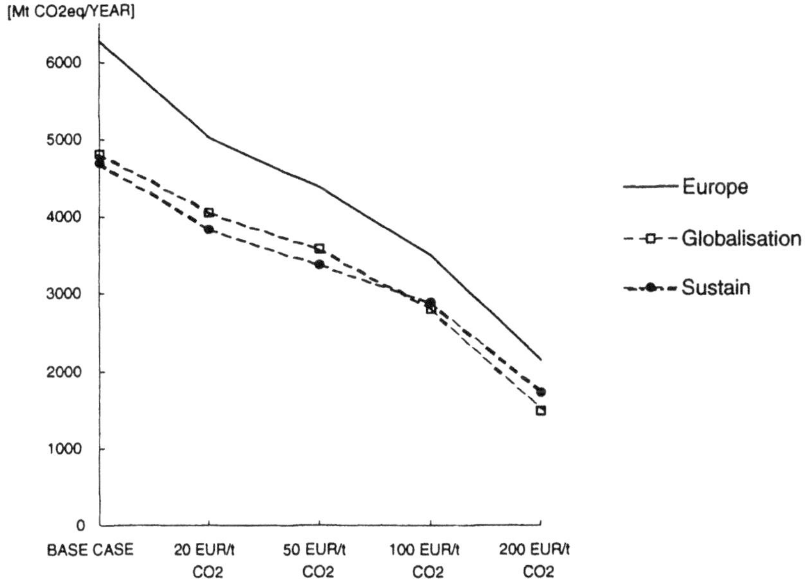

Fig. 2. CO2 emission reduction as a result of increasing GHG permit prices in 3 scenarios, 2030, Western European energy system (Gielen et al. 2000).

What is the effect of market-based instruments? In addition to technology policies, MARKAL can be used to examine market-based instruments (e.g. see also Fig. 2). Much speculation has occurred about the interaction between technology policies, energy price instruments (e.g., taxes, carbon permits), and economic growth. As the price of energy services falls in response to efficiency-induced increases in the supply of energy services, a concern is that many of the reductions from technology gains will be lost. This phenomenon is known as the "rebound" effect. As a result, it is important to design carbon mitigation or energy policies that incorporate technology and market based policies. Instruments such as the trade of emission permits (part of Kyoto protocol) have also been addressed with MARKAL, see Fig. 3 (Adapted from: Bahn et al. 1998). Emission permits can reduce the marginal cost of reducing emissions.

How to model technology dynamics and the impact of R&D? In a usual MARKAL model, technology characteristics may change in time but are exogenous parameters e.g. the investment cost a technology. The learning-by-doing curve concept can be used to model technology dynamics in a more consistent manner, but it introduces a non-linear relationship between model variables. MIP approximates the non-convex and non-linear objective function by piece-wise linear functions, and with use of a branch-and-bound algorithm a unique optimal solution can be found. PSI developed basic source code to incorporate technology learning in MARKAL (Barreto and Kypreos 1999; Seebregts et al. 2000). Fig. 4

shows an example of results obtained with this feature. In addition, two R&D variants are indicated: increase of R&D (re-allocation from research funds from nuclear to renewable energy) is assumed to have a beneficial effect on the progress ratio of solar PV and wind energy (faster decrease of investment cost). The progress ratio expresses the rate at which the cost declines each time the cumulative production (i.e. experience) doubles.

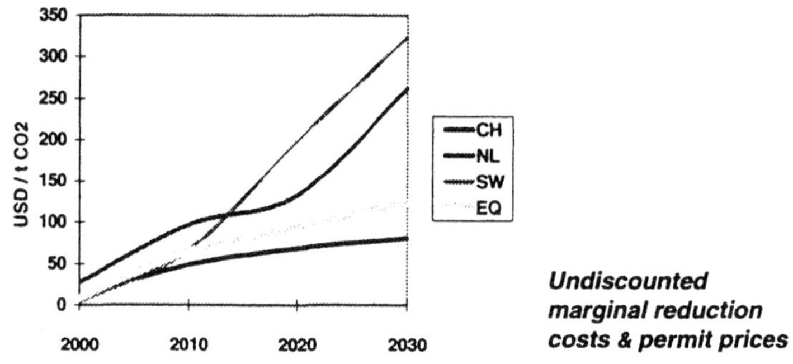

Scenario	Country	2000	2010	2020	2030
-20%	CH	-0.8	-1.5	-1.7	-3.7
	NL	-0.6	4.6	12.2	25.5
	SW	1.4	-3.2	-10.8	-22.2

Net export of CO_2 permits (Mt/year)

Fig. 3. Benefits of trading emission permits (Bahn et al. 1998) between the Netherlands (NL), Switzerland (CH) and Sweden (SW).

4 Recent Developments

TIMES (The Integrated MARKAL-EFOM System) is the evolutionary replacement for MARKAL. This modeling framework was introduced in 1999, expands the robustness with which MARKAL can address new application areas (ranging from local energy planning to technology-rich global modeling). Like MARKAL, TIMES is an optimization framework, which produces the least-cost solution subject to emissions or other constraints. The increased flexibility of the model allows for the analysis of a number of problems, which previously required undesirable compromises or were beyond the analytical limits of MARKAL. More details can be found at (ETSAP, 2001) and in (Goldstein et al. 2001).

Monte Carlo analysis is a method to analyse and propagate data uncertainties in models. It is a relatively time-consuming (i.e. computationally) method. However, with the current speed and memory capabilities of PC's, Monte Carlo analysis with complex MARKAL models now becomes feasible (Seebregts et al. 2001). Monte Carlo analysis can possibly be used complementary to conventional MARKAL practices as: Scenario analysis, sensitivity analysis, and stochastic programming.

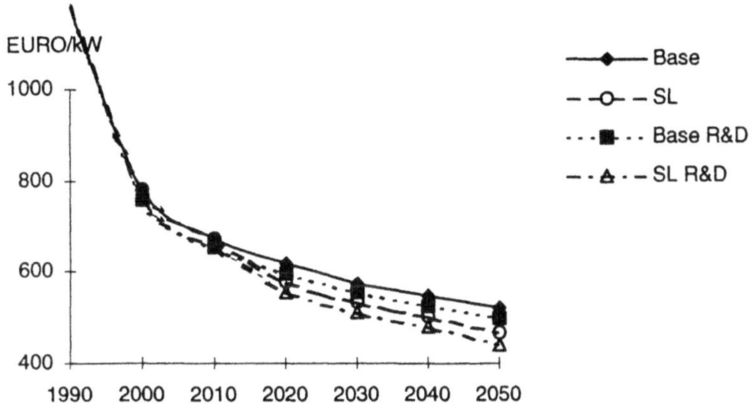

Fig. 4. Specific cost development of wind turbines. MARKAL-ETL was used, Western European energy system. 'Base' is a scenario without CO2 emission reduction; 'SL' denotes a emission reduction scenario which favors the deployment of wind turbines and hence, a faster decrease in costs.

Conclusion

The MARKAL family of models represents a series of extremely powerful tools for the analysis of energy planning with its associated environmental impacts. Through time, the model has evolved from a simple optimization framework used only by researchers, to a very sophisticated package with many potential applications to the analysis of energy/environment policy and planning questions. The Windows based "shell" ANSWER places the model framework within the reach of the user, who may not have knowledge of programming or optimization theory.

The MARKAL family provides a means of translating global commitments for the mitigation of GHG emissions into specific actions and projects. The cost effectiveness and benefits of these individual activities as well as the added benefits arising from cooperation opportunities need to be evaluated and quantified. The MARKAL family of models provides a flexible, well understood, proven, verifiable and evolving methodology that can contribute insights to assist with informed decision-making. TIMES will be the next generation member of the family.

References

1. Bahn O, Barreto L, Büeler B, Kypreos, S (1998) A Multi-regional MARKAL-MACRO Model to Study an International Market of CO2 Emission Permits: A detailed analysis of a burden sharing strategy among the Netherlands, Sweden and Switzerland. PSI technical paper, Villigen
2. Barreto L, Kypreos S (1999) Technological Learning in Energy Models: Experience and Scenario Analysis with MARKAL and the ERIS Model Prototype. PSI report Nr. 99-08, Villigen
3. Brooke A, Kendrick D. Meeraus A (1992) GAMS: A User's Guide, Release 2.25. Boyd & Fraser Publishing Company, Danvers, Massachusetts
4. ETSAP (2001) ETSAP website: http://www.ecn.nl/unit_bs/etsap
5. Gielen, DJ, Gerlagh T, Bos AJM (1998). MATTER 1.0 - A MARKAL Energy and Materials System - Model Characterisation. ECN report ECN-C—98-085, ECN, Petten, the Netherlands
6. Gielen, DJ, Bos AJM, Feber de MAPC, Gerlagh T (2000) Biomass for greenhouse gas emission reduction task 8 : optimal emission reduction strategies for Western Europe. ECN report ECN-C—00-001, ECN, Petten, the Netherlands
7. Goldstein GA, Remme U, Schellman U, Schlenzig C (2001) MESAP/TIMES - Advanced Decision Support for Energy and Environmental Planning. To be presented at this OR2001 symposium
8. Hamilton LD, Goldstein G, Lee JC, Manne A, Marcuse W, Morris SC, and Wene C-O (1992) MARKAL-MACRO: An Overview. Brookhaven National Laboratories #48377, November 1992
9. IPCC (2001) Intergovernmental Panel on Climate Change, Working Group I, Third Assessment Report, Summary for Policy Makers
10. Loulou, R, Lavigne D (1996) MARKAL model with Elastic Demands: Application to Greenhouse Gas Emission Control. In: C. Carraro, A. Haurie (eds.): Operations research and environmental management. Kluwers Academic Publishers, The Netherlands, pp 201-220
11. Regemorter D van, Goldstein G (1998) Development Of MARKAL - Towards a Partial Equilibrium Model. ETSAP Technical Paper, Dec 1998
12. Seebregts, AJ, Kram, T, Schaeffer, GJ, Bos AJM (2000) Endogenous learning of technology clusters in a MARKAL model of the Western European energy system. Int. J. Global Energy Issues, 14: 289-319
13. Seebregts, AJ, Smekens, K, Kram T (2001) Monte Carlo uncertainty analysis with MARKAL. In: Proceedings ETSAP Workshop, 15-17 May 2001, Venice, Italy.
14. Ybema, JR, Bos AJM, Seebregts AJ (1998) Incorporating the long term risk for deep emission reduction in near term CO2 mitigation strategies. Journal of Hazardous Materials 61: 217-227

Acknowledgment

The authors wish to acknowledge the support of the IEA ETSAP partners and colleagues. The IEA ETSAP community is a continuous driving force in further enhancing the MARKAL family of models, and in exchanging experience and information.

Section 3

Quantitative Methods in Marketing

Decision Trees with Parametric Enlarged Local Search

Andreas Hilbert, Henning Paul

Universität Augsburg, Lehrstuhl für Mathematische Methoden der Wirtschaftswissenschaften, D-86135 Augsburg

Abstract. Lately, decision trees became an important tool in marketing research, e.g. to classify potential customers in order to send individual ads to them. In contrast to many other situations here the classification should be done in such a manner that not the inhomogeneity of subsets is minimised but a cost orientated loss function. Unfortunately, split criteria based on cost orientated loss functions have the disadvantage to be too often indifferent between different splits. This is among others due to the local optimality of the split calculations ([1]) and due to the facts that a further improvement of the loss function could be better achieved if the best split in a "class of indifferent or nearly indifferent splits" is chosen. This paper presents a parametric enlargement of the local search for optimal splits using cost orientated loss functions.

1 Einführende Bemerkungen

Der Einsatzbereich von Entscheidungsbäumen umfasst u.a. rein deskriptive Strukturentdeckungen. In einer Gesamtmenge von Objekten sollen Teilmengen gesucht werden, die einen besonders hohen oder besonders niedrigen Anteil an einer interessierenden Klasse haben, z.B. die Kundenklassierung in potentielle Käufer und Nichtkäufer als Marketinginstrument. Dabei sollen aufgrund vorliegender Kundendaten Teilmengen mit höherem Käuferanteil und solche mit geringerem Käuferanteil entdeckt werden. Vergleichbare Fragestellungen gibt es auch in der medizinischen Diagnostik und Therapie ([2]). Derartige Anwendungsfälle haben oft zwei hier interessierende Charakteristika. Erstens können auch Teilmengen mit relativ geringen Anteilen für den Anwender wichtig sein, z.B. bei Direct Mailing Aktionen. Dabei werden i.a. diejenigen Teilmengen gesucht, deren potentieller Käuferanteil p oberhalb einer Break-Even-Schranke p* liegt. Weil hier Zielfunktionen in Abhängigkeit der Summe von Anteilswerten der interessierenden Klasse eingesetzt werden, kann es leicht vorkommen, dass mehrere unterschiedliche Splits zum selben Zielfunktionswert führen und erst der nachfolgende Split eine Verbesserung zeigt. Zum anderen sind interessante Teilmengen nicht immer durch eine sukzessive "optimale" Trennung mit Hilfe einzelner Merkmale zu erreichen, sondern nur durch den gleichzeitigen Einsatz einer geeigneten Kombination von Merkmalsausprägungen. Das hier dargestellte Verfahren ist ein Ansatz, diese Art von Problemen besser zu lösen.

2 Das Modell

In Anlehnung an [1] und [4] sollen die für Entscheidungsbaumverfahren grundlegenden Begriffe im Folgenden skizziert werden.

Für jedes Objekt i, $i \in \{1,\ldots,N\}$, wird ein Tupel $(\boldsymbol{x}_i, j)$ beobachtet. Dabei ist der Vektor $\boldsymbol{x}_i \in \mathcal{X}$ mit $\mathcal{X} \subset \mathbb{R}^M$ der Merkmalsvektor von Objekt i; der Skalar j, $j \in \{1,\ldots,J\}$, hingegen beschreibt die Klassenzugehörigkeit des Objektes, ausgedrückt durch die Merkmalsausprägung eines kategorialen Merkmals. Jeder Merkmalsvektor $\boldsymbol{x}_i$ besteht aus M Merkmalen, die qualitatives oder quantitatives Skalenniveau haben können. Die Menge aller N Beobachtungen wird mit $X = \{x_1,\ldots,x_N\}$, $X \subset \mathcal{X}$, bezeichnet, jedes $\boldsymbol{x}_i$ entspricht einem Punkt im Merkmalsraum $\mathcal{X}$.

Entscheidungsbäume haben nun die Aufgabe, jedem $\boldsymbol{x}_i \in X$ bzw. jedem $\boldsymbol{x} \in \mathcal{X}$ eine Klasse $j \in J$ zuzuordnen; d.h., jedes bekannte Objekt i bzw. jedes neue Objekt soll auf Basis seines zugehörigen Merkmalsvektors klassifiziert werden. Dazu wird eine *optimale* disjunkte Aufteilung, d.h. eine optimale Partition, $\mathcal{P}(X)$ der Beobachtungsmenge X erstellt bzw. eine solche des Merkmalsraumes $\mathcal{X}$ gesucht, die "möglichst wenig Fehler verursacht". Die daraus resultierende Funktion

$$d: X \longrightarrow \{1,\ldots,J\} \quad \text{bzw.} \quad d: \mathcal{X} \longrightarrow \{1,\ldots,J\} \tag{1}$$

wird dann als **Klassifikationsregel** bezeichnet ([3]).

Die sukzessive Erstellung dieser Partition $\mathcal{P}(X)$ ist Gegenstand der Entscheidungsbaumverfahren. Durch sukzessives Teilen der Menge X bzw. der entstehenden Teilmengen $X_t \subset X$, $t \in T$ (T bezeichnet dabei eine Indexmenge zur Verwaltung sämtlicher Teilmengen), werden weitere Teilmengen, auch **Knoten** des Entscheidungsbaums genannt, gebildet. Soll die Beziehung dieser entstehenden Teilmengen bzw. Knoten zu ihren Vorgängerknoten X_t ausgedrückt werden, so findet bei **binären Splits**, auf die sich die Darstellungen in dieser Arbeit beschränken, die Bezeichnungen X_{t_L} und X_{t_R} für die linke und die rechte Teilmenge Verwendung. Alle nicht weiter geteilten Knoten werden **Endknoten** genannt; ihre Indexmenge wird mit $\tilde{T}$ bezeichnet. Damit kann die zur Klassifikationsregel (1) äquivalente Partition $\mathcal{P}(X)$ durch die Teilmengen X_t mit $t \in \tilde{T}$ angegeben werden.

In Fig. 1 ist ein Entscheidungsbaum zu erkennen, bei dem die Beobachtungsmenge X, hier indiziert als X_0, auf den Split-Ebenen 1 und 2 binär unterteilt wird. Auf Ebene 1 erfolgt dabei eine Teilung der Grundmenge X in die Teilmengen bzw. Knoten X_1 und X_2. Im Weiteren kann dann auf Ebene 2 die Teilmenge X_2 noch einmal in die Teilmengen X_3 und X_4 gesplittet werden. Möchte man den Zusammenhang zwischen den letztgenannten Knoten verdeutlichen, wäre auch die Darstellung $X_3 = X_{2_L}$ und $X_4 = X_{2_R}$ möglich. Eine weitere Aufteilung von Knoten X_1 soll hier nicht vorgenommen werden, was dadurch angedeutet wird, dass eine Kopie von Knoten X_1 der Ebene 1 auf Ebene 2 dargestellt wird. Die oben genannten Mengen T und $\tilde{T}$ werden in diesem Beispiel durch die Indizes der Knoten des Baumes dargestellt. T umfasst dabei die Indizes sämtlicher Knoten, d.h., $T = \{1, \ldots, 5\}$, während die Menge $\tilde{T}$ sich als Menge der Indizes der Endknoten wie folgt ergibt: $\tilde{T} = \{1, 3, 4\}$. Daraus folgt auch sofort, dass die Partition, die die Menge X wiedergibt, sich wie folgt gestaltet: $\mathcal{P}(X) = \{X_1, X_3, X_4\}$

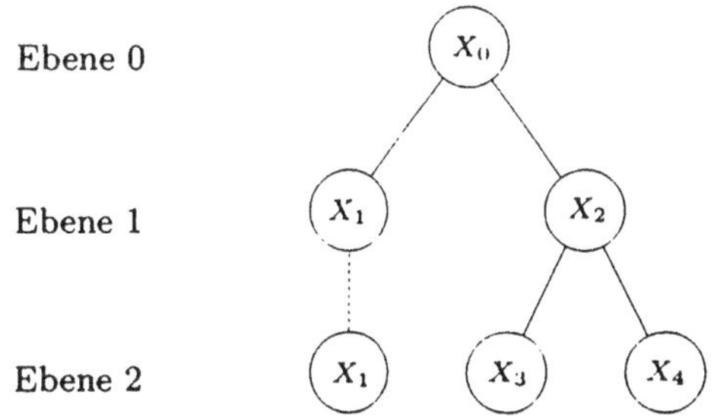

Fig. 1. Schematische Darstellung eines Entscheidungsbaumes

Des Weiteren müssen für die Menge X und die Teilmengen, d.h. Knoten, X_t noch einige absolute und relative Anteilswerte angegeben werden:

$N_j := |\{\boldsymbol{x}_i \in X : (\boldsymbol{x}_i, j)\}|$ ~ Anzahl beobachteter Objekte aus Klasse j

$N(t) := |\{\boldsymbol{x}_i \in X_t\}|$ ~ Anzahl beobachteter Objekte in Knoten t

$N_j(t) := |\{\boldsymbol{x}_i \in X_t : (\boldsymbol{x}_i, j)\}|$ ~ Anzahl beobachteter Objekte in Knoten t, die der Klasse j angehören

$p_j := \frac{N_j}{N}$ ~ Anteil beobachteter Objekte aus Klasse j

$p(t) := \frac{N(t)}{N}$ ~ Anteil beobachteter Objekte in Knoten t

$p_j(t) := \frac{N_j(t)}{N(t)}$ ~ Anteil beobachteter Objekte in Knoten t, die der Klasse j angehören

Um zu entscheiden, welche der möglichen Partitionen am geeignetsten für eine Aufgabenstellung ist, muss eine adäquate Bewertung eingeführt werden. Dazu wird einer Partition $\mathcal{P}(X)$ ein **Maß** $\mathcal{C}(\mathcal{P}(X))$ zugeordnet, das sich fast immer additiv aus den **Bewertungen** $C(X_t)$ der Endknoten X_t mit $t \in \tilde{T}$ zusammensetzt und im Folgenden stets möglichst klein sein soll:

$$\mathcal{C}(\mathcal{P}(X)) := \sum_{t \in \tilde{T}} C(X_t) \tag{2}$$

Das Splitten der Teilmengen bzw. Knoten X_t in die Teilmengen X_{t_L} und X_{t_R} erfolgt i.a. so, dass deren Bewertung besser wird als die des Knotens $C(X_t)$ und für den (binären) Splitwert $C_{SP}(X_t)$ gilt:

$$C(X_t) > C_{\mathrm{SP}}(X_t) := C(X_{t_L}) + C(X_{t_L}) \tag{3}$$

Der optimale Binärsplit $C^*_{\mathrm{SP}}(X_t)$ ergibt sich dann wie folgt:

$$C^*_{\mathrm{SP}}(X_t) = \min_{X_{t_L} \subset X_t} (C(X_{t_L}) + C(X_{t_R})) \quad \text{mit} \quad X_{t_R} = X_t \backslash X_{t_L} \tag{4}$$

Von allen bekannten Knotenbewertungsfunktionen seien hier nur die drei in ihrer Art sehr verschiedenen Funktionen C_1, C_2 und C_3 angegeben, die ihrerseits wiederum Basis der Partitionsmaße $\mathcal{C}_i(\mathcal{P}(X))$ $(i = 1, 2, 3)$ sind:

$$C_1(X_t) := -p(t) \cdot \sum_{j=1}^{J} p_j(t) \cdot \mathrm{ld}(p_j(t)) \tag{5}$$

$$C_2(X_t) := |\{(\boldsymbol{x}, j) : \boldsymbol{x} \in X_t, j \neq d(\boldsymbol{x})\}| \tag{6}$$

$$C_3(X_t) := \begin{cases} 0 & \text{für} \quad p_j(t) < p^* \\ -(E \cdot p_j(t) - c) \cdot N(t) & \text{für} \quad p_j(t) \geq p^* \end{cases} \tag{7}$$

Die Bewertungsfunktion C_1 basiert auf der **Shannonschen Entropie**, während C_2 sich als **Summe fehlklassierter Objekte** darstellt. C_3 hingegen ist eine **gewinnorientierte Bewertungsfunktion**, die zum Ausdruck bringt, dass eine mit Kosten c je Objekt behaftete Aktion nur dann stattfindet, wenn der Anteil $p_j(t)$ an einer bestimmten Klasse einen Break-Even-Punkt p^* überschreitet. Dann fallen je Objekt in dieser Klasse Einnahmen in Höhe von E an. Weitere Bewertungsfunktionen sind meist nur Abwandlungen dieser hier angegebenen Funktionen. Schließlich gelten bei diesen Bewertungsfunktionen die üblichen Konventionen wie z.B. $0 \cdot \mathrm{ld}(0) = 0$ oder $C_i(\emptyset) = 0$. Für Teilmengen X_t, die nur aus Objekten einer Klasse bestehen, gilt im Übrigen auch $C_i(X_t) = 0$.

3 Modellkritik

Der Aufgabe der Zuordnung von Objekten zu bekannten Klassen, d.h. die Klassifikation von Objekten, widmet sich eine ganzes Teilgebiet der Datenanalyse: Es sind dies die Verfahren der **Identifikation** von nominalen, abhängigen Merkmalen. Lineare Verfahren wie die binäre Regression, die logistische Regression bzw. die Diskriminanzanalyse, aber auch sehr rechenintensive, nichtlineare und daher schwierig zu interpretierende Ansätze wie die Künstlichen Neuronalen Netze sind zu diesem Teilgebiet zu zählen.

Entscheidungsbaumverfahren sind im Gegensatz dazu – wie alle heuristischen Verfahren – dadurch gekennzeichnet, dass sie in der Regel sehr rechengünstig implementiert werden können, und dass ihre Ergebnisse leicht zu interpretieren sind. Eine wesentliche Schwäche der Verfahren liegt aber in der lokal begrenzten Suche nach Optimallösungen, was sich z.B. darin äußert, dass eine optimale Klassifikation i.a. nicht über eine Folge von "besten", d.h. lokal-optimalen Splits zu erreichen ist. Insbesondere die auf Basis der Bewertungen C_2 bzw. C_3 abgeleiteten Verfahren sind oft sehr indifferent gegenüber verschiedenen, scheinbar gleichwertigen Splitmöglichkeiten. Außerdem gibt es viele Fälle, in denen die beste Partition nicht durch Befolgung der Bedingung (4) sondern über Splits erreicht wird, in denen Folgendes gelten darf:

$$C^*_{SP}(X_t) \leq C_{\mathrm{SP}}(X_t) = C(X_{t_L}) + C(X_{t_L}) \tag{8}$$

4 Parameter-gesteuerte Splits

Im Folgenden soll eine Verfahrensvariante vorgestellt werden, die versucht, den oben skizzierten Nachteil der Entscheidungsbaumverfahren zu umgehen. Dazu wird das **lokal begrenzte Suchen** nach Optimallösungen – hier sind es die optimalen Splits auf jeder Ebene – derart ausgeweitet, dass ein Knoten X_t nicht nur unter Berücksichtigung seiner binären Teilmengen X_{t_L} und X_{t_R} gesplittet wird, sondern dass vielmehr die Optimalität des gesamten entstehenden Teilbaums angestrebt wird. Da aber selbst bei binären Bäumen die Anzahl der zu untersuchenden Möglichkeiten sehr schnell mit der Anzahl der Ebenen e des Baumes wächst, muss die Tiefe der zu untersuchenden Teilbäume, auf deren Gesamtbewertung dann die Entscheidung zu Gunsten eines Splits erfolgt, begrenzt werden.

Eine rekursive Möglichkeit, dieses **Look-Ahead** auf l tiefer liegenden Ebenen umzusetzen, kann z.B. wie folgt geschehen:

$$C^*_{\mathrm{SP}}(X_t|l) := \begin{cases} \min\limits_{X_{t_L} \subset X_t} \{C_{\mathrm{SP}}(X_{t_L}|l-1) + C_{\mathrm{SP}}(X_{t_R}|l-1)\} & \text{für} \quad l \geq 1 \\ \min\limits_{X_{t_L} \subset X_t} \{C_{\mathrm{SP}}(X_t) = C(X_{t_L}) + C(X_{t_R})\} & \text{für} \quad l = 0 \end{cases}$$

X_{t_R} ist dabei wie oben definiert. Zu beachten ist hierbei, dass zwischen den **Knotenbewertungen** C und den **Splitbewertungen** C_{SP} unterschieden werden muss. Während die Splitbewertungen alleine zur Steuerung der Aufteilung der Beobachtungsmenge X bzw. ihrer Teilmengen X_t dienen, beschreiben die Knotenbewertungen die Güte der Knoten X_t auf Basis der bekannten Bewertungsfunktionen. In diesem Zusammenhang kommt es bei Entscheidungsbäumen mit e Ebenen zu interessanten Ergebnissen. Bezeichnet $\mathcal{P}_e(X)$ die sich ergebende Partition eines Entscheidungsbaumes, die auf Basis des klassischen Ansatzes konstruiert wurde und $\mathcal{P}_e^{\mathcal{L}}(X)$ die äquivalente Partition eines Entscheidungsbaumes, die mit Hilfe des Look-Ahead-Verfahrens unter Berücksichtigung der Sequenz $\mathcal{L}$ (sie steuert für jeden Knoten die Entscheidung bzgl. des Parameters l), so berechnet sich der Wert der erstellten Partition weiterhin nur aus den Knotenbewertungen $C(X_t)$ der Endknoten X_t. Dabei gilt:

$$\mathcal{C}(\mathcal{P}_e^{\mathcal{L}}(X)) = \sum_{X_t \in \mathcal{P}_e^{\mathcal{L}}(X)} C(X_t) \quad \leq \quad \sum_{X_t \in \mathcal{P}_e(X)} C(X_t) = \mathcal{C}(\mathcal{P}_e(X)) \tag{9}$$

Somit kann die **Dominanzaussage** formuliert werden, dass für jeden Split mit $l \geq 1$ der Look-Ahead-Split für die $(l+1)$-nachfolgenden Ebenen mindestens ebenso gut ist wie $l+1$ nacheinander ausgeführte *normale* lokaloptimale Splits. Des Weiteren ist zu beachten, dass für einen Zwischenknoten X_t des Look-Ahead-Entscheidungsbaumes, d.h. für alle $X_t \not\subset \mathcal{P}_e^{\mathcal{L}}(X)$, nicht die Optimalität bzgl. des klassischen Splitkriteriums $C_{\mathrm{SP}}(X_t)$ gilt:

$$C_{\mathrm{SP}}(X_t|l) \quad \geq \quad \min_{X_{t_L} \subset X_t} \{C(X_{t_L}) + C(X_{t_R})\} \qquad \forall X_t \not\subset \mathcal{P}_e(X) \tag{10}$$

Für den Verfahrensablauf bedeutet dies im Übrigen, dass man zumindest die letzte Split-Ebene, also die Endknoten, stets mit dem Parameter $l = 0$ erstellen muss. Wegen des mit wachsendem l stark anwachsenden zusätzlichen Rechenaufwandes scheint es sinnvoll zu sein, größere Werte für l nur in besonderen Fällen einzusetzen.

5 Ergebnisse

Zum Vergleich der verschiedenen Verfahren soll hier die Gesamtbewertung der jeweiligen Partition $\mathcal{P}(X)$, d.h. die Summe der Bewertungen der Endknoten X_t mit $t \in \tilde{T}$, herangezogen werden. Diese Größen ergeben sich aus den oben angegebenen Knoten- bzw. Splitwerten gemäß:

$$\mathcal{E}_0 := \mathcal{C}(\mathcal{P}_e(X)) = \sum_{X_t \in \mathcal{P}_e(X)} C(X_t) \tag{11}$$

$$\mathcal{E}_{10} := \mathcal{C}(\mathcal{P}_e^{\mathcal{L}}(X)) = \sum_{X_t \in \mathcal{P}_e^{\mathcal{L}}(X)} C(X_t) \qquad \text{mit } \mathcal{L} = (1, \ldots, 1, 0) \tag{12}$$

$$\mathcal{E}_{210} := \mathcal{C}(\mathcal{P}_e^{\mathcal{L}}(X)) = \sum_{X_t \in \mathcal{P}_e^{\mathcal{L}}(X)} C(X_t) \qquad \text{mit } \mathcal{L} = (2, \ldots, 2, 1, 0) \tag{13}$$

Die Sequenz $\mathcal{L} = (1, \ldots, 1, 0)$ bringt zum Ausdruck, dass auf den ersten $e - l - 1$ Ebenen Look-Ahead-Splits mit dem Parameter $l = 1$ durchgeführt wurden, während die Sequenz $\mathcal{L} = (2, \ldots, 2, 1, 0)$ angibt, dass auf den ersten $e - l - 2$ Ebenen Look-Ahead-Splits mit dem Parameter $l = 2$ und auf der drittletzten Ebene mit $l = 1$ durchgeführt wurden. Darüber hinaus wurde in beiden Fällen auf der letzten Ebene der Parameter $l = 0$ gewählt.

Table 1. Testergebnisse

Bewertung	C_1			C_2			$-C_3$		
Verfahren	$\mathcal{E}_0$	$\mathcal{E}_{10}$	$\mathcal{E}_{210}$	$\mathcal{E}_0$	$\mathcal{E}_{10}$	$\mathcal{E}_{210}$	$\mathcal{E}_0$	$\mathcal{E}_{10}$	$\mathcal{E}_{210}$
Split-Ebene	Ergebnisse								
0	0,89	-	-	34	-	-	70	-	-
1	0,88	-	-	32	-	-	72	-	-
2	0,84	0,79	-	31	30	-		76	-
3	0,77	0,71	0,68	30	28	28		84	84
4	0,62	0,61	0,60		26	26		88	84
5	0,54	0,56	0,54		24	20		91	98
6	0,50	0,48	0,51		21	19		98	105
7	0,44	0,42	0,48		20	18		100	106
8	0,41	0,36	0,45		17	17		103	
9	0,39								

Legende: keine Berechnung mit diesem Verfahren

leere Zelle keine Verbesserung erreichbar

Zur Demonstration der Auswirkungen des Look-Ahead-Verfahrens wurden das klassische Entscheidungsbaumverfahren und das Look-Ahead-Verfahren an einem Datensatz mit 100 Objekten und mit den drei oben angegebenen Zielfunktionen angewandt. Jedem Objekt waren ein binäres und zwei ordinale Merkmale mit 4 bzw. 9 Ausprägungen sowie seine Klassenzugehörigkeit zugeordnet. Die Festlegung der Merkmalsausprägungen bei jedem Objekt erfolgte durch einen Zufallszahlengenerator. Tabelle 1 zeigt die Resultate, die wie folgt interpretiert werden können:

- Wie zu erwarten war, ergeben sich bei der ein-, zwei- bzw. dreifachen Aufsplittung der Daten (Ebene 1, 2 bzw. 3) die besten Lösungen für die Verfahren $\mathcal{E}_0$, $\mathcal{E}_{10}$ bzw. $\mathcal{E}_{210}$. Möchte man z.B. den Datensatz bzgl. des Splitkriteriums $\mathcal{C}_1$ unterteilen und dies auf drei Ebenen durchführen, so erhält man den optimalen Split mit einer Bewertung von 0.68 unter Verwendung der Sequenz $\mathcal{L} = (2, 1, 0)$.
- Des Weiteren erkennt man, dass das Look-Ahead deutliche Stärken besitzt. Die Aufteilung der Daten auf drei Ebenen unter Verwendung der Bewertungsfunktion $\mathcal{C}_1$ zeigt dies eindrucksvoll. So ergibt die Verwendung des klassischen Splitverfahrens dort eine Aufteilung, die mit 0.77 bewertet werden kann. Die Verwendung des Look-Ahead-Ansatzes mit der Sequenz $\mathcal{L} = (1, 0)$ führt zu einer Splittung der Daten mit einem Maß von 0.71, während die Optimallösung dieses Problems – berechnet auf Basis der Sequenz $\mathcal{L} = (2, 1, 0)$ – mit 0.68 bewertet werden.
- Auch zeigt sich, dass das Look-Ahead-Verfahren bei den diskreten Maßen $\mathcal{C}_2$ und $\mathcal{C}_3$ mehr Vorteile bringt als beim quasi-stetigen Maß $\mathcal{C}_1$. Der Look-Ahead-Ansatz kann hier deutlich bessere Splits finden als das klassische Verfahren. Während $\mathcal{E}_0$ z.B. nach einem Split mit einer Bewertung von 72 abbricht, können die verschiedenen Varianten des Look-Ahead deutlich mehr und deutlich bessere Splits finden.

Man erkennt an diesem Beispiel insgesamt auch die Überlegenheit der Look-Ahead-Heuristik, auch wenn sie partiell nicht immer garantiert werden kann ($\mathcal{E}_{10}$ und $\mathcal{E}_{210}$ bzgl. $\mathcal{C}_1$ auf Ebene 6 - 8). Generell lässt sich aber sagen, dass zumindest für Probleme, die genau $e + 1$ Split-Ebenen aufweisen sollen, der hier vorgestellte Ansatz mit dem Parameter $l = e$ optimal ist. Inwieweit der deutlich erhöhte Rechenanwand für große l-Werte in der Praxis sinnvoll ist, müssen indes weitere Simulationsstudien zeigen. Generell dürfte nach ersten Tests aber gelten, dass eine Wahl von $l \geq 3$ in den meisten Problemfällen nicht gerechtfertigt ist.

References

1. Breiman, L.; Friedman, J.H.; Olshen, R.A.; Stone, C.J. (1984) Classification and Regression Trees. Wadsworth Inc., Pacific Grove, California.
2. Hippner, H.; Küsters, U.; Meyer, M.; Wilde, K. (2001) Handbuch Datamining im Marketing. Vieweg, Gabler, Braunschweig, Wiesbaden.
3. Kass, G.V. (1980) An Exploratory Technique for Investigating Large Quantities of Categorial Data. In: Applied Statistics. Vol. 29, No. 2, S. 119–127
4. Quinlan, J.R. (1986) Introduction of decision trees. In: Machine Learning 1(1), S. 81–106

Acquisition of High-Value Customers for Automotive Banks

Stefan Müller, Katja Gelbrich

TU Dresden, Dep. of Marketing, Mommsenstr. 13, 01062 Dresden

1 Rationale

Many companies still follow the principle "the more customers the better". However, the buyers' contributions to a firm's profit fluctuate evidently. Therefore, it is reasonable to attract "lucrative" persons. This idea has even been incorporated into a definition of marketing as the "art of attracting and keeping profitable customers" [1]. A powerful tool to assess an account is customer value, which is the sum of a purchaser's contributions to a company's well-being [2]. Contributions can be positive (benefits) or negative (costs) and monetary or non-monetary. Classical evaluations focus on current rather than on prospective accounts, because the prevailing marketing paradigm is the management of customer relations, i.e. it focuses on re-purchase [3]. However, in accordance with Keane and Wang [4] or Shepard [5], we suggest that any evaluation should start with the *acquisition*, because it is five to seven times as expensive as customer retention [6]. Moreover, a value oriented acquisition brings about two advantages:

- It empowers a company to spend its limited marketing budget on the "right" prospective purchasers. This will à priori positively bias its customer portfolio.
- The à priori evaluation enables the firm to segment the market. Each value segment can be addressed according to its supposed contributions to the firm: Low-value purchasers get a standardized treatment, while high-value accounts may enjoy special attention (e.g. face-to-face consultation). Thus, value orientation is the base for the "right" scope of customer orientation.

In reality, firms rarely evaluate customers, because so far, researchers have failed to develop an operational and commonly recognized concept of customer value. We introduce an operational model and apply it to a special branch – the *automotive credit business*. The model is new insofar as it incorporates the state-of-the-art of research in customer evaluation, is based on efficiency (ratio of benefits and costs) rather than on a contribution margin or a simple score, aggregates monetary and non-monetary components as well as benefits and costs to a single value score, and weights them objectively by using Data Envelopment Analysis (DEA).

2 Components of Customer Value

Reviewing the relevant publications of customer value, e.g. [2] and [6], and applying it to an automotive bank, one can derive the benefits and costs a customer brings about in this branch. The only *monetary benefit* is the money someone is going to spend on cars in his remaining life, discounted to the presence, i.e. Customer Lifetime Value (CLV). *Non-monetary benefits* are:

- The readiness to buy on credit (CRED) is a prerequisite for this business. If someone considers it to be unworthy to be in debt and rather pays his car in cash, only the automotive company will realize some turnover, the bank not.
- Cross-Selling Potential (CROSS) means, that a consumer purchases more than the core product from the same firm. Automotive banks originally offered only financing / leasing contracts. Nowadays, they face an increasing competition of retail, non- and near-banks. As a reaction, they are about to penetrate the banks' traditional core market by offering investments. Thus, cross-selling potential refers to the willingness to invest money in this particular case.
- Brand Loyalty equals the re-purchase of a brand. One has to distinguish between attitude and behavior. Attitude refers to the general readiness to stay with a brand if one is satisfied. This is called generic brand loyalty (GENER), because it does not refer to a particular product, but reflects a general brand affinity. Behavior cannot be observed when dealing with a future decision. It is therefore measured as the intention to re-purchase a particular brand. However, in the case of customer acquisition, it will be re-coded as intention to SWITCH, because new accounts are mainly switchers from competing brands.
- Opinion Leadership (LEAD) is due to the fact that people, who have a central position within a social group (social integration), are experts in a particular field, e.g. cars (expert knowledge) and extensively, and intensively communicate with others (social network), are often asked for advice [7].

A car buyer also brings about costs to the credit corporation: efforts for acquisition (advertising, provision for sales personnel), credit rating, customer care (customer retention, collections) and re-activation (mailings). Generally speaking, they should all be included, but most of them are either very low or do hardly fluctuate individually. The only important and customer specific cost driver is customer care, more specifically, collections. According to experts, 80% to 90% of the overall expenses of financial service providers occur if somebody does not pay back his rates duly. Thus, we solely include delinquent payment. Since prospective accounts are not yet debt-accounts, one has to predict it as the *probability of default* (PROB). To sum it up, one faces a number of rather heterogeneous value components: While costs are measured in per cent, benefits are either scores (e.g. LEAD), Likert scale items (e.g. CRED) or monetary units (e.g. CLV). The challenge is to aggregate them to one score (see chapter 3).

3 Principle of Data Envelopment Analysis

DEA compares the efficiency of "Decision Making Units" (DMUs) based on multiple in- (costs) and output (benefits) criteria, i.e. the seven components suggested above: Best performing will be those with the highest output/input ratio. We explain the principle by means of a simple example of only eight DMUs with one input and output for each [8]. Fig. 1 shows the DMUs in a two-dimensional space. For the reason of simplicity, inputs and outputs are displayed as ten-point scales. In this two-dimensional example which is only to explain the principle of DEA, the scaling is arbitrary. In case of multicriteria evaluation, all scales have to be standardized in order to make them comparable.

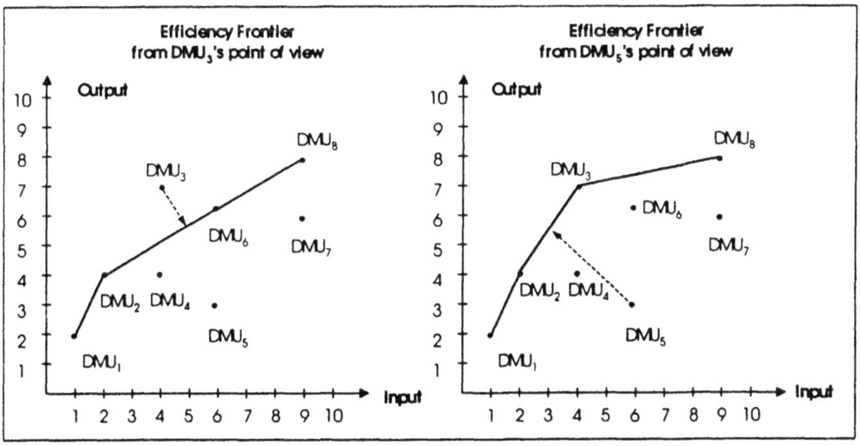

Fig. 1. Envelopment surfaces from different points of view

As for DMU_1, for instance, an input of 1 yields an output of 2; DMU_8 displays values of 9 and 8 for in- and output, respectively. The efficiency of a specific unit is assigned in terms of its *relative situation to all other objects*. In order to calculate the value of DMU_3 (left-hand side of Fig. 1), for instance, this particular unit is excluded from consideration [9]. The remaining batch of objects has a convex upper boundary, the so-called efficiency frontier line or envelopment surface. The units on the frontier – DMU_1, DMU_2, DMU_6 and DMU_8 – are called efficient. They are assigned a value of 1. The reason is that, compared with the DMUs beneath the boundary, units on the boundary would either produce the same output with less input (e.g. DMU_2 vs. DMU_4) or a higher output with the same input (e.g. DMU_8 vs. DMU_7). All units beneath the line are inefficient and score less (i.e. between 0 and < 1). DMU_3 is situated above the line showing it as *superefficient* from the standpoint of all others. Its efficiency score is > 1. For evaluating DMU_5 (right-hand side of Fig. 1), this unit is excluded yielding a different envelopment surface. According to this, DMU_5 is inefficient. The exact scores for each unit depend on their distances from the efficiency line, either from beneath (value < 1) or from above (value > 1). So far, the explanations have been limited to a simple, two-dimensional case with one in- and output. But our automotive bank faces multiple in- and outputs, which brings about the problem of *weighting* u_r and v_r in equation (1) [10]. In order to ensure the comparability of the multiple criteria, they have to be transformed into so-called z-values (see equation 2), i.e. the variables x and y in equation (1) are all standardized values.

$$\text{Efficiency of DMU}_j = \frac{u_1 y_{1j} + u_2 y_{2j} + \ldots + u_R y_{Rj}}{v_1 x_{1j} + v_2 x_{2j} + \ldots + v_I x_{Ij}} = \frac{\sum_{r=1}^{R} u_r y_{rj}}{\sum_{i=1}^{I} v_i x_{ij}} \qquad (1)$$

y	....	Output	x	.....	Input
u_r	....	Weighting coefficient for output r			
v_i	....	Weighting coefficient for input i			
R	....	Number of outputs	I	.....	Number of inputs
j	....	Number of DMU			

$$z_{ij} = \frac{x_{ij} - \bar{x}_i}{\sigma_i} \qquad (2)$$

z_{ij}	....	standardized value	σ_i	.....	standard deviation of i
x_{ij}	....	originial value	i	.....	variable
$\bar{x}_i$	....	arithmetical mean of i	j	.....	value of variable

The weighting problem in equation (1) is solved by "allowing" each DMU to present oneself in the *best light*: The weighting coefficients are determined so, that the efficiency of the concerning unit is maximized with respect to all others [10]. For a particular DMU_0 this means: Maximize the efficiency of DMU_0 under the condition that the efficiency of all other DMUs ≤ 1[1]. In order to make sure that all in- and outputs are included, all weighting coefficients have to be greater or equal to a small, positive value ε. Equation (3) describes this optimization problem mathematically with respect to our example object DMU_0. In order to solve equation (3), the system of equations is transferred into a linear form. For the *maximization*, the single in- and output values are of no interest; what matters is their ratio. Thus, the denominator is set constant just maximizing the numerator [10], as the equation system (4) displays. Solving it yields the relative efficiency of the object at hand (DMU_0) as well as the weights responsible for this value.

1 If superefficient DMUs are admitted, this condition will be applied to all objects. If superefficient DMUs are allowed (as here), this holds only for the other objects.

$$\max\ \text{eff}_0 = \frac{\sum_{r=1}^{R} u_r y_{r0}}{\sum_{i=1}^{I} v_i x_{i0}} \quad \text{under the condition that} \ldots \tag{3}$$

$$\frac{\sum_{r=1}^{R} u_r y_{rj}}{\sum_{i=1}^{I} v_i x_{ij}} \leq 1 \text{ for each unit } j \neq 0 \qquad \text{and } ur,\ vi \geq \varepsilon > 0$$

$$\max\ \text{eff}_0 = \sum_{r=1}^{R} u_r y_{r0} \quad \text{under the condition that} \ldots \tag{4}$$

$$\sum_{i=1}^{I} v_i x_{i0} = 1$$

$$\sum_{r=1}^{R} u_r y_{rj} - \sum_{i=1}^{I} v_i x_{ij} \leq 0 \qquad j = 1, 2, \ldots, n \quad \text{and} \quad ur,\ vi \geq \varepsilon > 0$$

4 Results of the Evaluation

The DEA score is one-dimensional and therefore might consider rather different DMUs as equal. Along with the fact that it is cost-reducing to address market segments rather than individuals, this weakness necessitates a data pre-processing: Prior to the DEA, the customers has to be segmented with respect to the seven value components. A cluster analysis in a data base of 5,432 German car drivers yielded a *10-cluster solution*. In order to evaluate the 10 clusters, we use the average input and output factors within each group (see Table 1). Table 2 reveals the results. There are two superefficient (score > 1), three efficient (score = 1), and five inefficient customer clusters (score < 1). According to this, cluster 7, 8, 6, 9, and 1 are promising target groups. In order to interpret the efficiency scores correctly, we have to recall the principle of DEA: The efficiency of a DMU is determined in that way, that it performs best in relation to all other units. Doing so, all factors with very low input or very high output values get very high weighting coefficients. In an extreme case, a unit is assigned as superefficient if it is superior to all the others with respect to only one output/input ratio, regardless of all other components. The results should therefore be called into question as long as we haven't had a closer look at the single components of the DMUs.

Table 1. Unstandardized[2] in- and output values for the DEA

CL	CLV	CRED	GENER	SWITCH	CROSS	LEAD	PROB
1	55,286	2.79	3.16	1.01	2.88	2.72	1.50
2	39,097	1.22	3.09	1.38	1.75	1.93	46.99
3	43,991	1.63	3.28	2.63	1.68	2.42	1.80
4	5,858	1.63	3.35	2.20	2.76	2.52	46.81
5	29,943	1.38	3.27	1.11	2.29	1.99	1.77
6	53,637	3.23	3.31	2.64	2.97	2.71	1.55
7	143,537	1.86	3.40	1.94	3.40	2.95	0.85
8	43,604	1.27	3.59	1.12	3.40	2.85	1.22
9	43,922	1.46	3.21	2.71	3.30	2.56	1.61
10	42,288	1.88	1.78	1.53	2.41	2.16	1.59

CL ... Cluster, CLV ... Customer Lifetime Value, GENER ... Generic Brand Loyalty, CROSS ... Cross-Selling, LEAD ... Opinion Leadership, PROB ... Probability of Default

Table 2. Ranking of customer clusters

CL	Score	Size*	CLV	CRED	GENER	SWITCH	CROSS	LEAD	PROB
7	2.399	7.6	++	+	+	+	++	++	++
8	1.188	17.0	-	-	++	-	++	++	++
6	1.000	7.9	O	++	+	++	O	+	+
9	1.000	16.2	-	--	O	++	++	O	+
1	0.999	13.9	+	++	O	--	O	+	+
10	0.997	8.5	-	+	--	-	O	-	+
3	0.996	9.3	-	-	+	++	--	O	O
4	0.974	0.8	+	-	+	+	O	O	--
5	0.968	18.2	--	-	O	-	-	--	+
2	0.693	0.6	-	-	-	-	--	--	--

Key: O average
++ much better or + better than average
-- much worse or - worse than average
* Size in percent of sample

Note: With reference to the probability of default (PROB), "+" means lower than average, a "-" means higher. Thus, the sign does not mean "higher/lower", but "better/worse".

Cluster 7 is by far most attractive, which is reasonable, because it is superior to the others with respect to any component, although the readiness to buy on credit is only slightly above average (1.86 vs. 1.82). In general, most ranks make sense: Cluster 2, for instance, scores lowest, because all its components are subordinate to the others. However, the effect described above brings about two *incongruities*:

Case 1. Cluster 3 scores higher than cluster 4, although the former is only below average with regard to three components (CLV, CRED, CROSS), the latter having only two inferior attributes (CRED, PROB). Moreover, cluster 3 exceeds the

[2] Prior to the DEA, the values were transofrmed into z-values (see equation 2).

average twice (GENER, SWITCH) while cluster 4 has three superior components (CLV, GENER, SWITCH). Taking a closer look at this incongruity, cluster 4 has an extremely high probability of default (46.81%) and – to reinforce this unfavorable situation – not a single output superior to all the other groups. On the contrary, cluster 3 has a by far better creditworthiness (1.8%). This provides those accounts with a favorable position: Due to the low input variable, all possible benefit/costs ratios will be better. This effect is even reinforced, by the variable SWITCH, which is much above average (2.63 vs. 1.73). Thus, SWITCH gets a high weighting coefficient boosting the rank of cluster 3. This effect of DEA described can be tolerated or is even desired in the problem at hand. The reason is, that creditworthiness is a so-called knock-out-criterion in the financing business. If a debtor does not pay back his rates duly, costs rise tremendously due to extensive debt collection. Since the probability of default is the only input in our value model, DEA implicitly provides outliers in terms of costs with high weighting coefficients. With regard to the financing business, this is not only sensible but necessary. The "disadvantage" turns out to be an advantage.

Case 2. A similar incongruity can be observed when comparing cluster 6 and 8. One could speculate, that these groups would be ranked vice versa, because cluster 6 does not have any below-average attribute while 8 has even three. However, the efficiency of 8 is that high, because its probability of default amounts only to 1.22%, cluster 6 having a higher one (1.55%). Moreover, four output factors receive a "++" while cluster 6 displays "++" only twice. This has to be seen in a more critical light. *First*, the probability of default, which is so important to the financing business, is very low in both segments (1.22% and 1.55%). *Second*, cluster 6 outstrips cluster 8 with respect to its average willingness to buy on credit, which puts the former into a positive light, because the readiness to be in debt is a prerequisite for a finance contract. Since cluster 6 scores higher on two core components of the business concerned, this group may be seen in a more positive light than DEA suggests. Moreover, the components by which cluster 8 outstrips the others (cross-selling, opinion leadership) are of a more derivative nature: Their effect is – while of importance – rather indirect. It should be borne in mind, however, that the DEA shows both of these clusters to be efficient.

To sum up, the results of the DEA meet the demand of *content validity*: No cluster scores (super) efficient, just because it has *one* superior output/input relation, all the others being extremely bad. This is due to the underlying data base which does not contain extreme patterns. Just in one case (cluster 8 and 6), do the assigned efficiencies contradict 'common sense'. Notwithstanding this particular case, DEA can be considered as a powerful base for a value-oriented acquisition of new accounts: The company concerned should focus its acquisition activities on the first five clusters (with special attention to the two top clusters). Due to the preceding clustering, the financial service provider even knows, that cluster 7, 8 and 9 will most likely react to an investment offer, while their message for clusters 6 and 1 should appeal to their readiness to buy on credit.

5 Guidelines for Future Research

As discussed before, content validity is given in general. However, the concept of validity is much broader, requiring *empirical validity*, which means finding another criterion that can corroborate with the result at hand. To check this, one could turn to experts (i.e. managers or employees of an automotive bank) and

- aks them to weight the seven value components according to their experiences and calculate a single efficiency score based on a scoring model (This approach, however begs the central question of which weighting to apply) or
- ask them directly to rank the ten clusters according to their subjective opinion.

The expert and DEA scores could then to be compared. If they match, the DEA results will be further evidenced. One could also check empirical validity by switching the sequence of analysis. As opposed to the introduced research steps, one could start with the DEA and then segment the market according to the individuals' efficiency scores (e.g. by ABC analysis). Both results – first segmentation, then DEA vs. first DEA, then segmentation – have to be compared.

References

[1] Kotler, P., Armstrong, G. (1996) Principles of Marketing. Englewood Cliffs/NJ

[2] Cornelsen, J. (2000) Customer Value Analysis in Relationship Marketing, Nürnberg

[3] Grönross, C. (1994): Paradigm Shift in Marketing. Management Decision 32: 4-20

[4] Keane, T. J., Wang, P. (1995) Applications for the Lifetime Value Model in Modern Newspaper Publishing. Journal of Direct Marketing 9: 59-66

[5] Shepard, D. (1990) The New Direct Marketing. Homewood/IL

[6] Johnson, M. D., Herrmann, et al. (1997) Customer Retention. Wiesbaden

[7] Chan, K. K., Misra, S. (1990) Opinion Leader. Journal of Advertising 19: 53-61

[8] Charnes, A., Cooper, W. W. et al. (1994) Data Envelopment Analysis. Boston/MA

[9] Keller, J., Paterson, I. et al. (2000) Multi-Criteria-Ranking. ECML. Barcelona

[10] Dyson, R. G., Thanassoulis, E. et al. (2000) A Data Envelopment Analysis Tutorial, Tutorial Paper in Operational Research, Warwick Business School

The authors thank Iain Paterson (IHS, Vienna) for a review of a former version of this paper, for his helpful expertise and fruitful discussions.

Quantitative Entscheidungsunterstützung für das Preismanagement von TK-Dienstleistungen

Cornelia Schön

Institut für Wirtschaftstheorie und Operations Research, Universität Karsruhe (TH) , D-76128 Karlsruhe, schoen@wior.uni-karlsruhe.de

1 Einleitung

Im wettbewerbsumkämpften Telekommunikationsmarkt stellt für Anbieter von klassischen festnetzbasierten Telefondienstleistungen ein gewinnorientiertes Preismanagement und die zielgruppenspezifische Marktbearbeitung durch attraktive Tarifgestaltung eines der wichtigsten Marketinginstrumente zur Sicherung des langfristigen Markterfolgs dar [9]. Dabei müssen alle relevanten Determinanten der Preisbildung im Hinblick auf ihre Gewinnwirkung berücksichtigt werden: Nachfragereaktionen, Kosten- und Konkurrenzaspekte sowie regulatorische Rahmenbedingungen [8]. Zudem erstreckt sich die Preisentscheidung i.a. nicht nur über eine eindimensionale Größe, sondern betrifft eine mehrdimensionale Preisstruktur, die verschiedene Preiskomponenten umfaßt. Mit dem Ziel, modell- und methodenbasierte Entscheidungsunterstützung für die gewinnorientierte Tarifgestaltung von festnetzbasierten TK-Dienstleistungen aus Sicht der Deutschen Telekom (DT) zu leisten, werden im folgenden die wesentlichen Determinanten des TK-Preismanagements diskutiert und Modell- und Lösungsansätze vorgestellt. Die Komplexität der Fragestellung gestattet dabei nur einen kleinen Einblick, der aufgrund des Datengeheimnisses gegenüber der DT stellenweise mehr als Ansatz generischer Natur zu verstehen ist.

2 Determinanten des TK-Preismanagements

2.1 Entscheidungsvariable "Preis"

Zur Verdeutlichung des mehrdimensionalen Entscheidungsgegenstandes und des Begriffs *Preis* bzw. *Tarif* seien einige am Markt real existierende Beispiele genannt, wie die Tarife „T-Net Aktiv Plus“ oder „Arcor ISDN Happy Sunday“. In Anlehnung an [12] verstehen wir unter einem *Tarif* eine Abbildung, die in Abhängigkeit einer oder mehrerer Preiskomponenten berechnet, wieviel ein Nutzer für eine bestimmte nachgefragte Menge einer definierten Leistung bezahlen muß. Bei dieser Leistung handelt es sich hier um festnetzbasierte Telefondienstleistungen, also ein mit dem Kunden vertraglich fixiertes Leistungs*paket*, das a) die Kernleistungen des festnetzbasierten Telefondienstes (Anschluß- und Verbindungsleistungen) und b) eventuelle Nebenleistungen als zusätzliche Leistungsmerkmale (z.B. Anrufweiterschaltung) umfaßt [4]. Ein Tarif kann demnach durch die mit ihm verbundenen Kern- und Nebenleistungen („*leistungsbezogene Tarifattribute*") und durch die für diese Leistungen erhobenen Preiskomponenten („*preisbezogene Tarifattribute*") beschrieben werden, wobei wir als Gegenstand der Tarifgestaltung i.e.S. nur die Optimierung der preisbezogenen Tarifattribut*ausprägungen* betrachten und davon ausgehen, daß die konkreten Leistungen der Tarife und die Tarifform bereits à priori festgelegt sind [12]. Klassische Preiskomponenten der Kernleistungen eines (zweiteiligen) Tarifs j sind der nutzungsintensitätsunabhängige monatliche Grundpreis p_j^G, sowie Nutzungspreise p_{jkl} pro Takteinheit einer Verbindung, die i.a. nach „Slots" kl, d.h. Kombinationen aus Entfernungszonen l (z.B. $l \in L$ = {Ort, Fern, Ausland, Mobilfunknetz} und Zeitfenstern k (z.B. $k \in K$ = {Peak, Offpeak}) differenziert sind. Weitere stets festzulegende Preiskomponente ist die Länge d_j des Zeittaktes bei Tarif j (Taktzeit). Andere innovativere Preiskomponenten wie Mindestumsatz, Freiminuten, Sockeldauer, Rabatte oder explizite Preiskomponenten für Nebenleistungen sind in beliebiger Form denkbar. Die möglichen Tarifformen reichen vom klassischen zweiteiligen Tarif, der zunächst Gegenstand der Betrachtung sein soll, über den Blocktarif bis hin zur kontinuierlichen Preisstruktur und zu multiattributiven Tarifen [10,12].

2.2 Zielkriterium

Die Angebotspalette von TK-Unternehmen (hier: DT) setzt sich i.a. im Sinne einer zielgruppenspezifischen Marktbearbeitung aus mehreren differenzierten, *optionalen Tarifen* zusammen [3,4,12], so daß substitutive Interdependenzen zwischen diesen in Form der „Produktlinienpreisbildung" [11] berücksichtigt werden sollten. In diesem Sinne konzentriert sich der Beitrag auf die Maximierung des Gesamtgewinns über alle angebotenen Tarife. Die Modellierung der Zielfunktion erfordert zunächst die Abbildung von Nachfrage und Kosten.

2.2.1 Nachfrage

Im Sinne einer zielgruppenspezifischen Marktbearbeitung gehen wir von einer A-priori-Segmentierung des Marktes in eine Menge I von Segmenten $i \in I$ aus. Bei der Untersuchung des Nachfrageverhaltens interessieren vier gewinnbestimmende Teilentscheidungen der Konsumenten hinsichtlich Wahl und Nutzung von TK-Dienstleistungen:

a) Wahl des Grundtarifs: Die Wahl des Grundtarifs j betrifft die Entscheidung eines jeden Konsumenten mit Festnetzanschluß, einen voreingestellten „Default"-

Tarif für die Kernleistungen des festnetzbasierten Telefondienstes (Netzanschluß und Verbindungsmöglichkeit zum vereinbarten Tarif). Zur Abbildung der Grundtarifwahl bieten sich insbesondere *Attraktivitätsmodelle*] (auch: *multiattributive Auswahlmodelle* [2]) an, die sich in jüngster Zeit zur Beschreibung des Auswahlverhaltens von Konsumenten bei optionalen Tarifen in Abhängigkeit erklärender Tarifattribute im Rahmen von Discrete Choice Analysen durchsetzen konnten [3]. Der Marktanteilen MA_{ij} eines Tarifs j im Segment i läßt sich schätzen durch:

$$MA_{ij} = \frac{A_{ij}}{\sum_{j \in J} A_{ij}}, \tag{1}$$

wobei $J = J^T \cup J^W$ die Menge der interessierenden (Telekom- und Wettbewerber-) Tarife am Markt ist und A_{ij} die *‚Attraktivität'* (oft auch: Nutzen) von Produkt j für Segment i beschreibt, die durch verschiedene preis- oder leistungsbezogene Produktattribute gesteuert wird. Zu den bekanntesten Modellen der Form (1) zählen das MNL- und MCI-Modell [2], das BTL- und das LOGIT-Markenwahlmodell [14]. Im Fall des empirisch häufig bestätigten MNL-Modells könnte sich die Attraktivität A_{ij} von Tarif j für Segment i in Abhängigkeit des Grundpreises p_j^G, ausgewählter Verbindungspreise p_{jkl} und der Taktdauer d_j bei zuschätzenden Parametern α_i, β_{ikl} und γ_i, z.B. folgendermaßen darstellen:

$$A_{ij} = \exp\left(\alpha_i p_j^G + \sum_{k \in K} \sum_{l \in L} \beta_{ikl} p_{jkl} + \gamma_i d_j \right). \tag{2}$$

b) Call-by-Call Tarifwahl: Die Call-by-Call Tarifwahl bildet die gesprächsweise Wahlentscheidung eines Konsumenten ab, über welchen Tarif eine einzelne Verbindung in einem bestimmten Slot geführt wird. Bei jeder Verbindung hat der Konsument die Möglichkeit, entweder über seinen voreingestellten Grundtarif j zu telefonieren oder durch Vorwahl 010XX einen anderen Verbindungsnetzbetreiber und damit Call-by-Call Tarif zu wählen. Zu schätzen ist aus DT-Sicht der entsprechende Wahlanteil $MA^V_{ijkl|j}$, mit der ein Segment i, vorausgesetzt dass im Rahmen der Grundtarifwahl die Entscheidung für Telekomtarif $j \in J^T$ gefallen ist, eine einzelne Verbindung im Slot kl auch über diesen und nicht einen Wettbewerbertarif per Call-by-Call führt. Als einfacher, durch Marktforschungsstudien begründbarer Modellansatz für die Call-by-Call Tarifwahl dient hier die Abbildung als konstanter Modellparameter mit

$$MA^V_{ijkl|j} = const., \tag{3}$$

der im Rahmen von Sensitivitätsanalysen oder Monte-Carlo-Simulationen unter Annahme seiner empirischen Verteilung systematisch variiert werden kann [10].

c) Verbindungsanzahl: Die Gesamtzahl an Verbindungen Q_{ijkl}, die ein Segment i mit Grundtarif j in einem Slot kl insgesamt in einer durchschnittlichen Periode (z.B. pro Monat) führt, kann i.a. mit einem Regressionsansatz geschätzt werden. Als geeignete funktionale Spezifikation hat sich dabei in verschiedenen empirischen Untersuchungen die multiplikative Preis-Absatz-Funktion erwiesen [10]. Preisliche Einflußgrößen ist z.B. der Verbindungspreis des Grundtarifs (und evtl. genutzter Call-by-Call Tarife) im entsprechenden Slot und in anderen Zeitfenstern zur Erfassung von Verkehrsverlagerungen zwischen Peak- und Off-Peak-Perioden. Eine exemplarische Modellspezifikation bei zwei Zeitfenstern k, k' mit zu schätzenden Parametern $Q_{0,ikl}$, ε_{ikl} und $\varepsilon_{ikk'l}$ sieht für $i \in I, j \in J, l \in L$ wie folgt aus:

$$Q_{ijkl} = Q_{0,ikl} \left(p_{jkl} \right)^{\varepsilon_{ikl}} \left(p_{jk'l} \right)^{\varepsilon_{ikk'l}} . \tag{4}$$

d) Gesprächsdauer: Die Dauer einer Telefonverbindung ist im allgemeinen sowohl aus Konsumentensicht als auch aus Sicht des gewählten Carriers zu Beginn des Gesprächs unbekannt. In diversen empirischen Untersuchungen zur Gesprächsdauer wird diese als Realisation einer exponential- oder lognormalverteilten Zufallsvariable aufgefaßt [5]. Unter Kenntnis der Gesprächsdauerverteilung ergibt sich die erwartete Anzahl tarifierter Takteinheiten für eine Verbindung von Segment i im Slot kl in Abhängigkeit der Taktdauer d_j bei Tarif j aus:

$$E(\#Takte) = \mu_{ijkl}^{Takt} = \sum_{\tau=1}^{\infty} \tau \int_{(\tau-1) \cdot d_j}^{\tau \cdot d_j} f_{ikl}(X_{GD}) dX_{GD} . \tag{5}$$

Dabei bezeichnet X_{GD} die Gesprächsdauer (Zufallsvariable) und f_{ikl} die Dichtefunktion der Gesprächsdauerverteilung von Segment i im Slot kl. Da die Berechnung der unendlichen Summe i.a. approximativ unter Annahme einer fixierten Taktzeit d_j erfolgt [5], setzen wir die Kenntnis der erwarteten Taktzahl einer Verbindung im folgenden voraus. Empirische Untersuchungen zur Gesprächsdauer auf Basis von Regressionsanalysen und Experteneinschätzungen stützen die Annahme eines vernachlässigbar geringen Einflusses des Preises auf die Verteilungsparameter [10].

2.2.2 Kostenaspekte und Gewinnbetrachtung

Entscheidungsrelevante Kosten in der Festnetzkommunikation, die durch Preismaßnahmen verursacht werden [8,9], sind zum einen *variable Kosten der Dienstenutzung* c_{ikl}^{V}, die durch die Inanspruchnahme einer im Tarif enthaltenen Verbindungsleistung durch den Kunden entstehen [DM/Verbindung], z.B. für Nutzung eigener und fremder Netzinfrastruktur (Interconnectionentgelte). *Variable Kosten des Tarifs* c_{ij}^{Tarif} (pro Periode) sind zwar unabhängig von der

Nutzung der mit dem Tarif verbundenen Dienste, fallen aber pro verkauftem Tarif (Vertrag) einmalig oder laufend an, z.B. für Akquisition & Bereitstellung (periodisiert), Customer Care, Rechnungsabwicklung, etc. Daneben sind hier zur Vereinfachung vernachlässigte *beschäftigungsfixe Kosten* der Preisentscheidung evtl. zu berücksichtigen (z.B. Markteinführungskosten) [10].

Unter Einbezug der variablen Kostengrößen und der Ergebnisse (1) - (5) ergibt sich der erwartet (Perioden-) Gewinn G_{ij} von Segment $i \in I$ mit Segmentstärke ω_i und Grundtarif $j \in J^T$ aus:

$$G_{ij} = \sum_{k \in K} \sum_{l \in L} \left(\mu_{ijkl}^{Takt} p_{jkl} - c_{ikl}^{V} \right) MA_{ijkl|j}^{V} Q_{ijkl} + \omega_i \left(p_j^G - c_{ij}^{Tarif} \right). \tag{6}$$

Der Gesamtgewinn pro Periode über alle angebotenen Tarife $j \in J^T$ und Segmente $i \in I$ berechnet sich folglich aus (6) mit (1) gemäß:

$$G_{ges} = \sum_{i \in I} \sum_{j \in J^T} G_{ij} MA_{ij}. \tag{7}$$

2.3 Konkurrenzaspekte und Preisregulierung

Zur Erfassung wechselseitiger Interdependenzen zwischen preislichen Maßnahmen konkurrierender Anbieter in der Gewinnformulierung dienen verschiedene strategische Szenarioanalysen. Gebräuchliche Hypothesen sind z.B. Nichtreaktion oder Konkurrenzpreise entsprechend der empirischen Preisverteilung, die Annahme einer einfachen empirischen Reaktionsfunktion oder rationales Verhalten im Sinne eines Bertrandwettbewerbs, der durch Aktions-/Reaktionssequenzen bzw. einen Tatonnementprozeß untersucht werden kann [10,14].

Im Gegensatz zu ihren Wettbewerbern unterliegt die DT als ehemaliger Monopolist der Preisregulierung. Im Rahmen der Price-Cap-Regulierung gibt die Regulierungsbehörde eine obere Schranke (Price-Cap-Index PCI) für das relative Preisniveau (Durchschnittspreisindex DPI) eines definierten Warenkorbs aus regulierten Telefondienstleistungen bzw. der entsprechenden Tarife vor. Dabei werden Privatkunden- und Geschäftskundentarife in zwei gesonderten Warenkörben Ω_{PK} und Ω_{GK} reguliert. In [10] wird gezeigt, dass sich die Price-Cap-Restriktion in bekannter Notation mit gegebenen Parametern $\mu_{jkl}^{Takt.ref}$, $\Theta_{jkl}^{V.ref}$, $\Theta_j^{Tarif.ref}$ und U_{ges}^{ref} eines „Referenzmengengerüsts" als lineare Ungleichung darstellen läßt:

$$\sum_{j \in \Omega_{PK}} \sum_{k \in K} \sum_{l \in L} \mu_{jkl}^{Takt.ref} p_{jkl} \Theta_{jkl}^{V.ref} + p_j^G \Theta_j^{Tarif.ref} \leq U_{ges}^{ref}. \tag{8a}$$

Eine entsprechende Ungleichung (8b) gibt es für die regulierten Tarife des Geschäftskunden-Warenkorbs. Alle Entgelte für Leistungen, die nicht oder noch nicht Bestandteil der Dienstekörbe, aber der Preisregulierung sind, unterliegen der Einzelpreisgenehmigung, die durch allgemeine Beschränktheit aller Preiskomponenten formuliert werden kann:

$$\breve{p}_{jkl} \leq p_{jkl} \leq \hat{p}_{jkl} \qquad j \in J^T, k \in K, l \in L \tag{9}$$

$$\breve{p}_j^G \leq p_j^G \leq \hat{p}_j^G \qquad j \in J^T. \tag{10}$$

3 Modell- und Lösungsansätze

Die abgebildeten Determinanten der Preisbildung lassen sich unmittelbar zu einem Optimierungsmodells für die Gestaltung zweiteiliger Tarife zusammensetzen mit:

Max. (7) u.d.N. (8a), (8b), (9), (10).

Für die typischen funktionalen Spezifikationen der Nachfrage in (7), die sich empirisch bestätigt haben, handelt es sich bei der Problemstellung um ein nichtlineares Optimierungsmodell mit nichtlinearer differenzierbarer Zielfunktion, zwei linearen Price-Cap Restriktionen und beschränkten stetigen Variablen. Zur Lösung schlagen wir ein Sequential Quadratic Programming (SQP)-Verfahren vor, das derzeit als effizientestes universell einsetzbares Verfahren für derartige Probleme kleiner bis mittlerer Größenordnung gilt [13].

Ein etwas anderer, flexiblerer Ansatz zur Annäherung an die Fragestellung, der auch innovative (insbesondere diskrete) Preiskomponenten und nahezu beliebige multiattributive Tarifformen berücksichtigen kann, stellt eine dem Problem der Produktliniengestaltung ähnliche Formulierung dar, die wir im folgenden noch kurz darstellen wollen [10]. Ein Tarif für Telefondienstleistungen wird weiterhin wie in Abschnitt 2.1 allgemein beschrieben durch nutzenbildende Tarifattribute beschrieben, nämlich preisbezogene Attribute $m = 1,\ldots,M$ und bereits in den Ausprägungen festgelegte nichtpreisliche leistungsbezogene Tarifattribute. Die Attributausprägung des preisbezogenen Attributs m des Tarifs j sei mit z_{jm} bezeichnet, die definierten Attributausprägungen der leistungsbezogenen Tarifattribute in Vektorform mit $\hat{\mathbf{z}}_j$. Ein ausgestalteter Tarif ist dann beschrieben durch eine Belegung $\mathbf{b}_j = (\hat{\mathbf{z}}_j; z_{j1}, \ldots, z_{jm}, \ldots, z_{jM})$ der kaufrelevanten Tarifattribute mit konkreten Attributausprägungen. Die Wahlwahrscheinlichkeit MA_{ij} von Segment i für Tarif j lässt sich wiederum mit einem Attraktivitätsmodell (1) als Funktion der Tarifattribute schätzen [6,12]. Gegeben seien nun $j = 1, \ldots, |J^T|$ verschiedene Tarife (unveränderbare bestehende, zu repositionierende und oder neu einzuführende), wobei es zu jedem Tarif j R_j zu testende Belegungen $\mathbf{b}_{jr}$ ($r = 1, \ldots, R_j$) durch Variation der preislichen Attributausprägungen gibt. Gegeben sei auch die segmentspezifische Attraktivität (Nutzen, Konsumentenrente) $A_{ijr} = f$

($\hat{\mathbf{z}}_j$, z_{jr1}, ..., z_{jrm}, ..., z_{jrM}) auf Produktebene in Abhängigkeit dieser Belegung $\mathbf{b}_{j,r}$ von Tarif j, sowie die aggregierte Attraktivität A_i^W aller Wettbewerbertarife. Ferner sei der erwartete Gewinn G_{ijr} pro Periode für Tarifkandidat jr (Tarif j mit Belegung $\mathbf{b}_{jr}$) bei Wahl von Segment i bekannt, der unmittelbar aus (6) durch entsprechende Anpassungen für weitere/andere Preisattribute hervorgeht.

Gesucht ist die nun Entscheidung, welche Tarifkandidaten am Markt tatsächlich angeboten werden sollen, so dass der Gewinn über gesamte Angebotspalette maximal wird. Dafür definieren wir für jeden Tarifkandidaten jr mit $j \in J^T$, $r = 1,\ldots,R_j$ eine Binärvariable mit $x_{jr} = 1$, falls Tarif j mit Belegung b_{jr} angeboten wird und 0 sonst. Mit dieser Definition können das Attraktivitätsmodell MA_{ij} zur Tarifwahl und der Gewinn G_{ij} formuliert werden als:

$$MA_{ij} = \left(\sum_{r=1}^{R_j} A_{ijr} x_{jr}\right) \Bigg/ \left(A_i^W + \sum_{j'=1}^{|J^T|} \sum_{r=1}^{R_j} A_{ij'r} x_{j'r}\right) \text{ bzw. } G_{ij} = \sum_{r=1}^{R_j} G_{ijr} x_{jr} . \tag{11}$$

Damit ergibt sich folgendes Optimierungsproblem:

$$\text{Max.} \left(\sum_{i \in I} \sum_{j \in J^T} G_{ij} MA_{ij} = \sum_{i \in I} \frac{\sum_{j=1}^{|J^T|} \sum_{r=1}^{R_j} G_{ijr} A_{ijr} x_{jr}}{A_i^W + \sum_{j'=1}^{|J^T|} \sum_{r=1}^{R_j} A_{ij'r} x_{j'r}} \right) \tag{12}$$

$$\sum_{r=1}^{R_j} x_{jr} \leq 1 \quad \text{für } j \in J^T \tag{13}$$

$$\varphi_u \leq \sum_{j=1}^{|J^T|} \sum_{r=1}^{R_j} x_{jr} \leq \varphi_o \tag{14}$$

$$x_{jr} \in \{0,1\} \quad \text{für } j \in J^T, \quad r = 1,\ldots,R_j . \tag{15}$$

Die Zielfunktion (12) stellt den Gewinn über die gesamte Produktlinie angebotener Tarife $j \in J^T$ dar. Bedingung (13) stellt sicher, daß ein Tarif j mit genau einer Belegung b_{ir} oder gar nicht angeboten. Mit (14) wird die Anzahl insgesamt angebotener Tarife nach unten durch φ_u und oben durch φ_o beschränkt und (15) definiert die Binärvariablen. Zur Einbeziehung von Price-Cap-Restriktion und beschäftigungsfixen Kosten einer Tarifmaßnahme vgl. [10]. Die Problemstellung ist ein binäres Sum-of-Ratios Problem mit linearen Nebenbedingungen, das durch Einführen zusätzlicher Variablen in ein lineares gemischt-binäres Problem umgewandelt werden kann [1]. Ferner bieten sich auch für die nichtlineare Problemstellung einige Heuristiken zur Produktliniengestaltung an [14]. Zu bemerken ist noch, dass bei Wegfall des Segmentindex das Problem ein binäres Quotientenoptimierungsproblem mit total unimodularer Koeffizientenmatrix ist, das bekanntlich effizient global lösbar ist [6,7].

Die dargestellten Modelle eröffnen entsprechen der Komplexität der Fragestellung eröffnet viele Möglichkeiten für Erweiterungen, die teilweise bereits im Text angesprochen wurden. So verdient die DT z.B. auch an Kunden mit Wettbewerbergrundtarifen $j \in J^w$ im Maße des T-Net Standardtarifs j_S, wenn ein Konsument entweder für einzelne Gespräche über Call-by-Call 01033 die DT wählt oder aber der Wettbewerbertarif ein sog. Pre-Selection Tarif ist, bei dem Anschluß- und Ortsnetzverbindungsleistungen weiterhin von der DT erbracht werden. Für eine weitergehende Diskussion dieser und weiterer Punkte hinsichtlich Modellschätzung, dynamischer Aspekte, etc. sei auf [10] verwiesen.

Literatur

[1] Chang, C.-T. (2001): On the polynomial mixed 0-1 fractional programming problem, EJOR 131, 224-227

[2] Decker, R. (1994): Analyse und Simulation des Kaufverhaltens auf Konsumgütermärkten, Peter Lang, Frankfurt a.M.

[3] Elixmann, D. (1997): Discrete Choice Analysis of Optional Tariff Demand Characteristics in Germany, presented at the 1997 ICFC Conference

[4] Gerpott, T.J. (1998): Wettbewerbsstrategien im Telekommunikationsmarkt, Schäffer-Poeschel Verlag, Stuttgart

[5] Giese, J. (2000a): Mathematisch-statistische Analyse von Kenngrößen des Kommunikationsverhaltens, Arbeitspapier, Philipps-Universität Marburg

[6] Hausman W.H. / Chen, K.D. (2000): Technical Note: Mathematical Properties of the Optimal Product Line Selection Problem, Management Science 46 (2), 327-332

[7] Neumann, K. / Morlock, M. (1993): Operations Research, Carl Hanser, München

[8] Schön, C. (2000): Relevante Einflußgrößen von Preisentscheidung in der TK, WIOR-Report-584, Universität Karlsruhe

[9] Schön, C. (2000): Preisbildung in der Telekommunikation, in: Arnold, Franz (Hrsg.): Handbuch der Telekommunikation, Deutscher Wirtschaftsdienst, Oktober 2000

[10] Schön, C. (2001): Modellorientierte Entscheidungsunterstützung für das Preismanagement von TK-Dienstleistungen, Arbeitspapier, Institut für Wirtschaftstheorie und Operations Research, Universität Karlsruhe

[11] Simon, H. (1992): Preismanagement; Gabler Verlag, Wiesbaden

[12] Skiera, B. (1999): Mengenbezogene Preisdifferenzierung bei Dienstleistungen, Gabler DUV, Wiesbaden (zugl. Habilitationsschrift, Universität Kiel)

[13] Spellucci, P. (1993): Numerische Verfahren der nichtlinearen Optimierung, Birkhäuser, Basel u.a.

[14] Steiner, W. (1999): Optimale Neuproduktplanung, Gabler DUV, Wiesbaden

The Estimation of Market Volumes

Manfred Schwaiger, Matthias Meyer

University of Munich, Chair for Empirical Research and Quantitative Business Administration, Kaulbachstr. 45, D-80539 Munich

Abstract. The estimation of market volumes is a crucial point in the business planning process, especially when considering start-up companies. Depending on the industry and/or the type of product and with respect to a product's position in the life cycle several methods are applicable, starting from common sense based qualitative analysis of consumer behavior up to sophisticated statistical methods. Among them, univariate time series analysis and diffusion models form one group, another consists of expansion procedures. The present paper will give a review of the different approaches, showing which conditions enable the user to apply a certain method and pointing out an algorithm for dealing with biased samples.

1 Market Analysis and Market Volumes

Due to increased competition, market information is becoming more and more significant when trying to design competitive products and services and to position them adequately ([17], p. 15). The necessity of an elaborated market analysis is out of the question. Therefore, market analysis is a constituent of every proper business plan ([11], p. 65), which has to be presented not only on formation of a company but in many other situations, too, e.g. if potential investors are to be convinced.

The main content of market analysis is referring to

- market structure,
- market size and growth and
- the identification of market segments.

For additional information we would like to refer to Schwaiger and Zinnbauer [24] and the references given there.

This paper focuses on the estimation of market volumes that may be systematized, similar to Gierl ([6], p. 432), as shown in table 1, but we omitted response functions, because their main application is to forecast sales or market shares within a single company depending on the marketing mix applied [9]. Nevertheless, in specific situations they might be useful estimation tools as well.

In section 2 we are going to discuss qualitative and in section 3 quantitative approaches, before we may draw some conclusions. But first we need to clear terms and definitions: strictly speaking the drawn sales of a company (sales volume) plus drawn sales of competitors are defined as **market**

	"Time Series Analysis"			"Sampling Theory"
	"Qualitative" Approaches	Polynomial in Time	Diffusion Model	Survey Data
Market Shape	saturated, commodities	monotonous	innovation market	arbitrary
Product Prices	nearly constant	nearly constant	w/o influence on adopters behav.	arbitrary

Table 1. Methods available to estimate market volumes

volume. Adding the free market volume looking at potential buyers (i.e. present non-buyers due to lack of information, leap-froggers, customers currently short of money a.s.o.) we get the **market potential**.

Generally speaking, all methods discussed below are able to estimate market volumes, whereas market potentials require expansion procedures applied to survey data to be calculated properly.

2 Qualitative Approaches

Screening literature one can find surveys (among sales force agents, retailers and wholesalers or customers) subsumed in qualitative approaches ([14], pp. 178ff.).

Aside from the fact that the accuracy of discrimination between qualitative and quantitative methods is subject to personal preferences, the attribute "qualitative" in the named cases is correct only if there is no extrapolation of collected data.

Other qualitative approaches ([10], pp. 217ff.) are

- expert interviews using the DELPHI method,
- conclusion by analogy and
- the use of indicators.

Drawing conclusions by analogy on the market volume in demand may be possible if we can analyze sales figures being descended from comparable business segments or from similar markets, say from foreign countries or from former periods.

Applying the indicator method means, we are looking for related facts and figures, that may serve as hints to our estimation problem. Just when expensive field research has to be dismissed because of costs or strategic considerations (nondisclosure!) this kind of qualitative estimation may provide valuable results allowing to come to a fair decision. An impressive example on the precision of common sense based calculation, based on simple ventilations, is given in ([11], p. 71). So, if applicable, qualitative methods may serve well, but we can imagine that there are situations where we have to switch to more sophisticated methods.

3 Quantitative Approaches

In this section we are going to discuss the estimation of market volumes using polynomials in time, diffusion models or methods of expansion (parameter extrapolation).

Using a polynomial in time we assume a coherence of the form $y_t = \beta_0 + \sum_{j=1}^{k} \beta_j t^j$, where y_t denotes sales volume or market volume in t and the coefficients β_j may be estimated by means of curve fitting techniques [8]. An example is given by Gierl ([6], p. 435), who describes the estimation of per capita amounts of mineral water consumption. We have to criticize that sales volumes only are to be considered whereas potentials not exhausted must be neglected, and that a dependence of the market volume from time only seems to be rather unlikely. That means, even achieving a good fit in the polynomial does not ensure the estimated market volume (which is a forecasting value in this case) to be proper. For this reason we would recommend this approach only very seldom.

3.1 Diffusion Models

In contrast to the just mentioned polynomial in time, diffusion models do not look at the chronological development of sales figures but on the amount of initial purchases (repeated and multiple purchases are ignored). So the focus is on innovators and imitators, and diffusion models are valuable tools if the relation between them is the critical parameter.

Basic assumption and leading thought of most diffusion models is, that the increase in sales Δy_t is following a normal distribution. In this case, the cumulated sales y_t (which equals the market volume for $t \to \infty$) can be expressed by a logistic growth function. The logistic model $y_t = \frac{\beta_0}{1+e^{\beta_1 - \beta_2 t}}$ can be linearized only (and then solved by LS-techniques) if the saturation level β_0 is given. As β_0 is exactly the figure we have to estimate, linearization has to drop out. For this reason there are only two more possibilities: If we have sufficient evidence on the reversal point, we may take advantage of the fact that the reversal point is crossed when exactly half the saturation level is reached ([15], p. 162), and directly calculate the market volume. Otherwise we have to apply algorithms like gradient techniques, Gauss-Newton-Algorithm, Newton-Raphson-Algorithm or Marquardt-Algorithm [27] to estimate the parameters b_j. Except from the gradient technique the named algorithms replace a nonlinear function as the given sales increase by a Taylor-row approximation of first or second order. Hence, estimation results are highly dependent on the starting values and confidence limits are dramatically spread. For this, simulations were carried out by us that show that a reliable estimation of market volumes using the just mentioned models is possible only if the present sales data series is long enough to reach the reversal point. In other words: a reliable estimation of market volumes is possible only if complex models are no longer needed. The same does apply to all generalized logistic functions of order x [12], [13], so we abstain from describing these models here. As opposed to this, the Bass-Model [2] can be linearized, so numeric problems do not appear and good results may be expected, if the shape of sales data available matches the shape of the density trace required in theory.

3.2 Total Value Estimation in Simple Random Samples

In the following we assume that we have collected data that enable us – from a substantial point of view – to infer on the market volume. Such data may be sales data collected or e.g. purchase intentions resulting from a valid sample design (see [20] on the topic of buying intentions). These data are stored in our survey variable Y, but probably there is additional information stored in so called auxiliary variables X_j $(j = 1, \ldots, k)$, we may exploit to increase precision of estimation results. These data are descended from a simple random sample of size n. Simple random samples are assumed to represent a scaled down, but otherwise veridical picture of the population. The process of expansion has to reverse the miniaturization of the population during the sample selection. That means, from answers of several market participants (e.g. potential customers) we have to infer about the market volume. In mathematical terms we have observed realizations $(y_1, x_{11}, \ldots, x_{1k}); \ldots; (y_n, x_{n1}, \ldots, x_{nk})$ on a survey variable Y and (probably) on one or more auxiliary variables X_j, and we are interested in estimating T_y, the total value of Y in population G.

Standard Expansion If there are no auxiliary variables available, we may use the standard expansion procedure according to which the total value $\hat{T}_Y$ is calculated by

$$\hat{T}_Y = N \cdot \frac{1}{n} \sum_{i=1}^{n} Y_i = N \cdot \bar{Y} \qquad \text{with a variance of} \tag{1}$$

$$\text{Var}(\hat{T}_Y) = N^2 \cdot \frac{\sigma_Y^2}{n} \frac{N-n}{N}, \tag{2}$$

where n, N and σ_Y^2 denote the size of the sample, the population size and the variance of Y respectively. As σ_Y^2 is unknown in practice, we use the unbiased estimator $\hat{\sigma}_Y^2 = \frac{1}{n-1} \sum_{i=1}^{n} (y_i - \bar{y})^2$. The last term in (2) is necessary to correct the finiteness of the population. In further variance equations we may omit this correction term and assume that the ratio between sample size and population size is small enough (less than 5 percent). To evaluate the goodness of our estimation we have a look at the spread of the confidence interval, that contains the total value or market volume sought after with a probability of $1 - \alpha$. Prevailing is ([1], p. 162ff.)

$$\left[\hat{T}_y - c \cdot \hat{\sigma}_{\hat{T}_y} \; ; \; \hat{T}_y + c \cdot \hat{\sigma}_{\hat{T}_y}\right] \tag{3}$$

for the lower and upper confidence limit, where $\hat{T}_y$ denotes the realization of the point estimator, $\hat{\sigma}_{\hat{T}_y}$ the estimated standard deviation of $\hat{T}_Y$ and c the $(1-\alpha/2)$-fractile of the standard normal distribution.

Linear Regression Estimation To use an auxiliary variable X_j to increase the precision of market volume estimation the expected value μ_j in G has to be known. In this case we compare μ_j to the corresponding mean value $\bar{x}_j$ in our sample. Given they differ from each other we correct the estimator in (1), because we assume that a bias detected in X_j might also occur in Y. This is plausible at least if there is a correlation between Y and X_j. Among the numerous expansion algorithms using auxiliary variables ([21], pp. 58ff.) we pick out **linear regression estimation**, which never can be worse than competing procedures ([25], pp. 119ff.). The total value estimator and his estimated variance are given in (4) and (5).

$$\hat{T}_Y^{\mathrm{LR}} = N \cdot \left(\bar{Y} + \hat{b}(\mu_X - \bar{X})\right) \quad \text{with } \hat{b} = \frac{\sum_{i=1}^{n}(y_i - \bar{y})(x_i - \bar{x})}{\sum_{i=1}^{n}(x_i - \bar{x})^2} , \quad (4)$$

$$\mathrm{Var}(\hat{T}_Y^{\mathrm{LR}}) = N^2 \cdot \frac{1}{n}(\sigma_Y^2 + \hat{b}^2\sigma_X^2 - 2\hat{b}\sigma_{XY}) \quad . \quad (5)$$

In some cases it may be possible to use more than one auxiliary variable. We may then transfer findings from **multiple linear regression** to our problem. The point estimator and its variance are given in ([21], pp. 82ff.)

Expansion Using Object-Weighting An algorithm to estimate total values from biased samples was proposed by Schwaiger ([21], p. 131ff.). The basic idea is simple: every sample object $i, i = 1, \ldots, n$ is given a special weight g_i instead of the (usual) weight $1/n$ so that the sum of squared deviations $\sum d_j^2$ with $d_j = \left|\sum_{i=1}^{n} g_i x_i^{(j)} - \mu_{X^{(j)}}\right|$, $j = 1, \ldots, k$, between weighted mean values of the k auxiliary variables and the corresponding expected values in the population is being minimized. The weights found are then applied to the m survey variables for estimating population values by using $\hat{T}^{G}_{Y^{(p)}} = N \cdot \left(\sum_{i=1}^{n} g_i Y_i^{(p)}\right)$. Once calculated, the weights allow multiple parameter estimation simultaneously. The determination of weights, estimation of variances, user's benefit and financial consequences of using the described methods are illustrated in Schwaiger [22], [23] by means of an example dealing with the business planning process of a publisher. It is shown that expected revenues can be increased dramatically.

3.3 Total Value Estimation Procedures in Stratified Samples

Frequently, we may notice that, in spite of using auxiliary variables, the precision of our estimation is insufficient, because variances are to big. The reason therefore may be, that the population is very heterogenous with respect to our survey variable Y. If there exists an auxiliary variable of arbitrary scale whose realizations are known for every object in G and which divides G into homogenous groups, we may stratify our sample. Properly, one is looking for an auxiliary variable correlated to our survey variable Y, because then the variance of Y in each stratum will be lower than in the population.

To estimate market volumes in stratified samples literature offers two versions of regression estimation: separate and combined regression estimation ([4], pp. 201ff.). Using **separate regression estimation**, each stratum is treated like a separate sample. According to (4) estimators are calculated ($\hat{b}_h$ denoting the regression coefficient in stratum h) and then aggregated by adding them weighted with the stratum share. Prerequisites are the knowledge of the expected value μ_{X_h} and a sample size $n_h > 30$ in each stratum h. The estimator is calculated according to

$$\hat{T}_Y^{\text{LRs}} = \sum_{h=1}^{H} N_h(\bar{Y}_h + \hat{b}_h(\mu_{X_h} - \bar{X}_h)) \quad \text{with} \tag{6}$$

$$\text{Var}(\hat{T}_Y^{\text{LRs}}) = \sum_{h=1}^{H} \frac{N_h^2}{n_h}(\sigma_{Y_h}^2 + \hat{b}_h^2\sigma_{X_h}^2 - 2\hat{b}_h\sigma_{X_hY_h}) \,. \tag{7}$$

If expected values of the auxiliary variables are known in G, but not in all H strata, or if $n_h > 30\,\forall h$ is violated, one can switch to combined regression estimation if at least $n > 30$ can be ensured. **Combined regression estimation** is very similar to separate regression as shown in (6), but we assume identical regression coefficients in all strata. The corresponding formula are given by (8) and (9), see ([4], p. 239), where $\bar{Y}_{\text{Ges}}$ and $\bar{X}_{\text{Ges}}$ denote arithmetic means in the total sample .

$$\hat{T}_Y^{\text{LRc}} = N \cdot \left(\bar{Y}_{\text{Ges}} + \hat{b}_c(\mu_X - \bar{X}_{\text{Ges}})\right) \quad \text{with} \tag{8}$$

$$\hat{b}_c = \frac{\sum_{h=1}^{H} \frac{N_h^2}{n_h}\frac{1}{n_h-1}\sum_{i=1}^{n_h}(y_{ih} - \bar{y}_h)(y_{ih} - \bar{x}_h)}{\sum_{h=1}^{H} \frac{N_h^2}{n_h}\frac{1}{n_h-1}\sum_{i=1}^{n_h}(x_{ih} - \bar{x}_h)^2} \,.$$

$$\text{Var}(\hat{T}_Y^{\text{LRc}}) = \sum_{h=1}^{H} \frac{N_h^2}{n_h}(\sigma_{Y_h}^2 + \hat{b}_c^2\sigma_{X_h}^2 - 2\hat{b}_c\sigma_{X_hY_h}). \tag{9}$$

3.4 Recent Developments

Instead of using $\hat{T}_Y = N \cdot (\bar{Y} + f(\mu_X - \bar{x}))$ we might consider a non-linear model $\hat{T}_Y = N \cdot (\bar{Y} + f(\mu_X) - f(\bar{x}))$ and transfer concepts of exponential regression, hyperbolic regression and/or polynomial regression ([16], pp. 27ff.) to our problem. The problems that have to be faced are primarily of a numerical type ([3], p. 345).

Another extension is competitive to expansion procedures shown in section 3.3: Münnich ([16], pp. 36ff.) shows how spline regression may help to achieve better results.

Rendtel [18], [19] published several papers on the parameter estimation from panel data, where he focuses on the variance of differences of two cross sectional surveys emerging from a panel. His basic idea is to divide the panel into random groups (subpanels) and then to estimate the variance of weighting results over all subpanels.

Due to space restrictions we had to omit several expansion techniques, mainly those dealing with expansion procedures in multistage samples. Details are given by [28] and [7]. Bookware containing an expansion software tool was presented by Schwieren [26].

4 Recommendations and Conclusions

What to do when in need of estimating market volumes? First of all, one might check, if qualitative approaches as given in section 2 are applicable. If the data needed can be found the user may save time and money. Else, we have to switch to quantitative methods.

Time series analysis may serve well, if we have enough data from the past showing a monotonous trend and the product prices are nearly constant (e.g. commodity markets). Diffusion models are appropriate in (already existing) innovation markets, if buying probability is proportionate to the present cumulated sales level. This is plausible, if the most important adoption motive is social approval or the pool of experience with the product, or if the present cumulated sales level determines application possibilities (facsimile, visual telephone).

In all other cases we have to use survey techniques, collect data via field research and apply expansion procedures. If possible, regression estimation should be applied both in simple and in stratified samples, where separate is superior to combined regression. Object weighting is indicated, if we are facing a biased sample. Schwaiger [21] shows in different examples for many expansion procedures how strong the gain in precision may be compared to standard procedures.

References

1. Bamberg, G., Baur, F. (2001) Statistik, 11th ed., Oldenbourg, München
2. Bass, F.M. (1969) A New Product Growth Model for Consumer Durables. Management Science **15** 215–227
3. Brachinger, H.W., Steinhauser, U. (1995) Rechentechnische Vorteile? Eine Anmerkung zum Verschiebungssatz. Allgemeines Statistisches Archiv **79** 341–346
4. Cochran, W.G. (1972) Stichprobenverfahren, Deutsche Übersetzung von Dr. Wulf Böing, Berlin

5. Cochran, W.G. (1977) Sampling techniques, 3rd ed., Wiley, New York
6. Gierl, H. (1995) Marketing, Kohlhammer, Stuttgart et al.
7. Hansen, M.H., Hurwitz, W.N., Madow, W.G. (1993) Sample Survey Methods and Theory, Vol. I + II. Wiley, New York
8. Höschel, H.-P., Spiridon, P. (1980) Least-squares curve-fitting for nonlinear models with errors in variables and globally convergent Gauss-Newton-procedures, Center for Operations Research and Econometrics, Discussion papers, Louvain
9. Hruschka, H. (1991) Marktreaktionsfunktionen mit Interaktion zwischen Marketing-Instrumenten. Zeitschrift für Betriebswirtschaft **61** 339–356
10. Hüttner, M. (1986) Prognoseverfahren und ihre Anwendung. De Gruyter, Berlin et al.
11. Kubr, T., Ilar, D., Marchesi, H. (1997) Planen, gründen, wachsen. Mit dem professionellen Businessplan zum Erfolg, Ueberreuter, Zürich
12. Lewandowski, R. (1974) Prognose- und Informationssysteme und ihre Anwendungen, Band I, Berlin
13. Lewandowski, R. (1980) Prognose- und Informationssysteme und ihre Anwendungen, Band II, Berlin
14. Meffert, H. (2000) Marketing. 9., überarb. und erw. Aufl.. Gabler, Wiesbaden
15. Mertens, P., Falk, J. (1993) Mittel- und langfristige Absatzprognose auf der Basis von Sättigungsmodellen. In: Mertens, P. (ed.) Prognoserechnung, 5th ed., Physica, Heidelberg, 157–194
16. Münnich, R. (1997) Gebundene Hochrechnung bei Stichprobenerhebungen mit Hilfe von Splines. Vandenhoeck & Rupprecht, Göttingen
17. Probst, G., Raub, St., Romhardt, K. (1999) Wissen managen. Wie Unternehmen ihre wertvollste Ressource optimal nutzen, 3rd ed., Gabler, Wiesbaden
18. Rendtel, U. (1990) Hochrechnung und Stichprobenfehler in Panelerhebungen. Arbeitspapier Nr. 321, Univ. Frankfurt am Main
19. Rendtel, U. (1991) Die Schätzung von Populationswerten in Panelerhebungen. Allgemeines Statistisches Archiv **75** 225–244
20. Schnedlitz, P. (1985) Einstellungen und soziale Beeinflussung als Bedingungen von Kaufabsichten, eine empirische Untersuchungsreihe. Lang, Frankfurt am Main
21. Schwaiger, M. (1993) Hochrechnungsverfahren im Marketing. Vahlen, München
22. Schwaiger, M. (1995a) Informationsgewinnung aus verzerrten Stichproben. Marketing Zeitschrift für Forschung und Praxis **17** 122–129
23. Schwaiger, M. (1995b) Finanzielle Konsequenzen des Einsatzes neuerer Stichprobenverfahren. Marketing Zeitschrift für Forschung und Praxis **17** 259–263
24. Schwaiger, M., Zinnbauer, M. (2001) Markt und Wettbewerb: Methoden und Kriterien zur Bestimmung des Zielmarktes und zur Marktanalyse, in: Hommel, U., Knecht, T. (eds.) Handbuch Wertorientiertes Management wachstumsstarker Start-up Unternehmen. Vahlen, München
25. Schwarz, H. (1975) Stichprobenverfahren. Oldenbourg, München
26. Schwieren, P. (1996) Computergestützte Stichprobenverfahren: objektorientierte Programmierung der Gebundenen Hochrechnung bei geschichteten Stichproben. Kovac, Hamburg
27. Seber, G.A., Wild, C.J. (1989) Nonlinear Regression, Wiley, New York
28. Sukhatme, P.V., Sukhatme, B.V. (1970) Sampling Theory of Surveys with Applications, Ames/Iowa

Section 4

Production, Logistics and SCM

Strategies for Capacity Planning in a Complex Production System

K. Alicke[1], T. Hanne[2,3], and T. Melo[2]

[1] ICON Industrie Consulting GmbH, Hertzstr. 16, D-76187 Karlsruhe, Germany
[2] Fraunhofer Institut für Techno- und Wirtschaftsmathematik (ITWM), Gottlieb-Daimler-Str. 49, D-67663 Kaiserslautern, Germany
[3] Corresponding author, e-mail: hanne@itwm.fhg.de

Abstract. We consider a lotsizing problem in a multi-stage production setting with capacity-constrained resources and a general production structure which may change during the planning horizon. Based on an earlier presentation [1], we focus on some novel concepts of capacity planning. We also present computational results for test problems obtained with the most recent version of our planning software.

1 Introduction

Material Requirements Planning (MRP) concepts have been developed as computer-based tools for generating a master schedule that specifies a production plan for each end item over a planning horizon. Each end item production plan induces secondary demand for its components. Taking into account each component's production, purchasing or transportation lead times and its current and planned stock, the MRP approach determines a synchronized production plan for all components stating in which periods and in which quantities each component should be produced. It is well-known that many aspects relevant to real-life multi-stage production settings are disregarded by most MRP systems with capacity constraints being the most notorious ones.

We developed a heuristic procedure for a multi-stage, multi-item dynamic capacitated lotsizing problem (MSCLP) arising from a manufacturer of electronic devices. This planning software belongs to the class of Enterprise Resource Planning (ERP) tools which succeed traditional MRP tools by including more realistic goals and production constraints, in particular the capacities of shared resources. ERP is widely used in industry to plan the production of goods over the short to medium term in a multi-stage environment.

In the following section we present a brief description of the considered planning problem. In Section 3 some concepts of the capacity planning heuristics included in our software are discussed. Computational issues and some results concerning running times are presented in Section 4. Finally, some concluding remarks are given in Section 5.

2 The Production Planning Problem

We consider a lotsizing problem in a multi-stage production setting with capacity-constrained resources and a general production structure which may change during the planning horizon. The production structure is described by a bill of materials (BOM) which includes production coefficients, lead times, and other relevant data regarding components (also called predecessors of a given product) needed for manufacturing a product (also denoted as successor of its components). The number of levels in the BOM is not limited and the production structure may be any graph free of cycles (production tree). Transportation relationships can be described in a similar way. Moreover, resources needed for production processes can be defined together with their capacities.

Since the problem is hard to solve as an optimization problem, it is formulated as a feasibility problem with various kinds of hard and soft constraints which arise in many practical settings and substantially complicate the planning problem. These include, among others, the following ones:

- due dates and priority classes for end product demands,
- forbidden production periods,
- reordering intervals,
- minimum lot sizes,
- target inventory levels,
- split relationships of given components.

Since usually it is not possible to fulfill all soft constraints, there exists a priority ranking defining which of them may be violated first. For instance, a target inventory level may rather be disregarded than a fulfillment of customer orders. The overall goal of the planning is to satisfy a large number of prioritized customer orders with a minimum of lateness, and to determine the corresponding production plans considering lead times and other restrictions.

It is also possible to model various alternatives in production, for instance, the usage of alternative components or alternative resources. Most of all planning and replanning procedures involve changes in the production structure. This is a consequence of short product life cycles and rapid technological changes which lead to the substitution of components by new ones during the planning horizon. Several types of such substitutions (called hard rolls or soft rolls) are considered in the planning software. A more detailed description of the problem can be found in our earlier paper [1].

3 Strategies for Capacity Planning

Since optimal solutions can only be obtained for special cases or small problems [5], most of the research on the multi-stage capacitated lotsizing problem focusses on the development of methods for simplified versions of the problem. Recent contributions include those by [2], [4], [5], and [6]. One group of promising approaches is based on the decomposition of the problem into a series of single-level, multi-item problems. In [6], capacity constraints are taken into account while solving each single-level problem with a modified version of the Dixon-Silver procedure [3] which considers level relationships. Other approaches (e.g. [2] and [4]) first disregard capacity constraints and generate a feasible solution to the uncapacitated problem by applying sequentially the Wagner-Whitin algorithm [7]. Afterwards, production is shifted from overloaded periods to earlier or later underloaded periods in order to attain capacity feasibility.

Following this basic approach, we developed a two-step greedy heuristic. In the first step, the capacity constraints are disregarded and the problem is decomposed in a similar way as done by [2] and [4] for a simplified version of the MSCLP. Starting at the end item level, the orders from the different priority classes are aggregated by end product type. Lotsizing and timing decisions are made for the end products taking into account their aggregated demands and the availability of direct predecessor components. Since lot sizes on a given level determine the secondary demand for components on lower levels, the procedure evolves iteratively by generating lot sizes level by level until the last BOM level has been reached. In each level, a forward-oriented procedure is applied which schedules items period by period starting with period 1 and ending with the last period in the planning horizon. Let us note that this simple top down approach already poses difficulties (e.g. when only a limited amount of a given part can be purchased), which may lead to the use of replanning routines similar to those applied in the second step of the heuristic as described below.

In the second step, capacity feasibility is attained, e.g., by shifting production of sub-product structures backwards and forwards in time. Specifically, it is determined in which periods the available capacity of each resource is exceeded. Products which consume the overloaded resource in such a period are determined and called critical parts. Next, production is replanned using different strategies for exploring alternatives in the production structure and, possibly, allowing the violation of various soft constraints. These strategies are summarized as follows:

1. Use of alternative resources.
A simple strategy to reduce the consumption of a resource is to try to produce the critical parts without the overloaded resource(s) or to use a combination of resources with a smaller consumption of the overloaded resource(s).

2. Use of alternative components.
In many cases, it is possible to use alternative components for manufacturing a product. If a product consuming an overloaded resource can be replaced by another part (using other resources for its production), then the consumption of the overloaded resource is reduced. The strategy requires an analysis of

the production structure above a critical part which creates demand for that part.

3. Earlier production.
This strategy is based on shifting the production backward (in time), i.e. producing earlier. This requires a replanning of the production structure below the critical part.

4. Use of interdepot transportation.
Possibly, a product consuming an overloaded resource, or some product in the production structure above it which creates demand in a critical period, can be replaced by an equivalent product which is available through transportation from another depot.

5. Inventory reduction.
An obvious way of reducing the consumption of an overloaded resource is to postpone the production of an item which is kept in stock for some time. Such additional production may, for instance, result from safety stock requirements which are soft constraints. Due to their low priority, it is preferred to violate them instead of the due dates of customer orders. The reduction of inventory concerns the critical products as well as parts in the production structure above them which create some demand in critical periods.

6. Creation of end demand backlog.
If none of the above strategies is sufficiently successful in reducing the consumption of the overloaded resource, then it is not possible to satisfy all end product demand on time. In that case, some (or the whole) production amount is shifted forward in time resulting in backlogs of customer orders. These orders belong to different priority classes. Low prioritized demand is first selected to be filled after its due date.

A replanning according to the above strategies requires capacity checking and the consideration of the availability of components in different parts of the production structure. Some of the procedures require an analysis of the production structure above a critical part, i.e. the critical path. In most of the strategies, some special consideration of predecessors of the critical path is necessary. The parts can, for instance, be kept in stock if their demand is shifted to a later period. Alternatively, the corresponding produced amounts can be reduced in the original periods and replanned in new (earlier or later) periods. The checking and replanning of upper or lower subtrees of critical parts is done in a recursive fashion.

In the following, we will discuss two of the recursive replanning methods. However, many details of the implemented routines are omitted due to space restrictions. Furthermore, to keep the exposition simple the following features are not considered in the algorithms presented below:

- production times,
- lead times,
- forbidden periods and other calendar/time restrictions,
- restrictions on minimum production quantities and lot sizes,

- desired split values for groups of alternative components,
- numerical issues.

For instance, let us consider a strategy for checking alternatives for producing an item that consumes the overloaded resource. This is carried out in the upper production tree (of successors). The goal is to replan production in order to reduce the resource consumption. The recursive pseudo code routine in Fig. 1 is first called with the critical part (product) and the desired production amount to be reduced as parameters.

use_alternative_components(product, amount):

```
summed_up_shifted_amount := 0;
for each successor S of product
   if S has at least 2 groups of alternative components then
      while there is at least one unconsidered group G of alternative components
         if product is not in G then
           up_to_amount := amount - summed_up_shifted_amount;
           shifted_amount := try_to_replan_production_with_group_G(product, up_to_amount);
           summed_up_shifted_amount := summed_up_shifted_amount + shifted_amount;
         end;
      end;
   end;
end;
while summed_up_shifted_amount < amount and
   if there is at least one predecessor P of product
      shifted_amount := coeff * use_alternative_components(P, (amount -
                          summed_up_shifted_amount)/coeff);
      summed_up_shifted_amount := summed_up_shifted_amount + shifted_amount;
   end;
end;
decrease production of product using the violated resource,
   or with the group of alternative components in which the predecessor
   in the recursion path (successor in the production tree) belongs;
return summed_up_shifted_amount;
```

Fig. 1. Algorithm for replanning production by using alternative components

In this algorithm, `coeff` is the production coefficient stating the amount of a component (`product`) necessary to produce one unit of the considered predecessor (`P`). The basic idea of the routine is first to look for production possibilities using alternative components for the current product. Afterwards, the upper tree is recursively checked for such replanning possibilities as long as the already shifted amount `summed_up_shifted_amount` is not sufficient. In the algorithm, the subroutine `try_to_replan_production_with_group_G` shown in Fig. 2 is necessary for replanning the production of the part with a specific group of alternative components.

Let us note that two different strategies for finding bottlenecks in production (sub)trees are used: Either the relevant (sub)tree is first checked concerning replanning possibilities (as indicated in the above routine), or a replanning is done immediately with the possibility of undoing the changes when tighter restrictions are detected in the (sub)tree.

try_to_replan_production_with_group_G(product, amount):

```
feasible_amount := feasible production of product;
possible_amount := min(amount, feasible_amount);
for each componenent C in G
   component_amount := C.coeff * possible_amount;
   determine possible_production_of_C with any of its groups of alternative components;
   if possible_amount > possible_production_of_C / C.coeff then
      possible_amount := possible_production_of_C / C.coeff;
   end;
end;
for each componenent C in G
   component_amount := C.coeff * possible_amount;
   for each group CG of alternative components of C
      if (component_amount > 0)
         replanned_component_amount :=
            try_to_replan_production_with_group_CG(C, component_amount);
         component_amount := component_amount - replanned_component_amount;
      end;
   end;
end;
increase production of product by possible_amount;
return possible_amount;
```

Fig. 2. Algorithm for replanning production of a product and its subtree

4 Computational Issues and Test Results

The considered MSCLP can be formulated as a mixed integer linear programming problem. This problem is NP-hard and real-life instances are typically very large. On the other hand, it is desired to solve the problem within a short time (say 20 minutes for a large-scale instance), possibly including features for interactively solving a modified problem starting from an existing solution (what-if-analysis).

Therefore, the average running time of the software should be almost linearly dependent on the number of products and the number of planning periods. The top down approach in step one of our heuristic procedure basically fulfills this requisite but the complexity of the replanning routines may have quite time consuming effects. Thus, for typical situations with overloaded resources requiring various repair routines, test problems were constructed that involve an intensive use of the replanning procedures.

A major strategy for tuning the algorithm was to store additional data concerning information on already explored and exhausted possibilities for shifting production. In a subsequent repair stage, these shifting possibilities do not need to be explored again. Let us clarify this case by means of a simple example: Assume that a resource has been overloaded by one product in any (feasible) planning period, and that no alternatives in production or transportation exist. Clearly, demand cannot be completely satisfied on time. As a result, backlogs will occur and grow monotonously until the end of the planning horizon. In that case, in order to reduce the consumption of the overloaded resources we only need to explore all possibilities in the first period. For the subsequent periods, we know that no production alternatives exist and so production will have to be shifted completely to the end of the planning horizon.

This and other strategies lead to a significant reduction of the CPU times as shown in Table 1. The columns 'RT old' and 'RT new' indicate the absolute CPU times in milliseconds (ms) before and after the tuning (measured on a Pentium III, 350 MHz with 128 MB RAM). All test problems except the 'real data' problem are artificially constructed and include various typical production scenarios.

Table 1. Running times for test problems before and after tuning.

test problem	RT old (ms)	RT new (ms)	speedup (%)
1	714	62	91.3
2	671	58	91.4
3	1413	63	95.5
4	49686	637	98.7
5	2831	89	96.9
6	1526	109	92.9
7	7327	396	94.6
8	5663	112	98.0
9	1592	90	94.3
10	5787	163	97.2
linear_small	60253	1811	97.0
linear	2500165	78348	96.9
real data	342700	5351	84.4

For most of the test problems the table shows a speedup of more than 90%. For the 'real data' problem the speedup is somewhat smaller. The main reason is that the latter problem does not contain many of the complex situations that were intentionally included in problems 1 up to 10. As a result, some time consuming replanning routines are not required. The large CPU time reported for this problem is due to the size of the problem: There are 26528 customer orders which are aggregated in 6501 end products, and 30430 BOM items (organized in 4 levels).

The 'linear' test problem represents a deep production structure with shared resources on different BOM levels and a very long planning horizon. A reduced version of this problem, 'linear_small', shows almost the same speedup, indicating the scalability of the speedup.

5 Conclusions

In this paper, we have considered a multi-stage lotsizing problem involving multiple capacitated resources which are shared by different products. The complexity of the problem is increased by various soft and hard constraints which model real-life planning situations. Since large instances of the problem, formulated as a large-scale mixed integer linear program, cannot be solved within acceptable computing time, a planning heuristic has been developed. During its greedy first stage, capacity constraints are disregarded. In the second stage, capacity violations are repaired using recursive replanning routines. Various experiments with small and large-scale test problems led to the tuning of the routines and to a considerable reduction of the computational requirements.

Future research should lead to a modification of the model towards an optimization problem. This involves developing a refined cost model for pricing the violation of different kinds of constraints. By using means of local optimization strategies, it should then be possible to improve the planning heuristics and the quality of the generated production plans.

References

[1] Alicke, K., Eley, M., Hanne, T., Melo, T. (2001): A heuristic approach for a multistage lotsizing problem with dynamic product structure. In: Fleischmann, B.; Lasch, R.; Derigs, U.; Domschke, W.; Rieder, U., (Eds.) Operations Research Proceedings 2000. Selected Papers of the Symposium on Operations Research (OR 2000), Dresden, September 9-12, 2000, 278–283

[2] Clark, A., Armentano, V. (1995): A heuristic for a resource-capacitated multistage lot-sizing problem with lead times. Journal of the Operational Research Society **46**, 1208–1222

[3] Dixon, P., Silver, E. (1981): A heuristic solution procedure for the multi-item, single-level, limited capacity lot-sizing problem. Journal of Operations Management **2**, 23–39

[4] França, P., Armentano, V., Beretta, R., Clark, A. (1997): A heuristic method for lot-sizing in multi-stage systems. Computers in Operations Research **24**, 861–874

[5] Tempelmeier, H., Derstroff, M. (1996): A Lagrangean-based heuristic for dynamic multilevel, multiitem constrained lotsizing with setup times. Management Science **41**, 738–757

[6] Tempelmeier, H., Helber, S. (1994): A heuristic for dynamic multi-item multilevel capacitated lotsizing for general product structures. European Journal of Operational Research **75**, 296–311

[7] Wagner, H., Whitin, T. (1958): Dynamic version of the economic lot size model. Management Science **5**, 89–96

A Simple Queueing Model for the Estimation of Man Machine Interference in Semiconductor Wafer Fabrication

Hermann Gold

Infineon Technologies AG, D-93049 Regensburg, Germany
email: hermann.gold@infineon.com

Abstract. In this paper a simple queueing model to deal with the man machine interference problem in semiconductor manufacturing is developed. The aim is to estimate the cycle time increase at a work center due to the fact that each lot entering a work center brings in not only work for the machines but also for the operators. The complex situation at a work center with forty and more machines brought upon by machine dedication, operator cross training, maintenance schedules etc. is mapped onto a double queue with synchronous arrivals for operators and machines. Modelling all operators as one server and doing alike for all machines requires some basic condition to be fulfilled which is known as resource pooling in queueing literature.

1 Introduction

The classical queueing model for the description of the man machine interference problem is the product form queueing network model as presented by Kelly in [4]. Kelly considers N machines under the care of a single operator. He assumes that the machines need attention at specified time instances during which operation of the machines is interrupted. At such a time instance a certain machine is stopped and waits for an operator. Once the operator is available the machine is to be served for some time by the operator before it can resume running. Both service time and running time of a machine are negative exponentially distributed. In this model machines are competing for the attention of the operator as soon as more than one machine is stopped and is waiting. This is the way machines interfere with each other. There is no similar interference amongst operators in this model, since only machines act as traffic sources and there is only one operator, i. e, the situation is deliberately biased. In semiconductor wafer fabrication however machinery is very expensive and therefore a number N of machines will be attended by a number $K > 1$ of operators. As a consequence the situation is more or less symmetric between operators and machines in the sense that operators might have to wait for a machine to finish processing so that a certain control measurement can be taken for the lot in process, a preventive maintenance, setup or tuning operation for the running machine can be carried out, etc..

Machines in turn might have to wait for an operator for loading or unloading of lots.

In [6] the classical machine interference model of Kelly has been changed so that the traffic source for operators are the lots and no longer the machines. The machine behaviour is built into the model by considering the first M (number of machines) places of the lot queue waiting for operators as machine instances. As a consequence it is not possible that one or more machines are busy at some time epoch where all operators are idle. This is not the case in environments considered in this paper.

In our modelling approach the traffic source for both operators and machines are the lots. The man machine interference is taken into account by considering a statistic which combines the load situation for the operators and for the machines at one time, i. e. the maximum of the unfinished jobs for operators and for machines at each time instance. Hereby we assume that operators prioritize jobs which have to be done immediately in order to keep machines running. In particular loading and unloading are high priority jobs. Furthermore we assume that operators work as a team in the sense that the work load to be done can be distributed homogeneously amongst the operators, i. e. statistically all operators have to work the same percentage of time. In a similar way the same should be true for the machines. Formally we require from both resource types, operators and machines, that the condition for resource pooling is fulfilled. The condition for resource pooling says, that whenever there is dedicated traffic λ_1 and λ_2 for each of two servers of the same type, operator or machine, in order that the two servers can be considered as one single resource pool there has to be an amount of at least $|\lambda_1 - \lambda_2|$ of descretionary traffic which can join the shorter one of the two queues (see e. g. [5] for more details on resource pooling).

2 Basic model and Markov chain state equations

In Fig. 1 the modelling of man machine interference for an example work center is shown. There are two machine groups with service rates $\mu_{m,1}$ and $\mu_{m,2}$, respectively, and two operator groups with service rates $\mu_{o,1}$ and $\mu_{o,2}$, respectively. Job type 1 with arrival rate λ_1 can be done on machine group 1 and needs assistance by operator group 1. Job type 2 with arrival rate λ_2 can be done on both machine group 1 or machine group 2 and can be assisted by operator group 1 or by operator group 2. Finally, Job type 3 with arrival rate λ_3 can be done on machine group 2 and needs assistance by operator group 2. Assuming that the resource pooling condition is fulfilled for both operators and machines, i.e. $\lambda_2 > |\lambda_1 - \lambda_3|$, the service rates μ_o and μ_m of the simplified model are given by $\mu_{o,1} + \mu_{o,2}$ and $\mu_{m,1} + \mu_{m,2}$, respectively. The arrival rate for the simplified model is $\lambda = \lambda_1 + \lambda_2 + \lambda_3$.

In the sequel we refer only to the simplified model. As a performance characteristic which expresses the main effects of man machine interference

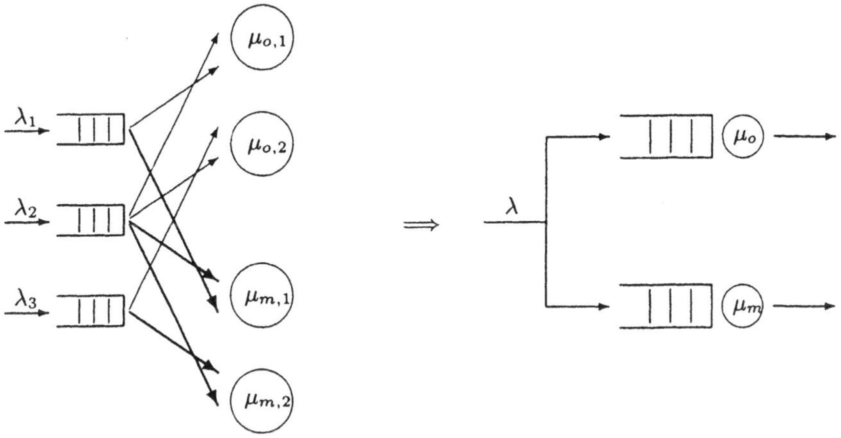

Fig. 1. Man Machine Interference Modelling for a Work Center

we consider the expected value of the maximum of jobs in each of the two queues. We define a two-dimensional random walk (X, Y) with the following meaning:

i : number of unfinished jobs for machines
j : number of unfinished jobs for operators

The joint probability $P(X = i, Y = j)$ is denoted by $p_{i,j}$. Let μ_o the altogether service rate of the operators as one team, μ_m the altogether service rate of the machines of the work center considered and λ the arrival rate of lots to the work center. Taking into account that each lot brings in work for operators as well as machines synchronously, the corresponding Markov chain equations are given by the following. For $i \geq 1$ and $j \geq 1$:

$$p_{i,j} \cdot [\alpha + \beta + 1] = p_{i-1,j-1} + p_{i+1,j}\,\alpha + p_{i,j+1}\,\beta \tag{1}$$

where $\alpha = 1/\rho = \mu_m/\lambda$ and $\beta = 1/\xi = \mu_o/\lambda$.

3 Analysis

In [2] the equilibrium equations (1) are converted into a functional equation for the generating function $P(z, w)$ of the joint probability of (X, Y). The derivation of important system characteristics for the Markov chain described can be considerably simplified through the definition of the following macro states.

$$\acute{p}_k = \sum_{i-j=k} p_{i,j} \qquad k = \ldots, -2, -1, 0, 1, 2, \ldots \tag{2}$$

Using the macro state equations

$$\acute{p}_k \mu_m = \acute{p}_{k-1}\mu_o - p_{k-1,0}\mu_o. \quad \text{for } k > 0 \tag{3}$$

and

$$\acute{p}_k \mu_o = \acute{p}_{k+1}\mu_m - p_{0,k-1}\mu_m, \quad \text{for } k < 0 \tag{4}$$

we get by recursive substitution

$$\acute{p}_k = \begin{cases} \left(\frac{\beta}{\alpha}\right)^k \acute{p}_0 - \sum\limits_{i=0}^{k-1} p_{i,0} \left(\frac{\beta}{\alpha}\right)^{k-i} & : \text{ for } k > 0 \\ \left(\frac{\alpha}{\beta}\right)^{-k} \acute{p}_0 - \sum\limits_{j=0}^{-(k+1)} p_{0,j} \left(\frac{\alpha}{\beta}\right)^{-(k+j)} & : \text{ for } k < 0. \end{cases} \tag{5}$$

Hence we get for $\acute{p}_0$

$$\acute{p}_0 = \left(\frac{\alpha}{\beta}\right)^k \acute{p}_k + \sum_{i=0}^{k-1} p_{i,0} \left(\frac{\alpha}{\beta}\right)^i \quad \text{for } k > 0 \tag{6}$$

and

$$\acute{p}_0 = \left(\frac{\beta}{\alpha}\right)^{-k} \acute{p}_k + \sum_{j=0}^{-(k+1)} p_{0,j} \left(\frac{\beta}{\alpha}\right)^j \quad \text{for } k < 0. \tag{7}$$

Since

$$\lim_{k \to \pm\infty} \acute{p}_k = 0 \tag{8}$$

we get as result for the macro state $\acute{p}_0$ consisting of all states where both queues have the same length that it equals the z-transform of the base line probabilities taken at α/β and β/α, respectively. Formally, the base line probabilities correspond to the marginal distributions of the joint distribution of the integral-valued random variables X and Y. Their generating functions have been derived in [2] and are denoted by $P(z,0)$ and $P(0,w)$, respectively. Hence we get

$$\acute{p}_0 = P(\alpha/\beta, 0) = P(0, \beta/\alpha). \tag{9}$$

This result is valuable for numerical evaluation since it captures an initial macro state comprising an infinite number of single states. In the sequel we use eqn. (5), $P(z,0)$ and $P(0,w)$ to calculate $\acute{p}_k$ for $|k| > 0$. The expectation of the maximum in both queues can now be yield in a simple way by the following manipulation.

$$E[\max(X,Y)] = \sum_{i=0}^{\infty}\sum_{j=0}^{\infty} \max(i,j) \cdot p_{i,j} \tag{10}$$

$$= \sum_{i=0}^{\infty}\sum_{j=0}^{i-1} i \cdot p_{i,j} + \sum_{i=0}^{\infty}\sum_{j=i}^{\infty} (i \cdot p_{i,j} + (j-i) \cdot p_{i,j})$$

$$= \sum_{i=0}^{\infty}\sum_{j=0}^{\infty} i p_{i,j} + \sum_{k=0}^{-\infty} |k|\, \acute{p}_k$$

$$= \frac{\rho}{1-\rho} + \sum_{k=0}^{-\infty} |k|\, \acute{p}_k$$

How many derivations do we have to calculate in order to get a given accuracy of the performance measure $E[\max(X,Y)]$? To answer this question we make use of the limiting tail probabilities for $p_{i,0}$ and $p_{0,j}$. Without loss of generality we assume that $\alpha < \beta$ as μ_o and μ_m can be interchanged if necessary. In [2] the limit for $p_{0,j}$, $j \to \infty$ has been given. The formula for the limit of $p_{i,0}$ as it has been given in [2] is wrong. It has to be substituted by the following.

$$p_{i,0} \approx \frac{c_1}{\alpha^i} \quad \text{for} \quad i \to \infty, \quad \text{with } c_1 = \frac{q \cdot \frac{u}{p} - v}{q} \cdot \frac{p-q}{u}, \quad \alpha = \frac{p}{q}, \quad \beta = \frac{u}{v}. \tag{11}$$

The asymptote for $p_{i,0}$ is approached much more quickly than the one for $p_{0,j}$. Since the asymptote is approached from below (and hence $\acute{p}_k$ is overestimated when using the limiting probability $p_{k-1,0}$), care has to be taken when only a small number of derivatives of the generating function $P(z,0)$ can be calculated, especially when α and β differ only slightly. As an upper bound for $E[\max(X,Y)]$ the value for a symmetric system with $\alpha = \min(\alpha,\beta)$ can be used. By using the successive derivations of $P(z,0)$ we get for the special case $\alpha = \beta$

$$\sum_{k=0}^{\infty} k \cdot \acute{p}_k = \lim_{S\to\infty} \left[\frac{S(S+1)}{2} \left[\frac{\alpha-1}{\alpha} - \left(\frac{\alpha-1}{\alpha}\right)^{\frac{3}{2}} - \right.\right. \tag{12}$$

$$\left.\left. \left(\frac{\alpha-1}{\alpha}\right)^{\frac{3}{2}} \sum_{i=1}^{S} \prod_{k=1}^{i} \frac{2k-1}{2k} \alpha^{-i} \right] + \left(\frac{\alpha-1}{\alpha}\right)^{\frac{3}{2}} \sum_{i=1}^{S} \frac{i(i+1)}{2} \prod_{k=1}^{i} \frac{2k-1}{2k} \alpha^{-i} \right]$$

Numerical analysis of this expression yields

$$\sum_{k=0}^{\infty} k \cdot \acute{p}_k = \frac{1}{\alpha-1} \cdot \left(\frac{1}{2} - \frac{1}{8\alpha}\right), \quad \text{for} \quad \alpha = \beta. \tag{13}$$

Therefore under symmetric load for operator and machines we get

$$E[\max(X,Y)] = \frac{1}{\alpha-1} \cdot \left(\frac{3}{2} - \frac{1}{8\alpha}\right) \quad \text{for} \quad \alpha = \beta. \tag{14}$$

So far we have considered the system under the stability condition $\rho, \xi < 1$. As soon as one of the two servers is overloaded, i. e. $\rho > 1$ or $\xi > 1$ the buffer spaces for lots have to be restricted to some finite value S. Jobs seeing either of the two queues full upon arrival are considered to be blocked. The value for the expectation of the maximum in both queues is then given by

$$E[\max(X,Y)] = S - \frac{1}{\max(\rho,\xi) - 1} \tag{15}$$

4 Numerical Analysis and Applications

The use of the model for man machine interference as presented in the preceding sections is twofold. One area of application is the so-called Operating Curve manufacturing line performance assessment methodology, or in short OP Curve (see [1]). The aim of this method is to characterize the behaviour of a factory by its Operating Curve graph, a plot of the cycle time multiplier X over the bottleneck utilization. Hereby X is given as the ratio of Actual Line Cycle Time to Raw Cycle Time (see [1] for details). X is also called Flow Factor (FF) and this term will be used in the sequel. In [1] the Operating Curve graph is related to the famous Pollazcek-Khintchine formula for M/G/1 systems. We deploy Kingman's approximation formula for G/G/1 systems which allows for generally distributed arrival processes as suggested by Hopp and Spearman (see [3]). According to this formula the Flow Factor is given by

$$FF = \alpha' \frac{\rho}{1-\rho} + 1, \tag{16}$$

where α' is a measure for the variability of material flows and service times of the factory (not to be confused with $\alpha = \mu_m/\lambda$ in the analysis of Sections 2 and 3). As utilization factor ρ the median of all machine utilization factors is taken (as opposed to the bottleneck utilization factor suggested in [1]).

The idea of OP Curve is that once a factory's OP Curve graph is established the effect of market changes and correspondingly raised or lowered production rates on manufacturing lead times can be estimated in advance. It is necessary that α keeps roughly the same for one and the same factory to make this work.

Of course OP Curve is also applied to smaller manufacturing areas of a factory. These smaller manufacturing areas should feature the property of a resource pool, each by itself, as described in Section 1. There are a couple of influences in a real world fab which make the G/G/1 approximation a somewhat unsatisfactory heuristic. Among these are machine downtimes (see e. g. [7]), unbalanced loading on the machines and man machine interference. Our studies have shown that the G/G/1 approximation can be much improved when man machine interference is taken into regard. We simply assumed that operator and machine utilization are roughly the same in the overall factory. Therefore as an alternative to eqn. (16) the formula

$$FF = \alpha'' \left(\frac{1}{2\rho} + \frac{7}{8} \right) \frac{\rho}{1-\rho} + 1, \tag{17}$$

is used to fit the factory's OP Curve, where α'' is the variability of the new system. This formula is easily derived from eqns. (14), which is based on Markovian queueing systems and $\alpha'' = 1$ in there. In the application it turned out that for a number of 55 historical operating points of a fab the sum of the squared errors with respect to the best fitted α' and α'' is reduced by 23.5% when using eqn. (17) in stead of eqn. (16).

A second area of application for the man machine interference model is simulation of the material flows of a factory. Since human beings can adapt themselves to real world situations in an excellent way it is very hard to incorporate operator behaviour into simulation models. Therefore simulation models are often build up by assuming that there are no operator constraints. Of course this modelling approach yields results which are too optimistic as compared to the real world behaviour of the factory. Experience shows that overall WIP in a real world factory is typically a 1.3 to 1.5 multiple as compared to that of a corresponding simulation model without operator constraints. This is in line with eqn. (14), which suggests that this multiplying factor is 1.375 for $\rho \to 0$ and 1.5 for $\rho \to 1$ under the assumption that man and machine are equally loaded. Hence the cycle time estimations provided by simulation models with no operators incorporated can be improved by using a mathematical plug-in which stretches the cycle times for each resource pool according to the man machine interference modelling results.

So far we have assumed, that man machine interference can be considered independently of the general variability of a factory. However this is not quite true. In Tab. 1 we give the Flow Factor as calculated from the maximum of a symmetrically loaded double queue with synchronous Poisson arrivals ($c_a^2 = 1$) and service time distributions according to a Gamma distribution with a squared coefficient of variation c_s^2 of 2. In the second and third columns of the table the Flow Factors as calculated from eqn. (17) can be compared to results from simulative analysis for corresponding utilization factors. It turns out that the above assumption leads to a certain moderate discrepancy in the performance results for systems with non-Markovian servers. In a number of studies which are not shown in the table we investigated that under medium and high load the development of the G/G/1 approximation to the double queue produces too optimistic results when the coefficient of variation of the service time is > 1; on the other hands results are too pessimistic when the coefficient of variation of the service time is < 1.

5 Conclusion and Outlook

The main results of our work are the following. The ability to estimate lead times on the factory level can be much improved using eqn. (17) instead of

Table 1. Validity of Flow Factor calculation for non-Markovian systems

U	FF (17)	FF (sim.)
0.1	1.979	1.857
0.2	2.266	2.178
0.3	2.634	2.593
0.4	3.125	3.130
0.5	3.813	3.859
0.6	4.844	4.945
0.7	6.563	6.804
0.8	10.000	10.160
0.9	20.313	22.382

the well-known G/G/1 approximation. On the tool set level the interference of man and machine can be compensated for by adjusting lead times or target flow factors according to the maximum of the two lead times in a system with two parallel queues created by arrivals with two demands. Hereby the load in the two queues need not be balanced. In the analysis macro state definitions and limiting probabilities have been developed which make numerical evaluation much more efficient as compared to dealing solely with transforms. Future work will develop the concept and application of resource pooling when machine dedication and operator cross training is more intricate than in the example of section 2. Hereby connectivity of nodes in graphs and some sort of nested resource pools would play an important role.

References

[1] Aurand, S., Miller, p. (1997) The Operating Curve: A method to measure and benchmark manufacturing line productivity. IEEE/SEMI Advanced Semiconductor Manufacturing Conference, 391-397

[2] Flatto, L., Hahn, S. (1984) Two parallel queues created by arrivals with two demands I. SIAM J. Appl. Math. 44 (5), 1041-1053

[3] Hopp, W. J., Spearman, M. L. (1996) Factory Physics - Foundations of Manufacturing Management. Irwin, Chicago

[4] Kelly, F. P. (1979) Reversability and stochastic networks, Wiley

[5] Kelly, F. P., Laws, C. N. (1993) Dynamic routing in open queueing networks: Brownian models, cut constraints and resource pooling. Queueing Systems 13, 47-86

[6] Meyersdorf, Biron, O., Ozelkan E. C., Fowler. J. W. (1997) Staffing analysis tool for operator-machine-lot interference in semiconductor manufacturing. IEEE/SEMI Advanced Semiconductor Manufacturing Conference, 335-340

[7] Rose, O., Duemmler M., Schoemig, A. (2000) On the validity of approximation formulae for machine downtimes. Proceedings of the International Conference on Modeling and Analysis of Semiconductor Manufacturing (MASM 2000), 188-193

Optimal Coordination of Manufacturing and Remanufacturing Decisions in Case of Product Substitution

Karl Inderfurth[1]

Otto-von-Guericke University Magdeburg, Faculty of Economics and Management, P.O.Box 4120, D-39016 Magdeburg, Germany
e-mail: inderfurth@ww.uni-magdeburg.de

1 Decision Problem

Due to both economic incentives and legal pressure more and more companies are going to be engaged in the product recovery business which refers to activities for regaining materials and value added of used products. A very important field of product recovery is remanufacturing which refers to those activities that bring used products or major modules back to a condition that remanufactured items are almost or even just as good as new. This is widely found for high-valued industrial products like copiers, computers, vehicle engines, or medical equipment. In many industries original equipment manufactures are active in the remanufacturing business themselves. In this case coordination problems arise.

If manufacturing and remanufacturing processes are using the same productive resources, capacity allocation and production scheduling is a major task. If remanufactured products are as good as new from the custumers' perspective, two interchangeable sources of supply exist which create specific problems for a demand-driven production and materials planning. The majority of quantitative research in the area of remanufacturing is dealing with these problems under deterministic as well as under stochastic conditions (for an overview see [1]). If remanufactured products are significantly different from new ones, e.g. because of downgrading, they are sold in different markets to different customers at different (i.e. lower) prices. In this case there seems to be no need for coordination, given that jointly used resources do not exist. The manufacturing and the remanufacturing process can be managed independently. However, things are different if a company is willing to offer its customers of remanufactured items a higher-valued original one as substitute in an out-of-stock situation. This for instance happens in the automotive business with overhauled engines. In this case of possible downward substitution of products an interconnection between both markets exists which makes it necessary to coordinate manufacturing and remanufacturing decisions. This two-product decision problem has not been dealt with up to now for hybrid systems with remanufacturing.

2 Model Formulation

The model to be formulated will be used to provide some insight in the structure of an optimally coordinated manufacturing and remanufacturing policy for the above problem under product substitution. Starting point will be a single-stage single-period problem with stochastic demands for both product types. The returns of used products are also allowed to be stochastic. Returned items which are not remanufactured will be disposed of. Serviceable products are sold at a type-specific price. Products left over after the demand period earn specific salvage values. An excess stock of manufactured serviceables can be used to satisfy excess demands for remanufactured products. Manufacturing, remanufacturing, and disposal are charged with a respective cost per unit. The objective is to maximize the expected profit from all activities, i.e. from determining the manufacturing and the remanufacturing order quantity given an arbitrary starting inventory of serviceable products. Stochastic returns arrive during some return period after which the process of remanufacturing can start. The outcome of this process is restricted by the number of returned items. By this formulation this model is some kind of product substitution variant of the standard stochastic remanufacturing models found in literature (e.g. see [2] and [3]). Related inventory models with substitution, but without the remanufacturing environment are treated in [4] and [5].

In the sequel we use the following notation where index i stands for manufacturing ($i = M$) and remanufacturing ($i = R$), while manufactured and remanufactured serviceable products are abbreviated by MP and RP, respectively :

x_i : MP/RP inventory *before* manuf./reman.
y_i : MP/RP inventory *after* manuf./reman.
q : quantity of product substitution
R : returns of used products (stochastic)
D_i : MP/RP demand (stochastic)
p_i : per unit sales price for MPs/RPs
s_i : per unit salvage value for MPs/RPs
c_i : per unit manuf./reman. cost
c_D : per unit disposal cost for returned products
λ_i : leadtime for manuf./reman.

The model formulation has to be different for two different leadtime cases. Hereby, it is assumed that both manufacturing and remanufacturing will start in time such that they are completed when the demand period begins. In case 1, when the manufacturing leadtime is shorter or equal to the remanufacturing one (i.e. $\lambda_M \leq \lambda_R$) both production decisions will be made simultaneously because there is no difference in the state of information. In case 2, when the remanufacturing leadtime is shorter (i.e. $\lambda_M > \lambda_R$), at the time of the manufacturing decision the return uncertainty will not yet have been revealed completely. So a sequential decision making under uncertainty is necessary.

In the following analysis we will concentrate on case 1 ($\lambda_M \leq \lambda_R$) and start with deriving the objective function. By the end of the demand period - not yet regarding possible product substitution - revenues from selling and salvaging MPs and RPs are found to be

$$r_i(y_i, D_i) = p_i \cdot \min[y_i, D_i] + s_i \cdot \max[y_i - D_i, 0] \quad \text{with} \quad i \in \{M, R\}.$$

Expected revenues are denoted by

$$L_i(y_i) = \mathop{\mathrm{E}}_{D_i} \{r_i(y_i, D_i\} \quad \text{with} \quad i \in \{M, R\}.$$

After demand realisation as much RP demand will be substituted by MP leftover inventory as necessary and possible. This results in the following substitution quantity q:

$$q(y_M, y_R, D_M, D_R) = \begin{cases} D_R - y_R & \text{if} \quad y_M - D_M \geq D_R - y_R > 0 \\ y_M - D_M & \text{if} \quad D_R - y_R > y_M - D_M > 0 \\ 0 & \text{else} \end{cases}$$

The expected amount of substitution is :

$$Q_M(y_M, y_R) = \mathop{\mathrm{E}}_{D_M, D_R} \{q(y_M, y_R, D_M, D_R\}$$

Subtracting manufacturing, remanufacturing and disposal costs from the expected revenues before substitution and from expected net profit by substitution leads to the following expected total profit TP :

$$\begin{aligned} TP(y_M, y_R) = {} & L_M(y_M) + L_R(y_R) + (p_R - s_M) \cdot Q(y_M, y_R) \\ & - c_M \cdot (y_M - x_M) - c_R \cdot (y_R - x_R) - c_D(R + x_R - y_R) \end{aligned}$$

Here $p_R - s_M$ is the amount of net profit which can be earned by substituting one unit MP servicable for one unit RP item.

3 Policy Analysis

Using total profit TP as objective function and regarding the restrictions for increasing the MP and RP inventories by manufacturing and remanufacturing results in the following optimization problem:

$$\underset{y_M, y_R}{\text{Max}} \quad TP(y_M, y_R)$$

$$\text{s.t. } y_M \geq x_M \text{ and } \quad R + x_R \geq y_R \geq x_R$$

Under reasonable revenue and cost parameter restrictions (e.g. $p_R > s_M$, to let substitution be profitable) it can be shown (see [6]) that objective function $TP(y_M, y_R)$ is jointly concave in y_M and y_R. This concavity property results in optimality conditions $\partial TP/\partial y_M = 0$ and $\partial TP/\partial y_R = 0$ which lead to order-up-to-*functions* $y_M^+ = S_M(y_R)$ and $y_R^+ = S_R(y_M)$ as well as to order-up-to-*levels* S_M^+ and S_R^+. With these parameters the structure of the optimal manufacturing and remanufacturing policy can be described as follows (see [6]):

Table 1. Optimal policy structure

RP-Inventory	MP-Inventory	Reman. Decision	Manuf. Decision
$x_R + R < S_R^+$	$x_M \leq S_M(x_R + R)$	$y_R^* = x_R + R$	$y_M^* = S_M(x_R + R)$
	$x_M > S_M(x_R + R)$		$y_M^* = x_M$
$S_R^+ - R \leq x_R \leq S_R^+$	$x_M \leq S_M^+$	$y_R^* = S_R^+$	$y_M^* = S_M^+$
	$x_M > S_M^+$	$y_R^* = S_R^+(x_M)$	$y_M^* = x_M$
$x_R > S_R^+$	$x_M \leq S_M(x_R)$	$y_R^* = x_R$	$y_M^* = S_M(x_R)$
	$x_M > S_M(x_R)$		$y_M^* = x_M$

Thus we see that the optimal coordination of manufacturing and remanufacturing in case of product substitution leads to a medium-simple policy which can be characterised by two simple order-up-to-levels, but also has to take into consideration two more complex order-up-to-functions. In case 2 ($\lambda_M > \lambda_R$) the manufacturing decision has to be made before the number of available returns is known with certainty while for the remanufacturing decision the information on returns is deterministic. In this situation both decision problems have to be solved sequentially because a two-stage stochastic optimization problem is in question. The optimal manufacturing policy has to be determined using results of conditional optimization (given possible realisations of stochastic returns) of the remanufacturing problem. It turns out (see [6]) that under these conditions also a fairly simple policy structure is optimal which can be characterized by types of order-up-to-functions for manufacturing and remanufacturing,different from those in case 1.

The analysis presented here is open to possible extensions. First of all, it seems worthwhile to investigate if the evaluated policy structure also remains optimal in a multi-period setting. Fustherone, from an application point of view it should be tested if a simple order-up-to-level (S_M^+, S_R^+) policy can be used as a good approximation of the more complicated policy with additional order-up-to-functions presented in Table 1.

References

1. Fleischmann, M., Bloemhof-Ruwaard, J.M., Dekker, R., van der Laan, E., van Nunen, J.A.E.E. and L.N. van Wassenhove (1997): Quantitative Models for Reverse Logistics: A Review. European Journal of Operational Research 103, 1-17

2. Inderfurth, K. (1997): Simple Optimal Replenishment and Disposal Policies for a Product Recovery System with Leadtimes. OR Spektrum 19, 111-122.

3. van der Laan, E., Salomon, M., Dekker, R. and L. van Wassenhove (1999): Inventory Control in Hybrid Systems with Remanufacturing. Management Science 45, 733-747.

4. Pasternack, B.A. and Z. Drezner (1991): Optimal Inventory Policies for Substitutable Commodities with Stochastic Demand. Naval Research Logistics 38, 221-240.

5. Bassok, Y., Anupindi, R. and R. Akella (1999): Single-Period Multiproduct Inventory Models with Substitution. Operations Research 47, 632-642.

6. Inderfurth, K. (2001): Optimal Policies in Hybrid Manufacturing/Remanufacturing Systems with Product Substition. FWW-Working Paper, University of Magdeburg, Germany.

Planning Sales Territories - A Facility Location Approach

Jörg Kalcsics[1], Teresa Melo[1], Stefan Nickel[1], and Hartmut Gündra[2]

[1] kalcsics@itwm.fhg.de, Fraunhofer Institut für Techno- und Wirtschaftsmathematik, Gottlieb-Daimler-Str. 49, D-67663 Kaiserslautern, Germany

[2] Geomer GmbH – Geoinformatik & Ressourcenmanagement, Redtenbacherstr. 5, D-69126 Heidelberg, Germany

Abstract. Planning sales territories may be viewed as the problem of grouping small geographic sales coverage units (e.g. counties, zip code areas, company trading areas) into larger geographic clusters called sales territories in such a way that the latter are acceptable according to managerially relevant criteria (e.g. minimal/maximal/average sales potentials or travel times). In contrast to the currently available set–partitioning oriented models, we propose a facility location approach for planning sales territories. The integer programming formulation we developed for this approach is a discrete capacitated m–median problem with the additional requirement that sales territories have to be connected. As this problem is NP–hard, we devised an efficient, interchange based heuristic procedure for solving large scale problems. The algorithms are embedded in the so–called *BusinessManager*, a software tool developed by *Geomer GmbH*, Germany.

1 Introduction

Due to fast changing markets or the introduction of new products, sales territory planning decisions have frequently to be re–evaluated. Especially for large sales territories this is a lengthy process and therefore an algorithmic approach for expediting the process is often desired. Well–planned decisions enable an efficient market penetration and lead to decreased costs and improved customer service.

Another advantage of an algorithmic approach is the possibility of performing sensitivity analysis to find out how robust the obtained solutions are. One aspect concerns the possibly inaccurate forecasts of relevant data. Questions which commonly arise in this context are for example: How much do sales forecasts influence the optimal solution? Do we obtain the same solution if sales figures were 10% higher/lower (worst-case, best-case)?

In spite of the fact that planning sales territories is an area where management science techniques are likely to be accepted and implemented, it is also an area in which marketing science literature does not propose a generalized optimization approach to the problem. Several heuristic methods exist (see e.g. [4]) but all have the major shortcoming of being trial and error procedures which rely on an adjust–and–evaluate mechanism to arrive at reasonable territory configurations. Two types of mathematical programming models have been employed. In [5] the problem is formulated as a set–partitioning model. Alternatively, in [2] so-called geographic sales coverage unit (SCU) assignment models are introduced, and subsequently refined in [7]. The latter model takes multiple criteria into account and eliminates the problems of disconnected sales regions encountered in the first model by introducing a hierarchical SCU–adjacency tree. Both models have the major drawback that they require the centers of the sales territories, i.e. the location of the sales person(s), to be fixed in advance.

In this paper we report on the development of a tool to support sales territory planning, where the centers of the sales territories can but need not be fixed in advance. In the latter case, the optimal locations of sales persons within the territories are also determined. The developed algorithms are embedded in the so–called *BusinessManager*, a software extension developed by *Geomer GmbH, Germany* for the Geographic Information System ArcView GIS[1].

In the next section we will formally introduce the problem. Section 3 is dedicated to solution techniques that assist managers in planning sales territories. In Section 4 we present the numerical results obtained for randomly generated test problems. Some concluding remarks are given in Section 5.

2 Problem Formulation

We are given a set of n sales coverage units. These SCUs may be just points (locations), but may also represent geographical areas (see Fig. 1). In case of 2-dimensional objects these are represented by the coordinates of the center of the area. Counties, zip code areas, company trading areas and other predetermined prospect clusters are some examples of SCUs. Furthermore, a sales figure is assigned to each SCU allowing to formulate additional planning criteria. These figures may be combinations of different values. Typical examples range from demographical data, e.g. number of inhabitants, purchasing power, to company specific values like turnover figures.

Each sales territory (ST) consists of a subset of SCUs. In addition, a sales person (SP) is assigned to each ST and sited in one of the SCUs belonging to the ST. The generation of sales territories must fulfill certain planning criteria. Typical examples are the following:

- The sales figure of each ST has to be within certain limits, i.e. a minimal or maximal ST size is required. The sales figure of the ST is simply the sum of the figures of its SCUs. For example, we would like to plan STs in such a way that the sales person assigned to each neither has to serve too many or too less SCUs.

[1] ArcView® GIS is a registered trademark of Environmental Systems Research Institute, Inc. (ESRI), USA.

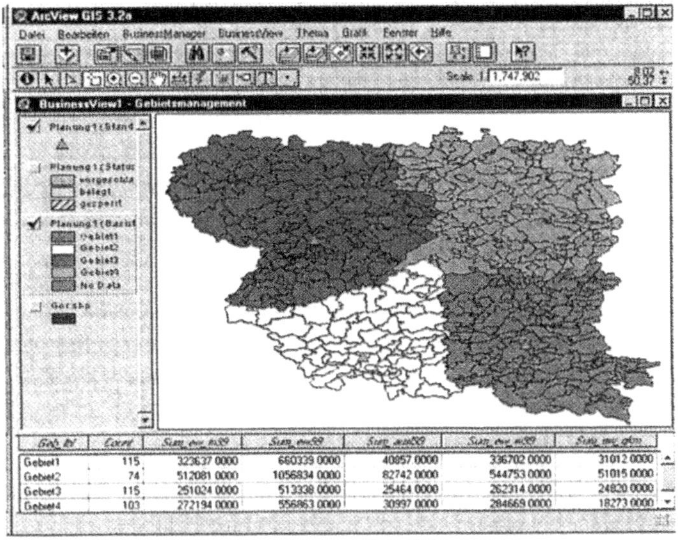

Fig. 1. Graphical display of sales territories consisting of zip–code areas in the *Geomer – BusinessManager*.

- A maximal distance/travel time from the SCU where a sales person is located to all SCUs served by him/her. For example, the distance a sales person can drive in one day starting from his/her home location may be used as a planning criterium.

Furthermore, STs have to be disjoint, i.e. an SCU belongs to exactly one ST, and connected (see the four shaded areas in Fig. 1). Finally, we are given a matrix indicating the distance between the centers of any two SCUs. The entries of the matrix may be measured in kilometers based on Euclidean distances or correspond to distances on an underlying road–network. Alternatively, travel times or costs can be considered. Where required, the distances can be weighted by additional factors which may differ from the SCU sales figures.

The objective is to find a given number m of new sales territories minimizing the sum of distances from the SCUs to the locations of the sales persons within the corresponding ST to which the SCUs belong. The planning can either be performed from scratch or in the presence of already existing STs or even locations of sales persons. In the former case, it is allowed to add additional SCUs to an existing ST. In the latter case, a new sales territory will be built around the location of the SP. Moreover, the number m of STs needs not be fixed in advance. Instead, the algorithm will choose it in such a way that the planning criteria are fulfilled. A last option is that some SCUs may remain unassigned if the maximal allowed ST size is too restricting.

3 A Facility Location Approach

In contrast to the previously known set–partitioning oriented models, we propose a facility location approach for planning sales territories. The problem is modeled as a discrete capacitated m–facility location problem. SCUs correspond to existing sites, while locations of sales persons correspond to new facilities, whose optimal locations are sought. Candidate sites for these facilities are (a subset of) the SCUs.

The assignment of SCUs to a sales person is based on the minimum distance criterium. The set of SCUs assigned to the same candidate site now defines a sales territory. This means that the catchment area of an SP contains all sales coverage units that are closer to this sales person than to all other SPs. The main advantages of this approach are:

- Locations of sales persons can but need not be fixed in advance.
- Connected STs are obtained as a result of assigning each SCU to the closest selected candidate site, in the case that the distance matrix fulfills the triangular inequality.
- No neighborhood information (of SCUs) is needed for the planning process.
- A broad range of techniques for solving discrete facility location problems is available.

In addition, the above facility location approach always yields the optimal solution, if no size constraints are specified for the STs, i.e. we have an unconstrained planning problem. The drawback of this approach is that for very size restrictive problems a feasible solution may not be found although one exists.

3.1 Integer Programming Formulation

The integer programming (IP) formulation we developed for this approach is a discrete capacitated m–median model with the additional requirement that SCUs are always allocated to the closest selected candidate location.

We denote by $SCU_1, \ldots, SCU_n$ the sales coverage units, and by SF_i the sales figure of SCU_i, $i \in \mathcal{I} := \{1, \ldots, n\}$. Let d_{ij} be the (weighted) distance from SCU_i to SCU_j, $i, j \in \mathcal{I}$. Furthermore, let $MinSF$ and $MaxSF$ denote the minimal and maximal sales limits of a sales territory, respectively. Finally, the following decision variables are defined:

$$y_j = \begin{cases} 1 \text{ if a sales person is located at } SCU_j \\ 0 \text{ otherwise} \end{cases}$$

$$x_{ij} = \begin{cases} 1 \text{ if } SCU_i \text{ is assigned to a sales person located at } SCU_j \\ 0 \text{ otherwise} \end{cases}$$

The following formulation includes constraints on the upper and lower sales figures of the sales territories. Moreover, it is assumed that planning is carried out from scratch.

$$
\begin{array}{llll}
\text{MIN} & \displaystyle\sum_{i=1}^{n}\sum_{j=1}^{n} d_{ij}x_{ij} & & \\
\text{s.t.} & \displaystyle\sum_{j=1}^{n} x_{ij} = 1 & \forall i = 1,\dots,n & (1) \\
& \displaystyle\sum_{j=1}^{n} y_j = m & & (2) \\
& x_{ij} \leq y_j & \forall i,j = 1,\dots,n & (3) \\
& \displaystyle\sum_{i=1}^{n} SF_i\, x_{ij} \geq MinSF\, y_j & \forall j = 1,\dots,n & (4) \\
& \displaystyle\sum_{i=1}^{n} SF_i\, x_{ij} \leq MaxSF\, y_j & \forall j = 1,\dots,n & (5) \\
& y_k(d_{ij} - d_{ik}) \leq (1 - x_{ij})(d_{ij} + d_{ik}) & \forall i,j,k = 1,\dots,n & (6) \\
& x_{ij},\, y_j \in \{0,1\} & \forall i,j = 1,\dots,n & (7)
\end{array}
$$

The objective is to minimize the sum of (weighted) distances from the SCUs to the new sites. Constraints (1) insure that an SCU is assigned to exactly one new site, i.e. sales person. By (2) we select exactly m new sites and by (3) SCUs can only be allocated to SCU sites where an SP is located. Furthermore, constraints (4) and (5) insure that the sales figures of STs are within the desired limits. By (6), if SCU_i is assigned to a new facility located at SCU_j, i.e. $x_{ij} = 1$, then for every other SP located at some SCU_k, $k \neq j$, the distance from i to k is not smaller than the distance from i to j, i.e. SCU_i is assigned to the closest SP.

Note that we have $O(n^2)$ binary decision variables and $O(n^3)$ constraints.

3.2 Heuristic Procedure

Observe that our problem is NP–hard since it contains as a special case the discrete m–median problem. Hence, to solve large scale problems, we developed an efficient heuristic procedure which is based on a variable neighborhood search (VNS) technique [1] and utilizes as local search method the interchange algorithm of [6].

An initial site selection of sales persons is obtained by solving a continuous m-facility location problem. This is done using the heuristic method described in [3], which can also take already existing SPs into account. For each site generated for a sales person by this procedure, the closest SCU site is taken as the actual SP location.

Based on the initial site selection, a combined interchange and variable neighborhood search heuristic is applied in order to determine a better solution. In our case, we slightly modified the local exchange step of the interchange procedure [6]. Namely, the current SP site is only exchanged by another SCU site, if the latter belongs to the catchment area, i.e. the sales territory, of this SP. This simplification of the basic interchange procedure is applied, since from our computational experience we could observe that the initial site selection already provides very good solutions (see Table 1).

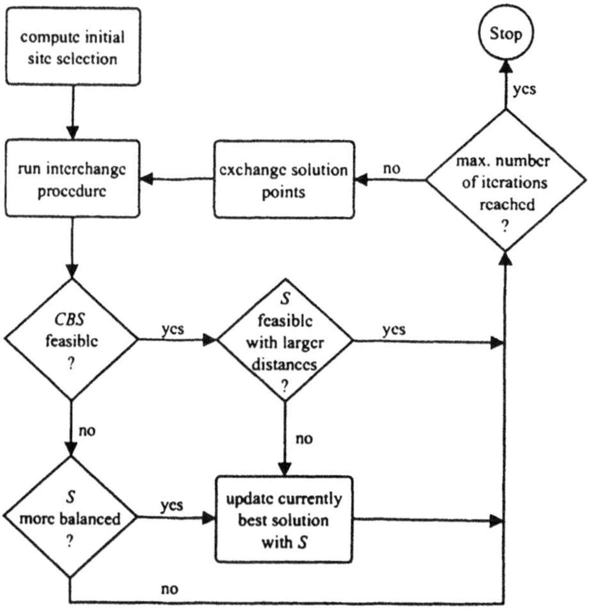

Fig. 2. Outline of the heuristic.

Fig. 2 gives an outline of the heuristic. Using the initial solution, we run the modified interchange procedure and store the new site selection as currently best solution, CBS. Afterwards, we exchange a certain number of SP sites of the CBS by SCU sites which belong to the sales territories of the corresponding SPs. The exchange number hereby depends on the number of new SPs, m. Using this modified site selection, we run again the interchange procedure and obtain a modified solution S. We accept this new site selection S as currently best solution either if CBS is infeasible and S is more balanced, i.e. the min. and max. sales figure limits for the STs are less violated, or both site selections are feasible but S has a lower total distance. Next, we iterate this process of exchanging SP sites and evaluating the new solution. The algorithm stops after a certain, predetermined, number of iterations.

4 Computational Results

In this section we evaluate the quality of our heuristic procedure on a set of test problems.

The coordinates of SCUs were drawn randomly from a uniform distribution between 0 and 100, and the sales figures between 4 and 8. For a fixed number of SCUs, we generated 20 instances and solved every instance 5 times, each with a different number of new sales persons. For unconstrained planning problems, input data was generated for six problem types ranging from 100 up to 600 SCUs. The number of new sales persons ranged from 4 to 25 depending on the size of the problem.

Table 1. Computational results for unconstrained problems

	Initial			VNS			Exact
SCUs	A–Gap	W–Gap	A–CPU	A–Gap	W–Gap	A–CPU	A–CPU
100	2.77	8.82	0.04	0.39	4.85	0.32	4.26
200	3.39	10.19	0.06	0.50	4.08	0.90	35.03
300	2.90	10.23	0.13	0.31	3.06	2.53	128.01
400	3.66	11.86	0.17	0.39	2.00	4.37	351.90
500	3.61	7.80	0.33	0.43	2.67	10.14	779.73
600	3.92	7.80	0.38	0.36	1.92	15.35	1649.92

Table 1 summarizes the results obtained for the 600 test problems. We compare the initial solution given by the procedure in [3] ('Initial') and our heuristic approach ('VNS') with the optimal solution ('Exact'). The first column of the table indicates the total number of SCUs in each problem category. The 'A–Gap' and 'W–Gap' columns present the average and worst, respectively, relative percentage gap with respect to the optimal objective function values. All tests were performed on a Pentium III, 600 MHz with 1 GB RAM, and the CPU times (in seconds) are given in the columns 'A–CPU'. Finally, all problems were solved optimally with CPLEX 7.0.

By comparing the gaps of the initial solution with the VNS approach, we observe that the latter are considerably better and are obtained at the expense of an acceptable increase in the computational time. As expected, solving a problem to optimality involves a considerable computational burden, even for relatively small instances. In contrast, the VNS approach yields very good solutions for large scale problems in reasonable time.

For the constrained planning case, input data was generated for problems with 50 and 75 SCUs with the above settings. The minimal and maximal sales figure limits, $MinSF$ and $MaxSF$, for the STs are 90% and 110%, respectively, of the sum of the sales figures of the SCUs divided by the number of new SPs. Table 2 shows the results obtained.

Table 2. Computational results for constrained planning problems

	VNS			Exact
SCUs	A-Gap	W-Gap	A-CPU	A-CPU
50	2.46	12.06	0.54	11819
75	1.54	4.07	0.76	60617

Constrained problems are much more difficult to solve optimally as the computational times show. Therefore, the use of heuristic procedures for solving real life problems is necessary. Although the average gap of the VNS approach worsens compared to the unconstrained case, it is still acceptable.

5 Conclusions and Further Research

In this paper we presented an efficient, interchange based heuristic procedure for planning sales territories. Some issues require further investigation. One topic is the problem that in constrained planning it is often desired by managers that the STs not only fulfill the sales figure restrictions but should also be of equal size, i.e. somehow balanced. Another yet un–modeled issue is a multi–criteria approach which allows the user to specify more than one sales figure for each SCU and therefore, also more than one planning restriction on the STs. Concerning the IP formulation, polyhedral properties will be investigated.

References

1. Hansen, P., Mladenovic, N., Taillard, E. (1998) Heuristic solution of the multisource Weber problem as a p-median problem. Oper. Res. Lett. **22**, 55–62
2. Hess, S. W., Samuels, S. A. (1971) Experiences with a sales districting model: criteria and implementation. Management Sci. **4**, Part II, 41–54
3. Kalcsics, J., Melo, T., Nickel, S., Schmid-Lutz, V. (1999) Facility location decisions in supply chain management. In *Operations Research Proceedings 1999.* Inderfurth, K., Schwödiauer, G., Domschke, W., Juhnke, F., Kleinschmidt, P., Wäscher, G. (ed.). Berlin: Springer-Verlag (2000), 467–472
4. Lodish, L. M. (1975) Sales territory alignment to maximize profit. J. Marketing Res. **12**, 30–36
5. Shanker, R. J., Turner, R. E., Zoltners, A. A. (1975) Sales territory design: an integrated approach. Management Sci. **3**, 309–320
6. Teitz, M. B., Bart, P. (1968) Heuristic methods for estimating generalized vertex median of a weighted graph. Oper. Res. **16**, 955–961.
7. Zoltners, A. A., Sinha, P. (1983) Sales territory alignment: a review and model. Management Sci. **29**, 1237–1256

Eine spieltheoretische Analyse von Zulieferer-Abnehmer-Beziehungen auf Basis des JELS-Modells

Eric Sucky

Seminar für Logistik und Verkehr, Johann Wolfgang Goethe-Universität, Frankfurt am Main, esucky@wiwi.uni-frankfurt.de

1 Problemstellung

Die ganzheitliche Analyse, Planung, Steuerung und Kontrolle unternehmensübergreifender Wertschöpfungsprozesse stellt die zentrale Aufgabe des Supply Chain Management dar (vgl. [3], S. 1-2). Im Rahmen der Analyse einer gemeinsamen Bestell- und Produktionspolitik eines Zulieferers und eines Abnehmers entwickelte *Banerjee* [1] das Joint Economic-Lot-Size Model (JELS-Modell). *Banerjee* [1] zeigt, dass eine gemeinsame Bestimmung der Bestell- und Produktionspolitik zu einer Minimierung der Summe der entscheidungsrelevanten Kosten führt. Verfügt jedoch einer der Akteure über die Marktmacht, seine individuell optimale Bestell- bzw. Produktionspolitik gegenüber dem anderen durchzusetzen, so besteht für ihn kein Anreiz, von seiner optimalen Politik abzuweichen. Eine gemeinsame Bestell- und Produktionspolitik kann daher nur das Ergebnis von Verhandlungen sein. In diesem Beitrag werden auf Basis des JELS-Modells Verhandlungslösungen zwischen Abnehmer und Zulieferer für den Fall analysiert, dass der Zulieferer nur unvollständig über Ausprägungen relevanter Merkmale des Abnehmers informiert ist.

2 Bestell- und Produktionspolitiken im JELS-Modell

2.1 Individuelle Bestell- und Produktionspolitiken

Dem JELS-Modell liegen folgende Annahmen zu Grunde: Es wird ein Abnehmer (A) und ein Zulieferer (P) eines Gutes betrachtet, für das eine im Zeitverlauf gleichbleibende Nachfrage in konstanter Höhe besteht. (A) und (P) bestimmen ihre Bestellmenge bzw. Losgröße unter dem Ziel der Minimierung der individuellen entscheidungsrelevanten Kosten je Periode, welche für (A) die Bestell- und Lagerkosten umfassen und sich für (P) aus den Rüst- und Lagerkosten zusammen setzen. (A) und (P) ermitteln die optimale Bestellmenge bzw. Losgröße mit der klassischen Bestellmengenformel bzw. optimalen Losgrößenformel (vgl. [1], S. 293-294). Die Entscheidungssituation von (A) stellt sich wie folgt dar: Der bekannte Gesamtbedarf, gemessen in Mengeneinheiten je Periode, beträgt b [ME/PE]. Bei einer Periodenlänge von τ

Zeiteinheiten erfolgt der Lagerabgang mit der konstanten Nachfragerate von $\frac{b}{T}$ [ME/ZE]. Der Lagerzugang erfolgt unendlich schnell in gleich großen Losen x_A [ME]. Es treten keine Fehlmengen auf; die Bestellmenge entspricht der Liefermenge. Der durchschnittliche Lagerbestand beträgt $\frac{x_A}{2}$ [ME]. Mit dem Lagerhaltungskostensatz h_A [GE/ME ZE] und den Bestellkosten B [GE] ergeben sich die relevanten Kosten von (A) je Periode in Abhängigkeit von x_A. Der Entscheidungssituation von (P) liegen folgende Annahmen zu Grunde: Die Produktionskapazität von (P) beträgt d [ME/PE]. Der Lagerzugang erfolgt mit der konstanten Produktionsrate je Zeiteinheit $\frac{d}{T}$ [ME/ZE]. Es gilt $d \geq b$, da für $d < b$ die Nachfrage nicht befriedigt werden kann. (P) verfolgt eine lot-for-lot-production-strategy, d.h., die Liefermenge entspricht dem Produktionslos x_P [ME]. Der durchschnittliche Lagerbestand beträgt $\frac{x_P}{2}\frac{b}{d}$ [ME]. Mit dem Lagerhaltungskostensatz h_P und den Rüstkosten R [GE] ergeben sich die relevanten Kosten von (P) je Periode in Abhängigkeit von x_P. *Tab. 1* zeigt die optimalen Bestell- und Produktionspolitiken x_A^* und x_P^* sowie die resultierenden minimalen Kosten von (A) und (P) (vgl. [7], S. 966).

Tabelle1. Individuelle Bestell- und Produktionspolitiken

	Abnehmer (A)	Zulieferer (P)
Kostenfunktion	$K^A(x_A) = B\frac{b}{x_A} + \frac{x_A}{2}h_A$	$K^P(x_P) = R\frac{b}{x_P} + \frac{x_P}{2}\frac{b}{d}h_P$
Bestellmenge/Losgröße	$x_A^* = \sqrt{\frac{2Bb}{h_A}}$	$x_P^* = \sqrt{\frac{2Rd}{h_P}}$
Minimale Kosten	$K^A(x_A^*) = \sqrt{2Bbh_A}$	$K^P(x_P^*) = b\sqrt{\frac{2Rh_P}{d}}$

2.2 Gemeinsame Bestell- und Produktionspolitik

Zur Ermittlung der gemeinsamen Bestell- und Produktionspolitik ist die Bestellmenge und Losgröße $x_G = x_A = x_P$ so zu wählen, dass die Summe der relevanten Kosten von (A) und (P) minimiert wird (vgl. [1], S. 299). Es gilt:

$$K^G(x_G) = (B+R)\frac{b}{x_G} + \frac{x_G}{2}(h_A + \frac{b}{d}h_P) \qquad (1)$$

Im Minimum der entscheidungsrelevanten Kosten von (A) und (P) gilt:

$$\frac{dK^G(x_G)}{dx_G} = \frac{1}{2}(h_A + \frac{b}{d}h_P) - (B+R)\frac{b}{x_G^2} = 0 \qquad (2)$$

Im Minimum der Summe der Kosten von (A) und (P) entspricht die Kostensteigerung bei (A) bzw. (P) bei Änderung der Bestell- bzw. Produktionsmenge genau der daraus resultierenden Kostensenkung bei (P) bzw. (A):

$$\frac{1}{2}\frac{b}{d}h_P - R\frac{b}{x_G^2} = -\frac{1}{2}h_A + B\frac{b}{x_G^2} \quad \Rightarrow \quad \frac{dK^P(x_G)}{dx_G} = -\frac{dK^A(x_G)}{dx_G} \qquad (3)$$

Die gemeinsame, kostenminimale Bestellmenge und Losgröße x_G^* sowie die daraus resultierenden Gesamtkosten betragen:

$$x_G^* = \sqrt{\frac{2b(B+R)}{h_A + \frac{b}{d}h_P}} \quad \Rightarrow \quad K^G(x_G^*) = \sqrt{2b(B+R)(h_A + \frac{b}{d}h_P)} \tag{4}$$

2.3 Vergleich der Bestell- und Produktionspolitiken

Verfolgt (P) eine lot-for-lot-production-strategy, so muss das Produktionslos der Bestellmenge entsprechen ($x_P = x_A = x_G$). Im Weiteren ist zu analysieren, wann diese Bedingung erfüllt ist. Das Verhältnis der Rüst- zu den Bestellkosten sowie das Verhältnis der Lagerkosten von (P) und (A) in Abhängigkeit einer bestimmten Bestellmenge und Losgröße x kann mit den Parametern α und β ausgedrückt werden (vgl. [1], S. 299):

$$\alpha = \frac{R\frac{b}{x}}{B\frac{b}{x}} = \frac{R}{B} \quad \text{und} \quad \beta = \frac{\frac{x}{2}\frac{b}{d}h_P}{\frac{x}{2}h_A} = \frac{bh_P}{dh_A} \tag{5}$$

Mit den Parametern α und β ergeben sich folgende Beziehungen:

$$x_A^* = \sqrt{\frac{\beta}{\alpha}}x_P^*, \quad x_P^* = \sqrt{\frac{\alpha}{\beta}}x_A^*, \quad x_G^* = \sqrt{\frac{1+\alpha}{1+\beta}}x_A^* = \sqrt{\frac{1+\frac{1}{\alpha}}{1+\frac{1}{\beta}}}x_P^* \tag{6}$$

Nur für $\alpha = \beta$ gilt $x_A^* = x_P^* = x_G^*$. Für $\alpha \neq \beta$ gilt $x_G^* \in]x_A^*, x_P^*[$ und es ergibt sich nur dann eine Lösung, wenn sich (P) an (A) anpasst ($x_P = x_A^* = x_G$), (A) sich an (P) anpasst ($x_A = x_P^* = x_G$) oder (A) und (P) eine gemeinsame Menge $x_A = x_P = x_G$ wählen. *Tab. 2* zeigt die Konsequenzen der Bestell- und Produktionspolitiken x_A^*, x_P^* und x_G^* für (A) und (P) (vgl. [6], S. 189).

Tabelle2. Kostenwirkungen von Bestell- und Produktionspolitiken

x_G	Kosten bei Abnehmer (A)	Kosten bei Zulieferer (P)
x_A^*	$K^A(x_A^*) = \sqrt{2Bbh_A}$	$K^P(x_A^*) = \frac{1}{2}\left(\sqrt{\frac{\alpha}{\beta} + \frac{\beta}{\alpha}}\right) K^P(x_P^*)$
x_P^*	$K^A(x_P^*) = \frac{1}{2}\left(\sqrt{\frac{\alpha}{\beta} + \frac{\beta}{\alpha}}\right) K^A(x_A^*)$	$K^P(x_P^*) = b\sqrt{\frac{2Rd}{h_P}}$
x_G^*	$K^A(x_G^*) = \frac{1}{2}\left(\sqrt{\frac{1+\alpha}{1+\beta}} + \sqrt{\frac{1+\beta}{1+\alpha}}\right) K^A(x_A^*)$	$K^P(x_G^*) = \frac{1}{2}\left(\sqrt{\frac{1+\frac{1}{\alpha}}{1+\frac{1}{\beta}}} + \sqrt{\frac{1+\frac{1}{\beta}}{1+\frac{1}{\alpha}}}\right) K^P(x_P^*)$

Für $\alpha \neq \beta$ und $\alpha, \beta > 0$ gilt:

$$\frac{1}{2}\left(\sqrt{\frac{\alpha}{\beta}} + \sqrt{\frac{\beta}{\alpha}}\right) > \frac{1}{2}\left(\sqrt{\frac{1+\alpha}{1+\beta}} + \sqrt{\frac{1+\beta}{1+\alpha}}\right) > 1 \tag{7}$$

$$\frac{1}{2}\left(\sqrt{\frac{\alpha}{\beta}}+\sqrt{\frac{\beta}{\alpha}}\right)>\frac{1}{2}\left(\sqrt{\frac{1+\frac{1}{\alpha}}{1+\frac{1}{\beta}}}+\sqrt{\frac{1+\frac{1}{\beta}}{1+\frac{1}{\alpha}}}\right)>1 \tag{8}$$

Ein Abweichen von der individuell optimalen Bestell- bzw. Produktionspolitik ist mit einem Anstieg der relevanten Kosten beim Abweichenden verbunden. Verhalten sich (A) und (P) individuell rational im Sinne der verfolgten Zielsetzungen, so wählen sie ihre jeweils optimalen Mengen x_A^* und x_P^*. Verfügt (A) bzw. (P) über die Marktmacht, seine individuell optimale Menge gegenüber (P) bzw. (A) durchzusetzen, so besteht für (A) bzw. (P) kein Anreiz, eine andere, als die individuell optimale Menge x_A^* bzw. x_P^* zu wählen. Der schwächere Akteur muss von seiner individuell optimalen Menge abweichen und seine Bestell- bzw. Produktionspolitik an die des stärkeren Akteurs anpassen. Sind Verhandlungen möglich, so kann der schwächere Marktteilnehmer versuchen, z.B. durch Kompensationszahlungen, den stärkeren Marktteilnehmer zur Wahl einer anderen als der individuell optimalen Menge zu bewegen. Für den Fall $\alpha \neq \beta$ soll analysiert werden, welche Menge $x_G = x_A = x_P$ realisiert wird, wenn (A) und (P) über x_G verhandeln.

3 Verhandlungen bei unvollständiger Information

3.1 Die Verhandlungssituation bei unvollständiger Information

Es wird angenommen, dass (A) über die Marktmacht verfügt, seine optimale Bestellmenge x_A^* gegenüber (P) durchzusetzen. (P) möchte (A) ein Angebot unterbreiten, eine gemeinsame Menge x_G gegen eine Kompensationszahlung z zu wählen. Zum Zeitpunkt der Festlegung des Vertragsangebots verfügt (P) über unvollständige Information derart, dass (P) die Funktion der relevanten Kosten von (A) nicht mit Sicherheit kennt, d.h., (P) besitzt unvollständige Informationen über den Abnehmertyp (vgl. [5], S. 467). (P) nimmt zwei alternative Funktionen der relevanten Kosten $K_1^A(x)$, $K_2^A(x)$ und damit zwei mögliche Bestellmengen $x_{A,1}^*$, $x_{A,2}^*$ von (A) an. (P) muss für jeden Abnehmertyp (A_1) und (A_2) ein Angebot ermitteln, das diese zu akzeptieren bereit sind (vgl. [5], S. 467). Außerdem müssen die Angebote für (A_1) und (A_2) anreizverträglich sein, d.h., die Angebote sind so zu gestalten, dass jeder Abnehmertyp genau das Angebot akzeptiert, das für ihn bestimmt ist (vgl. [2], S. 561-562). Diese Form der Angebotsgestaltung wird als Selbstwahlmechanismus oder Screening bezeichnet (vgl. [4], S. 79-80).

3.2 Das Screening-Modell zur Ermittlung optimaler Angebote

(P) legt seinen Überlegungen folgende Annahmen zu Grunde: Es existieren zwei mögliche Abnehmertypen (A_1) und (A_2). Mit einer Wahrscheinlichkeit von $p_1 > 0$ liegt der Abnehmertyp (A_1) vor; mit der Gegenwahrscheinlichkeit von $p_2 = 1 - p_1 > 0$ handelt es sich um Abnehmertyp (A_2). Die Akteure sind risikoneutral. Das Optimierungsproblem von (P) lautet somit:

$$minE[K^P(x_1,z_1,x_2,z_2)] = p_1(K^P(x_1)+z_1)+p_2(K^P(x_2)+z_2) \quad u.d.N. \tag{9}$$

$$K_1^A(x^*_{A,1}) - K_1^A(x_1) + z_1 \geq 0 \text{ und } K_2^A(x^*_{A,2}) - K_2^A(x_2) + z_2 \geq 0 \tag{10}$$

$$K_1^A(x_2) - z_2 - K_1^A(x_1) + z_1 \geq 0 \text{ und } K_2^A(x_1) - z_1 - K_2^A(x_2) + z_2 \geq 0 \tag{11}$$

$$x_1, x_2, z_1, z_2 \geq 0 \tag{12}$$

Die Nebenbedingungen (9) stellen die Teilnahmebedingungen dar: Für jeden Abnehmertyp (A_1) und (A_2) dürfen die aus der Annahme des Angebots von (P) resultierenden Kosten höchsten so hoch sein, wie bei Realisierung der individuell optimalen Bestellmengen $x^*_{A,1}$ bzw. $x^*_{A,2}$. Die Bedingungen (10) stellen die Anreizverträglichkeit sicher. Jeder Abnehmertyp (A_1), (A_2) muss ein Interesse daran haben, genau das Angebot anzunehmen, das für ihn bestimmt ist, d.h. es darf sich nicht lohnen, den jeweils anderen Abnehmertyp zu imitieren. Aus der Lagrange-Funktion

$$\begin{aligned} L(x_1, z_1, x_2, z_2) &= p_1(K^P(x_1) + z_1) + p_2(K^P(x_2) + z_2) \\ -\lambda_1(K_1^A(x^*_{A,1}) &- K_1^A(x_1) + z_1) - \lambda_2(K_2^A(x^*_{A,2}) - K_2^A(x_2) + z_2) \\ -\mu_1(K_1^A(x_2) - z_2 &- K_1^A(x_1) + z_1) - \mu_2(K_2^A(x_1) - z_1 - K_2^A(x_2) + z_2) \end{aligned} \tag{13}$$

lassen sich die Karush-Kuhn-Tucker- (KKT-) Bedingungen herleiten:

$$\frac{\partial L}{\partial x_1} = p_1\frac{\partial K^P(x_1)}{\partial x_1} + \lambda_1\frac{\partial K_1^A(x_1)}{\partial x_1} + \mu_1\frac{\partial K_1^A(x_1)}{\partial x_1} - \mu_2\frac{\partial K_2^A(x_1)}{\partial x_1} \geq 0 \tag{14}$$

$$x_1\frac{\partial L}{\partial x_1} = x_1\left(p_1\frac{\partial K^P(x_1)}{\partial x_1} + \lambda_1\frac{\partial K_1^A(x_1)}{\partial x_1} + \mu_1\frac{\partial K_1^A(x_1)}{\partial x_1} - \mu_2\frac{\partial K_2^A(x_1)}{\partial x_1}\right) = 0 \tag{15}$$

$$\frac{\partial L}{\partial x_2} = p_2\frac{\partial K^P(x_2)}{\partial x_2} + \lambda_2\frac{\partial K_2^A(x_2)}{\partial x_2} - \mu_1\frac{\partial K_1^A(x_2)}{\partial x_2} + \mu_2\frac{\partial K_2^A(x_2)}{\partial x_2} \geq 0 \tag{16}$$

$$x_2\frac{\partial L}{\partial x_2} = x_2\left(p_2\frac{\partial K^P(x_2)}{\partial x_2} + \lambda_2\frac{\partial K_2^A(x_2)}{\partial x_2} - \mu_1\frac{\partial K_1^A(x_2)}{\partial x_2} + \mu_2\frac{\partial K_2^A(x_2)}{\partial x_2}\right) = 0 \tag{17}$$

$$\frac{\partial L}{\partial z_1} = p_1 - \lambda_1 - \mu_1 + \mu_2 \geq 0 \text{ und } z_1\frac{\partial L}{\partial z_1} = z_1\,(p_1 - \lambda_1 - \mu_1 + \mu_2) = 0 \tag{18}$$

$$\frac{\partial L}{\partial z_2} = p_2 - \lambda_2 + \mu_1 - \mu_2 \geq 0 \text{ und } z_2\frac{\partial L}{\partial z_2} = z_2\,(p_2 - \lambda_2 + \mu_1 - \mu_2) = 0 \tag{19}$$

$$-\frac{\partial L}{\partial \lambda_1} = K_1^A(x^*_{A,1}) - K_1^A(x_1) + z_1 \geq 0 \text{ und } \lambda_1\big(K_1^A(x^*_{A,1}) - K_1^A(x_1) + z_1\big) = 0 \tag{20}$$

$$-\frac{\partial L}{\partial \lambda_2} = K_2^A(x^*_{A,2}) - K_2^A(x_2) + z_2 \geq 0 \text{ und } \lambda_2\big(K_2^A(x^*_{A,2}) - K_2^A(x_2) + z_2\big) = 0 \tag{21}$$

$$-\frac{\partial L}{\partial \mu_1} = K_1^A(x_2) - z_2 - K_1^A(x_1) + z_1 \geq 0 \text{ und } \mu_1(K_1^A(x_2) - z_2 - K_1^A(x_1) + z_1) = 0 \quad (22)$$

$$-\frac{\partial L}{\partial \mu_2} = K_2^A(x_1) - z_1 - K_2^A(x_2) + z_2 \geq 0 \text{ und } \mu_2(K_2^A(x_1) - z_1 - K_2^A(x_2) + z_2) = 0 \quad (23)$$

$$x_1, x_2, z_1, z_2, \lambda_1, \lambda_2, \mu_1, \mu_2 \geq 0 \quad (24)$$

Die Analyse der möglichen Kombination der Ausprägungen für $\lambda_1, \lambda_2, \mu_1, \mu_2$ ergibt, dass für $x_1, x_2, z_1, z_2 > 0$ in sechs Fällen die KKT-Bedingungen erfüllt werden können. In Abhängigkeit der Kostenfunktionen von (A_1), (A_2) und (P) ergeben sich sechs mögliche Angebotsmenus.

Angebotsmenu 1: Im Fall $(\lambda_1 > 0, \lambda_2 > 0, \mu_1 = 0, \mu_2 = 0)$ ergibt sich:

$$x_1 = \sqrt{\frac{2b(B_1 + R)}{h_{A,1} + \frac{b}{d}h_P}} \text{ mit } z_1 = K_1^A(x_1) - K_1^A(x^*_{A,1}) \quad (25)$$

$$x_2 = \sqrt{\frac{2b(B_2 + R)}{h_{A,2} + \frac{b}{d}h_P}} \text{ mit } z_2 = K_2^A(x_2) - K_2^A(x^*_{A,2}) \quad (26)$$

Diese Lösung ist optimal, wenn weder (A_1) noch (A_2) einen Anreiz haben, den jeweils anderen zu imitieren, da sie sich bei Annahme des für den jeweils anderen bestimmten Angebots schlechter stellen würden. Die Zahlungen z_1 und z_2 kompensieren genau die Kostenanstiege bei (A_1) und (A_2), die durch die Übergänge von $x^*_{A,1}$ auf x_1 bzw. von $x^*_{A,2}$ auf x_2 induziert werden.

Angebotsmenu 2: Im Fall $(\lambda_1 > 0, \lambda_2 = 0, \mu_1 = 0, \mu_2 > 0)$ ergibt sich:

$$x_1 = \sqrt{\frac{(p_1 R + p_1 B_2 + B_1 - B_2)b}{\frac{p_1}{2}\frac{b}{d} + \left(\frac{p_1}{2} - \frac{1}{2}\right)h_{A,2} + \frac{1}{2}h_{A,1}}} \text{ mit } z_1 = K_1^A(x_1) - K_1^A(x^*_{A,1}) \quad (27)$$

$$x_2 = \sqrt{\frac{2b(B_2 + R)}{h_{A,2} + \frac{b}{d}h_P}} \text{ mit } z_2 = K_2^A(x_2) - K_2^A(x_1) + K_1^A(x_1) - K_1^A(x^*_{A,1}) \quad (28)$$

Dies kann optimal sein, wenn (A_2) einen Anreiz hat (A_1) zu imitieren, d.h., (A_2) sich durch Annahme des Angebots für (A_1) besser stellt. Die Zahlung z_2 setzt sich aus der Kompensation der Kosten, die aus dem Übergang von $x^*_{A,2}$ auf x_2 resultieren und einer Imitationsverzichtsprämie zusammen.

Angebotsmenu 3: Im Fall $(\lambda_1 = 0, \lambda_2 > 0, \mu_1 > 0, \mu_2 = 0)$ folgt:

$$x_1 = \sqrt{\frac{2b(B_1 + R)}{h_{A,1} + \frac{b}{d}h_P}} \text{ mit } z_1 = K_1^A(x_1) - K_1^A(x_2) + K_2^A(x_2) - K_2^A(x^*_{A,2}) \quad (29)$$

$$x_2 = \sqrt{\frac{(p_2 R + p_2 B_1 + B_2 - B_1)b}{\frac{p_2}{2}\frac{b}{d} + \left(\frac{p_2}{2} - \frac{1}{2}\right)h_{A,1} + \frac{1}{2}h_{A,2}}} \text{ mit } z_2 = K_2^A(x_2) - K_2^A(x^*_{A,2}) \quad (30)$$

Dies kann optimal sein, wenn (A_1) einen Anreiz hat (A_2) zu imitieren.
Angebotsmenu 4: Im Fall $(\lambda_1 > 0, \lambda_2 > 0, \mu_1 = 0, \mu_2 > 0)$ ergibt sich:

$$x_{1_{1.2}} = \frac{K_1^A(x_{A,1}^*) - K_2^A(x_{A,2}^*)}{h_{A,2} - h_{A,1}} \pm \sqrt{\left(\frac{K_1^A(x_{A,1}^*) - K_2^A(x_{A,2}^*)}{h_{A,2} - h_{A,1}}\right)^2 - \frac{2(B_2 + B_1)b}{h_{A,2} - h_{A,1}}}$$

$$\text{mit} \quad z_1 = K_1^A(x_1) - K_1^A(x_{A,1}^*) \qquad (31)$$

$$x_2 = \sqrt{\frac{2b(B_2 + R)}{h_{A,2} + \frac{b}{d} h_P}} \quad \text{mit} \quad z_2 = K_2^A(x_2) - K_2^A(x_{A,2}^*) \qquad (32)$$

In diesem Fall wird die Imitation von (A_1) durch (A_2) durch eine gegenüber x_G^* veränderten Angebotsmenge für (A_1) verhindert.
Angebotsmenu 5: Im Fall $(\lambda_1 > 0, \lambda_2 > 0, \mu_1 > 0, \mu_2 = 0)$ ergibt sich:

$$x_1 = \sqrt{\frac{2b(B_1 + R)}{h_{A,1} + \frac{b}{d} h_P}} \quad \text{mit} \quad z_1 = K_1^A(x_1) - K_1^A(x_{A,1}^*) \qquad (33)$$

$$x_{2_{1.2}} = \frac{K_2^A(x_{A,2}^*) - K_1^A(x_{A,1}^*)}{h_{A,1} - h_{A,2}} \pm \sqrt{\left(\frac{K_2^A(x_{A,2}^*) - K_1^A(x_{A,1}^*)}{h_{A,1} - h_{A,2}}\right)^2 - \frac{2(B_1 + B_2)b}{h_{A,1} - h_{A,2}}}$$

$$\text{mit} \quad z_2 = K_2^A(x_2) - K_2^A(x_{A,2}^*) \qquad (34)$$

In diesem Fall wird die Imitation von (A_2) durch (A_1) durch eine gegenüber x_G^* veränderten Angebotsmenge für (A_2) verhindert.
Angebotsmenu 6: Im Fall $(\lambda_1 > 0, \lambda_2 > 0, \mu_1 > 0, \mu_2 > 0)$ ergibt sich:

$$x_{1_{1.2}} = \frac{K_1^A(x_{A,1}^*) - K_2^A(x_{A,2}^*)}{h_{A,2} - h_{A,1}} \pm \sqrt{\left(\frac{K_1^A(x_{A,1}^*) - K_2^A(x_{A,2}^*)}{h_{A,2} - h_{A,1}}\right)^2 - \frac{2(B_2 + B_1)b}{h_{A,2} - h_{A,1}}}$$

$$\text{mit} \quad z_1 = K_1^A(x_1) - K_1^A(x_{A,1}^*) \qquad (35)$$

$$x_{2_{1.2}} = \frac{K_2^A(x_{A,2}^*) - K_1^A(x_{A,1}^*)}{h_{A,1} - h_{A,2}} \pm \sqrt{\left(\frac{K_2^A(x_{A,2}^*) - K_1^A(x_{A,1}^*)}{h_{A,1} - h_{A,2}}\right)^2 - \frac{2(B_1 + B_2)b}{h_{A,1} - h_{A,2}}}$$

$$\text{mit} \quad z_2 = K_2^A(x_2) - K_2^A(x_{A,2}^*) \qquad (36)$$

3.3 Beispiel zur Ermittlung eines optimalen Angebotsmenus

Die abgeleiteten Ergebnisse werden an einem Beispiel verdeutlicht. *Tab. 3* zeigt die Daten des Beispiels, mit $p_1 = p_2 = 0,5$, $d = 15000$ und $b = 10000$. Verfügt (A) über die Marktmacht, $x_{A,1}^*$ bzw. $x_{A,2}^*$ gegenüber (P) durchzusetzen, so ergibt sich mit $K^P(x_{A,1}^*) = 9750$, $K^P(x_{A,2}^*) = 15000$ ohne Verhandlungen: $E\left[K^P(200, 0, 100, 0)\right] = 12375$. Angebotsmenu 5 ist optimal: $x_1 = 242,38, z_1 = 185,25, x_2 = 141,42, z_2 = 606,57$ mit $E\left[K^P\right] = 10832,31$.

Tabelle3. Daten des Beispiels

Zulieferer (P)	Abnehmertyp (A_1)	Abnehmertyp (A_2)
$R = 135, h_p = 45$	$B_1 = 100, h_{A,1} = 50$	$B_2 = 50, h_{A,2} = 100$
$x_P^* = 300$	$x_{A,1}^* = 200$	$x_{A,2}^* = 100$
$K^P(x_P^*) = 9000$	$K_1^A(x_{A,1}^*) = 10000$	$K_2^A(x_{A,2}^*) = 10000$

4 Schlussbetrachtung

Im vorliegenden Beitrag wurde auf der Grundlage JELS-Modells eine gemeinsame Bestell- und Produktionspolitik als Verhandlungslösung entwickelt. Verfügt der Abnehmer über die Marktmacht, seine optimale Bestellpolitik gegenüber dem Zulieferer durchzusetzen und hat der Zulieferer nur unvollständige Informationen über die Funktion der relevanten Kosten des Abnehmers, so lässt sich das optimale Angebotsmenu ermitteln, wenn der Zulieferer die Kostenfunktion des Abnehmers hinreichend genau abschätzen kann. Verhält sich der Abnehmer rational im Sinne seiner Zielsetzung, so nimmt er genau das für ihn bestimmte Angebot an.

Literatur

1. Banerjee, A., A Joint Economic-Lot-Size Model for Purchaser and Vendor, in: Decision Sciences, Vol. 17, 1986, S. 292-311.
2. Illing, G., Private Information as Transaction Costs: The Coase Theorem Revisted, in: Zeitschrift für die gesamte Staatswissenschaft, Vol. 148, 1992, S. 558-576.
3. Knolmayer, G./Mertens, P./Zeier, A., Supply Chain Management auf Basis von SAP-Systemen, Berlin u.a., 2000.
4. Schenk-Mathes, H. Y., Effiziente Gestaltung von Lieferverträgen bei unvollständiger Information, in: Dyckhoff, H./Derigs, U./Salomon, M./Tijms, H. C. (Hrsg.), Operations Research Proceedings 1993, Berlin u.a., 1994, S. 467-473.
5. Schenk-Mathes, H. Y., Gestaltung von Lieferbeziehungen bei Informationsasymmetrie, Wiesbaden, 1999.
6. Toporowski, W., Logistik im Handel, Heidelberg, 1996.
7. Toporowski, W., Unternehmensübergreifende Optimierung der Bestellpolitik - das JELS-Modell mit einem Intermediär, ZfbF 51, 1999, S. 963-989.

Section 5

Finance, Banking and Insurance

Ein bedingtes Mehrfaktorenmodell zur Quantifizierung der Renditen von Bankaktien auf dem deutschen Kapitalmarkt

Wolfgang Bessler, Heiko Opfer

Justus-Liebig-Universität Gießen

Zusammenfassung:
Das Ziel dieser Studie ist es, die Renditen von Bankaktien auf dem deutschen Kapitalmarkt mit Mehrfaktorenmodellen zu untersuchen. Dazu werden sowohl unbedingte Modelle mit zeitkonstanten als auch bedingte Mehrfaktorenmodelle mit zeitvariablen Betakoeffizienten verwendet. Als Schätzverfahren kommt der klassische OLS-Ansatz sowie die "Generalized Method of Moments" im Rahmen eines Instrumentalvariablenansatzes zum Einsatz. Im Rahmen der Untersuchung finden sich bei beiden Schätzmethoden - für jeweils fünf Modellspezifikationen - Hinweise auf eine Zeitvariabilität der Betakoeffizienten.

1 Einführung

Die Untersuchung der Einflussfaktoren auf Aktienrenditen nimmt in der empirischen Kapitalmarktforschung seit langem einen herausragenden Stellenwert ein. Neben den traditionellen Gleichgewichtsansätzen, wie dem CAPM und der APT sind verschiedene Faktormodelle entwickelt worden.[1] Sie beruhen im Wesentlichen auf der Annahme, dass die an Wertpapiermärkten beobachtbaren Renditebewegungen durch eine begrenzte Anzahl von ökonomischen Variablen (Faktoren) erklärt werden können. In den meisten Studien werden unbedingte Faktormodelle genutzt. Darüber hinaus beschränken sich die empirischen Untersuchungen oftmals auf die Aktienrenditen von Industrieunternehmen auf dem US-amerikanischen Kapitalmarkt.[2] Zielsetzung dieses Beitrages ist der Vergleich von unbedingten Faktormodellen mit bedingten Faktormodellen. Dabei konzentriert sich die empirische Untersuchung auf den deutschen Kapitalmarkt und hier speziell auf den Vergleich von Bankaktien mit Aktien aus anderen Branchen.

---

1 Vgl. zu Faktormodellen Alexander/Sharpe/Bailey (1999).

2 Vgl. dazu die Arbeiten von Chen/Roll/Ross (1986) und Fama/French (1992).

2 Literaturüberblick

Bei der Analyse von Bankaktien ist der Einfluss von Zinsänderungen von besonderer Bedeutung. Banken sollten aufgrund des Zinsgeschäftes und der Fristentransformation eine über den normalen Umfang hinausgehende Sensitivität zu diesem Faktor aufweisen. Die Untersuchungen im Bereich von Mehrfaktormodellen für Finanzintermediäre konzentrieren sich deshalb primär auf den zusätzlichen Einfluss von Zinssätzen und Zinssatzänderungen. Das erste Mehrfaktorenmodell, das den Einfluss des Zinssatzes explizit betrachtet, ist das Modell von Stone (1974). In einer Untersuchung von Flannery/James (1984) wird der Einfluss von Zinssatzänderungen auf Bankaktien analysiert und ein signifikanter Zusammenhang zwischen der Laufzeitstruktur der Bankbilanz und der Reaktion des Aktienkurses ermittelt. Neben einem Zinsfaktor verwenden Choi/Elyasiani/Kopecky (1992) zusätzlich einen Wechselkursfaktor und ermitteln dafür einen signifikanten Einfluss, was darauf hindeutet, dass die Währungspositionen der untersuchten Banken nicht vollständig abgesichert waren.

Für den deutschen Kapitalmarkt liegen bisher nur wenige Untersuchungen mit Faktormodellen vor. So untersuchen Oertmann/Rendu/Zimmermann (2000) den Einfluss nationaler und internationaler Zinssätze auf europäische Banken und Versicherungen. Nur eine kleine Gruppe von Studien verwendete bisher bedingte Faktormodelle. Dazu zählen z.B. Flannery/Hameed/Harjes (1997), die mit Hilfe eines GARCH-Modells zeitvariable Risikoprämien zur Analyse von Bankrenditen in den USA spezifizieren, sowie Bessler/Booth (1994), die mit einem GARCH-Modell den Einfluss eines Zinsfaktors auf Bankaktien in den USA und der BRD vergleichen.[3]

3 Modell

In dieser Untersuchung wird neben der klassischen OLS-Methode ein Instrumentalvariablenansatz verwendet, um zwischen den verschiedenen Einflussfaktoren auf Aktienrenditen von Banken zu differenzieren und deren Signifikanz zu testen. Das OLS-Modell ist folgendermaßen strukturiert:

$$r_{i,t} = \alpha_i + \sum_{k=1}^{K} \beta_k \cdot f_{k,t} + \upsilon_t \qquad (1)$$

[3] Einen Überblick über weitere Literatur geben Choi/Elyasiani/Kopecky (1992) und Bessler/Booth (1994).

Die Rendite $r_{i,t}$ des Wertpapiers i im Zeitpunkt t setzt sich aus einer Linearkombination der k-Faktorausprägungen $f_{k,t}$ im Zeitpunkt t, gewichtet mit den jeweiligen faktorspezifischen Betakoeffizienten β_k zusammen. Zusätzlich existiert eine wertpapierspezifische Konstante α_i sowie ein Fehlerterm v_t. Um eine zeitvariable Spezifikation der Parameter eines Faktormodells zu erreichen, wird auf einen Instrumentalvariablenansatz zurückgegriffen.[4] Dieser Ansatz basiert auf der Idee, die im Zeitpunkt t für t+1 erwarteten Parameter des Modells in Abhängigkeit von der gesamten öffentlich verfügbaren Informationsmenge im Zeitpunkt t darzustellen. Da sich diese Informationsmenge jedoch nicht adäquat empirisch abbilden lässt, beschränkt man sich zur Spezifikation auf einen Set von h Instrumentalvariablen im Vektor $\mathbf{z}_t$. In der Untersuchung wird eine Modellstruktur mit zeitvariablen Betakoeffizienten verwendet [in Anlehnung an Ferson/Harvey (1993), Ferson/Harvey (1997)]. Die erwartete Rendite setzt sich aus der Kombination der k erwarteten Betakoeffizienten und den entsprechenden erwarteten Faktorausprägungen zusammen.

$$E_t(r_{i,t+1}) = \sum_{k=1}^{K} E_t\left[\beta_{k,t+1}(\mathbf{z}_t) \cdot f_{k,t+1}(\mathbf{z}_t)\right] \tag{2}$$

Die erwarteten Betakoeffizienten und Faktorausprägungen lassen sich als Linearkombination der h Instrumentalvariablen in $\mathbf{z}_t$ darstellen

$$\beta_{k,t+1}(\mathbf{z}_t) = \sum_{j=1}^{h} \varphi_j \cdot z_{j,t} \quad \text{(3a)} \qquad \text{bzw.} \quad f_{k,t+1}(\mathbf{z}_t) = \sum_{j=1}^{h} \gamma_j \cdot z_{j,t}\,. \quad \text{(3b)}$$

Als Schätzverfahren bietet sich ein "Generalized Method of Moments"-Schätzer von Hansen (1982) an. Die Schätzungen basieren auf folgendem Gleichungssystem:[5]

$$u_{1it} = r_{it} - \mathbf{z}_{t-1}^{A,F} \cdot \boldsymbol{\delta}_i \quad \text{(4a)} \qquad \text{bzw.} \quad u_{4it} = \mu_i - \mathbf{z}_{t-1}^{A,F} \cdot \boldsymbol{\delta}_i \quad \text{(4b)}$$

$$\mathbf{u}_{2it} = \mathbf{f}_t - \mathbf{z}_{t-1}^{A} \boldsymbol{\Gamma}_i \quad \text{mit } \boldsymbol{\Gamma} = (\gamma_{kj}) \text{ für } j = 1....h; k = 1....K \tag{5}$$

$$\mathbf{u}_{3it} = \left[\mathbf{u}_{2it}' \mathbf{u}_{2it} \left(\mathbf{z}_{t-1}^{A,F} \boldsymbol{\Phi}_i\right)' - \mathbf{f}_t' \, u_{1it}\right]' \text{ mit } \boldsymbol{\Phi} = (\varphi_{kj}) \text{ für } j = 1....h; k = 1....K \tag{6}$$

4 Vgl. zu Instrumentalvariablenmodellen Harvey/Kirby (1996).

5 Vgl. dazu Harvey/Kirby (1996).

$$u_{5it} = (-\alpha_i + \mu_i) - \mathbf{z}_{t-1}^{A,F} \mathbf{\Phi}_i \left(\mathbf{z}_{t-1}^{A} \mathbf{\Gamma}_i\right)' . \tag{7}$$

Gleichung (4a) stellt die Abweichung zwischen der Rendite des Wertpapiers und einer durch die mit $\boldsymbol{\delta}$ gewichteten Instrumentalvariablen geschätzten Rendite dar. Gleichung (4b) ermittelt den mittleren Fehler der Renditeschätzung aus Gleichung (4a), wobei μ_i die durchschnittliche Aktienrendite über den betrachteten Zeitraum darstellt. Analog ist Gleichung (5) aufgebaut, welche die Abweichung zwischen den k Faktorausprägungen und den durch einen Subset $\mathbf{z}_{t-1}^{A}$, der h Instrumentalvariablen in $\mathbf{z}_{t-1}^{A,F}$, erklärten Faktorausprägungen enthält. Die bedingten Betakoeffizienten sind in Gleichung (6) enthalten und setzen sich aus einer Kombination der Gewichte in der Matrix $\mathbf{\Phi}_i$ und der Instrumentalvariablen zusammen. In Gleichung (7) wird mit α_i der mittlere Fehler zwischen der Aktienrendite und der mit Hilfe des Mehrfaktorenmodells geschätzten Rendite ermittelt.

4 Daten

Grundlage der empirischen Untersuchung bilden fünf gleichgewichtete Branchenindices. Dabei handelt es sich um einen Index der Universalbanken (12 Werte), einen Index der Hypothekenbanken (7 Werte) und einen Index für alle Banken (19 Werte). Zusätzlich wird ein Index der Versicherungen (9 Werte) und ein Index der Industrieunternehmen (13 Werte) verwendet. In den ersten beiden Indices sind alle Aktien enthalten, die über den gesamten Zeitraum von Januar 1974 bis Dezember 1997 an einer Börse in Deutschland im amtlichen Handel notiert waren. Der Index der Industrieunternehmen enthält alle DAX-Werte (Basis Juni 2001), für die im Zeitraum von Januar 1974 bis Dezember 1997 Kurs-notierungen existieren. Ausgenommen davon sind die Bank- und Versicherungs-werte des DAX. Als Faktoren werden die Renditen des DAFOX abzüglich eines risikofreien Zinssatzes[6] (Marktrisikoprämie) sowie die monatlichen Differenzen der Spot-Rates mit einer Restlaufzeit von einem Jahr und die monatlichen Differenzen der Spot-Rates mit einer Restlaufzeit von zehn Jahren verwendet.[7] Bei der Auswahl der Instrumentalvariablen ist darauf zu achten, dass sie mindestens in derselben Frequenz wie die zu erklärenden Variablen vorliegen. Im Folgenden werden

---

6 Als risikofreier Zinssatz wird die Umlaufrendite von Bundesanleihen mit einer Restlaufzeit von einem Jahr verwendet.

7 Die Renditen der Aktien und des DAFOX wurden freundlicherweise von der Karlsruher Kapitalmarktdatenbank zur Verfügung gestellt. Die übrigen Zeitreihen stammen von der Deutschen Bundesbank und dem IWF.

monatliche Daten verwendet. Um eine ökonomische Fundierung zu erreichen, werden Regressionen zwischen den Instrumental-variablen und den Faktoren durchgeführt. Auf deren Basis kommen die folgenden sechs Instrumentalvariablen zum Einsatz: monatliche Rendite des Dow Jones Index, Veränderung des Wechselkurses DM/US$, Veränderung der saisonbe-reinigten Geldmenge M3, monatliche Veränderung des IWF-Verbraucherpreis-index für Deutschland sowie die Spot-Rates mit einer Restlaufzeit von einem bzw. zehn Jahren. Alle Instrumentalvariablen werden entsprechend dem Modell mit einem time-lag von einem Monat verwendet. Dadurch verkürzt sich der entsprechende Untersuchungszeitraum effektiv auf März 1974 bis Dezember 1997.[8]

5 Ergebnisse

Mit Hilfe des OLS-Verfahrens werden zunächst fünf verschiedene Faktormodelle für jeweils fünf Indices getestet.[9] Der DAFOX-Faktor ist in allen untersuchten Modellen auf dem 1%-Niveau signifikant. Die geringste Sensitivität des Marktfaktors weist dabei der Index der Hypothekenbanken auf, während im Industrie- und im Versicherungsindex die höchsten Sensitivitäten existieren.

Betrachtet man die beiden Einfaktormodelle mit den Spot-Rates mit einer Restlaufzeit von einem bzw. zehn Jahren als Faktor, so erhält man durchgehend negative Betas. Die stärkste Sensitivität bezüglich der langfristigen Spot-Rates weisen die Universalbanken auf, während die Hypothekenbanken in etwa auf dem selben Niveau wie die Industrieunternehmen liegen. Die Versicherungen weisen die geringste Zinssensitivität auf. Ein differenzierteres Bild ergibt sich bei Verwendung der kurzfristigen Spot-Rates. Die Versicherungen und Industrieunternehmen besitzen in diesem Fall kein bzw. ein nur schwach signifikantes Beta. Die Betas für die drei Bankindices sind alle auf dem 1%-Niveau signifikant und unterscheiden sich nur geringfügig.

Im Fall der Zweifaktormodelle kommt es zu differenzierteren Ergebnissen. So entsprechen die DAFOX-Betas in etwa denen der Einfaktormodelle. Dagegen sinkt das Niveau der Zinsbetas ab. Im Bereich der Versicherungen und Industriewerte weisen diese Betas positive Vorzeichen auf. Die stärkste Zinssensitivität im langfristigen Bereich existiert bei den Hypothekenbanken, während der Index der Universalbanken ein höheres negatives Beta bei Veränderungen in den kurz-

---

8 Da die Zeitreihe des IWF-Verbraucherpreisindex ist erst ab Ende Januar 1974 verfügbar ist.

9 Bei den Modellen handelt es sich jeweils Einfaktormodell mit den in Abschnitt 4 dargestellten Faktoren und zwei Zweifaktormodellen mit dem DAFOX Faktor und je einem kurzfristigen oder langfristigen Zinsfaktor. Alle Ergebnisse sind auf Anfrage von den Autoren erhältlich und konnten aus Platzgründen nicht abgedruckt werden.

fristigen Spot-Rates aufweist. Allerdings sind die Aussagen bezüglich der Zinssensitivität in den Zweifaktormodellen nur beschränkt statistisch unterlegt, da nicht alle Betakoeffizienten signifikant sind. Betrachtet man die Bestimmtheitsmaße der Regressionen, so ergeben sich Unterschiede bezüglich der Branchen und der verwendeten Faktoren. Die Spot-Rates besitzen in den Einfaktormodellen eine höhere Aussagekraft bezüglich der Renditen in den Bankindices. Wird der DAFOX-Faktor verwendet, so finden sich die höchsten Bestimmtheitsmaße bei den Industriewerten. Die geringste Erklärungskraft besitzt der DAFOX-Faktor für den Index der Hypothekenbanken. Die zusätzliche Verwendung der Spot-Rates als Faktor neben dem DAFOX-Faktor führt bei den drei Bankindices zu einem stärkeren Anstieg der Bestimmtheitsmaße als in den übrigen Indices. Um Problemen vorzubeugen, die aus einer eventuell vorhandenen Multikollinearität in den Zeitreihen der drei Faktoren resultieren, werden Variance-Inflation-Factors berechnet. Nach Chatterjee/Price (1991) weisen diese auf nachhaltige Multikollinearitätsprobleme hin, wenn Sie einen Wert von 10 übersteigen. Für die in diesem Zusammenhang verwendeten Faktorkombinationen wird in keinem Fall ein Variance-Inflation-Factor größer als 1,69 erreicht. Auf eine Orthogonalisierung der Zeitreihen wird daher verzichtet.

Um die Zeitkonstanz des zugrundeliegenden OLS-Ansatzes zu testen, sind für alle verwendeten OLS-Gleichungen Kusum-Tests auf eine Zeitkonstanz der Betakoeffizienten durchgeführt worden. Dabei konnte die Nullhypothese, dass zeitkonstante Betakoeffizienten vorliegen, auf dem 5%-Signifikanzniveau in 7 von 25 Modellen abgelehnt werden. Der Index der Versicherungsunternehmen weist dabei in allen Modellspezifikationen auf zeitvariable Betakoeffizienten hin.

Im weiteren Verlauf der Untersuchung wird nun mit Hilfe des dargestellten Instrumentalvariablenansatz ein Modell implementiert, das die Zeitvariabilität der Betakoeffizienten darstellen soll. In allen Einfaktormodellen ist der Alpha-Wert, der einen Hinweis auf eine mögliche Fehlspezifikation des Modells liefert, mindestens auf dem 5%-Niveau signifikant, während die Zweifaktormodelle anscheinend einen besseren Fit besitzen, da nur in vier von zehn Fällen der Alpha-Wert signifikant ist. Der Vergleich der Alpha-Werte deutet auf eine bessere Spezifikation durch das OLS-Modell hin. Bei den getesteten Einfaktormodellen ist nur ein Betakoeffizient nicht signifikant von Null verschieden und die Hypothese zeitvariabler Betakoeffizienten, die mit Hilfe eines Wald-Tests durchgeführt wird, kann in keinem weiteren Fall verworfen werden. Bei den Zweifaktormodellen kann die Hypothese zeitvariabler Betakoeffizienten für den DAFOX Faktor in fünf von zehn Fällen und für die Zinsfaktoren in zwei von zehn Fällen verworfen werden. Ein J-Test weist bei einem 5%-Signifikanzniveau in sieben Modellen auf eine Überspezifikation im Rahmen des verwendeten zugrundeliegenden GMM-Gleichungssystems hin.[10] Bei einem Vergleich der Ergebnisse für die verschiedenen Indices weist der Index der Universalbanken in allen Modelle eine höhere

[10] Vgl. zu J-Test Hansen (1982).

Zinssensitivität auf als der Index der Hypothekenbanken. Dabei sind die kurzfristigen Spot-Rates im Index der Hypothekenbanken nur im Zweifaktormodell signifikant. In allen Modellen sind die Betakoeffizienten für die Zinsfaktoren im Industrieindex signifikant, aber in keinem Fall erreicht die Zinssensitivität das Niveau der Universalbanken.

6 Schlussfolgerung

Im Rahmen dieser Untersuchung wurden die Renditen von fünf Branchenindices bezüglich verschiedener ökonomischer Einflussfaktoren untersucht sowie auf zeitvariable Betakoeffizienten getestet. Sowohl im Rahmen der OLS-Analyse als auch des Instrumentalvariablenansatzes ergaben sich Hinweise darauf, dass die drei Bankindices eine höhere negative Zinssensitivität besitzen als die Indices der Versicherungen und der Industrieunternehmen. Auch die Erklärungskraft der Spot-Rates als Faktoren, dargestellt durch das Bestimmtheitsmaß in den OLS-Gleichungen, ist bei den Bankindices am höchsten. Die Universalbanken besitzen in fast allen Modellspezifikationen eine höhere negative Zinssensitivität gegenüber der Veränderung der beiden Spot-Rates als die Hypothekenbanken. Vergleicht man den Index der Universalbanken mit dem der Hypothekenbanken, so deutet das höhere Bestimmtheitsmaß bei den Universalbanken aus ökonomischer Sicht auf ein Portfolio von nicht verbrieften Forderungstiteln (Kredite) und Beteiligungen hin, die dem gesamtwirtschaftlichen Risiko stärker ausgesetzt sind. Die Ergebnisse der durchgeführten Kusum-Tests deuten auch auf eine Zeitvariabilität der Betakoeffizienten hin. Bestätigt werden diese Ergebnisse durch die Anwendung des Instrumentalvariablenansatzes. Dort finden sich in der Mehrzahl der Fälle Hinweise auf eine zeitvariable Betaspezifikation. Die höhere Sensitivität der Universalbanken im Vergleich mit den Hypothekenbanken gegenüber den kurzfristigen Spot-Rates könnte durch eine stärkere Bedeutung des Zinsüberschussgeschäftes für die Universalbanken erklärt werden. Im Vergleich mit den Hypothekenbanken betreiben die Universalbanken eine positive Fristentransformation und sind daher auch empfindlicher gegenüber Zinssatzänderungen im langfristigen Bereich.

Literaturverzeichnis

Alexander, G. J., Sharpe, W. F., Bailey, J. V. (1999): Investments, 6th Edition. Prentice Hall, Upper Saddle River.

Bessler, W., Booth, G. G. (1994): Interest Rate Sensitivity of Bank Stock Returns in a Universal Banking System. Journal of International Financial Markets, Institutions and Money 3, 117-136.

Chatterjee, S., Price, B. (1991): Regression Analysis by Example, 2nd Edition. Wiley, New York.

Chen, N. F., Roll, R., Ross, S. A. (1986): Economic Forces and the Stock Market. Journal of Business 59, 383-403.

Choi, J. J., Elyasiani, E., Kopecky, K. J. (1992): The Sensitivity of Bank Stock Returns to Market, Interest and Exchange Rate Risk. Journal of Banking & Finance 16, 983-1004.

Fama, E. F., French, K. R. (1992): The Cross-Section of Expected Stock Returns. The Journal of Finance 47, 427-465.

Ferson, W. E., Harvey, C. R. (1993): The Risk and Predictability of International Equity Returns. The Review of Financial Studies 6, 527-566.

Ferson, W. E., Harvey, C. R. (1997): Fundamental Determinants of National Equity Market Return: A Perspective of Conditional Asset Pricing. Journal of Banking & Finance 21, 1625-1665.

Flannery, M. J., James, C. M. (1984): The Effect of Interest Rate Changes on the Common Stock Returns of Financial Institutions. The Journal of Finance 39, 1141-1154.

Flannery, M. J., Hameed, A. S., Harjes, R. H. (1997): Asset-Pricing, Time Varying Risk Premia and Interest Rate Risk. Journal of Banking & Finance 21, 315-335.

Hansen, L. P. (1982): Large Sample Properties of Generalized Method of Moments Estimators. Econometrica 50, 1029-1054.

Harvey, C. R., Kirby, C. (1996): Instrumental Variables Estimation of Conditional Beta Pricing Models. In: Maddala, G. S., Rao, C. R.: (Eds.): Statistical Methods in Finance. North-Holland, Amsterdam, 35-60.

Oertmann, P., Rendu, C., Zimmermann, H. (2000): Interest Rate Risk of European Financial Corporations. European Financial Management 6, 459-478.

Stone, B. K. (1974): Systematic Interest-Rate Risk in a Two-Index Model of Returns. Journal of Financial and Quantitative Analysis 9, 709-721.

Optimization of European Double-Barrier Options via Optimal Control of the Black-Scholes-Equation

Michael H. Breitner[1] and Tobias Burmester[2]

[1] Fachbereich Mathematik und Informatik, AG Wissenschaftliches Rechnen, Technische Universität Clausthal, Erzstraße 1, D-38678 Clausthal-Zellerfeld, Germany, breitner@math.tu-clausthal.de

[2] Systeme, Anwendungen und Produkte in der Datenverarbeitung (SAP) AG, Supply Chain Management (SCM 2), Neurottstraße 16, D-69190 Walldorf, Germany, tobias.burmester@sap.com

Abstract. In comparison to future contracts todays options have advantages and disadvantages. DAX-future contracts, e. g. the EUREX FDAX, have a constant profit/loss per DAX point, whereas the profit/loss per DAX point of common options varies undesirably with the DAX. An advantage of buying options results from the immediate payment of the option premium. There are no further payments during the life of the option. Similarly margin requirements force a FDAX buyer/seller to pay an initial margin plus a safety margin immediately. But in the event of adverse DAX movements the buyer/seller has to meet calls for substantial additional margin and may be forced to liquidate his position prematurely. Optimized European double-barrier options can combine the advantages of futures and options. The option expires if the DAX either hits the upper knock-out barrier DAX_{max} or the lower knock-out barrier DAX_{min} or if the option has matured. We present two approaches to optimize the cash settlements at expiration either analytically or numerically.

Keywords. Financial derivatives, options, futures, hedging tactics, Black-Scholes-model, optimal control, optimization, partial differential equation.

1 Real Life Case Study

A large German insurance company has a DAX-like stock portfolio worth 1 billion Euro (= 10^9 Euro). The DAX is 6000 points. "DAX" is used for the XETRA spot rate of the German stock index DAX 30. The company predicts decreasing stock values and wishes to immunize the portfolio, i. e. to generate a risk-free portfolio. Tax issues, market conditions or regulations inhibit the company from selling stocks.

2 Hedging with DAX Futures

The insurance company has the option to sell DAX futures at the EUREX, i. e. to open FDAX-short positions. One FDAX-short position has a profit/loss

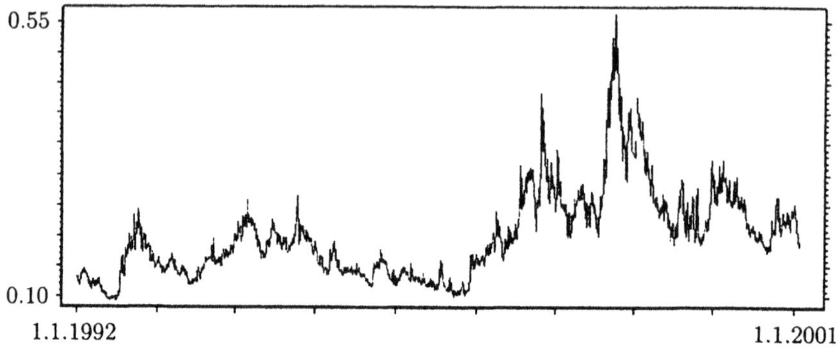

Fig. 1. Implied three-months at the money volatility (EUREX VDAX) of EUREX-options on the DAX. Source: Market-Maker (Release 1.40).

Δ_{FDAXs} per DAX point of 25 Euro. Thus the company has to sell $10^9/(25 * 6000) \approx 6667$ DAX futures to immunize the stock portfolio. The EUREX clearing agency demands an initial margin plus a safety margin immediately. The total margin per FDAX-short varies with the implied DAX volatility (EUREX VDAX) and is, e. g., about 5000 Euro for a low volatility of 0.17 and about 20000 Euro for a high volatility of 0.45. The total margin is between 3.33 % and 13.33 % of the stock portfolio value. In the event of an increasing DAX the company has to meet calls for additional margin, i. e. 166675 Euro per DAX point, and may be forced to liquidate its position prematurely. Moreover, the permanent margin adaption at the EUREX clearing agency is quite troublesome. Another disadvantage is the margin variation dependent on the VDAX, which oscillated between 0.10 and 0.55 during the last decade, see Fig. 1.

3 Hedging with DAX Calls with Constant Δ

The solution $C(t, p; r, \sigma)$ of the well known Black-Scholes-equation

$$\frac{\partial}{\partial t}C + r\,p\,\frac{\partial}{\partial p}C + \frac{1}{2}\sigma^2 p^2 \frac{\partial^2}{\partial p^2}C - r\,C = C_t + r\,p\,C_p + \frac{1}{2}\sigma^2 p^2 C_{pp} - r\,C \equiv 0, \quad (1)$$

which is a linear, homogeneous, parabolic partial differential equation of second order, yields a sufficiently accurate approximation of European DAX option values, see Hull (2000) and Redhead (1997). The time is denoted by $t \in [0, T]$, $p \in [\text{DAX}_{\text{min}}, \text{DAX}_{\text{max}}]$ denotes the DAX, $r > 0$ denotes the risk-free interest rate per year for the maturity period, $\sigma > 0$ denotes the implied volatility of the DAX and $T > 0$ denotes the initial maturity period at $t = 0$. Note that t and p are the independent variables in the linear, homogeneous, parabolic partial differential equation (1) of second order, whereas r and σ are constant parameters. For the profit/loss Δ per DAX point there holds $\Delta = C_p$. Assuming $\Delta \equiv \Delta_{\text{opt}}$, $\Delta_{\text{opt}} \geq 0$, for all

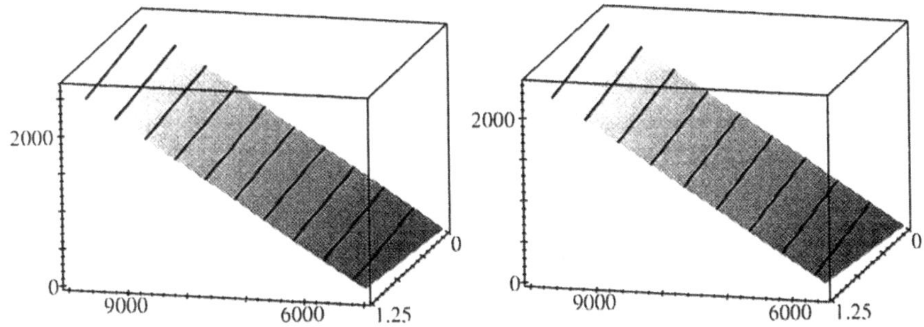

Fig. 2. Black-Scholes-values of *optimal* European double-barrier DAX calls with $\Delta_{oEdbc} \equiv 0.5$ for all DAX in $[5000, 10000]$ (left, strike $DAX_{min} = 5000$) and in $[5500, 10000]$ (right, strike $DAX_{min} = 5500$) and for all maturity periods of $[0, 1.25]$ years. Contour lines represent values of 250 Euro, 500 Euro, ...

$(t,p) \in [0,T] \times [DAX_{min}, DAX_{max}]$ we get $C_{pp} \equiv 0$. Equation (1) simplifies to a partial differential equation of first order with the solution

$$C_{\Delta_{opt}}(t,p;r) = \Delta_{opt}\left(p - DAX_{min}\, e^{-r\,(T-t)}\right), \tag{2}$$

which doesn't depend on σ, i.e. the anticipated future DAX volatility has no influence on the DAX call value. Solution (2) is unique solution of equation (1) if the initial/boundary conditions are adjusted to the analytic solution

$$C_{\Delta_{opt}}(t=T,p;r)=\Delta_{opt}(p-DAX_{min}) \text{ for } p\in[DAX_{min}, DAX_{max}], \tag{3}$$

$$C_{\Delta_{opt}}(t,p=DAX_{min};r)=\Delta_{opt}\, DAX_{min}\left(1-e^{-r\,(T-t)}\right) \text{ for } t\in[0,T[\text{ and} \tag{4}$$

$$C_{\Delta_{opt}}(t,p=DAX_{max};r)=\Delta_{opt}\left(DAX_{max}-DAX_{min}\, e^{-r\,(T-t)}\right) \text{ for } t\in[0,T[. \tag{5}$$

Dealing with an *European double-barrier* DAX call these initial/boundary conditions correspond to the cash settlements at the option's expiration. The option expires if either one of the knock-out barriers is hit, i.e. $p = DAX_{max}$ at the upper barrier or $p = DAX_{min}$ at the lower barrier, or if the expiration date is reached, i.e. $t = T$, see Fig. 2. Considering the case study the insurance company can sell $10^9/(0.5 * 6000) \approx 333333$ optimal DAX calls with $\Delta \equiv 0.5$ (covered call writing). The company accepts the obligation to pay the cash settlement (3), (4) or (5) at the option's expiration. The company's premium per written call at $t = 0$ is 651.47 Euro for a strike $DAX_{min} = 5000$ and 416.61 Euro for a strike $DAX_{min} = 5500$ if there holds $p = 6000$, $r = 0.05$ and $T = 1.25$ years, see Fig. 2. There are no further payments during the life of the option. The buyer of 50 options can easily synthesize 1 FDAX-long position costing 32573.50 Euro or 20830.50 Euro, respectively. Selling 50 options the insurance company easily synthesizes 1 FDAX-short position. In comparison a plain vanilla European DAX call has a profit/loss $\Delta_{pvEc}(t,p;r,\sigma)$ per DAX point, which strongly depends on p and σ and moderately on t and r, see

Fig. 3 for a strike $s = 7000$. The company has to pay the cash settlement $\max(p(T) - s, 0)$ Euro only at $t = T$. Due to the variation of Δ_{pvEc} the hedging of the DAX-like stock portfolio is tedious. The number of options required for a risk-free portfolio changes permanently with the German stock market conditions (DAX *and* VDAX).

4 Hedging with DAX Calls with Almost Constant Δ

Today European double-barrier options are well accepted at the financial markets. However, the non-constant cash settlements (4) and (5) at the knock-out barriers are quite extraordinary. Usually the payments at the barriers are constant, i. e.

$$C(t, p = \text{DAX}_{\text{min}}; r, \sigma) \equiv 0 \text{ and } C(t, p = \text{DAX}_{\text{max}}; r, \sigma) \equiv C_{\text{up}} \text{ with } C_{\text{up}} > 0. \quad (6)$$

In contrast a more complicate payment $C(t = T, p; r, \sigma) = C_T(p)$ is acceptable. Thus C_{up} and $C_T(p)$, $p \in [\text{DAX}_{\text{min}}, \text{DAX}_{\text{max}}]$, are optimized to get an almost constant, predefineable profit/loss $\Delta_{\text{aoEdbc}}(t, p; r, \sigma) \approx \Delta_{\text{opt}}$ per DAX point for all relevant p, σ and t, see Hoffmann (1999) for an overview of optimal control problems governed by partial differential equations.

In order to optimize $\Delta_{\text{aoEdbc}}(t, p; r, \sigma)$ the Black-Scholes-equation (1) has to be solved numerically. Here the Crank-Nicholson method is used, which is a combination of an explicit and an implicit finite difference method, see Ames (1992), Hull (2000), Prisman (2000), Seydel (2000) and Thomas (1998). With a time discretization Δt and DAX discretization Δp the Crank-Nicholson method converges robustly, i. e. the absolute error of the numerical solution decreases by $\mathcal{O}(\Delta t^2) + \mathcal{O}(\Delta p^2)$ for $\max(\Delta t, \Delta p) \to 0$. The computing time increases by $\mathcal{O}(1/(\Delta t * \Delta p))$ for $\max(\Delta t, \Delta p) \to 0$. We consider options with $t \in [0, 1.25]$ years, $p \in [4000, 10000]$, $T = 1.25$ years, $r = 0.05$, $s = 7000$ and $\sigma \in [0.17, 0.45]$. The parameterized initial/boundary conditions are

$$C_T(p) = \begin{cases} 0 & \text{for } p \in [4000, 7000], \\ x_1\,(p - 7000)/600 & \text{for } p \in]7000, 7600], \\ (x_1\,(8200 - p) + x_2\,(p - 7600))/600 & \text{for } p \in]7600, 8200], \\ (x_2\,(8800 - p) + x_3\,(p - 8200))/600 & \text{for } p \in]8200, 8800], \\ (x_3\,(9400 - p) + x_4\,(p - 8800))/600 & \text{for } p \in]8800, 9400], \\ (x_4\,(10000 - p) + x_5\,(p - 9400))/600 & \text{for } p \in]9400, 10000], \end{cases} \quad (7)$$

$$C\,(t, p = 4000; r, \sigma) = 0 \quad \text{for} \quad t \in [0, 1.25[\quad \text{and} \quad (8)$$

$$C\,(t, p = 10000; r, \sigma) = x_5 \quad \text{for} \quad t \in [0, 1.25[, \quad (9)$$

where $C_T(p) = C(t = 1.25, p; r, \sigma)$ and $\boldsymbol{x} = (x_1, x_2, \ldots, x_5)^T$ is the vector of parameters to optimize with $0 \leq x_i \leq 3000$ for all i. We consider the relevant implied DAX volatilities $\sigma_j = 0.17 + 0.04\,(j - 1)$ for $j = 1, 2, \ldots, 8$ and compute approximations $C_{\text{CN}}(t, p; 0.05, \sigma_j, \boldsymbol{x})$ on a 1200×1200 grid, i. e.

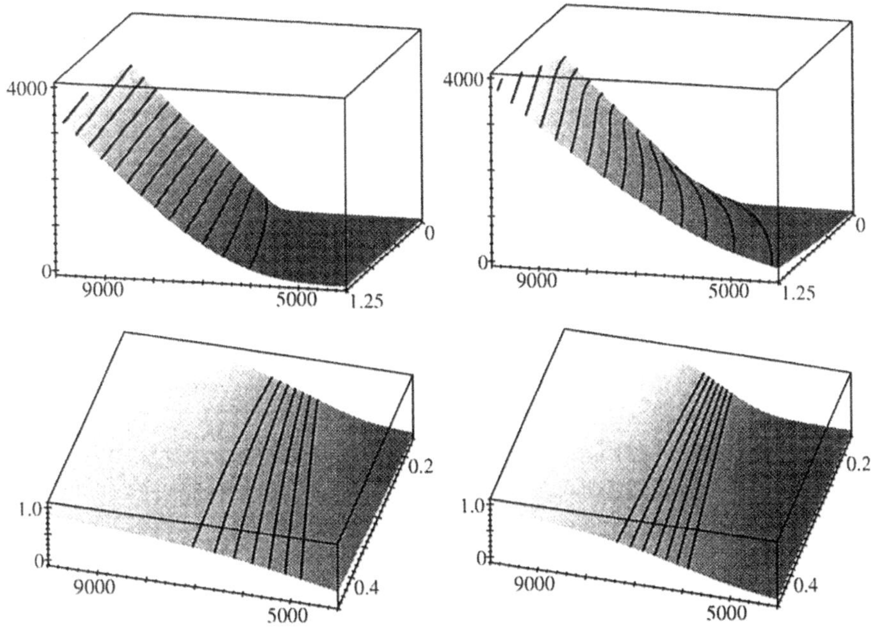

Fig. 3. Black-Scholes-value of a plain vanilla European DAX call dependent on the DAX in [4000, 10000] and the maturity period in [0, 1.25] years for an implied DAX volatility of 0.17 (upper left) and 0.45 (upper right) and Δ_{pvEc} of this call dependent on the DAX and the implied DAX volatility in [0.17, 0.45] for a maturity period of 1.25 years (lower left) and 0.417 years (lower right). Contour lines represent $\Delta_{\mathrm{pvEc}} \equiv 0.35$, $\Delta_{\mathrm{pvEc}} \equiv 0.40$, ... , $\Delta_{\mathrm{pvEc}} \equiv 0.65$.

with $\Delta t = 1.25/1200$ and $\Delta p = (10000 - 4000)/1200$. With $C_{j,k,l}(\boldsymbol{x}) := C_{\mathrm{CN}}(k\,\Delta t, 4000 + l\,\Delta p; 0.05, \sigma_j, \boldsymbol{x})$, $w_1 = w_2 = w_3 = 5$, $w_4 = w_5 = 2$ and $w_6 = w_7 = w_8 = 1$ we minimize the Euclidian norm error function

$$\varepsilon(\boldsymbol{x}) := \sum_{j=1}^{8} w_j \sum_{k=0}^{960} \sum_{l=100}^{1100} \left(\Delta_{\mathrm{opt}} - \frac{C_{j,k,l+1}(\boldsymbol{x}) - C_{j,k,l-1}(\boldsymbol{x})}{2\,\Delta p} \right)^2 \qquad (10)$$

w.r.t. $\boldsymbol{x}$. Only the relevant $(t, p) \in [0, 1.0] \times [4500, 9500]$ are considered and the more important volatilities have larger weights w_j. The profit/loss Δ_{aoEdbc} per DAX point of the simplified optimizeable European double-barrier DAX call is computed with the symmetric difference approximation $(C_{j,k,l+1}(\boldsymbol{x}) - C_{j,k,l-1}(\boldsymbol{x}))/(2\,\Delta p)$ for the first partial derivative of $C(k\,\Delta t, 4000 + l\,\Delta p; 0.05, \sigma_j, \boldsymbol{x})$ w.r.t. p. The error function $\varepsilon(\boldsymbol{x})$ is minimized with the sequential quadratic programming (SQP) method NPSOL, see Gill et al. (2000). The total computing time for a Pentium III/1 GHz PC is less than 5 minutes. Various computations have shown that $\varepsilon(\boldsymbol{x})$ has very many local minima. Thus computations should be made for different initializations for $\boldsymbol{x}$ to find a *good* local or the global minimum. For the case study

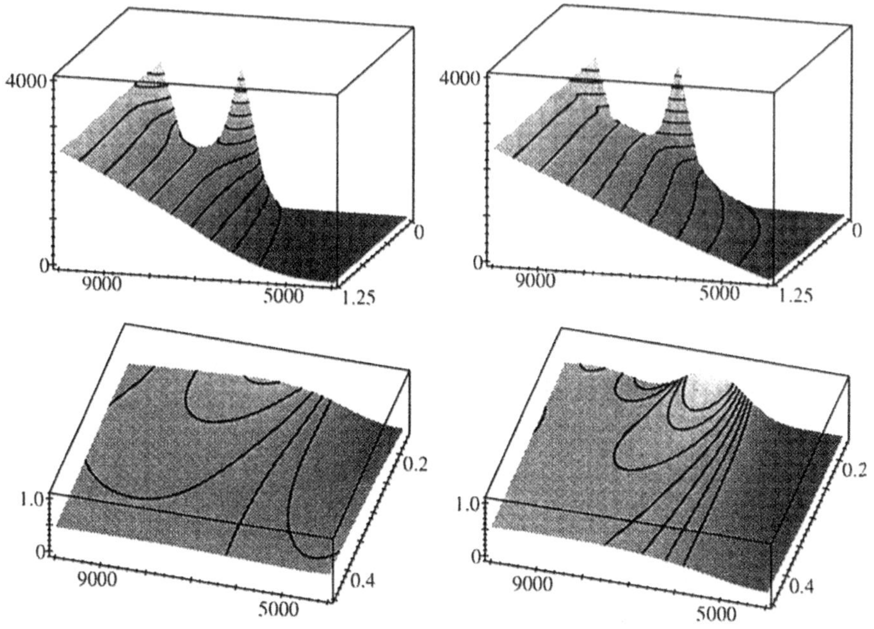

Fig. 4. Black-Scholes-value of an *optimized* European double-barrier DAX call ($\Delta_{\text{aoEdbc}} \approx 0.5$, strike 7000) dependent on the DAX in $[4000, 10000]$ and the maturity period in $[0, 1.25]$ years for an implied DAX volatility of 0.17 (upper left) and 0.45 (upper right) and Δ_{aoEdbc} of this call dependent on the DAX and the implied DAX volatility in $[0.17, 0.45]$ for a maturity period of 1.25 years (lower left) and 0.417 years (lower right). Contour lines represent $\Delta_{\text{aoEdbc}} \equiv 0.35$, $\Delta_{\text{aoEdbc}} \equiv 0.40$, ... , $\Delta_{\text{aoEdbc}} \equiv 0.65$.

and $\Delta_{\text{opt}} = 0.5$ the best vector found is

$$\boldsymbol{x}^* = (3000.00, 688.459, 865.610, 3000.00, 2433.73)^T, \tag{11}$$

see Fig. 4 and compare Fig. 4 to Figs. 2 and 3. The insurance company can sell optimized DAX calls at $t = 0$ with $\Delta \in [0.38, 0.46]$ dependent on the actual implied DAX volatility $\sigma \in [0.17, 0.45]$ (covered call writing), see Fig. 4. The company accepts the obligation to pay the cash settlement (7), (8) or (9) at the option's expiration. In comparison to the plain vanilla DAX call the optimized DAX call has a widely almost constant $\Delta_{\text{aoEdbc}} \approx \Delta_{\text{opt}}$. Both buyer and seller of the optimized DAX call can benefit from this desirable behaviour. The actual Beta-version of the WARRANT-PRO-2 software handles an user definable number of parameters x_i to discretize the initial/boundary conditions (7) and (9). Computations with up to 25 parameters for (7) and up to 10 parameters for (9) have converged. However, computing times are very long and the large number of bad local minima makes it difficult to find good optimized European double-barrier DAX calls. So far it turned out that the optimal $C_{\text{up}}{}^*(t)$ tends to be almost constant and that the optimal

$C_T{}^*(p)$ tends to be more and more oszillating for an increasing number of parameters.

5 Outlook

An user friendly Windows/Linux version of WARRANT-PRO-2 with graphical user interface is under development in cooperation with the Dresdner Bank AG, Frankfurt. The software will provide a framework for professional option and warrant design. The gradient of $\varepsilon(\boldsymbol{x})$ needed for the minimization of $\varepsilon(\boldsymbol{x})$ currently is computed via numerical differentiation. The gradient's accuracy is only 2 – 3 correct decimal digits due to the low accuracy of $\varepsilon(\boldsymbol{x})$ (4 – 7 digits). The gradient computation will be implemented with the help of automatic/algorithmic differentiation (accuracy 4 – 6 digits), see Griewank (2000). Additionally the improved, analytic gradient computation will cut down the computing times significantly. Moreover, ongoing research of the first author is dedicated to the replacement of the deterministic Black-Scholes-model by more accurate heuristic option pricing models, see Breitner (2000).

Acknowledgements

The authors gratefully appreciate support by Prof. Dr. P. E. Gill, University of California San Diego, providing his excellent SQP optimization method NPSOL, by the Market Maker Software AG, Kaiserslautern, and by the Dresdner Bank AG, Frankfurt.

References

1. Ames, W. F. (1992): Numerical Methods for Partial Differential Equations. Academic Press, Boston
2. Breitner, M. H. (2000): Heuristic Option Pricing with Neural Networks and the Neuro-computer SYNAPSE 3. Optimization 47, 319–333
3. Chriss, N. A. (1997): Black-Scholes and Beyond. McGraw-Hill, London
4. Gill, P. E., Murray, W., and Wright, M. H. (2000): Practical Optimization. Academic Press, London
5. Griewank, A. (2000): Evaluating Derivatives: Principles and Techniques of Algorithmic Differentiation. SIAM, Philadelphia
6. Hoffmann, K.-H. (Ed.) (1999): Optimal Control of Partial Differential Equations (International Conference in Chemnitz, Germany, April 20 – 25, 1998). Birkhäuser, Basel
7. Hull, J. C. (2000): Options, Futures, and other Derivatives (with microdisk). Prentice Hall, New York
8. Prisman, E. Z. (2000): Pricing Derivative Securities: An Interactive Dynamic Environment with Maple V and Matlab (with CD-ROM). Academic Press, San Diego

9. Redhead, K. (1997): Financial Derivatives — An Introduction to Futures, Forwards, Options and Swaps. Prentice Hall, London
10. Seydel, R. (2000): Einführung in die numerische Berechnung von Finanz-Derivaten (Computational Finance). Springer, Berlin
11. Thomas, J. W. (1998): Numerical Partial Differential Equations (Part 1): Finite Difference Methods. Springer, New York

How to Incorporate Estimation Risk into Markowitz Optimization

Alexander Kempf, Klaus Kreuzberg, and Christoph Memmel

University of Cologne, Department of Finance, Germany. Please address all correspondence to the first author. E-mail: kempf@wiso.uni-koeln.de

Abstract. This paper presents a new approach to incorporate estimation risk into mean-variance portfolio selection. The key contribution of our analysis is that we model the estimation risk as a second, independent source of risk. We show that widely used portfolio strategies (naive diversification, global minimum variance investment and classical two-step portfolio optimization) are nested in our model. In the second part of the paper we empirically test these strategies using G7-countries MSCI index return data. Our model leads to substantially superior performance as compared to several benchmark strategies. We find evidence that the estimation risk depends on the inhomogeneity of a market.

Portfolio theory is a cornerstone of modern financial theory. It provides efficient portfolio weights *given* the distribution parameters of stock returns. To implement the model, a two-step approach is typically used. In the first step, the distribution parameters are estimated from return data. In the second step, the estimated values are used as input variables in the Markowitz [6] portfolio model.

However, this two-step approach often leads to poor portfolio performance. This is caused by the fact that the estimates of the distribution parameters differ largely from the true distribution parameters. This problem is most pronounced when estimating expected returns since expected returns can be estimated only with a low precision (Merton [7]) and have a tremendous impact on portfolio holdings and performance (Chopra/Ziemba [2]).

In this paper we focus on estimation risk with respect to expected returns and neglect the less important issue of estimation risk in variances and covariances.[1] We provide a model which uses an optimal estimator for expected returns and takes the remaining estimation risk into account. This model is tested using international index return data and found to lead to a better performance than several benchmark models.

Our paper relates to two strands of literature. One deals with the question of how to estimate expected returns in an optimal way. The seminal work here is Stein [9] who shows that the arithmetic average is not the optimal estimator for expected returns given that they are multidimensionally normal distributed. He finds that the estimation error can be substantially reduced by

[1] For methods of estimating variances and covariances see, e.g. Chan et al. [1] and Ledoit [5].

smoothing the estimate. The second strand of literature deals with the question of how to incorporate estimation risk into the Markowitz framework. Jorion [4] contributes to this literature by providing a Bayesian approach to portfolio selection. Under the restrictive assumption that estimation risk enlarges the return risk by a constant factor, he proves the resulting expected return estimate to belong to Stein's class of superior estimators. Within a similar setting Frost/Savarino [3] perform simultaneous mean and covariance estimation under the restrictive Jorion [4] assumption regarding the estimation risk.

In this paper we present a model (Section 1) which is close to the Jorion approach but avoids the restrictive assumption regarding estimation risk. We provide analytical solutions for optimal portfolio weights and show under which conditions heuristic trading strategies (investing in the Global Minimum Variance Portfolio GMVP or in the Equally Weighted Portfolio EWP) are optimal. In Section 2 we test the model using G7-countries index return data. Section 3 concludes.

1 The Model

We consider an economy with N stocks. The excess return of stock i at time t is denoted by $r_{i,t}$, the N-vector of excess returns by r_t. Excess returns can be decomposed into a mean μ and innovations ε_t, i.e. $r_t = \mu + \varepsilon_t$. Innovations are assumed to be serially independent and normally distributed $\varepsilon_t \sim N(0, \Sigma)$. Σ denotes the $N \times N$-variance-covariance matrix of the innovations. We assume that Σ is common knowledge to all investors.

To allow for estimation risk we assume that investors do not know expected stock returns μ, but know only that the mean vector is time-invariant and normally distributed with $\mu \sim N(\omega, \Omega)$.

1.1 General Model of Estimation Risk

We assume that there are T historical excess return observation vectors $r_{t-T}, \ldots, r_{t-1}$. Without any loss of information, time series data can be summarized into the historical average $\bar{r}_t(T) := \frac{1}{T}\sum_{s=1}^{T} r_{t-s}$.[2] Using standard regression theory one can show the conditional return distribution - given time series information ($\bar{r}_t(T)$) and prior information ($\mu \sim N(\omega, \Omega)$) - to be normal, too:

$$r_t|\bar{r}_t(T) \sim N(\mu_t(T), \Sigma(T)), \tag{1}$$

$$\mu_t(T) = \omega + \Omega\left(\Omega + \frac{1}{T}\Sigma\right)^{-1}(\bar{r}_t(T) - \omega), \tag{2}$$

$$\Sigma(T) = \Sigma + \Omega - \Omega\left(\Omega + \frac{1}{T}\Sigma\right)^{-1}\Omega. \tag{3}$$

[2] See Press [8], pp. 161–163.

Equation (3) shows the impact of estimation risk on total portfolio risk. Portfolio risk in the presence of estimation risk always exceeds the portfolio risk without estimation risk ($\Sigma(T) > \Sigma$). The additional estimation risk can shown to be the lower, the longer the history of time-series data T (see the appendix for a proof). Furthermore, for long estimation horizons ($T \to \infty$) the conditional covariance matrix $\Sigma(T)$ converges to the covariance matrix Σ.
Based on the Separation Theorem of Tobin [10] we calculate the weights of the optimal stock portfolio using equations (2) and (3) as

$$w_t(T) = \frac{\Sigma(T)^{-1}\mu_t(T)}{\underline{1}'\Sigma(T)^{-1}\mu_t(T)}, \tag{4}$$

where $\underline{1}$ denotes the N-vector of ones. To implement the model one has to specify the distribution parameters (ω, Ω) of the prior. We will discuss the Jorion specification in Section 1.2 and our more general specification in Section 1.3.

1.2 Specification of Jorion (1986)

The specification of Jorion is characterized by two assumptions. First, the prior mean is identical across all N stocks:

$$\omega = \phi\underline{1}. \tag{5}$$

Thus, Jorion assumes an investor without stock specific information. For such an investor it seems reasonable to treat all stocks as being equal with respect to their expected returns.
The second and more critical assumption of Jorion is that the estimation risk is proportional (with a factor $\frac{1}{\lambda}$) to the innovation risk:

$$\Omega = \frac{1}{\lambda}\Sigma. \tag{6}$$

At first view this assumption looks harmless, but it implies that innovations and estimation errors are one *single* source of risk. Inserting equations (6) and (5) into equations (2) and (3) leads to the following conditional return distribution parameters of Jorion:

$$\mu_t^J(T) = \frac{\lambda}{\lambda+T}\phi\underline{1} + \frac{T}{\lambda+T}\bar{r}_t(T), \tag{7}$$

$$\Sigma^J(T) = \Sigma\left(1 + \frac{1}{T+\lambda}\right). \tag{8}$$

The estimator (7) has the typical properties of a shrinkage estimator as introduced by Stein [9]: The estimate $\mu_t^J(T)$ is a weighted average of the time series estimator $\bar{r}_t(T)$ and a prior $\phi\underline{1}$, where the scalar λ is a measure of precision of prior information. The greater λ, the more precise the prior information is and consequently the larger the faith in prior information.

1.3 Generalized Specification

Like Jorion we assume that the prior mean is identical across all N stocks ($\omega = \phi \underline{1}$). However, our model differs from his approach by the way of modeling estimation risk. Instead of assuming estimation risk to be a linear function of innovation risk we model estimation risk as a *second* source of risk. We assume that the estimation risk is determined by the inhomogeneity of the market (different management skills, economic prospects, ...). Thus, we model the covariance matrix of expected returns as

$$\Omega = \tau^2 I, \tag{9}$$

where τ characterizes the inhomogeneity of the market and I denotes the identity matrix. The larger τ the more inhomogeneous the market and the larger the variety of expected returns are. Combining equations (2), (3), (5) and (9) yields the parameters of the conditional return distribution:

$$\mu_t^G(T) = \left[\Sigma + T\tau^2 I\right]^{-1} \Sigma \phi \underline{1} + \left[\Sigma + T\tau^2 I\right]^{-1} T\tau^2 \bar{r}_t(T), \tag{10}$$

$$\Sigma^G(T) = \Sigma \left[I + \tau^2 \left(T\tau^2 I + \Sigma\right)^{-1}\right]. \tag{11}$$

The structure of the formulas is similar to those of Jorion. The estimation risk leads to an increase of the total risk which the investor faces. The conditional expected return is a weighted sum of the prior and the historical average. However, since Ω is no longer a multiple of Σ, the weighting factors are no longer scalars but matrices.
Stock markets can be classified by two characteristics in our model: The length of their return history (T) and their inhomogeneity (τ). Based on these market characteristics we analyze four scenarios and derive the optimal trading strategies. The results are summarized in Table 1. In Scenarios

Table 1. Market Classification by T and τ

Case	Time series	Inhomogeneity	Optimal Strategy
I	$T = 0$	$\tau = 0$	GMVP
II	$T = 0$	$\tau \to \infty$	EWP
III	$T \to \infty$	$\tau = 0$	GMVP
IV	$T \to \infty$	$\tau \to \infty$	Two Step Approach

I and III the market is perfectly homogeneous and, consequently, there is no estimation risk. Therefore, the investor relies entirely on the prior ϕ which is identical across all stocks ($\mu_i = \phi$, $i = 1, \ldots, N$). Since all stocks have identical expected returns in this setting, the expected returns of all portfolios are identical and the investor will chose the portfolio weights to minimize the risk of the portfolio. He will hold the Global Minimum Variance Portfolio (GMVP). In Scenario IV stocks differ extremely in expected returns ($\tau \rightarrow \infty$), but a long stock return history is available. Due to the extreme noise in prior information and the high precision of the historical average, the classical two-step implementation of the Markowitz model proves to be optimal. In Scenario II neither reliable prior information nor historical stock returns are available.[3] In this setting estimation risk is huge. The estimation risk is optimally diversified by choosing the Equally Weighted Portfolio (EWP).

Thus, the existence of estimation risk can serve as a foundation for trading strategies which are heavily used in practice. However, these trading strategies will be optimal only in extreme market situations. In general, a portfolio composition based on the parameters (10) and (11) is optimal.

2 Empirical Study

2.1 Data and Implementation

Our tests are based on monthly data of MSCI total return indices of the G7 countries (Canada, France, Germany, Italy, Japan, UK, USA). All national MSCI indices are calculated on US-Dollar basis and we use 3-month US T-bill-rate when calculating excess returns. All data are obtained from Thomson Financial Datastream.

We analyze the empirical performance of the five strategies examined above. We consider (i) a portfolio based on the classical two-step portfolio optimization, (ii) the Global Minimum Variance Portfolio (GMVP), (iii) the Equally Weighted Portfolio (EWP), (iv) an optimal portfolio using the assumptions of Jorion[4] and (v) an optimal portfolio using our model.

Every month the parameters are estimated using monthly return observations of the past T years. Based on those estimates we determine the optimal portfolio weights and hold this portfolio for one month. Then we reestimate the parameters and adjust the portfolio weights. This portfolio strategy is implemented over the 15-year period from January 1986 to December 2000. For each strategy we obtain a 15-year return time series on a monthly basis. To judge the performance of the strategies we compute their empirical Sharpe-Ratios. Robustness of our results is tested by examining five year subperiods and by using different estimation periods T.

2.2 Results

The Sharpe-Ratios of the different strategies are reported in Table 2 for the total period and for three 5-year subperiods. We find four main results: First,

[3] For technical reasons we assume $T\tau^2 \rightarrow 0$ in Scenario II.

[4] The shrinkage estimator is implemented following Press [8], pp. 170–171.

Table 2. Performance of Different Strategies ($T = 10$)

Investment Period	Two Step	GMVP	EWP	Jorion	Gen. Prior
1/1986-12/2000	0.465	0.466	0.438	0.416	0.530
Subperiods					
1/1986-12/1990	0.212	0.343	0.449	0.329	0.331
1/1991-12/1995	0.491	0.659	0.325	0.610	0.740
1/1996-12/2000	0.728	0.507	0.534	0.421	0.612

our model leads to substantially superior performance as compared to the four benchmark models. This holds for the whole research period. In the subperiods, our model is always among the best strategies. Second, our generalized specification of prior beliefs outperforms the restrictive Jorion specification. This indicates that estimation risk should be treated as an additional separate source of risk. Third, taking estimation risk into account generally leads to superior results as compared to global minimum variance or equally weighted investments. This suggests that stocks are inhomogeneous, but inhomogeneity is limited. Fourth, the classical two-step portfolio optimization provides poor results in most cases. Thus, allowing for prior beliefs leads to improved investment strategies.

To check robustness in our results, we vary the testing period from 8 to 12 years. In Table 3 the performance of the five strategies during the total testing period from January 1986 to December 2000 is reported for different estimation horizons. We find that our generalized prior specification outperforms

Table 3. Performance of Different Strategies for Varying Estimation Period (Investment Period: 1/1986-12/2000)

Estimation Period	Two Step	GMVP	EWP	Jorion	Gen. Prior
$T = 8$	0.482	0.425	0.438	0.501	0.545
$T = 9$	0.400	0.436	0.438	0.496	0.488
$T = 10$	0.465	0.466	0.438	0.416	0.530
$T = 11$	0.441	0.500	0.438	0.465	0.591
$T = 12$	0.410	0.485	0.438	0.354	0.500

both the restrictive Jorion specification and the three heuristic strategies (GMVP, EWP and classical two-step optimization) for all estimation periods T. Thus, we conclude that our results are insensitive to the chosen estimation period.

3 Conclusion

This paper presents a portfolio model which incorporates estimation risk with respect to expected returns. We assume that estimation risk is an additional, independent source of risk and depends on the inhomogeneity of a market. This assumption differentiates our approach from Jorion [4] who assumes that estimation risk is a multiple of the innovation risk.
In the empirical part we compare the performance of our model with the performance of four benchmark strategies. We find that our model generally outperforms the benchmark strategies for the estimation and testing periods analyzed. From this result we conclude that (i) estimation risk is an important factor one has to recognize in the portfolio selection process and that (ii) the estimation risk seems to depend on the inhomogeneity of a market.

Appendix

Proof of the properties of $\Sigma(T)$

Let $\Delta(T) \equiv \Sigma(T) - \Sigma(T+1)$. After a few arrangements and using equation (3) $\Delta(T)$ can shown to be

$$\Delta(T) = \Omega \left(\left(\Omega + \frac{1}{T+1}\Sigma \right)^{-1} - \left(\Omega + \frac{1}{T}\Sigma \right)^{-1} \right) \Omega, \tag{12}$$

where $(\Omega + \frac{1}{T+1}\Sigma)$ and $(\Omega + \frac{1}{T}\Sigma)$ are both positive definite. Using matrix algebra (see e.g. Press [8], pp. 35-36), we can prove $(\Omega + \frac{1}{T+1}\Sigma)^{-1} - (\Omega + \frac{1}{T}\Sigma)^{-1}$ to be positive definite, too. Because Ω is nonsingular, $\Delta(T)$ is positive definite. Using $\lim_{T\to\infty} \Sigma(T) \to \Sigma$, we find $\Sigma(T) - \Sigma$ also to be positive definite.

References

1. Chan, L., Karceski, J., and Lakonishok, J. (1999) On Portfolio Optimization: Forecasting Covariances and Choosing the Risk Model, The Review of Financial Studies **12**, 937–974.
2. Chopra, V.K., and Ziemba, W.T. (1993) The Effect of Errors in Means, Variances, and Covariances on Optimal Portfolio Choice, Journal of Portfolio Management **19**, 6–11.
3. Frost, P., and Savarino, J. (1986) An Empirical Bayes Approach to Efficient Portfolio Selection, Journal of Financial and Quantitative Analysis **21**, 293–305.
4. Jorion, P. (1986) Bayes-Stein Estimation for Portfolio Analysis, Journal of Financial and Quantitative Analysis **21**, 279–292.
5. Ledoit, O. (1999) Improved Estimation of the Covariance Matrix of Stock Returns with an Application to Portfolio Selection, Working Paper, University of California Los Angeles.

6. Markowitz, H.M. (1952) Portfolio Selection, The Journal of Finance **7**, 77–91.
7. Merton, R.C. (1980) On Estimating the Expected Return on the Market: An Exploratory Investigation, Journal of Financial Economics **8**, 323–361.
8. Press, J.S. (1972) Applied Multivariate Analysis, New York.
9. Stein, C. (1955) Inadmissibility of the Usual Estimator for the Mean of a Multivariate Normal Distribution, Proceedings of the 3rd Berkely Symposium on Probability and Statistics **1**, Berkeley.
10. Tobin, J. (1958) Liquidity Preference as Behaviour Towards Risk, The Review of Economic Studies **25**, 65–86.

Risk-/Return-orientierte Optimierung des Gesamtbank-Portfolios unter Verwendung des Conditional Value at Risk

Ursula-A. Theiler

Risk Training, Carl-Zeiss-Str. 11, D-83052 Bruckmühl

Abstract. In einem sich verschärfenden Wettbewerb sind die Banken gezwungen, Rentabilitäts- und Risikomanagement eng miteinander zu verzahnen und die Risiko-/Ertrags-Struktur des Gesamtbank-Portfolios zu optimieren. In diesem Beitrag wird ein Überblick über ein Optimierungsmodell gegeben, das den erwarteten Gesamtbank-Return maximiert und dabei die Verlustrisiken gleichzeitig aus interner und aus aufsichtrechtlicher Sicht begrenzt und dadurch die zur Verfügung stehenden Kapitalressourcen des ökonomischen und des regulatorischen Kapitals bestmöglich nutzt. Die Optimierung basiert auf einem neuen Ansatz der internen Risikomessung mittels des Conditional Value at Risk, der eine integrierte Betrachtung von Marktpreis- und Kreditrisiken erlaubt, ein konvexes Risikomaß darstellt und die Lösbarkeit der betrachteten Portfolio-Optimierungsaufgabe sichert.

1 Problemstellung

Das Umfeld der Banken ist geprägt durch einen sich verschärfenden Wettbewerb, der zu sinkenden Ergebnismargen im Bankgeschäft führt. Gleichzeitig sehen sich die Banken steigenden Risiken ausgesetzt, die aus einer zunehmenden Komplexität der gehandelten Finanzinstrumente sowie steigenden Volatilitäten und zunehmenden Verflechtungen der Finanzmärkte resultieren. So wird es für die Banken überlebensnotwendig, Rentabilitäts- und Risikomanagement enger miteinander zu verzahnen und das Gesamtbank-Portfolio Risk-/Return-orientiert zu steuern.

Ziel dieses Beitrags ist es, einen Überblick über einen Optimierungsansatz für das Gesamtbank-Portfolio zu geben, der die Risk-/Return-Struktur des Gesamtbank-Portfolios optimiert, indem der erwartete Return des Gesamtbank-Portfolios maximiert wird und dabei die Verlustrisiken gleichzeitig aus interner und aus aufsichtrechtlicher Sicht begrenzt werden. Das Optimierungsmodell hat die folgende Grundstruktur:

(P) Maximierung des erwarteten Gesamtbank-Return		(1)
unter den **Nebenbedingungen**		
NB 1	Nach internen Verfahren gemessenes Verlustrisiko ≤ Risikotragfähigkeitsgrenze (Ökonomisches Kapital)	
NB 2	Kapitalstrukturbegrenzungen für das aufsichtsrechtliche Kapital, Nach aufsichtsichtrechtlichen Regeln gemessenes Verlustrisiko ≤ Aufsichtsrechtliches Eigenkapital (Grundsatz I-Regeln)	
NB 3	Volumenbegrenzungen	

2 Grundlegendes Risk-/Return-Optimierungsmodell

Eine Hauptschwierigkeit liegt in der Formulierung der Nebenbedingung NB 1 der Optimierungsaufgabe (P). Gegenstand dieses Kapitels ist daher die Darstellung eines grundlegenden Risk-/Return-Optimierungsmodells, das den erwarteten Gesamtbank-Return maximiert und dabei eine Obergrenze für das interne Verlustrisiko einhält (NB 1) und nur zulässige Portfolios betrachtet (NB 3). Im Kapitel 3 wird skizziert, wie dieses Grundmodell zu einem Optimierungsmodell für das Gesamtbank-Portfolio erweitert werden kann, indem zusätzlich die aufsichtsrechtlichen Kapitalstruktur- und Risikobegrenzungsregeln (NB 2) berücksichtigt werden.

2.1 Festlegung eines Risikomaßes für das Gesamtbank-Portfolio

Zur Risikomessung des Gesamtbank-Portfolios wird ein einseitiges oder Downside-Risikomaß verwendet, welches das Verlustrisiko entsprechend dem betriebswirtschaftlichen Risikobegriff als die *Gefahr der negativen Abweichung des realisierten von einem geplanten Ergebnis* misst (vgl. z. B. [2], S. 865). Eine besondere Klasse der Downside-Risikomaße bilden die quantilsabhängigen Risikomaße; hierzu zählen der Value at Risk, kurz VaR, und der Conditional Value at Risk, kurz CVaR. Während der VaR den Verlust misst, der innerhalb einer bestimmten Haltedauer mit einer vorgegebenen Wahrscheinlichkeit nicht überschritten wird, gibt der CVaR an, welcher Verlust bei Eintritt dieses Extremfalls, d. h. bei Überschreitung des VaR, zu erwaren ist.

Der CVaR besitzt wünschenswerte Eigenschaften für die Portfoliorisiko-Steuerung. Er stellt ein kohärentes Risikomaß dar. Unter dem Begriff der Kohärenz werden vier Eigenschaften für Risikomaße axiomatisch gefordert, die eine konsistente Risikomessung im Portfolioverbund garantieren [1]. Ein kohärentes Risikomaß erfüllt die Eigenschaften der positiven Homogenität, Subadditivität, Translationsinvarianz und Monotonie und erlaubt eine für die Portfolio-Steuerung entscheidende Risiko-Aggregation verschiedener Teilportfolios über mehrere Hierarchiestufen.

Aus der positiven Homogenität und der Subadditivität des CVaR folgt weiterhin, dass der CVaR ein konvexes Risikomaß bildet. Dies stellt sicher, dass eine Risk-/Return-Optimierung des Portfolios stets lösbar ist, sofern eine konkave Return-Funktion verwendet wird und der zulässige Bereich konvex und nicht leer ist. Weiterhin lässt sich zeigen, dass aus entscheidungstheoretischer Sicht eine Risk-/Return-Optimierung unter Verwendung des CVaR im Einklang mit einer Erwartungsnutzen-Maximierung risikoaverser Investoren, d. h. Anteilseigner der Bank, steht. Insofern liefert diese Optimierungsaufgabe aus betriebswirtschaftlicher Sicht fundierte und aussagekräftige Ergebnisse.

In der Praxis hat sich der Value at Risk weithin als Risikomaß zur Messung der Marktpreis-Risiken etabliert. Dieses Risikomaß liefert konsistente Informationen der Portfoliorisiko-Messung, wenn normalverteilte Portfolio-Returns vorliegen. Falls diese Normalverteilungssnnahme nicht erfüllt ist, geht jedoch die Eigenschaft der Subadditivität des VaR und damit dessen Konvexität und Kohärenz verloren. Daher ist der VaR nicht zur Risikosteuerung des Gesamtbank-Portfolios geeignet, das zum überwiegenden Teil aus Kreditpositionen besteht und dessen Returns keinesfalls als normalverteilt angenommen werden können.

Da der CVaR grundlegende Eignungsmerkmale der Risikosteuerung des Gesamtbank-Portfolios besitzt, wird dieses Risikomaß im Optimierungsmodell zur Abbildung der Nebenbedingung NB 1 der Verlustrisiko-Begrenzung nach internen Regeln verwendet.

Nachfolgend werden wesentliche **Modellannahmen** und **Definitionen** für das Optimierungsmodell erläutert. Ein Portfolio wird als Vektor $\mathbf{x}$ der Nominalvolumina der Einzelpositionen x_i, $i=1,...,n$, dargestellt, $\mathbf{x}=(x_1,...,x_n)'$. Dabei bedeuten im Folgenden fett gedruckte Buchstaben Vektoren. Zufallsabhängige Größen sind die Marktpreise der einzelnen Positionen zum Prognosehorizont. Jeder Position x_i ist der Marktpreis y_i, $i=1,...,n$, zugeordnet. Die Marktpreise werden im Vektor $\mathbf{y}=(y_1,...,y_n)'$ zusammengefasst. Der Marktwert des Portfolios $MW_{Portfolio}$ berechnet sich als

$$MW_{Portfolio} = \mathbf{y}'\mathbf{x} = \sum_{i=1}^{n} y_i x_i \,. \tag{2}$$

Die Existenz einer Dichtefunktion für den Marktpreisvektor ist dabei keine notwendige Bedingung für die rechnerische Umsetzung des Optimierungsmodells. Vielmehr wird dazu eine gemeinsame Stichprobe der Marktpreis-Realisationen $\mathbf{y}_1,...,\mathbf{y}_J$ benötigt (vgl. Gleichung (6) in Kapitel 2.2).

Die **Verlustfunktion** $f(\mathbf{x},\mathbf{y})$ misst den *Portfolioverlust als negative Abweichung des realisierten von dem erwarteten Portfoliowert* und wird definiert als

$$f(\mathbf{x},\mathbf{y}) = E[\mathbf{y}]'\mathbf{x} - \mathbf{y}'\mathbf{x} = (E[\mathbf{y}] - \mathbf{y})'\mathbf{x} = \sum_{i=1}^{n} (E[y_i] - y_i)x_i \,. \tag{3}$$

Dabei bezeichnet E[y] den Erwartungsvektor der Marktpreise zum Prognosehorizont. Im Folgenden wird von einem Konfidenzniveau der Höhe α, z. B. $\alpha = 99\%$, ausgegangen. Der **VaR** und **CVaR** der Verlustfunktion f(**x**,**y**) zum Konfidenzniveau α, die mit $\mathrm{VaR}_\alpha(\mathbf{x})$ und $\mathrm{CVaR}_\alpha(\mathbf{x})$ bezeichnet werden, lauten dann [5], [7]:

$$\mathrm{VaR}_\alpha(\mathbf{x}) = \inf\{z \in \Re \mid P((f(\mathbf{x},\mathbf{y})) \le z) \ge \alpha\}, \quad (4)$$
$$\mathrm{CVaR}_\alpha(\mathbf{x}) = E[f(\mathbf{x},\mathbf{y}) \mid f(\mathbf{x},\mathbf{y}) \ge \mathrm{VaR}_\alpha(\mathbf{x})].$$

Die folgende Abbildung zeigt die Verlustfunktion f(**x**,**y**) und die zugehörigen Risikomaße VaR und CVaR zum Konfidenzniveau α:

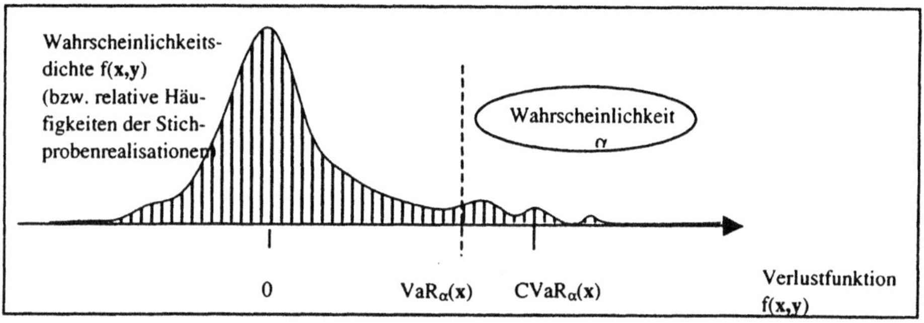

Abbildung 1: Wahrscheinlichkeitsdichte der Verlustfunktion f(**x**,**y**) und zugehörige Risikomaße VaR und CVaR

2.2 Formulierung des Risk-/Return-Optimierungsmodells

Im Folgenden wird das **grundlegende Risk-/Return-Optimierungsmodell** formuliert [7]. Es lässt sich auf Portfolios mit beliebigen Assets, beispielsweise Aktien-, Options- oder Kreditpositionen, anwenden, d. h. es werden keine Bedingungen an die zu Grunde liegenden Verteilungen der Assel-Returns gestellt. Es basiert auf dem Risikomaß des CVaR. Allgemein lautet die Optimierungsaufgabe:

(P1) Maximiere den erwarteten Portfolio-Return (5)
Unter den Nebenbedingungen:
NB 1 CVaR des Portfolios ≤ Verlustrisiko-Obergrenze,
NB 3 Betrachtung nur zulässiger Portfolios.

Die CVaR-Nebenbedingung wird sukzessive modifiziert, um das Ausgangsproblem (P1) in ein rechnerisch einfacher zu lösendes Optimierungsproblem zu transformieren: Das Problem (P1) wird in ein äquivalentes Problem überführt, das eine Hilfsfunktion für den CVaR verwendet. Anschließend wird diese Hilfsfunktion für den CVaR durch eine Schätzfunktion approximiert und die CVaR-Nebenbedingung durch ein System linearer Nebenbedingungen ersetzt. Es resultiert das folgende Optimierungsproblem (P2), wobei $\mu(\mathbf{x})$ eine lineare Return-Funktion darstellt.

(P2) Maximiere $\mu(\mathbf{x})$ (6)

Unter den Nebenbedingungen

(i)	$q + \frac{1}{(1-\alpha)} \cdot \frac{1}{J} \sum_{j=1}^{J} z_j$	$\leq$	$Riskap\_max$	
(ii)	$f(\mathbf{x}, \mathbf{y}_j) - q$	$\leq$	z_j,	$j = 1, \dots, J$,
(iii)	$-z_j$	$\leq$	0,	$j = 1, \dots, J$,
(iv)	$\mathbf{x}$	$\in$	X,	
(v)	q	$\in$	$\Re$.	

Das Optimierungsproblem (P2) stellt ein lineares Optimierungsproblem dar, das mittels Simplex-Algorithmus relativ einfach zu lösen ist. Die Nebenbedingungen (i) bis (iii) bilden die CVaR-Nebenbedingung ab, wobei die freie Variable q im Optimum eine Näherung für den Portfolio-VaR liefert. Die Nebenbedingung (iv) legt den zulässigen Bereich fest und entspricht der Bedingung NB 3 in der Gleichung (1) (vgl. [7] sowie [6], S. 117 ff.).

3 Risk-/Return-Optimierungsmodell für das Gesamtbank-Portfolio

Im Folgenden werden wesentliche Aspekte der Erweiterung des grundlegenden Optimierungsansatzes zu einem Optimierungsmodell für das Gesamtbank-Portfolio aufgezeigt ([5], S. 147 ff.). Es werden wesentliche Modellannahmen dargestellt, welche die besonderen Risiko-/Ertrags-Strukturen des Gesamtbank-Portfolios abbilden, und die weiteren Nebenbedingungen zur Begrenzung des Verlustrisikos aus aufsichtsrechtlicher Sicht (NB 2) betrachtet.

Das **Gesamtbank-Portfolio** wird als Vektor der Nominalvolumina der Einzelgeschäfte abgebildet, die nach verschiedenen Merkmalen aggregiert werden, die zur Ergebnis- und Risikobewertung notwendig sind, beispielweise nach Restlaufzeiten, Währungsarten, Ratingklassen und nach der Profit Center-Zugehörigkeit. Dabei ist letztere für die Aggregation der Risk-/Return-Kennzahlen auf Profit Center-Ebene erforderlich. Als **Entscheidungsvariable des Optimierungsmodells** für das Gesamtbank-Portfolio wird das Neugeschäft des betrachteten Prognosezeitraumes modelliert, das sowohl positive als auch negative Werte annehmen kann und dadurch Geschäftszuwachs und -abschmelzung berücksichtigt.

Es werden die **Wertveränderungen jeder Einzelgeschäftsart** auf Grund von Marktpreis- und Bonitätsveränderungen isoliert erfasst, indem jede Einzelgeschäftsart doppelt dargestellt wird, nämlich jeweils einmal pro betrachteter Risikoart. Der Portfolio-Vektor der Gesamtbank verdoppelt sich dadurch gegenüber seiner ursprünglichen Länge. Damit erhalten der Entscheidungsvektor $\mathbf{x}_{Ges}$ und der Marktpreisvektor $\mathbf{y}_{Ges}$ des Gesamtbank-Portfolios die folgende Gestalt:

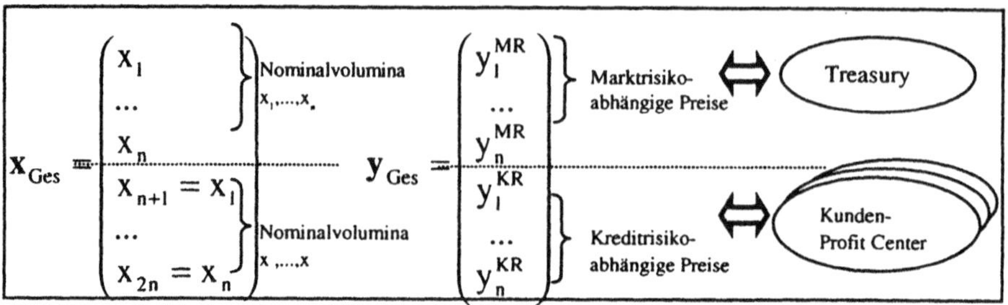

Abbildung 2: Marktpreis- und Kreditrisiko-Komponenten des Gesamtbank-Vektors

Damit werden in der Verlustfunktion $f(\mathbf{x}_{Ges},\mathbf{y}_{Ges})$ in den ersten n Summanden die Verluste auf Grund von Marktpreisveränderungen, in den zweiten n Summanden die Verluste auf Grund von Bonitätsveränderungen berechnet.

Entsprechend ist die **Return-Funktion** strukturiert. Gemäß dem Prinzip der Marktzinsmethode werden in den ersten n Komponenten die erwarteten Strukturbeiträge, die der Treasury-Einheit zugerechnet werden, in den zweiten n Komponenten die erwarteten Konditionsbeiträge erfasst, die den Profit Centern zugeordnet werden ([4], S. 72 ff., [5], S. 153 ff.]. Neben den erwarteten Margenbeiträgen sind weiterhin die erwarteten Wertveränderungen des Neugeschäfts zu berücksichtigen, so dass sich die folgende Zielfunktion ergibt:

„Maximiere die Summe der erwarteten Marktwertänderungen und der erwarteten Ergebnisbeiträge des Neugeschäfts." (7)

Insgesamt ergibt sich die folgende Erfassung von Risiken und Erträgen im Optimierungsmodell:

Profit Center-Zurechnung	**Risiko (CVaR)**	**Return (Zielfunktion)**	
		Wertveränderungen +	Margenbeitrag
Kunden-Profit Center	Kreditrisiko	Wertveränderungen aufgrund von Bonitätsänderungen	Konditionsbeitrag
Treasury	Marktrisiko	Wertveränderungen aufgrund von Marktpreisänderungen	Strukturbeitrag

Tabelle 1: Erfassung von Risken und Erträgen im Optimierungsmodell

Neben der internen Risikonebenbedingung für den CVaR werden zusätzlich die **aufsichtsrechtlichen Verlustrisiko-Begrenzungsregeln** abgebildet. Unter der Nebenbedingung NB 2 werden die aufsichtsrechtlichen Eigenkapitalstruktur-Regeln, sowie die beiden Risikobegrenzungsregeln des Grundsatzes I für das Ausfallrisiko des Anlagebuchs und für das Marktrisiko des Handelsbuchs zusammengefasst [3]. Die **Eigenkapitalunterlegung der Risikoaktiva** in der ersten Bedingung des Grundsatz I wird durch eine lineare Nebenbedingung dargestellt. Wird zur Begrenzung der **Marktrisiken des Handelsbuchs** ein internes Modell verwendet ([3], §§ 32 ff.), so kommt weitere CVaR-Nebenbedingung als obere Abschätzung des Handelsbuch-VaR hinzu. Beim Einsatz der Standardmethode ergibt sich eine weitere lineare Nebenbedingung für die Handelsbuch-Positionen.

4 Anwendungsbeispiel

Die Wirkungsweise der Risk-/Return-Optimierung des Gesamtbank-Portfolios lässt sich anhand eines Anwendungsbeispiels verdeutlichen (vgl. [5], S. 207 ff.). Es soll das Planportfolio einer Beispiel-Bank festlegt werden. Das aufsichtsrechtliche Eigenkapital ist derzeit nahezu ausgelastet, und im folgenden Geschäftjahr ist keine Kapitalerhöhung möglich. Um zusätzliches Geschäftspotential zu schöpfen, erwägt der Vorstand eine Erhöhung des derzeit eingesetzten ökonomischen Kapitals um 17 Einheiten, kurz mit E bezeichnet. Weiterhin soll im Planjahr ein Mindest-RORAC von 14,00% erreicht werden, der definiert ist als das erwartete Gesamtbank-Ergebnis, dividiert durch den CVaR des Gesamtbank-Portfolios. Für die Unternehmensleitung stellen sich die Fragen: Soll das aktuelle Risikoniveau beibehalten werden (Szenario 1) oder erhöht werden (Szenario 2)? Welche Risk-/Return-Plankennzahlen ergeben sich daraus für das Gesamtbank-Planportfolio?

Zur Lösung des Problems werden verschiedene Optimierungen unter alternativen internen Risikoniveaus, d. h. Obergrenzen für das ökonomische Kapital, durchgeführt. Das derzeitige Ausgangsportfolio wird mit PF 0 bezeichnet.

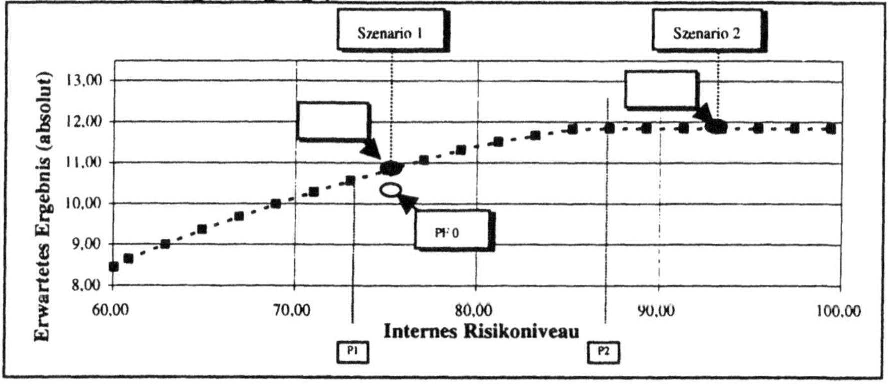

Abbildung 3: Effizienzlinie, Risk-/Return-optimale Portfolios der Beispiel-Bank

Die verschiedenen Optimierungen zur Berechnung der Effizienzlinie der Risk-/ Return-optimalen Planportfolios zeigen, dass links des Risikoniveaus P1 nur die interne Risikonebenbedingung, rechts des Risikoniveaus P2 nur die aufsichtsrechtliche Restriktion aktiv ist. Im Intervall zwischen P1 und P2 sind beide, die interne und die aufsichtsrechtlichen Nebenbedingungen aktiv. Das bedeutet, dass nur in diesem Intervall beide Kapitalressourcen maximal ausgenutzt werden.

Ein Vergleich der optimalen Portfolios PF 1 und PF 2 der betrachteten Szenarien zeigt, dass eine Erhöhung des ökonomischen Kapitals im Szenario 2 zwar zu einem zusätzlichen Ergebnis von 0,88 E bzw. 8,03% gegenüber Szenario 1 führt, dass jedoch der RORAC von 13,57% die geforderte Mindest-Risikorentabilität von 14,00% nicht erreicht und außerdem niedriger ist als derjenige des Portfolios PF 1 (14,39%) und des Ausgangsportfolios PF 0 (13,98%). Weiterhin wird das interne

Risikokapital im Szenario 2 nicht voll ausgelastet. Als Entscheidungsempfehlung für die Jahresplanung ergibt sich daher: Szenario 1 des derzeitigen Risikoniveaus wird befürwortet. Das Risk-/Return-optimale Planportfolio PF 1 führt gegenüber dem aktuellen Portfolio PF 0 bei gleichem Risikoniveau zu einer absoluten Ergebnissteigerung von 0,31 E, zu einem Plan-RORAC von 14,39% und lastet die Ressourcen des ökonomischen und des regulatorischen Kapitals maximal aus.

5 Fazit

Das dargestellte Optimierungsmodell erlaubt eine Risk-/Return-orientierte Portfolio-Optimierung des Gesamtbank-Portfolios. Es basiert auf dem CVaR, einem konvexen und kohärenten Risikomaß, das zur Risikomessung des Gesamtbank-Portfolios geeignet ist. Mit dem Optimierungsmodell lassen sich Portfolio-Optimierungen unter alternativen Rahmenbedingungen durchführen, wobei sowohl aufsichtsrechtliche als auch interne Risikobegrenzungsregeln gleichzeitig eingehalten werden und beide Kapitalressourcen bestmöglich genutzt werden. Das Optimierungsmodell liefert fundierte Entscheidungsempfehlungen für eine Risk-/Return-orientierte Gesamtbank-Steuerung und kann zur Sicherung der Wettbewerbsposition des Unternehmens beitragen.

6 Literaturverzeichnis

[1] Artzner, Ph., Delbaen, F., Eber, J.-M., Heath, D.(1999): Coherent Measures of Risk. In: Mathematical Finance,Vol. 9, No. 3 (July 1999), S. 203-228.

[2] Büschgen, H. (Bankbetriebslehre, 1998): Bankbetriebslehre: Bankgeschäfte und Bankmanagement. 5. Auflage Gabler Wiesbaden.

[3] Bundesaufsichtsamt für das Kreditwesen (1997): Bekanntmachung über die Änderung und Ergänzung der Grundsätze über das Eigenkapital und die Liquidität der Kreditinstitute. Berlin.

[4] Schierenbeck, H. (1997): Ertragsorientiertes Bankmanagement: Band 1: Grundlagen, Marktzinsmethode und Rentabilitäts-Controlling. 5. Auflage Gabler Wiesbaden.

[5] Tasche, D. (1999): Risk Contributions and Performance Measurement. Working Paper. Zentrum Mathematik, Technische Universität München Juni 1999.

[6] Theiler, U. (2001): Integriertes Risk-/Return-Steuerungsverfahren für das Gesamtbank-Portfolio. Gabler Wiesbaden, erscheint 2001.

[7] Uryasev, S., Palmquist, J. Krokhmal, P. (1999): Portfolio Optimization with Conditional Value at Risk Objective and Constraints. Research Report No. 99-14. Center for Applied Optimization, Department of Industrial and Systems Engineering, University of Florida, Gainesville, November 1999.

Section 6

Scheduling and Project Planning

Optimal Properties for Scheduling Deteriorating Jobs for the Total Completion Time Minimization

Aleksander Bachman[1], Adam Janiak[1], and Wieslaw Wajs[2]

[1] Institute of Engineering Cybernetics, Wrocław University of Technology, Janiszewskiego 11/17, 50-372 Wrocław, Poland
[2] Department of Automatics, University of Mining and Metallurgy, Mickiewicz Avenue 30, 30-059 Kraków, Poland

Abstract. The paper deals with a problem of scheduling deteriorating jobs on a single machine to minimize the total completion time. Deterioration of a job means that its processing time is a function of its execution start time. We present some optimal properties for three problems, which correspond to three different decreasing linear functions, whose increasing counterparts have been studied in the literature. Based on these properties, two of these problems are solved in $O(n \log n)$ time. We present also a comparison between the problems with job processing times being decreasing and increasing linear functions of their start times.

1 Introduction

Problems of scheduling jobs with start time dependent processing times (deteriorating jobs) attract the attention of many researchers, recently. The scientists introduce new models to describe different aspects of this phenomenon and solve different problems for various criteria (see [1] and [8]). However, in spite of many efforts, there are still some unsolved problems. The problem of minimizing the total completion time for the increasing linear model belongs, for a long time, to the unsolved one. In this model, each job processing time contains two parts. One part is a constant (*normal processing time*) and the other part is start time dependent. The dependence on the start time is described by an *increasing rate*. Mosheiov [6], who considered this problem, assumed that all the jobs are characterized by a common positive value of the normal processing time. On the basis of this assumption, he proved that the optimal solution is symmetric and has a V-shape with respect to the increasing rates. Unfortunately, the complexity status of the problem remained an open question. In another paper, Mosheiov [7] considered the linear model with the values of the normal processing times equal to zero. For this case, he found the optimal solution. The optimal solution was also found for the linear model with increasing rates proportional to the normal processing times [2]. Apart from the increasing linear model for the job processing times, there is also a decreasing linear model. This model was introduced by Ho, Leung and

Wei [5]. It also contains two parts, but in contrast to the increasing model, the start time dependent part is described by a *decreasing rate.*

Some interesting relations between the linear models with decreasing and increasing start time dependent parts have been presented by Cheng and Ding [3]. The results obtained by Cheng and Ding were an inspiration of this paper.

The remaining part of the paper is organized as follows. Section 2 contains the precise formulation of the problem. Section 3 is devoted to the properties of the optimal solution found for three special cases of the problem under consideration. Based on these properties, in Section 4, we present the comparison of the problems with increasing and decreasing linear models of job processing times.

2 Problem formulation

There are given a single machine and a set $J = \{1, \ldots, n\}$ of n independent and non-preemptive jobs immediately available for processing. The processing time p_i of job i is given as a decreasing linear function dependent on its execution start time S_i:

$$p_i(S_i) = a_i - b_i S_i, \tag{1}$$

where $a_i > 0$ denotes the *normal processing time* of job i and b_i denotes its *decreasing rate.* It is assumed that the decreasing rates satisfy the following conditions: $0 < b_i < 1$ and $b_i(\sum_{j=1}^{n} a_j - a_i) < a_i$. The first condition ensures that the decrease of each job processing time is less than one unit for every unit delay in its starting moment. The second one ensures that all job processing times are positive in any schedule (see [5] for some additional explanations).

Define for convenience that $B_i = 1 - b_i$. The objective is to find a schedule π for which the value of the total completion time (TC) is minimized:

$$TC = \sum_{i=1}^{n} C_{\pi(i)} = \sum_{i=1}^{n} \left[\sum_{j=1}^{i} a_{\pi(j)} \prod_{k=j+1}^{i} B_{\pi(k)} \right], \tag{2}$$

where $C_{\pi(i)}$ denotes the completion time of the job placed on the ith position in the schedule π and $C_{\pi(0)} \triangleq 0$. The problems considered in the paper are denoted according to the three-field notation $\alpha \mid \beta \mid \gamma$ introduced by Graham *et al.* [4].

3 Optimal properties for some special cases

This section presents some properties of the optimal solution found for the special cases of the problem under consideration.

Property 1. The schedule obtained by the non-decreasing order of the normal processing times a_i is optimal for the problem $1 \mid p_i(S_i) = a_i - bS_i \mid \sum C_i$.

Proof. Let two permutations π and π' be given. The permutation π' is obtained from the permutation π by interchanging the jobs from the ith and the i+1st positions of π. Assume that the permutation π, in which $a_{\pi(i)} > a_{\pi(i+1)}$, is optimal. The difference between the values of TC obtained for both permutations is equal to

$$\sum_{j=1}^{n} C_{\pi(j)} - \sum_{j=1}^{n} C_{\pi'(j)} = \left(a_{\pi(i)} - a_{\pi(i+1)}\right)(1-b)^{n-i}.$$

Since $a_{\pi(i)} > a_{\pi(i+1)}$, the result obtained above is positive. Thus, the non-decreasing order of the normal processing times is optimal for the considered problem. □

Property 2. Scheduling jobs according to non-decreasing order of their normal processing times a_i gives optimal solution for the problem $1 \mid p_i(S_i) = a_i - ka_iS_i \mid \sum C_i$, where $k(\sum_{j=1}^{n} a_j - a_{min}) < 1$ and $a_{min} = \min_{i=1,\ldots,n} a_i$.

Proof. Assume that two permutations π and π' are given. The permutation π' has been obtained from the permutation π by interchanging jobs from the ith and the $i+1$st positions. Assume also that the permutation π in which $a_{\pi(i)} > a_{\pi(i+1)}$ is optimal. Note that for the job processing times given above, the completion time of the job placed on the ith position in the permutation π is given by the following expression: $C_{\pi(i)} = -\frac{1}{k}\prod_{j=1}^{i}(1 - ka_{\pi(j)}) + \frac{1}{k}$. Therefore, for both permutations only the completion times of the jobs scheduled in the ith position are different. The difference between the values of the TC obtained for both permutations is equal to

$$\sum C_{\pi(j)} - \sum C_{\pi'(j)} = \left(a_{\pi(i)} - a_{\pi(i+1)}\right)\prod_{j=1}^{i-1}(1 - ka_{\pi(j)}).$$

Since $a_{\pi(i)} > a_{\pi(i+1)}$, then the result obtained above is positive. Thus, the schedule in which the jobs are arranged according to non-decreasing order of their normal processing times a_i is optimal. □

Property 3. For the problem $1 \mid p_i(S_i) = a - b_iS_i \mid \sum C_i$ any optimal schedule has a Λ-shape with respect to the values of the job decreasing rates. It means that the jobs scheduled before the job with the largest decreasing rate and the jobs scheduled after it, are arranged in non-decreasing and non-increasing order of their decreasing rates, respectively.

Proof. Assume that the following three permutations are given

$$\pi = (\pi(1), \pi(2), \ldots, \pi(i-2), \pi(\mathbf{i-1}), \pi(\mathbf{i}), \pi(\mathbf{i+1}), \pi(i+2), \ldots, \pi(n)),$$

$$\pi 1 = (\pi(1), \pi(2), \ldots, \pi(i-2), \pi(\mathbf{i}), \pi(\mathbf{i-1}), \pi(\mathbf{i+1}), \pi(i+2), \ldots, \pi(n)),$$

and

$$\pi 2 = (\pi(1), \pi(2), \ldots, \pi(i-2), \pi(\mathbf{i-1}), \pi(\mathbf{i+1}), \pi(\mathbf{i}), \pi(i+2), \ldots, \pi(n)).$$

Assume also that the permutation π is optimal and that the following conditions are satisfied: $\Delta_1 = b_{\pi(i-1)} - b_{\pi(i)} > 0$ and $\Delta_2 = b_{\pi(i+1)} - b_{\pi(i)} > 0$. Let TC, $TC(1)$ and $TC(2)$ denote the values of the total completion time obtained for the permutations π, $\pi 1$ and $\pi 2$, respectively. We will show now that $TC(1)$ or $TC(2)$ is smaller than TC, which means that the permutation π cannot be optimal. Let's compare the following two differences $TC - TC(1)$ and $TC - TC(2)$.

$$TC - TC(1) = a\Delta_1 \left[B_{\pi(i+1)} \left[1 + \sum_{j=i+2}^{n} \prod_{k=i+2}^{j} B_{\pi(k)} \right] - \sum_{j=2}^{i-2} \prod_{k=j}^{i-2} B_{\pi(k)} \right], \quad (3)$$

$$TC - TC(2) = a\Delta_2 \left[B_{\pi(i-1)} \left[1 + \sum_{j=2}^{i-2} \prod_{k=j}^{i-2} B_{\pi(k)} \right] - \sum_{j=i+2}^{n} \prod_{k=i+2}^{j} B_{\pi(k)} \right]. \quad (4)$$

Define $X = \sum_{j=2}^{i-2} \prod_{k=j}^{i-2} B_{\pi(k)}$ and $Y = \sum_{j=i+2}^{n} \prod_{k=i+2}^{j} B_{\pi(k)}$. The expressions (3) and (4) are equal to

$$TC - TC(1) = a\Delta_1 \left[B_{\pi(i+1)}(1+Y) - X \right], \quad (5)$$

$$TC - TC(2) = a\Delta_2 \left[B_{\pi(i-1)}(1+X) - Y \right]. \quad (6)$$

Since the permutation π is optimal, both expressions (5) and (6) should be negative. Therefore, the following inequality

$$1 - b_{\pi(i+1)}(1+Y) < (X - Y) < -1 + b_{\pi(i-1)}(1+X)$$

which is equivalent to

$$2 < b_{\pi(i+1)}(1+Y) + b_{\pi(i-1)}(1+X)$$

should be satisfied. Since the following estimations $X < i-2$ and $Y < n-i-1$ are true and from the model requirements follows that $b_i < (n-1)^{-1}$, the final result is given as follows

$$2 < (n-1)^{-1}(n-1) = 1.$$

It contradicts with the assumption about the optimality of the permutation π. Thus the schedule, which is Λ-shaped with respect to the values of the job decreasing rates, is optimal. □

Property 4. For the problem $1 \mid p_i(S_i) = a - b_i S_i \mid \sum C_i$, the values of TC obtained for the schedule $\pi = (\pi(1), \pi(2), \ldots, \pi(n))$ and the reversed schedule $\pi' = (\pi(1), \pi(n), \pi(n-1), \ldots, \pi(2))$ are equal.

Proof. The value of TC for the schedule π is equal to:

$$\sum_{i=1}^{n} C_{\pi(i)} = an + a\sum_{j=2}^{n} B_{\pi(j)} + a\sum_{j=2}^{n-1} B_{\pi(j)}B_{\pi(j+1)}$$
$$+ \ldots + a\sum_{j=2}^{2} B_{\pi(j)} \ldots B_{\pi(j+n-2)}.$$

The value of TC obtained for the schedule π' is given by:

$$\sum_{i=1}^{n} C_{\pi'(i)} = an + a\sum_{j=2}^{n} B_{\pi(n+2-j)} + a\sum_{j=2}^{n-1} B_{\pi(n+2-j)}B_{\pi(n+1-j)}$$
$$+ \ldots + a\sum_{j=2}^{2} B_{\pi(n+2-j)} \ldots B_{\pi(4-j)}.$$

Since both expressions obtained above are equal, the property is true. □

It follows from Property 3 and Property 4 that the job with the smallest value of the decreasing rate is scheduled either on the first or on the last position. Using the jobs interchanging strategy the following property can be proved.

Property 5. For the problem $1 \mid p_i(S_i) = a - b_i S_i \mid \sum C_i$, the job with the smallest value of the decreasing rate is scheduled in the first position.

4 Conclusions

In this paper, we investigated the problem of scheduling jobs with start time dependent processing times for the total completion time minimization. We established the properties for some special cases of the general problem. Collecting the results from the literature and this paper, we can notice that there is some relation between the problems with non-decreasing and non-increasing models of start time dependent job processing times. Namely, the non-decreasing order of the normal processing times is optimal for the problems $1 \mid p_i(S_i) = a_i - bS_i \mid \sum C_i$ (Property 1) and $1 \mid p_i(S_i) = a_i - ka_i S_i \mid \sum C_i$ (Property 2). The optimal solution for the corresponding problems with non-decreasing processing times, i.e., $1 \mid p_i(S_i) = a_i + bS_i \mid \sum C_i$ and $1 \mid p_i(S_i) = a_i + ka_i S_i \mid \sum C_i$, is also given by the non-decreasing order of the normal processing times [2]. The optimal solution for the problem $1 \mid p_i(S_i) = a - b_i S_i \mid \sum C_i$ is symmetric with respect to the job decreasing rates (Property 4) and is Λ-shaped (Property 3). The optimal solution for the problem $1 \mid p_i(S_i) = a + b_i S_i \mid \sum C_i$ is also symmetric (with respect to the increasing rates), but it is V-shaped [6]. For the problem $1 \mid p_i(S_i) = a - b_i S_i \mid \sum C_i$, the first position in the optimal schedule is occupied by the job with the smallest value of the decreasing rate (Property 5). In the optimal solution of the corresponding problem with non-decreasing model, the first position is reserved for the job with the largest value of the increasing rate [6]. The last similarity of these two problems deals with the status of their computational complexity which, in spite of many efforts, is still an open question.

References

1. Alidaee B., Womer N. K. (1999) Scheduling with time dependent processing times: Review and extensions. Journal of the Operational Research Society 50, 711–720
2. Bachman A., Janiak A. (1997) Scheduling jobs with special type of start time dependent processing times. Wroclaw University of Technology, Report PRE 34/97
3. Cheng T. C. E., Ding Q. (1998) The complexity of scheduling starting time dependent tasks with release dates. Information Processing Letters 65, 75–79
4. Graham R. L., Lawler E. L., Lenstra J. K., Rinnooy Kan A. H. G. (1979) Optimization and approximation in deterministic sequencing and scheduling: a survey. Annals of Discrete Mathematics 3, 287–326
5. Ho K. I-J., Leung J. Y-T., Wei W-D. (1993) Complexity of scheduling tasks with time-dependent execution times. Information Processing Letters 48, 315–320
6. Mosheiov G. (1991) V-shaped policies for scheduling deteriorating jobs. Operations Research 39/6, 979–991
7. Mosheiov G. (1994) Scheduling jobs under simple linear deterioration. Computers and Operations Research 21/6, 653–659
8. Wajs W., Grabowski P. (1996) Studies in Automatics, University of Mining and Metallurgy, Krakow

Bicriterion Approach to a Single Machine Time-Dependent Scheduling Problem

Stanisław Gawiejnowicz, Wiesław Kurc, and Lidia Pankowska

Adam Mickiewicz University, Faculty of Mathematics and Computer Science, ul. Matejki 48/49, 60-769 Poznań, POLAND
E-mail: {stgawiej,wkurc,lpankow}@amu.edu.pl

Abstract. In the paper a single machine time-dependent scheduling problem with simultaneous minimization of two criteria is considered. Processing time p_j of each task is described by a linear function of the starting time t of the task, $p_j = 1 + \alpha_j t$, where $\alpha_j > 0$ for $j = 0, 1, 2, \ldots, n$. The criterion of optimality of a schedule is a convex combination of norms $\|\cdot\|_1$ and $\|\cdot\|_\infty$ in the form of $\lambda\|\cdot\|_1 + (1-\lambda)\|\cdot\|_\infty$, where $\lambda \in [0, 1]$ is an arbitrary real number and $\|\cdot\|_p$ is Hölder's norm l_p, $1 \le p \le \infty$. The main result of the paper is theorem saying that there exist numbers $\lambda_1 > \lambda_0 > 0$ such that for all $\lambda \in [0, \lambda_0]$ the problem is solvable in $O(n \log n)$ time and for all $\lambda \in [\lambda_1, 1]$ the optimal schedule for the problem has V-shape.

Keywords: single machine; time-dependent scheduling; deteriorating tasks; bicriteria scheduling

1 Introduction

The classical scheduling theory has been formulated in late fifties in order to solve the problems encountered in an industrial practice. Among all assumptions of the theory particularly two have been played important role in development of different scheduling models: 1° the processing times of tasks are fixed values and 2° there is only one criterion used to evaluate a schedule. In this paper we consider a scheduling problem in which both these assumptions are modified: bicriteria scheduling with time-dependent processing times of tasks.

The problem can be formulated as follows. We are given a single machine and a set of $n + 1$ tasks to be processed on the machine. Processing times of the tasks are in the form of $p_j(t) = 1 + \alpha_j t$, where t and $\alpha_j > 0$ denote, respectively, the starting time of the task and the deterioration rate, $j = 0, 1, 2, \ldots, n$. We are interested in minimizing the norm $\|C\|_{(\lambda)}$, where $C = [C_0, C_1, C_2, \ldots, C_n]^\top$ denotes a vector of completion times of the tasks, and $\|C\|_{(\lambda)} = \lambda\|C\|_1 + (1-\lambda)\|C\|_\infty$ for any C and fixed $\lambda \in [0, 1]$. The optimization is carried over all permutations of a given sequence α, where $\alpha = (\alpha_0, \alpha_1, \ldots, \alpha_n)$ is a vector of deterioration rates of the tasks.

2 Previous research

In this section we briefly describe both interesting us research areas: bicriteria scheduling and time-dependent scheduling. Recall that in the first case the processing times of tasks are known in advance, fixed values (i.e. integer or rational numbers) while in the second one the processing times are described by functions of time.

Bicriteria scheduling. In practical problems rarely we deal with only *one* criterion of optimality of a schedule. For instance, usually we wish to minimize costs of production and, at the same time, to maximize profits. Similarly, we can be interested in schedules for which not only the total completion time does achieve the minimum value but also the tasks are not delayed. Thus scheduling problems with *bicriteria* and *multiple* criteria play important role in modern scheduling theory, especially from applications point of view, see Hoogeveen [6]. For more details on bicriteria single machine scheduling we refer the reader to papers by Dileepan and Sen [3] and Lee and Vairaktarakis [9]. Review of literature concerning both, bicriteria and multiple criteria scheduling, one can found in Nagar et al. [12].

Time-dependent scheduling. There are many situations in which executed tasks have *variable* processing times. For example, medical procedures take different amounts of time to complete, computer programs need variable amount of time to execute sucessfully etc. One of possible approaches to the problem of task's processing time variability is scheduling with *deteriorating* processing times of tasks: the processing time of a task increases as the task is waiting for processing, and this dependency is described by a nondecreasing function of time. More details on a single machine time-dependent scheduling one can find in review by Alidaee and Womer [1]. There are also known results concerning time-dependent scheduling in multi-processor environment, for example Chen [2], Kononov [7], Mosheiov [11] (parallel identical machines) or Kononov and Gawiejnowicz [8] (parallel dedicated machines). Note also that time-dependent scheduling can be considered as a special case of so-called *discrete-continuous scheduling*, in which also other parameters of a problem, e.g. ready times of tasks or speeds of processors, are described by continuous functions. Problems of this kind are discussed in detail by Gawiejnowicz [4].

3 Preliminary results

In this section we present strict formulation of the problem as well as we state some auxiliary results for the ones in subsequent section. Due to space limitations we present our results without proofs only.

First note that due to the form of processing times of tasks the following recurrence equations hold:

$$C_j = C_{j-1} + p_j(C_{j-1}) = 1 + (1 + \alpha_j)C_{j-1} \tag{1}$$

where $j = 1, 2, \ldots, n$ and $C_0 = 1$.

The formula (1) can be written for $j = 0, 1, 2, \ldots, n$ as a superposition $C_j = f_j(f_{j-1}(\ldots(f_0(t))\ldots))$ of the functions $f_j(t) = t + p_j(t) = 1 + a_j t$, where $a_j = 1 + \alpha_j$, t denotes the starting time of a task and the first task is starting at $t = 0$.

Expanding recurrence relations (1) for $j = 0, 1, 2, \ldots, n$ and rewriting them in the matrix form we obtain the following:

$$\begin{bmatrix} 1 & 0 \ldots & 0\;0 \\ -a_1 & 1 \ldots & 0\;0 \\ 0 & -a_2 \ldots & 0\;0 \\ \vdots & \ldots & \vdots \\ 0 & 0 \ldots & -a_n\;1 \end{bmatrix} \begin{bmatrix} C_0 \\ C_1 \\ C_2 \\ \vdots \\ C_n \end{bmatrix} = \begin{bmatrix} 1 \\ 1 \\ 1 \\ \vdots \\ 1 \end{bmatrix},$$

i.e. $A(a)C(a) = d(1)$, where $d(1) = [1, 1, \ldots, 1]^\top$ and $C = [C_0, C_1, \ldots, C_n]^\top$ both are $(n+1)$-element vectors.

Recall that the l_p-type norm for $C = C(a) \in \mathbb{R}^{n+1}$ is given by

$$\|C(a)\|_p = \begin{cases} (\sum_{i=0}^{n} |C_i(a)|^p)^{\frac{1}{p}} , & 1 \le p < \infty, \\ \max_{0 \le i \le n} \{|C_i(a)|\} , & p = +\infty. \end{cases}$$

Our criterion $\| \cdot \|_{(\lambda)}$ is a convex combination of the norm $\| \cdot \|_1 = \sum_{j=0}^{n} C_j \equiv \sum C_j$ and the norm $\| \cdot \|_\infty = \max_{0 \le j \le n} \{C_j\} \equiv C_{\max}$, where $\lambda \in [0, 1]$ is arbitrary but fixed.

Let $a(a_q \leftrightarrow a_r)$ denote vector a with elements a_q and a_r interchanged. We begin with the following useful result.

Lemma 1. *Let $b = a(a_q \leftrightarrow a_{q+1})$, where $q = 1, \ldots, n-1$. Then for $n > 2$ and $i = 1, \ldots, n$ there holds the equality*

$$C_i(b) - C_i(a) = \begin{cases} (a_q - a_{q+1}) \cdot a_{q+2} \cdots a_i & , 1 \le q < i, \\ (a_{q+1} - a_q) \cdot \sum_{j=0}^{q-1} a_{j+1} \ldots a_{q-1} & , q = i, \\ 0 & , q > i. \end{cases}$$

Moreover, $C_0(b) - C_0(a) = 0$.

Summing differences $C_i(b) - C_i(a)$, for $i = 0, 1, \ldots, n$, we get immediately a formula concerning the behaviour of the norm $\|C(b)\|_1$ under transpositions $b = a(a_q \leftrightarrow a_{q+1})$.

Lemma 2. *Let $b = a(a_q \leftrightarrow a_{q+1})$. Then there holds the equality*

$$\|C(b)\|_1 - \|C(a)\|_1 = (a_{q+1} - a_q) \cdot \left(\sum_{j-0}^{q-1} a_{j+1} \ldots a_{q-1} - \sum_{i-q+1}^{n} a_{q+2} \ldots a_i \right),$$

where $q = 1, 2, \ldots, n-1$.

Since $\|C(a)\|_\infty = C_n(a)$, letting $i = n$ in Lemma 1, we obtain a formula concerning the behaviour of the norm $\|C(b)\|_\infty$ under transpositions $b = a(a_q \leftrightarrow a_{q+1})$.

Lemma 3. *Let* $b = a(a_q \leftrightarrow a_{q+1})$. *Then there holds the equality*

$$\|C(b)\|_\infty - \|C(a)\|_\infty = (a_q - a_{q+1}) \cdot a_{q+2} \ldots a_n,$$

where $q = 1, 2, \ldots, n-1$.

Finally, note that if we start at $t = 0$, the coefficient a_0 has no effect on the value of C_j for each j. Moreover, C_j depends on a_i $(i = 1, \ldots, j)$ in a monotonous, nondecreasing way. Therefore (cf. Mosheiov [10]), given any ordering of the sequence $(a_0, a_1, \ldots, a_n)$ the best strategy is to set as a_0 the maximal element in this sequence. Hence, we will assume in the sequel that the element a_0 has been established according to the above observation and consider the vector $(a_1, a_2, \ldots, a_n)$ only.

4 Main results

In this section we present our main results concerning properties of an optimal schedule for different values of the parameter $\lambda \geq 1$. We will also assume that $n > 2$, i.e. the sequences of coefficients $a = (a_1, a_2, \ldots, a_n)$ contain at least three elements such that $a_i \geq 1$ for $i = 1, 2, \ldots, n$.

Theorem 1. *Let* $b = a(a_q \leftrightarrow a_{q+1})$, *where* $a = (a_1, a_2, \ldots, a_n)$. *Then for* $q = 1, 2, \ldots, n-1$ *there holds the equality*

$$\|C(b)\|_{(\lambda)} - \|C(a)\|_{(\lambda)} =$$

$$= (a_{q+1} - a_q) \cdot \left(\lambda \left(\sum_{j=0}^{q-1} a_{j+1} \ldots a_{q-1} - \sum_{j=q+1}^{n-1} a_{q+2} \ldots a_j \right) - a_{q+2} \ldots a_n \right).$$

Let $\lambda \in [0, 1]$ be fixed but arbitrary and let $q = 1, 2, \ldots, n-1$. Define the function $\Lambda_q(\lambda)$ as follows

$$\Lambda_q(\lambda) \equiv \lambda \left(\sum_{j=0}^{q-1} a_{j+1} \ldots a_{q-1} - \sum_{j=q+1}^{n-1} a_{q+2} \ldots a_j \right) - a_{q+2} \ldots a_n.$$

Behaviour of the function $\Lambda_q(\lambda)$ is crucial for our purposes. We begin with the necessary condition for the sequence $a = (a_1, a_2, \ldots, a_n)$ to be optimal with respect to bicriterion measure $\| \cdot \|_{(\lambda)}$.

Theorem 2. *Let* $b = a(a_q \leftrightarrow a_{q+1})$ *and let the sequence* $a = (a_1, a_2, \ldots, a_n)$ *be optimal with respect to the norm* $\|C(a)\|_{(\lambda)}$. *Then for* $q = 1, 2, \ldots, n-1$ *there holds the inequality*

$$0 \leq \|C(b)\|_{(\lambda)} - \|C(a)\|_{(\lambda)} = (a_{q+1} - a_q) \cdot \Lambda_q(\lambda).$$

In view of Theorem 2, it is important to know the behaviour of the sign of the function $\Lambda_q(\lambda)$ for $q = 1, 2, \ldots, n-1$ and $\lambda \in [0, 1]$.

Lemma 4. *Let* $\lambda \in [0, 1]$ *be fixed but arbitrary. Then* $\Lambda_1(\lambda) < 0$ *and there hold inequalities*

$$\Lambda_1(\lambda) \leq \Lambda_2(\lambda) \leq \ldots \leq \Lambda_{n-1}(\lambda).$$

In view of these results, given the sequence $a = (a_1, a_2, \ldots, a_n)$, a fundamental problem is to determine numbers λ_0 and λ_1 $(0 < \lambda_0 \leq \lambda_1 < 1)$ such that $\Lambda_q(\lambda) \leq 0$ for $\lambda \in [0, \lambda_0]$ and $q = 1, 2, \ldots, n-1$, and $\Lambda_{n-1}(\lambda) \geq 0$ for $\lambda \in [\lambda_1, 1]$. Let $\overline{a} = \max\{a_1, a_2, \ldots, a_n\}$ and $\underline{a} = \min\{a_1, a_2, \ldots, a_n\}$. Define the numbers λ_0 and λ_1 as follows

$$\lambda_0 = \frac{\bar{a} - 1}{\bar{a}^{n-1} - 1} \tag{2}$$

and

$$\lambda_1 = \frac{\underline{a} - 1}{\underline{a}^{n-1} - 1}. \tag{3}$$

Lemma 5. *Let be given the sequence* $a = (a_1, a_2, \ldots, a_n)$. *Then* $\Lambda_q(\lambda) \leq 0$ *for* $q = 1, 2, \ldots, n-1$, *where* $\lambda \in [0, \lambda_0]$ *and* λ_0 *is defined by* (2).

Lemma 6. *Let be given the sequence* $a = (a_1, a_2, \ldots, a_n)$. *Then for each* $\lambda \in [\lambda_1, 1]$ *there holds* $\Lambda_{n-1}(\lambda) \geq 0$, *where* λ_1 *is defined by* (3).

As a corollary from Theorem 2, Lemma 5 and Lemma 6 we immediately obtain the main result of the paper.

Theorem 3. *Let the sequence* $a = (a_1, a_2, \ldots, a_n)$ *be optimal with respect to the norm* $\| \cdot \|_{(\lambda)}$ *and let* $0 < \lambda_0 \leq \lambda_1 < 1$ *be defined by* (2) *and* (3), *respectively. Then the following implications hold true:*
(a) *if* $\lambda \in [0, \lambda_0]$, *then* a *is nonincreasing and*
(b) *if* $\lambda \in [\lambda_1, 1]$, *then* a *has a* V*-shape.*
Moreover, if the sequence a *contains only distinct elements, then* $\lambda_0 < \lambda_1$.

Recall that, according to Mosheiov [10], the sequence $a = (a_1, a_2, \ldots, a_n)$ has a V-shape, if there exists an index $1 \leq m \leq n$ such that a_i is nonincreasing for $1 \leq i \leq m$, and a_i is nondecreasing for $m \leq i \leq n$.

There is also another version of Theorem 3, giving more precise conditions for the monotonicity and the V-shapeness of the optimal sequence a. This requires, however, $O(n \log n)$ additional operations to determine the respective values of λ_0 and λ_1. We obtained also a sufficient condition for the sequence $a = (a_1, \ldots, a_n)$ to be optimal. Below we present these two results.

For a given sequence $a = (a_1, a_2, \ldots, a_n)$, let us define the number $\lambda(a)$ as follows

$$\lambda(a) = \frac{1}{\sum_{j=0}^{n-2} a_{j+1} \cdots a_{n-2}}.$$

Moreover, let $\lambda(a_*) = \min_{\pi \in \Pi_n} \lambda(a_\pi)$ and $\lambda(a^*) = \max_{\pi \in \Pi_n} \lambda(a_\pi)$, where Π_n denotes the set of all permutations of the set $\{1, 2, \ldots, n\}$.

Theorem 4. *Let the sequence $a = (a_1, a_2, \ldots, a_n)$ be optimal with respect to the norm $\|\cdot\|_{(\lambda)}$, and let λ_0 and λ_1 be defined by (2) and (3), respectively. Then the following implications hold true:*
(a) *if $\lambda \in [0, \lambda(a_*)]$, then a is nonincreasing and*
(b) *if $\lambda \in [\lambda(a^*), 1]$, then a has a V-shape.*
Moreover, $\lambda_0 \le \lambda(a_*)$ and $\lambda(a^*) \le \lambda_1$, and the inequalities are strict whenever the sequence a contains only distinct elements.

Theorem 5. *A sufficient condition for the sequence $a = (a_1, \ldots, a_n)$ to be optimal with the respect to the norm $\|\cdot\|_{(\lambda)}$, where $\lambda \in [0, 1]$, is that a is nonincreasing and $0 \le \lambda \le \lambda(a_*)$.*

Now we present a few results concerning possible relations between the norm $\|\cdot\|_{(\lambda)}$ and norms $\|\cdot\|_1$ and $\|\cdot\|_\infty$.

Theorem 6. *For any $C = [C_0, C_1, C_2, \ldots, C_n]^\top$ and fixed $\lambda \in [0, 1]$ there hold inequalities:*

$$0 \le \frac{\|C\|_{(\lambda)} - \|C\|_\infty}{\|C\|_\infty} \le \lambda(n-1).$$

Similar estimation can be obtained for the norm $\|\cdot\|_{(\lambda)}$ in relation to the norm $\|\cdot\|_1$.

Theorem 7. *For any $C = [C_0, C_1, C_2, \ldots, C_n]^\top$ and fixed $\lambda \in [0, 1]$ there hold inequalities:*

$$0 \le \frac{\|C\|_1 - \|C\|_{(\lambda)}}{\|C\|_1} \le (1 - \lambda)\frac{n-1}{n}.$$

As a corollary from Theorems 6 and 7 we obtain the following general result.

Theorem 8. *For any $C = [C_0, C_1, C_2, \ldots, C_n]^\top$ and fixed $\lambda \in [0, 1]$ there hold inequalities:*

$$\|C\|_{(\lambda)} \le \|C\|_1 \le \frac{n}{1 + \lambda(n-1)} \|C\|_{(\lambda)}.$$

Using Theorems 6, 7 and 8 we obtained a number of different lower and upper bounds for the ratio of $\|\cdot\|_{(\lambda)}$ and $\|\cdot\|_1$ ($\|\cdot\|_{(\lambda)}$ and $\|\cdot\|_\infty$). For instance, one can show that hold the following two inequalities:

$$0 \leq \frac{\|C\|_{(\lambda)} - \|C\|_\infty}{\|C\|_\infty} \leq \frac{n-1}{n+1}$$

and

$$0 \leq \frac{\|C\|_1 - \|C\|_{(\lambda)}}{\|C\|_1} \leq \frac{n-1}{n+1}.$$

Moreover, one can show that holds the following equality:

$$\frac{\|C\|_{(\lambda)} - \|C\|_\infty}{\|C\|_1 - \|C\|_{(\lambda)}} = \frac{\lambda}{1-\lambda}.$$

This equality measures the relative distance of the norm $\|\cdot\|_{(\lambda)}$ from the norms $\|\cdot\|_1$ and $\|\cdot\|_\infty$.

At the end of this section we present a theorem that relates these original norms with the l_p-norms. This leads to another approach (see Gawiejnowicz et al. [5] for details) to get an approximate $O(n \log n)$ time solution of the problem for $\lambda = 1$ via considering monocriterion optimality measures $\|\cdot\|_p$, $1 \leq p \leq +\infty$.

Theorem 9. *Given sequence $a = (a_1, a_2, \ldots, a_n)$ and the vector $C(a)$ such that $A(a)C(a) = d(1)$, there holds inequality*

$$\|C(a)\|_p \leq \|C(a)\|_1^{\frac{1}{p}} \cdot \|C(a)\|_\infty^{1-\frac{1}{p}},$$

fulfilled for all $1 \leq p \leq +\infty$, and the equality holds only for limit cases $p = 1$ and $p = +\infty$.

5 Concluding remarks

The presented case of norm $\|\cdot\|_{(\lambda)}$ is general enough for simultaneous minimization of norms $\|\cdot\|_1$ and $\|\cdot\|_\infty$. Indeed, because

$$\alpha\|\cdot\|_1 + \beta\|\cdot\|_\infty = (\alpha+\beta)\left(\frac{\alpha}{\alpha+\beta}\|\cdot\|_1 + \frac{\beta}{\alpha+\beta}\|\cdot\|_\infty\right),$$

then letting $\lambda = \frac{\alpha}{\alpha+\beta}$ we can apply the approach presented in the paper for any criterion in the form of $\alpha\|\cdot\|_1 + \beta\|\cdot\|_\infty$, where $\alpha, \beta > 0$ are arbitrary real numbers.

There are, of course, some open problems. The first such open problem is hierarchical minimization of $\|\cdot\|_1$ and $\|\cdot\|_\infty$ criteria. We conjecture that the problem is at least NP-hard in an ordinary sense. The second, very intriguing problem is the time complexity of the problem in interval $[\lambda_0, \lambda_1]$.

Acknowledgments

The research has been partially supported by scientific grant no. GN-01/2001 of the Faculty of Mathematics and Computer Science, Adam Mickiewicz University, Poznań, Poland.

The research of the first author has been also partially supported by KBN (Polish Scientific Committee) grant no. 8T 11A 01618.

References

1. Alidaee, B., Womer, N.K. (1999) Scheduling with Time Dependent Processing Times: Review and Extensions. J. Optl Res. Soc. **50**, 711–720.
2. Chen, Z-L. (1996) Parallel Machine Scheduling with Time Dependent Processing Times. Discr. Appl. Math **70**, 81–93 (erratum: Discr. Appl. Math. **75**, 103).
3. Dileepan, P., Sen, T. (1998) Bicriterion Static Scheduling Research for a Single Machine. Omega **16**, 53–59.
4. Gawiejnowicz, S. (1996) Brief Survey of Continuous Models of Scheduling. Found. Comput. Dec. Sci. **21**, 81–100.
5. Gawiejnowicz, S., Kurc, W., Pankowska, L., Suwalski, C. (2001) Approximate Solution of a Time-dependent Scheduling Problem for l_p-norm-based Criteria. In: Operations Research Proceedings 2000, Fleischmann, B. et al. (eds.), Springer, Berlin, 372–377.
6. Hoogeveen, H. (1992) Single-Machine Bicriteria Scheduling. Ph.D. thesis, CWI, Amsterdam.
7. Kononov, A. (1997): Scheduling Problems with Linear Processing Times. In: Operations Research Proceedings 1996, Zimmermann, U. et al. (eds.), Springer, Berlin, 208–212.
8. Kononov, A., Gawiejnowicz, S. (2001) NP-hard Cases in Scheduling Deteriorating Jobs on Dedicated Machines. J. Optl Res. Soc. **52**, 708–717.
9. Lee, C.-Y., Vairaktarakis, G. (1993) Complexity of Single Machine Hierarchical Scheduling: A Survey. In: Complexity in Numerical Optimization, Pardalos, P.M. (ed.), World Scientific, Singapore, 269–298.
10. Mosheiov, G. (1991) V-shaped Policies to Schedule Deteriorating Jobs. Oper. Res. **39**, 979–991.
11. Mosheiov, G. (1998) Multi-machine Scheduling with Linear Deterioration. Infor **36**, 205–214.
12. Nagar, A., Haddock, J., Heragu, S. (1995) Multiple and Bicriteria Scheduling: A Literature Survey. Euro. J. Optl Res. **81**, 88–104.

Multi-machine Scheduling Problem with Optimal Due Interval Assignment Subject to Generalized sum Type Criterion

Adam Janiak, Marcin Marek

Institute of Engineering Cybernetics, Wroclaw University of Technology
Janiszewskiego 11/17, 50-372 Wroclaw, Poland

Abstract. In this paper, we consider a multi-machine scheduling problem, in which to each job a due interval should be assigned. In the scientific literature of scheduling theory, only an assignment of due date was considered. However in our case, to each job a due interval should be assigned. Due interval is a generalization of classical due date and describes a time interval, in which a job should be finished. For the considered problem, we should find a schedule of jobs and a common due interval such that the sum of: the total tardiness, the total earliness and due interval parameters is minimized. We proved an ordinary NP-completeness of the problem using the well known NP-complete partition problem. A heuristic algorithm solving the problem with its experimental analysis is also presented.

1 Introduction

In the scientific literature of scheduling theory, only problems with due date assignment were considered (see [1], [2] and [6]). In this paper we present a new class of scheduling problems in which a due interval assignment appears. The due interval is a generalization of classical due date and describes a time interval, in which a job should be finished. In the classical approach we should find an exact moment of job's completion. In our new approach, it is enough if a job completes in some moment from a given interval. Due interval assignment is a completely new approach and to our knowledge has not been investigated in the scientific literature. An application of the considered problems appears in some manufacturing systems in which the negotiation, with respect to due intervals, between a producer and a customer occurs. Especially, this problem occurs in periodical aircraft service.

2 Problem formulation

There is given a set of n independent and non-preemptive jobs to be scheduled on identical parallel machines. At the moment, each machine can process only one job. We assume that machines execute the jobs without idle times. However, we allow a time gap between the start of the optimization process and the execution of the first job on any machine. For each job its processing time p_j and a due interval $\langle d'_j = k_1; d"_j = k_2 \rangle$ are given, where k_1 and k_2 ($k_1 \leq k_2$) denote common (for all the jobs) due interval parameters. The problem is to find a schedule π and such values of the parameters k_1 and k_2, which minimize the following criterion:

$$f(\pi, k_1, k_2) = \sum_j E_j + \sum_j T_j + (k_2 - k_1), \quad (1)$$

where:
$E_j = \max(0, d'_j - C_j)$ - earliness of job j,
$T_j = \max(0, C_j - d''_j)$ - tardiness of job j,
C_j - the completion time of job j.

Using the three-field classification [5], our problem (denoted by PCS) can be characterized as follows:

$$P| < d'_j = k_1, d''_j = k_2 > | \sum_j E_j + \sum_j T_j + (k_2 - k_1).$$

An optimal solution of the single machine version of PCS has been constructed in [6].

3 Computational complexity

In this section we show that the decision version of the considered problem (PCS) is NP-complete. In order to prove that, we will use the NP-complete Partition Problem (PP) [3] defined as follows:
PP: Given a set X = $\{x_1, \ldots, x_m\}$ of m positive integers, for which $\sum_{i=1}^{m} x_i = 2B$; does there exist a partition of the set X into two disjoint subsets X_1 and X_2, such that $\sum_{x_i \in X_1} x_i = \sum_{x_i \in X_2} x_i = B$?

Theorem 1 *The problem* $P| < d'_j = k_1, d''_j = k_2 > | \sum_j E_j + \sum_j T_j + (k_2 - k_1)$ *is NP-hard even for the case with two identical parallel machines.*

Proof. Given an instance of PP, we construct the following instance of PCS. There are $n = m+2$ jobs. Among them, there are m partition jobs $i = 1, ..., m$, and 2 enforcer jobs e_1, e_2 with the following parameters:

$$p_i = x_i \text{ for } i = 1, ..., m$$

$$p_{e1} = p_{e2} = 2B.$$

The value y is equal to B.

The transformation given above is polynomial. The proof is trivial, therefore we omitted it.

We show that PP has a solution if and only if there exists a solution to the constructed instance of PCS with the value $\sum E_j + \sum T_j + (k_2 - k_1) \leq y = B$.

$(PP \Rightarrow PCS)$ Assume that subsets Y and X\Y are a solution of the problem PP. It means that $\sum_{x_i \in \Upsilon} x_i = \sum_{x_i \in X\setminus\Upsilon} x_i = B$. Based on the solution of PP, we construct the following solution for the problem PCS. On each machine the sequence starts with a single enforcer job which is followed by the partition jobs from one of the subsets Y or X\Y. The partition jobs are scheduled in an arbitrary order. Values of parameters and are given as follows:

$$\begin{aligned} k_1 &= p_{m+1} = 2B, \\ k_2 &= p_{m+1} + \sum_{x_i \in \Upsilon} x_i = 2B + B = 3B. \end{aligned}$$

For the values of the parameters and and a sequence of jobs π' described above, the criterion value is equal to:

$$f(\pi', k_1, k_2) = \sum E_j + \sum T_j + (k_2 - k_1) = (k_2 - k_1) = (3B - 2B) = B = y,$$

as it was required. It means that PP has a solution, if PCS has also one.

Before we will start with the second part of the proof, we will show that the enforcer job has to be processed as first on each machine.

Assume that at least one job with the processing time equal to ε is executed before the enforcer job, where ε is some positive integer. Without the loss of generality, we can assume that this situation occurs on one machine (see Fig. 1).

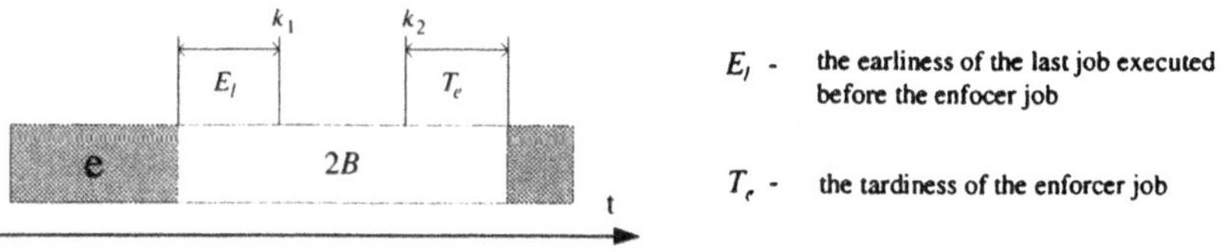

Fig. 1. The schedule where the enforcer job is not processed as first

For any given values of the parameters and , the criterion value (1) can be estimated as follows:

$$f(\pi, k_1, k_2) = \sum E_j + \sum T_j + (k_2 - k_1) \geq E_l + T_e + (k_2 - k_1) \geq 2B > y.$$

Since the criterion value (1) is greater than the required value y, in the further considerations, we will focus on schedules in which the enforcer jobs are processed as first on each machine.

$(\overline{PP} \Rightarrow \overline{PCS})$ Assume now that problem PP has no solution. It means that the sums of elements in both subsets are different, i.e., $\sum_{x_i \in \Upsilon} x_i \neq \sum_{x_i \in X\setminus\Upsilon} x_i$.

There are two cases to be considered, namely:

a) $\sum_{x_i \in \Upsilon} x_i = B + \varepsilon$ and $\sum_{x_i \in X \setminus \Upsilon} x_i = B - \varepsilon$,
b) $\sum_{x_i \in \Upsilon} x_i = B - \varepsilon$ and $\sum_{x_i \in X \setminus \Upsilon} x_i = B + \varepsilon$.

The cases a) and b) can be raduced to the case in which the sum of partition jobs is equal to $B + \varepsilon$ only on one machine (see Fig. 2).

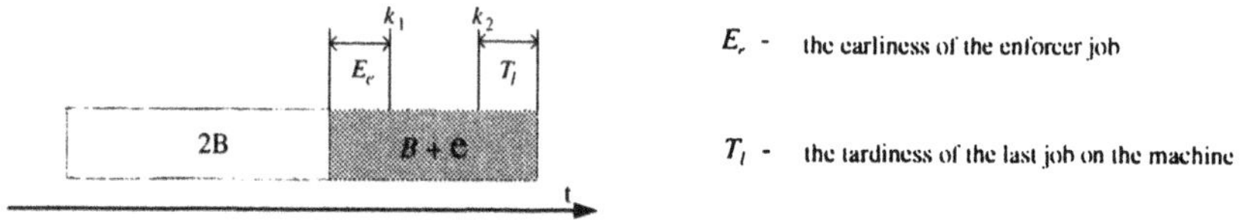

Fig. 2. The schedule where the enforcer job is processed as first

For any given values of the parameters and , for the case characterized on Fig. 2, the criterion value (1) can be estimated as follows:

$$f(\pi, k_1, k_2) = \sum E_j + \sum T_j + (k_2 - k_1) \geq E_e + T_l + (k_2 - k_1) \geq B + \varepsilon > y.$$

It means the no solution for the problem PP gives no solution for the problem PCS.

We showed that PP has a solution if and only if there exists a solution to the constructed instance of PCS. □

4 Heuristic algorithm

Since the considered problem is NP-hard, we do not expect to find its optimal solution in polynomial time. Therefore, to solve the problem we construct an heuristic algorithm. In further considerations we will use the following notations:
$C_{max} = \max_j C_j$,
q – the number of machines,
$S_{i[j]}$ – a starting time of the j-th job on the i-th machine,
$C_{i[j]}$ – a completion time of the j-th job on the i-th machine,
$p_{i[j]}$ – a processing time of the j-th job on the i-th machine.

Algorithm *APCS*

- **Step 1:** Rename the jobs according to the non-increasing order of their processing times.

- **Step 2:** Assign first q jobs to the machines (in increasing order of machine indices).

- **Step 3:** For the first q jobs calculate their starting times on appropriate machines according to the following expression: $S_{i[1]} = p_{1[1]} - p_{i[1]}$ for $i = 1, \ldots, q$.

- **Step 4:** Assign next jobs to the machines as they become free.

- **Step 5:** Set the values of the parameters: $k_1 = C_{1[1]}$ and $k_2 = C_{max}$.

- **Stop:** The obtained schedule π and the values of the parameters k_1 and k_2 are the solution for the problem PCS.

The computational complexity of Algorithm *APCS* is equal to $O(n \log n)$.

The following theorem describes the result of the worst case analysis of Algorithm *APCS*.

Theorem 2 *The worst case ratio of Algorithm APCS can be estimated as follows* $R_{APCS} \leq \frac{4}{3} - \frac{1}{3q}$.

Proof. In order to show that $R_{APCS} \leq \frac{4}{3} - \frac{1}{3q}$, we can use Graham's approach [4] which deals with the *LPT* Algorithm for the problem $P||C_{max}$. Note that the completion time of the first job on each machine is equal to $C_{1[1]}$. The next jobs are assigned to the machines according to the *LPT* rule starting from $C_{1[1]}$. It follows from the calculated values of parameters k_1 and k_2 that there are neither early nor tardy jobs. Therefore, the criterion considered for the problem PCS reduces to the makespan minimization of the problem $P||C_{max}$. □

5 Conclusion

We considered the problem of scheduling jobs on identical parallel machines and a common due interval assignment. The objective was to minimize the sum of the total earliness, the total tardiness and the due interval size. We proved that the problem is NP-hard. To solve the problem, we constructed a heuristic algorithm and showed its worst case ratio.

The results obtained are very promising and motivating to work with the other scheduling problems from this new class.

References

1. Baker K.R., Scudder G.D. (1990) Sequencing with earliness and tardiness penalties: a review. Operations Research 30, 22–36

2. Cheng T.C.E., Gupta M.C. (1989) Survey of scheduling research involving due date determination decision. European Journal of Operational Research 38, 156–166
3. Garey M. R., Johnson D. S. (1979) Computers and Intractability: A Guide to the Theory of NP-completeness. Freeman, San Francisco
4. Graham R.L. (1969) Bounds on multiprocessing timing anomalies. SIAM J. Appl. Math. 17, 263–369
5. Graham R. L., Lawler E. L., Lenstra J. K., Rinnooy Kan A. H. G. (1979) Optimization and approximation in deterministic sequencing and scheduling: a survey. Annals of Discrete Mathematics 3, 287-326
6. Janiak A., Marek M. (2000) Single machine scheduling problem with optimal due intervals assignment. Lecture Notes of Silesian University, Automation 129, 145–155 (in Polish)

Storage Problems in Batch Scheduling

Christoph Schwindt, Norbert Trautmann

Universität Karlsruhe, Institut für Wirtschaftstheorie und Operations Research,
D-76128 Karlsruhe, Germany
e-mail: {schwindt,trautmann}@wior.uni-karlsruhe.de

Abstract. This paper is concerned with the modelling of storage configurations for intermediate products in process industries. Those models form the basis of algorithms for scheduling chemical production plants. Different storage capacity settings (unlimited, finite, and no intermediate storage), storage homogeneity settings (dedicated and shared storage), and storage time settings (unlimited, finite, and no wait) are considered. We discuss a classification of storage constraints in batch scheduling and show how those constraints can be integrated into a general production scheduling model that is based on the concept of cumulative resources.

1 Process Scheduling

In process industries, e.g. the chemical or food industry, final products result from successive chemical or physical transformations of raw materials on scarce production resources (cf. [1]). The problem under consideration consists of scheduling a given set of operations on a production plant such that a given objective function is minimized. Typical objectives for process scheduling problems are the minimization of makespan, lateness, or setup cost.

Each operation needs a processing unit and operating staff for execution. In addition, each operation consumes a given amount of one or several input products and produces a given amount of one or several output products. At the end of an operation, the output products are unloaded from the processing unit. Certain intermediate products can then be buffered in storage facilities of given capacity. Other intermediate products cannot be stored and thus must be consumed without any delay. Backlogging is not allowed. For the storable intermediate products, a nonnegative safety stock is prescribed. Since one and the same product may be produced by several operations and may be consumed by several operations, it is generally not possible to establish a-priori precedence relationships between producing and consuming operations. If the plant is operated in batch production mode, i.e., the material flow is discontinuous, the input products of an operation must be available at its start and the output products arise at its completion. If the plant is operated in continuous production mode, the products are produced and consumed at constant rates. Production plants are driven in batch production mode if small amounts of a large number of products are required. In this case, the plant is configured according to the required final products. Continuous production mode is typical of mass production like oil refinement. In the sequel, we concentrate on batch production mode.

In chemical engineering literature (cf. e.g. [2], [3], or [4]), the following *storage capacity settings* for a given intermediate product are considered:

1. Unlimited intermediate storage UIS: The storage facility has unlimited capacity.
2. Finite intermediate storage FIS: The storage facility has limited capacity only.
3. No intermediate storage NIS: There is no storage capacity available.

Depending on whether for a given intermediate product, the stock-keeping facility is a single-product or a multi-product storage, we distinguish between two *storage homogeneity settings*:

1. Dedicated intermediate storage DIS: The product is stocked in a homogenous storage.
2. Shared intermediate storage SIS: The product is stocked together with other products in a heterogenous storage.

For what follows, we assume that the storage facilities are replenished and depleted based on a first-in first-out strategy. The *storage time settings* for a given intermediate product can then be classified according to:

1. Unlimited wait UW: The product can be consumed immediately after production and can be stored for an unlimited time.
2. Finite wait FW: A quarantine time $q \geq 0$ says that the product can be consumed q units of time after its production at the earliest. Symmetrically, a shelf life time $s \geq 0$ implies that the product must be consumed s units of time after its production at the latest.
3. Zero wait ZW: The product has to be consumed immediately after production.

2 Model

In this section we present a basic model for scheduling batch processes. In Section 3 we explain how to model the different storage capacity, storage homogeneity, and storage time settings. Furthermore, we show that the use of renewable resources like manpower or processing units is contained as a special case.

Let $\mathcal{O}$ be the set of operations to be scheduled. The duration of operation $i \in \mathcal{O}$ is denoted by p_i and the start time by S_i. The vector $S = (S_i)_{i \in \mathcal{O}}$ is called (production) schedule.

$\mathcal{C}$ is the set of cumulative resources (cf. [5] and [7]) stocking the intermediates produced and consumed by the operations $i \in \mathcal{O}$. For each cumulative resource, there is a prescribed minimum inventory level $\underline{R}_k$ and a prescribed maximum inventory level $\overline{R}_k \geq \underline{R}_k$. As the plant is operated in batch production mode, production and consumption rates are infinite, i.e., depletions occur at start times of operations, and replenishments occur at completion times of operations. r_{ik} denotes the—possibly negative—increase in the inventory level of resource k caused by operation i and will be referred to as the demand of operation i for resource k. If $r_{ik} > 0$, operation i replenishes resource k, and if $r_{ik} < 0$, operation i depletes resource k. For a given schedule S, the inventory level in resource $k \in \mathcal{R}^\gamma$ at time $t \geq 0$ is

$$r_k(S,t) := \sum_{i \in \mathcal{O}: r_{ik} < 0, S_i \leq t} r_{ik} + \sum_{i \in \mathcal{O}: r_{ik} > 0, S_i + p_i \leq t} r_{ik}$$

The inventory must not fall below the minimum nor exceed the maximum inventory level at any point in time, i.e., $\underline{R}_k \leq r_k(S,t) \leq \overline{R}_k$ ($k \in \mathcal{R}^\gamma$, $t \geq 0$).

In addition, we have a set $E \subset \mathcal{O} \times \mathcal{O}$ of operation pairs (i,j) with associated weights δ_{ij}. A nonnegative weight δ_{ij} corresponds to a minimum time lag of $d_{ij}^{min} = \delta_{ij}$ units of time between the starts of operations i and j. A negative weight δ_{ij} implies a maximum time lag of $d_{ji}^{max} = -\delta_{ij}$ units of time between the starts of operations j and i. The temporal constraints given by minimum and maximum time lags can be written as $S_j - S_i \geq \delta_{ij}$ ($(i,j) \in E$).

With $f(S)$ denoting the objective function, the production scheduling problem can be formulated as follows:

$$\begin{array}{lll}
\text{Minimize} & f(S) & \\
\text{subject to} & \underline{R}_k \leq r_k(S,t) \leq \overline{R}_k & (k \in \mathcal{C},\ t \geq 0) \quad (1) \\
& S_j - S_i \geq \delta_{ij} & ((i,j) \in E) \quad (2) \\
& S_i \geq 0 & (i \in \mathcal{O}) \quad (3)
\end{array}$$

Solution procedures for this problem are discussed in [5] and [6].

3 Applications

In Subsections 3.1, 3.2, and 3.3 we show how to formulate the different storage capacity, storage homogeneity, and storage time settings, respectively, of Section 1 on the basis of the model presented in Section 2. Subsection 3.4 is concerned with renewable resources like manpower and processing units.

3.1 Storage Capacity Settings

1. UIS. Each intermediate product corresponds to one cumulative resource $k \in \mathcal{C}$ with infinite maximum inventory level $\overline{R}_k$ and the minimum inventory level $\underline{R}_k$ being equal to the safety stock of the product. The demand r_{ik} of an operation i for resource k equals the (negative) amount of the intermediate produced (consumed).

2. FIS. The case of finite intermediate storage differs from the UIS case in that the maximum inventory level $\overline{R}_k$ of the corresponding cumulative resource k coincides with the capacity of the respective storage facility.

3. NIS. The case of no intermediate storage corresponds to the FIS case with $\underline{R}_k = \overline{R}_k = 0$.

3.2 Storage Homogeneity Settings

1. DIS. Each cumulative resource k corresponds to one intermediate product to be stocked. The storage capacity $\overline{R}_k$ coincides with the maximum inventory of the product.

2. SIS. We consider the case of a heterogeneous intermediate storage facility keeping $\pi > 1$ different products. This setting is modelled by $\pi+1$ cumulative resources. The resources $k = 1, \ldots, \pi$ correspond to one intermediate product each. The minimum inventory levels $\underline{R}_k$ of resources $k = 1, \ldots, \pi$ equal the safety stocks for the respective products. The maximum inventory levels $\overline{R}_k$ of those resources k are infinite. The capacity of the storage facility is taken into account by the maximum inventory level $\overline{R}_{\pi+1}$ of resource $\pi + 1$. The minimum inventory level $\underline{R}_{\pi+1}$ equals 0. The production of an intermediate replenishes the respective resource k as well as resource $\pi + 1$. Analogously, the consumption of an intermediate depletes the respective resource k as well as resource $\pi + 1$.

3.3 Storage Time Settings

1. UW. If an intermediate product can be consumed immediately after production and can be stored for an unlimited waiting time, no additional constraints are required.

2. FW. We first consider the case of an intermediate product with a quarantine time of $q \geq 0$ units of time. This can be modelled using two cumulative resources k and k' (cf. left-hand part of Fig. 1). Resource k represents the storage facility, and thus the minimum and maximum inventory levels $\underline{R}_k$ and $\overline{R}_k$ are chosen according to the storage capacity settings for the product. For resource k', we set zero minimum and infinite maximum inventory level, i.e., $\underline{R}_{k'} = 0$ and $\overline{R}_{k'} = \infty$. For each operation i producing the intermediate, we introduce a fictitious operation i' with duration $p_{i'} = p_i$. Operations i and i' are linked by a minimum time lag of q units of time, i.e., $d_{ii'}^{min} = q$. The demand of fictitious operation i' for resource k' equals the replenishment of resource k by operation i, i.e., $r_{i'k'} = r_{ik}$. Each consuming operation j depletes resources k and k' by the same amount, i.e., $r_{jk'} = r_{jk}$. Since the replenishment of resource k' occurs q units of time after the replenishment of resource k at the earliest, any consuming operation j, which depletes both

k and k', cannot be started earlier than q units of time after the production of the intermediate (otherwise, the inventory in resource k' would fall below the minimum inventory level of 0). Thus, there is a minimum time lag of q units of time between the production and the consumption of the product.

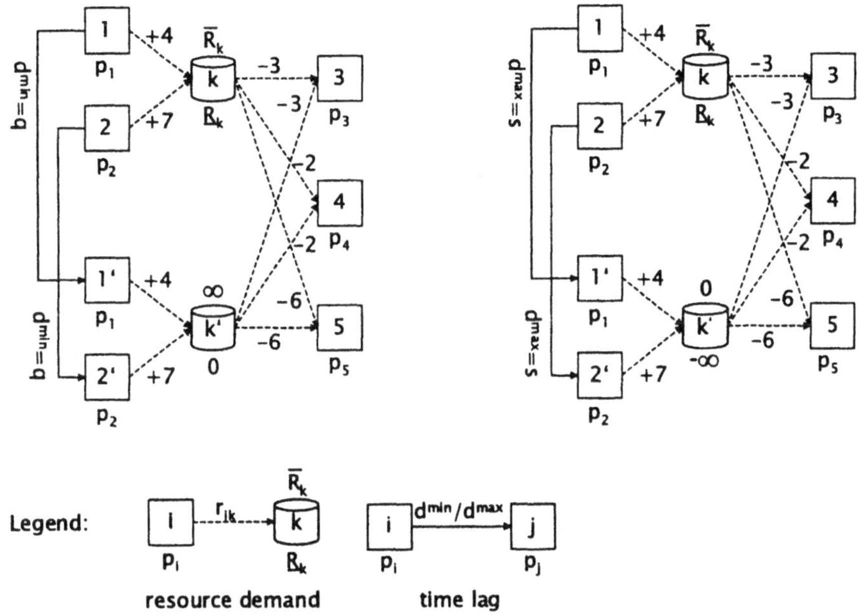

Fig. 1. Quarantine and shelf life times

Next, we deal with the case of an intermediate product with a shelf life time of $s \geq 0$ units of time. Analogously to the case of a quarantine time, we introduce two cumulative resources k and k' (cf. right-hand part of Fig. 1). The minimum inventory level of resource k' is now $\underline{R}_{k'} = -\infty$, and the maximum inventory level is $\overline{R}_{k'} = 0$. For each operation i producing the intermediate, a fictitious operation i' with duration $p_{i'} = p_i$ and resource demand $r_{i'k'} = r_{ik}$ is introduced. Operations i and i' are now linked by a maximum time lag of s units of time, i.e., $d_{ii'}^{max} = s$. Again, each consuming operation j depletes resources k and k' by the same amount, i.e., $r_{jk'} = r_{jk}$. Since the replenishment of resource k' occurs s units of time after the replenishment of resource k at the latest, any consuming operation j must be started no later than s units of time after the production of the intermediate (otherwise, the inventory in resource k' would exceed the maximum inventory level of 0). Thus, there is a maximum time lag of s units of time between the production and the consumption of the product.

3. ZW. The zero wait setting corresponds to both a quarantine and a shelf life time of 0, which can be modelled as described above. Note that instead of defining an additional resource k' it is sufficient to set $\overline{R}_k = 0$. Hence, the ZW and the NIS settings are equivalent.

3.4 Discrete Resources

Workers of same qualification and processing units of the same type are grouped to a pool. A pool corresponds to a (renewable) resource whose availability is independent of its previous utilization. Each pool is modelled as a cumulative resource k with a maximum inventory level $\overline{R}_k$ equal to the number of workers or processing units in the pool. The minimum inventory level is $\underline{R}_k = 0$. Each operation i is replaced by two dummy operations i' and i'' of duration 0. i' represents the start and i'' the completion of operation i. The duration p_i is modelled by a minimum and a maximum time lag $d_{i'i''}^{min} = d_{i'i''}^{max} = p_i$ saying that the completion of i occurs precisely p_i units of time after its start. The replenishment of k by operation i' and the depletion of k by operation i'' coincide with the number of workers required.

References

1. Applequist, G., Samikoglu, O., Pekny, J., Reklaitis, G. (1997): Issues in the use, design and evolution of process scheduling and planning systems. ISA Transactions **36**, 81–121
2. Ha, J.-K., Chang, H.-K., Lee, E.S., Lee, I.-B., Lee, B.-S., Yi, G. (2000): Intermediate storage tank operation strategies in the production scheduling of multiproduct batch processes. Computers & Chemical Engineering **24**, 1633–1640
3. Kim, M., Jung, J. H., Lee, I.–B. (1996): Optimal scheduling of multiproduct batch processes for various intermediate storage policies. Industrial & Engineering Chemistry Research **35**, 4058–4066
4. Kim, S.B., Lee, H.-K., Lee, I.-B., Lee, E.S., Lee, B. (2000): Scheduling of non–sequential multipurpose batch processes under finite intermediate storage policy. Computers & Chemical Engineering **24**, 1603–1610
5. Neumann, K., Schwindt, C. (1999): Project scheduling with inventory constraints. Report WIOR-572, University of Karlsruhe
6. Neumann, K., Schwindt, C., Trautmann. N. (2001): Scheduling material flows in process industries. Report WIOR-610, University of Karlsruhe
7. Schwindt, C., Trautmann, N. (2000): Batch scheduling in process industries: An application of resource-constrained project scheduling. OR Spektrum **22**, 501–524

Section 7

OR in the Health Service

Eine Entscheidung auf Leben und Tod: Decision Support für AIDS-Kontroll-Programme in Ostafrika

Steffen Fleßa

Evangelische Fachhochschule Nürnberg, Bärenschanzstraße 4, 90429 Nürnberg, Tel. 0911-27 25 38 86, Fax. 0911- 27 25 38 82, E-Mail: Steffen.Flessa@evfh-nuernberg.de

1 Epidemiologie

Die erworbene Immunschwächekrankheit AIDS (Acquired Immune Deficiency Syndrome) stellt die größte Bedrohung der Gesundheit der Bevölkerung Ostafrikas dar. In manchen Regionen ist diese Krankheit bereits die Haupttodesursache, in vielen Krankenhäusern überfordert die Behandlung der AIDS-Patienten längst die knappen Ressourcen. Die Gesundheitspolitik muss diese Veränderungen antizipieren und entsprechend rechtzeitig planen. Hierzu sind möglichst verlässliche Prognosedaten unentbehrlich. Aus diesem Grunde wurde ein System Dynamics Model der AIDS-Ausbreitung entwickelt und mögliche Interventionsstrategien wurden bewertet.

Nach Schätzungen der UNAIDS (*UNAIDS* 2000b, S. 6; *UNAIDS/WHO* 2001, S. 3) lebten Ende 2000 weltweit 36,1 Millionen Menschen mit AIDS bzw. einer HIV-Infektion. Die Zahl der Neuinfektionen im Jahre 2000 wird mit 5,3 Millionen und die Zahl der Todesopfer mit 3,0 Millionen angegeben. Seit Beginn der Epidemie bis zum 31. Dezember 2000 sind schätzungsweise 21,8 Millionen Menschen an dieser Krankheit gestorben. Besonders schwer betroffen sind die Länder Afrikas südlich der Sahara, wo am Ende des Jahres 2000 wahrscheinlich 25,3 Millionen Menschen infiziert waren (70,1 % der weltweiten Fälle) und ungefähr 3,8 Millionen neue Fälle zu verzeichnen waren (*UNAIDS/WHO* 2001, S. 5). Auf keinem anderen Kontinent wächst die Zahl der Infizierten so dramatisch wie in **Afrika**. Die am schlimmsten betroffenen Länder sind Swaziland, Malawi, Sambia, Zimbabwe, Namibia und Botswana, wo bis zu 36 % (Botswana) der Erwachsenen HIV-positiv sind.

Das Modell der AIDS-Ausbreitung simuliert die AIDS-Epidemie in einer Modellregion in Ostafrika. Tansania, Uganda und Kenia waren früh von der Epidemie betroffen und haben bis heute hohe Prävalenzen. So schätzt *UNAIDS* (2000b, S. 6) die Zahl der Infizierten Ende 1998 auf 1.300.000 (Tansania),

820.000 (Uganda) sowie 2.100.000 (Kenia). Die entsprechenden Prävalenzen unter der erwachsenen Bevölkerung betragen 8,09 %, 8,30 % bzw. 13,95 %. Weiterhin geht *UNAIDS* davon aus, dass bis Ende 1999 3,5 Millionen Kinder in diesen drei Ländern ihre Eltern durch AIDS verloren haben.

Diese Daten weisen intensiv auf die große menschliche und wirtschaftliche Tragödie hin, die AIDS für afrikanische Entwicklungsländer südlich der Sahara darstellt. Beim derzeitigen Stand des medizinischen Fortschrittes sowie der knappen Ressourcen in Entwicklungsländern führt jede Infektion automatisch zum Tod. Entsprechend müssen AIDS-Kontroll-Programme mit großer Sachkenntnis und mit modernsten Methoden geplant und durchgeführt werden, denn sie entscheiden über Leben und Tod der Betroffenen. Hierzu stellt ein System Dynamics Modell der AIDS-Ausbreitung ein gutes Instrument dar.

2 Modell

Für Entwicklungsländer sind bislang nur wenige detaillierte Modelle verfügbar. Das hier vorgestellte Modell, welches in Grundzügen auf einem System Dynamics Modell von *Heidenberger & Fleßa* (1993, S. 167-176) aufbaut, kann nicht einfach eine Transferierung eines europäischen oder amerikanischen Modells sein, sondern muss der speziellen Epidemiologie eines Entwicklungslandes gerecht werden.

Die Bevölkerung wird in 19.200 Subpopulationen eingeteilt. Auf der ersten Ebene erfolgt eine Gliederung nach Wohnsitz, Beruf und Herkunft in 8 Gruppen: „Männer am Land", „Frauen am Land", „Männer in der Stadt", „Frauen in der Stadt", „Berufsprostituierte", „Gelegenheitsprostituierte", „Entwicklungshelfer" und „Touristen". Jedes Compartment wird dann wiederum in 80 Altersklassen gegliedert, so dass die altersspezifischen Risiken berücksichtigt werden können. Auf der nächsten Ebene erfolgt eine Schichtung nach dem Gesundheitszustand, d.h. „Nicht-Infiziert", „HIV-positiv durch perinatale Infektion", „HIV-positiv durch Circumcision", „HIV-positiv durch Bluttransfusion", „HIV-positiv mit langer Inkubationszeit" sowie „AIDS-krank". Die Compartments der Infizierten und AIDS-Kranken werden wiederum in fünf Subpopulationen aufgeteilt, um den Zeitraum seit der Infektion bzw. dem Eintritt ins Stadium AIDS aufzuzeichnen.

Das demographische System besteht aus den Geburten, der Alterung, der natürlichen Mortalität und der Mobilität. Das epidemiologische System unterscheidet Infektionen durch Partnerschaften, promiskuitive Beziehungen, die Wiederverwendung von Injektionsnadeln, Circumcission sowie perinatale Infektionen. Intervenöser Drogenmissbrauch sowie Homosexualität spielen in Afrika keine Rolle. Infektionen durch traditionelle Beschneidungen von Mädchen und Jungen sowie die Übertragung des HI-Virus im Gesundheitswesen sind jedoch speziell zu modellieren. Hier kann auf kein bestehendes Modell zurückgegriffen werden.

3 Prävention

Mit Hilfe dieses Modells kann der erwartete Verlauf der AIDS-Epidemie in der Musterregion sowie die Wirksamkeit verschiedener Interventionsmaßnahmen untersucht werden. Es zeigt sich ,dass eine effektive **Impfung** gegen das HI-Virus die beste Möglichkeit der Bekämpfung wäre. Hierzu werden verschiedene Szenarien gerechnet: Der Standardlauf gibt die Fallzahlen ohne Impfung wieder. Das zweite Szenario entspricht der Situation, die sich ergibt, wenn die Impfung am 1. Januar 2001 durchgeführt wurde und lebenslang zuverlässig gegen AIDS schützt. Eine weitere Simulation zeigt den Verlauf der Fallzahlen, der sich ergibt, wenn die Impfung nur eine Wirksamkeit von 50 % hat. Es sei angenommen, dass diese beschränkte Wirksamkeit lebenslang anhält. Ein Szenario gibt den Fall wieder, dass die Impfung zwar voll wirksam ist, jedoch nach 5 Jahren ihren Impfschutz verliert und keine Wiederholungsimpfung erfolgt. Das letzte Szenario entspricht der Annahme, dass eine lebenslang und 100 %ig wirksame Impfung erst am 1. Januar 2006 verfügbar ist.

Abb. 1: Konsequenzen einer Impfung zum 1.1.2001

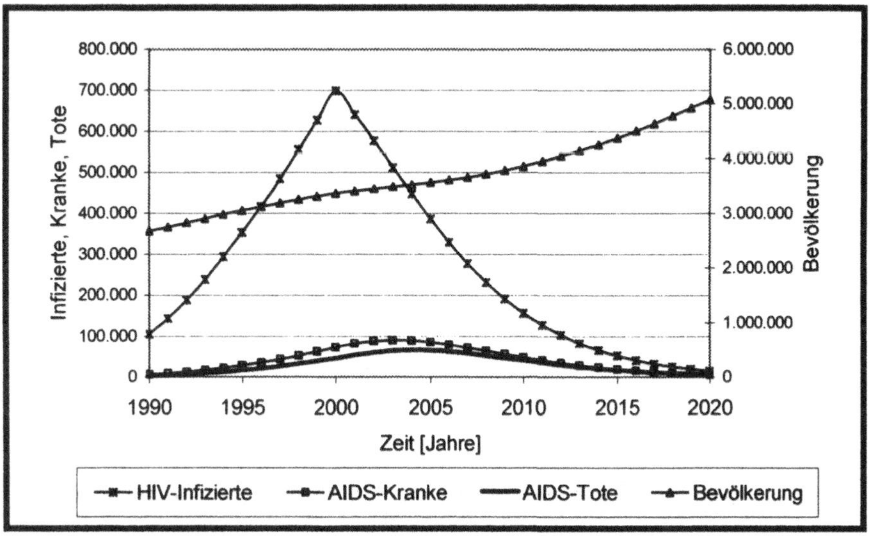

Abb. 2: Impfungen gegen AIDS: verschiedene Szenarien

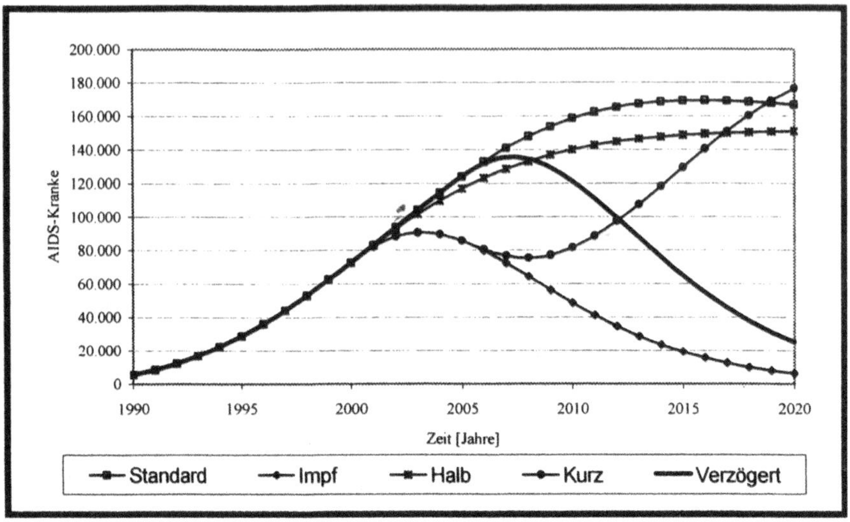

Wie zu erwarten, führen sowohl eine verspätete als auch eine Impfung mit eingeschränkter Wirksamkeit zu steigenden Fallzahlen. Da die Prävalenz progressiv steigt, ist langfristig ein Impfstoff, der erst verspätet zur Verfügung steht, jedoch hoch wirksam ist, einer Impfung mit geringer Wirksamkeit vorzuziehen. Besonders interessant ist der Verlauf der Fallzahlen, wenn die Impfung nur fünf Jahre wirksam ist. Ab 2006 steigt die Zahl der Neuinfektionen so stark an, dass im Jahr 2020 mehr Menschen an AIDS leiden als für den Fall, dass überhaupt keine Impfung durchgeführt worden ist. Dies ist darauf zurückzuführen, dass die Menschen, die fünf Jahre lang vor der Infektion mit HIV geschützt waren, anschließend ein großes Infektionsrisiko haben.

Ein Impfstoff, der keinen lebenslangen Schutz gewährt, kann folglich in Entwicklungsländern zu einer Akzelleration der demographischen und ökonomischen Konsequenzen von AIDS führen, wenn keine regelmäßigen Wiederholungsimpfungen durchgeführt werden. Deshalb müssen Impfkampagnen nachhaltig sein, sonst verstärken sie die ökonomische und epidemiologische Katastrophe, die AIDS für Entwicklungsländer bedeutet.

Die Entscheidung für oder gegen eine Impfung hängt von der Rate ab, mit der zukünftige Kosten und gewonnene Menschenleben diskontiert werden. Eine hohe Gegenwartsorientierung impliziert, dass zukünftige Ausgaben von geringer Bedeutung sind, so dass Kostenreduktionen, die weit in der Zukunft liegen, kaum berücksichtigt werden. Dementsprechend wird bei einer hohen Zeitpräferenz der Betrag, den eine Impfung kosten darf, geringer sein. Gesellschaften mit starker Gegenwartsorientierung haben ceteris paribus eine geringe Impfquote.

Tab. 1: Maximalkosten pro Geimpften, bei denen sich eine Impfung monetär rentiert

Diskontsatz	Impfung [US$]	Halb [US$]	Kurz [US$]	Verzögert [US$]
0 %	39,00	5,56	15,09	22,21
1 %	34,41	4,94	13,82	19,18
2 %	30,45	4,41	12,68	16,61
3 %	27,03	3,95	11,66	14,42
4 %	24,07	3,54	10,74	12,54
5 %	21,50	3,19	9,91	10,94
6 %	19,26	2,88	9,16	9,56
7 %	17,30	2,61	8,48	8,38
8 %	15,58	2,36	7,88	7,35
9 %	14,08	2,15	7,32	6,47
10 %	12,75	1,96	6,81	5,71
20 %	5,427	0,89	3,60	1,81
30 %	2,82	0,48	2,16	0,68
40 %	1,70	0,30	1,42	0,29
50 %	1,13	0,20	1,00	0,14

Mit dieser Methodik wird auch die Verhaltensprävention analysiert. Es zeigt sich, dass die bevölkerungsweite Verteilung (Szenario „Standard") von Kondomen eine geringere Effizienz hat als die Distribution an Prostituierte (Szenario „Prost") bzw. zum Schutz bei promiskuitiven Kontakten (Szenario „Promis"). Auch die Verwendung der Kondome für 50 % der promiskuitiven und 33 % der Partnerschaftskontakte (Szenario „Teil") hat keine hohe Effizienz. Dieses überraschende Ergebnis weist darauf hin, dass auch in Entwicklungsländern AIDS-Kontroll-Programme zielgruppenorientiert (z. B. Prostituierte) sein müssen. Die Verteilung von Kondomen an Prostituierte, Fernfahrer, Touristen und für Einmalkontakte hat eine hohe Kosten-Wirksamkeit.

Abb. 3: Verhaltensprävention: verschiedene Szenarien

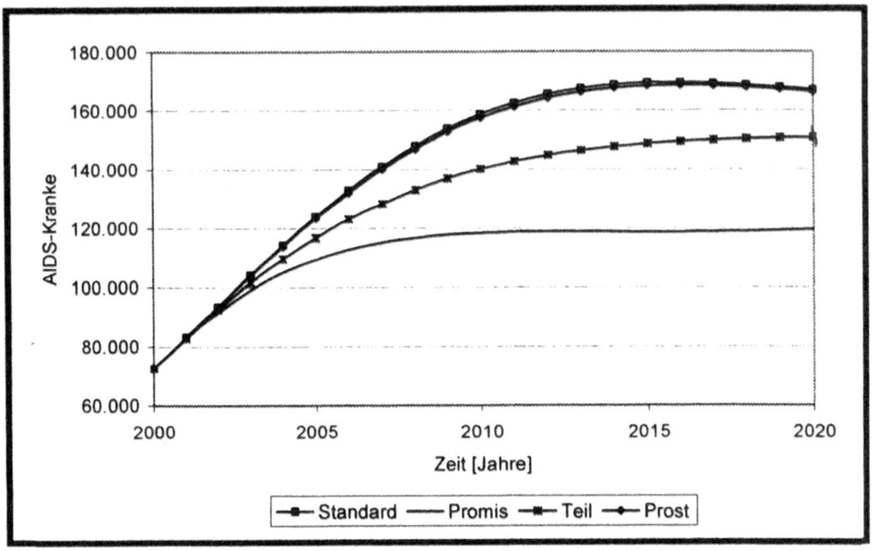

Abb. 4: Kostenersparnis durch Kondomverwendung[1]

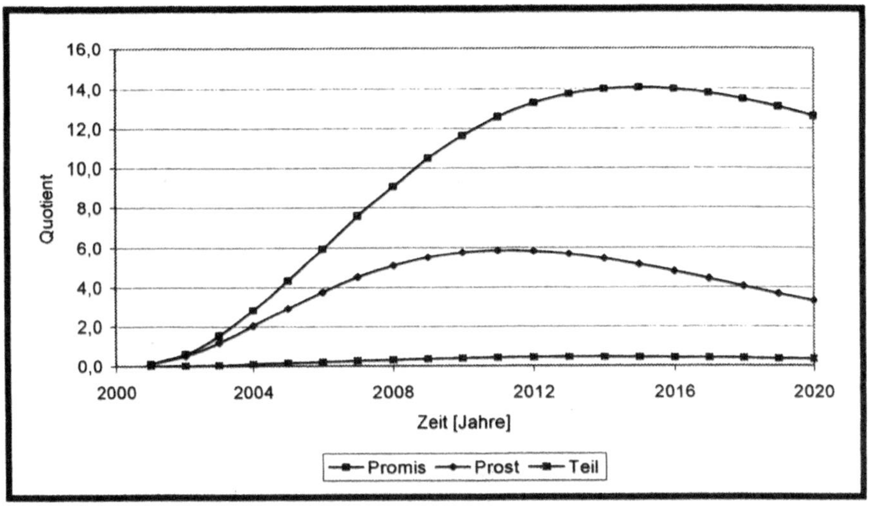

Die Unterbrechung der Übertragung perinataler Infektionen durch anti-retrovirale Medikamente erweist sich als ineffizient, wenn man monetäre Werte als alleinigen Maßstab des Handelns ansetzt. Die Szenarien gehen davon aus, dass die Übertra-

1 Eingesparte direkte Behandlungskosten geteilt durch Kosten der Kondome.

gungswahrscheinlichkeit von der Mutter auf das Kind 25 % (Szenario „Standard“), 12,5 % (Szenario „Halb“), 6,25 % (Szenario „Dreiviertel“) oder 0 % (Szenario „Null“) beträgt.

Tab. 2: Infizierte von 2001 bis 2020, verschiedene Szenarien perinataler Infektion

Infektionen	Szenarien			
	Standard	Halb	Dreiviertel	Null
Perinatal	461.938	231.861	116.151	0
Initiierungsriten	46	28	19	0
Einmalkontakte	879.446	890.211	895.543	900.844
Partnerschaftskontakte	1.488.034	1.504.231	1.512.249	1.520.214
Summe:	2.829.464	2.626.331	2.523.962	2.421.058

Abb. 5: Bevölkerung, verschiedene Szenarien der perinatalen Infektion

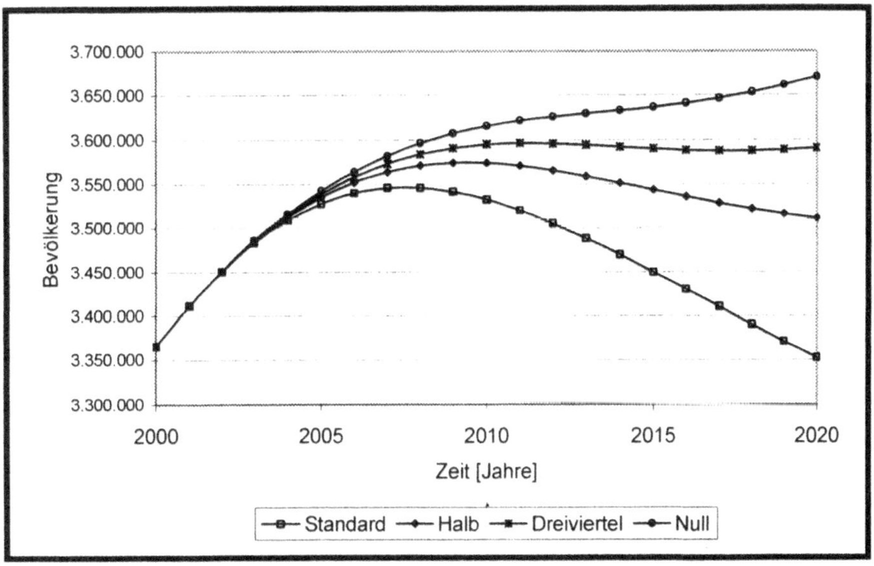

Die Modellrechnungen deuten darauf hin, dass bestehende AIDS-Kontroll-Programme auf Effizienz zu untersuchen sind, so dass die knappen Mittel in die Interventionsstrategien investiert werden, die die höchste Wirksamkeit haben. Die AIDS-Epidemie ist begrenzbar, jedoch können Least Developed Countries die nötigen Gesundheitsressourcen derzeit nicht aufbringen, um diese Interventionen nachhaltig zu implementieren. Hier ist die internationale Entwicklungshilfe gefordert, effiziente Präventionsmaßnahmen zu finanzieren.

Literatur:

UNAIDS (2000): Report on the global HIV/AIDS epidemic, June 2000. New York

UNAIDS/WHO (2001): AIDS epidemic update: January 2001. Genf

Heidenberger, K.; Fleßa, S. (1993): A system dynamics model for AIDS policy support in Tanzania. European Journal of Operational Research, Vol. 70, S. 167-176

Facilities Layout for Social Institutions

Horst W. Hamacher[1], Stefan Nickel[2], and Dagmar Tenfelde-Podehl[1]

[1] Universität Kaiserslautern, Fachbereich Mathematik, Postfach 3049, 67653 Kaiserslautern, Deutschland
[2] Fraunhofer Institut für Techno- und Wirtschaftsmathematik (ITWM), Gottlieb-Daimler-Straße, Geb. 49, 67663 Kaiserslautern, Deutschland

Abstract. Whenever a social institution plans new buildings or a modification of buildings (e.g. a hospital or a nursing home) the question arises how to arrange the rooms and functional units. One criterion is certainly the total walking distance of staff members, patients/clients and visitors.
Most of the existing approaches to the facilities layout problem either assume only one objective or they deal with several criteria but end up with only a single solution. Our approach focuses on determining Pareto-solutions which are presented to the decision makers. On that basis, it is possible to make a sound decision about the arrangement - by potentially incorporating those implicit criteria, which exist but have not been formulated before.
Furthermore, input data is most often assumed to be completely known which is in general not true. We show how uncertainty could be taken into account such that finally a robust solution is attained.

1 Introduction

In order to evaluate the quality of a layout for a social facility, we consider the total walking distance as a measure to be minimized. One has to decide, whether it is appropriate to minimize over all groups of persons (staff, patients/clients and visitors) at a time or whether it is better to differentiate between them? The same question could be posed regarding the period of time considered for minimization. In some cases it could be worthwhile to look separately at the data collected, for example, on a Monday morning and for data corresponding to a Sunday afternoon.
The question about optimality can also be viewed from another point of view: how is it possible to decide optimality when all data we work with are only estimations? They are possibly correct for the past, but more often they are only results of crude and subjective guesses. Furthermore, data about frequencies (how often does somebody go from facility i to facility j?) change over time. Hence these should be stochastic - and more needs to be done than only optimization with respect to the expected value of the data.
So what is an optimal solution? First of all it has to be reliable: reliable in terms of being a good solution for different possible values of the frequency matrix and also in terms of being as good as possible for several criteria.
One approach is to consider several criteria simultaneously and determine the so-called Pereto-optimal solutions, which are defined as follows:

Definition 1. A feasible solution x^* of (MOP) is called *Pareto optimal* if there does not exist another feasible solution x which is at least as good as x^* with respect to all criteria and strictly better for at least one objective. Let $\mathcal{X}_{Par}$ be the set of all Pareto optimal solutions. If x^* is Pareto optimal, $f(x^*)$ is called *efficient* or *nondominated.* The set of efficient points is denoted by $Y_{eff} := f(\mathcal{X}_{Par})$.

As an example we consider the Multicriteria Quadratic Assignment Problem (MCQAP) defined by Q frequency matrices $f^q = (f^q_{ik})$ and Q distance matrices $d^q = (d^q_{jl})$. (Here f^q_{ik} is the frequency of travel between facility i and facility k and d^q_{jl} is the distance between locations j and l, both with respect to criterion q.)With this denotation (MCQAP) can be stated as:

$$\min \begin{pmatrix} \sum_{i,j,k,l=1}^{n} f^1_{ik}\, d^1_{jl}\, x_{ij}\, x_{kl} \\ \vdots \\ \sum_{i,j,k,l=1}^{n} f^Q_{ik}\, d^Q_{jl}\, x_{ij}\, x_{kl} \end{pmatrix} \tag{1}$$

$$\text{s.t.} \sum_j x_{ij} = 1 \ \forall i = 1, \dots, n \tag{2}$$

$$\sum_i x_{ij} = 1 \ \forall j = 1, \dots, n \tag{3}$$

$$x_{ij} \in \{1, 0\} \ \forall i, j = 1, \dots, n. \tag{4}$$

where the problem is to find an assignment x defined as

$$x_{ij} = \begin{cases} 1, & \text{if facility } i \text{ is assigned to location } j \\ 0, & \text{otherwise,} \end{cases} \tag{5}$$

such that the total walking distances for all of the criteria is as small as possible. In Section 2 we propose an exact algorithm to compute the Pareto-optimal solutions of an MCQAP for any number of criteria.
The second approach, investigated in Section 3, deals with stochastic input data and again we consider the Quadratic Assignment Problem (QAP) as a basis for our investigations. We assume that the frequency matrix is stochastically distributed (distances between locations can be measured and hence are assumed to be known). Our goal is to obtain solutions which are as good as possible for both the expected value and the variance of the given objective function. Whereas minimizing the expected value can be done by solving a deterministic QAP with the expected frequency matrix, minimization of the variance is much more difficult: the objective is no longer a quadratic but a quartic term. For discrete distributions the corresponding program is a biquadratic assignment problem and can be solved e.g. by reformulating it using the ideas of Kaufman and Broeckx (cf. Burkard et al., 1994 [3]). Unfortunately the number of new (binary!) variables and constraints is large. To overcome this deficiency we combine ideas from Kaufman and Broeckx (1978)

[9] as well as from Adams and Sherali (1986) [2] and get a new reformulation yielding a mixed integer convex quadratic program, where the number of binary variables is not raised in comparison to the original problem.

2 An Exact Algorithm for the MCQAP

In this section we propose an algorithm to compute the Pareto solutions of an MCQAP lying inside a bounding box given by the decision maker. To state the algorithm we need the notion of level sets and level curves as given in the following definition:

Definition 2. Given a feasible set $\mathcal{F}$, an objective function $f = (f^1, \ldots, f^Q)$ and a real number z.
The *level set* of f^q at value z is $L^q_{\leq}(z) := \{x \in \mathcal{F} : f^q(x) \leq z\}$, and the *level curve* of f^q at value z is $L^q_{=}(z) := \{x \in \mathcal{F} : f^q(x) = z\}$.

The basic idea of the algorithm is valid for any multicriteria optimization problem (MOP) since it only relies on the next theorem which holds for a continuous as well as a combinatorial MOP. But in the following we will see how the special structure of the QAP can be exploited to make this general approach more efficient.
To determine Pareto-optimal solutions we use the fact (cf. Ehrgott et al , 1997 [5]) that a solution x^* is Pareto-optimal if and only if the intersection of the level sets coincides with the intersection of the level curves, i.e.

Theorem 1.

$$x^* \textit{ Pareto optimal} \Leftrightarrow \bigcap_{q \in \{1,\ldots,Q\}} L^q_{\leq}(z_q) = \bigcap_{q \in \{1,\ldots,Q\}} L^q_{=}(z_q),$$

where $z_q = f^q(x^*)$ *for all* $q = 1, \ldots, Q$.

Hence what we first need is a procedure to compute the level sets and level curves of a combinatorial optimization problem. For this purpose Hamacher and Queyranne developed in 1985 the so called *Binary Search Tree (BST) Algorithm* [8]. The idea of the BST Algorithm is the following:

- Partition the set of all feasible solutions $\mathcal{X}$ of a combinatorial problem in iteration k into k disjoint subsets, i.e. $\mathcal{X} = \mathcal{X}_1 \dot\cup \mathcal{X}_2 \dot\cup \ldots \dot\cup \mathcal{X}_k$
- In each set $\mathcal{X}_i$ we know a best solution X_i which is one of the k best solutions found so far
- With the exception of two sets $\mathcal{X}_p$ and $\mathcal{X}_k$ we also know a 2nd best solution in $\mathcal{X}_i$, $i \neq p, k$
- Then determine a 2nd best solution in $\mathcal{X}_p$ and $\mathcal{X}_k$
- After that, a comparison of all 2nd best solutions yields the $(k+1)^{st}$ best solution X_{k+1}

One of the main ingredients of the BST algorithm is hence a method to compute the second best solution of a combinatorial optimization problem. In the case of the QAP this can be done by excluding the known best solution $\tilde{x}$, e.g. by adding the constraint $\sum_{(i,j)\,:\,x_{ij}=1\,in\,\tilde{x}} x_{ij} \leq n-2$ to the original QAP. The meaning of this constraint is that at least two variables x_{ij} that are fixed to one in the optimal solution $\tilde{x}$ have to be changed to zero.

2.1 Determining the K Best Solution of the QAP: The Multiple Search Tree (MST) Algorithm

The MST Algorithm is a method for finding the K best solutions of the QAP, based on the idea of the BST Algorithm. But in contrast to the BST algorithm, MST fixes variables only to one and no longer to zero. In doing so, the dimension of each of the QAPs reduces by one. Although this approach makes it necessary to solve $n-1$ problems with variables fixed to one, first computational experiments (cf. Figure 1) show that avoiding variable-fixing to zero speeds up the procedure considerably.

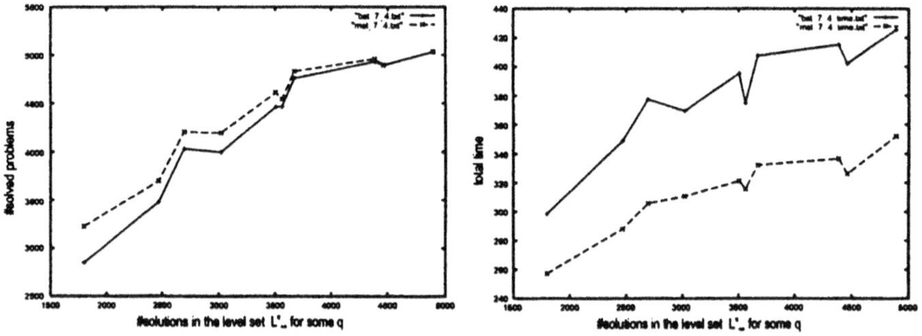

Fig. 1. 10 randomly generated MCQAP with seven facilities and four criteria

3 The Stochastic QAP (STQAP)

The problem we now address is uncertainty: the analysis of the frequencies often yields just estimations of the true numbers hence a stochastic inspection is interesting. Up to now not much has been done on stochastic quadratic assignment problems (STQAP). One of the few papers in this area, Li and MacGregor Smith in 1994 [10], is based on queueing systems. Assume now the entries of the frequency matrix are randomly distributed. Let $(\Omega, \mathcal{A}, P)$ be the probability space, $f_{ik} = f_{ik}(\omega)$, where $\omega \in \Omega$.

The two parameters most often used in stochastic investigations are the expectation and the variance. While the determination of the solution minimizing the expected objective is nothing else than solving a deterministic QAP with expected frequencies the problem of minimizing the variance is more complicated: The problem at hand is

$$\min_x \left\{ \text{var} \left[\sum_{ijkl}^{n} f_{ik}(\omega) d_{jl} x_{ij} x_{kl} \right] |(2) - (4) hold \right\} \tag{6}$$

which has no longer a quadratic objective function but is in fact a biquadratic assignement problem which was introduced by Burkard et. al. [3].

Lemma 1. *The problem (6) can be reformulated as a biquadratic assignment problem (VQAP) in the following form:*

$$\min_x \sum_{ijkl}^{n} \sum_{mpst}^{n} cov(f_{ik}(\omega), f_{ms}(\omega))\, d_{jl} d_{pt} x_{ij} x_{mp} x_{kl} x_{st} \quad (7) \qquad (VQAP)$$
$$s.t. \quad (2) - (4)$$

3.1 Possible Linearizations

In [3] two linearizations are proposed to solve a general biquadratic assignment problem, both straight forward translations of the well known linearizations for the (QAP). The first one yields a linear integer program by introducing $O(n^8)$ new integer variables $y_{ijmp} = x_{ij}x_{mp}$, $z_{ijklmpst} = y_{ijmp}y_{klst}$. To reduce the number of (integer) variables and the number of constraints Bukard et al. combine this idea (introducing n^4 binary variables $y_{ijkl} = x_{ij}x_{kl}$) with the idea of Kaufman and Broeckx [9] defining the variables $w_{ijkl} = y_{ijkl}(D_{ijkl}Y)$ where $D_{ijkl}Y = \sum_m^n \sum_s^n \sum_p^n \sum_t^n d_{mspt}y_{mspt}$ and $D_{ijkl} = \{\text{cov}\,(f_{ik}(\omega), f_{ms}(\omega))\, d_{jl}d_{pt}\}_{m,s,p,t=1,\ldots,n}$. The resulting program formulation has $n^4 + n^2$ binary variables, n^4 real variables and $2n^4 + 2n + 2$ constraints. Burkard et al. proposed a Branch & Bound procedure inducing the necessity of calculating lower bounds. Due to the linearization of a power-four function they could only report on rather weak bounds.

In the following section we introduce a different kind of simplification for the power-four objective function of the $(VQAP)$. Our goal is twofold: first, we want to avoid errors by applying a linearization, and secondly, we do not want to increase the number of binary variables.

3.2 A Quadratic Objective

Besides the linearization of Padberg and Rijal [11] there are mainly two other linearizations known for the QAP which introduce only real variables (for an overwiew over linearizations see for example [4]). One of these linearizations is the one of Kaufman and Broeckx as mentioned in the last section, the other linearization is based on the ideas of Frieze and Yadegar [6] and Adams and Sherali [2] and was proposed by Adams and Johnson [1] in 1994. The main idea here is to introduce real variables $y_{ijkl} = x_{ij}x_{kl}$.

Proposition 1. *The (VQAP) given in (7) and (2) - (4) is eqivalent to the following mixed integer quadratic program (MIQP) with n^2 boolean variables, n^4 real variables and $2n^3 + n^4/2 + 2n$ constraints:*

$$\min_{x,y} \sum_{ijkl}^{n} \sum_{mpst}^{n} cov(f_{ik}(\omega), f_{ms}(\omega))\, d_{jl} d_{pt} y_{ijkl} y_{mpst} \tag{8}$$

$$s.t. \quad (2), (3), (4)$$

$$\sum_{i \neq k}^{n} y_{ijkl} = x_{kl} \;\forall j, k, l = 1, \ldots, n \tag{9}$$

$$\sum_{j \neq p}^{n} y_{ijkl} = x_{kl} \;\forall i, k, l = 1, \ldots, n \tag{10}$$

$$y_{ijkl} = y_{klij} \;\forall i, j, k, l = 1, \ldots, n, i < k \tag{11}$$

$$y_{ijkl} \geq 0 \;\forall i, j, k, l = 1, \ldots, n \tag{12}$$

The proof of the proposition is analogous to the proof for the ordinary QAP.

Remark 1. The number of real variables could of course be reduced but due to easier notations it is more convenient to work with formulation (8) - (12). Unfortunately the objective function (8) of the latter ($MIQP$) is in general not convex but the advantage of this approach is that it yields for the QAP a linearization where the relaxed problem gives good lower bounds.

3.3 Applying the Idea of Kaufman and Broeckx

Up to now we did not fix any details about the distribution of the matrix entries. But to apply the idea of Kaufman and Broeckx in the case of nonindependent matrix entries we have to focus on discrete distributions. Therefore let $f_{ik}(\omega)$ take on values in $\{f_{ik}^1, \ldots, f_{ik}^{v_{ik}}\}$ with probabilities p_{ik}^b, $b = 1, \ldots, v_{ik}$ and for the joint distribution of $f_{ik}(\omega), f_{ms}(\omega)$ the set of possible values is $\{(f_{ik} f_{ms})^1, \ldots, (f_{ik} f_{ms})^{v_{ikms}}\}$, each of them taken with probability p_{ikms}^d, $d = 1, \ldots, v_{ikms}$. In addition we consider the case where the probabilities are corresponding to the entire matrix and not to the entries, i.e. $p_{ik}^b = p^b \;\forall\; (i, k)$. Then for the joint probability of $f_{ik}(\omega) f_{ms}(\omega)$ taking on the value $(f_{ik} f_{ms})^{bc}$ there holds $p_{ikms}^{bc} = \begin{cases} 0, & \text{if } b \neq c \\ p^b, & \text{if } b = c \end{cases}$.
Then the problem of minimizing the variance for this special discrete distribution can be written as

$$\min_x \sum_{b}^{v} p^b \left[\sum_{ijkl}^{n} \left(f_{ik}^b - \mathcal{E}(f_{ik})\right) d_{jl} x_{ij} x_{kl} \right]^2 \tag{13} \qquad (DVQAP)$$

$$\text{s.t.} \quad (2) - (4)$$

For applying the idea of Kaufman and Broeckx to $(VQAP)$ (or $(DVQAP)$) it is necessary that all coefficients are positive. As can be easily seen in $(VQAP)$ this is in general not the case due to the covariance involved in the problem. Therefore we define a constant C such that

$$C \begin{cases} \geq -\min_{i,j,k,l,b}\left[\left(f_{ik}^b - \mathcal{E}(f_{ik})\right) d_{jl}\right] & \text{if } \min_{i,j,k,l,b}\left[\left(f_{ik}^b - \mathcal{E}(f_{ik})\right) d_{jl}\right] < 0 \\ = 0 \text{ otherwise.} \end{cases}$$

Then there holds

Proposition 2. *Define* $c_{ij}^b = \sum_{kl}^n \left[\left(f_{ik}^b - \mathcal{E}(f_{ik})\right) d_{jl} + C\right] \ \forall\ i,j\ \forall\ b$. *Then the original problem* $(DVQAP)$ *is equivalent to the mixed integer program* $(MIQAP)$ *with* n^2 *Boolean variables,* vn^2 *real varialbes,* $4vn^2 + 2n$ *constraints, and a convex objective function:*

$$\min_{x,w} \sum_{b=1}^v p^b \left(\left[\sum_{ij}^n w_{ij}^b \right]^2 - 2Cn^2 \sum_{ij}^n w_{ij}^b \right)$$

$$s.t. \quad (2)-(4)$$

$$c_{ij}^b x_{ij} + \sum_{kl}^n \left[\left(f_{ik}^b - \mathcal{E}(f_{ik})\right) d_{jl} + C\right] x_{kl} - w_{ij}^b \leq c_{ij}^b \ \forall\ i,j\ \forall\ b \quad (14)$$

$$w_{ij}^b \leq x_{ij} c_{ij}^b \ \forall\ i,j\ \forall\ b \quad (15)$$

$$w_{ij}^b \leq \sum_{kl}^n \left[\left(f_{ik}^b - \mathcal{E}(f_{ik})\right) d_{jl} + C\right] x_{kl} \ \forall\ i,j\ \forall\ b \quad (16)$$

$$w_{ij}^b \geq 0 \ \forall\ i,j\ \forall\ b \quad (17)$$

Remark 2. Even if the lower bound obtained by solving the relaxation of $(MIQAP)$ is not the smallest possible, in any case it is bad.

Summing up: Applying just one of the Propositions 2 and 1 will yield a program with either a convex objective function with a relaxed problem giving bad lower bounds or with a QAP-relaxation giving good lower bounds but an objective function that is in general nonconvex. Thus for getting a problem with nice properties, we can combine both ideas:

Proposition 3. *Defining* C *as before Proposition (2) the following mixed integer quadratic problem with* n^2 *binary variables,* $n^4 + n^2$ *real variables and* $2n^3 + n^4/2 + vn^2 + 2n$ *constraints addutionally to the nonnegativity constraints is equivalent to the* $(DVQAP)$*:*

$$\min_{x,y,w} \sum_{b=1}^v p^b \left(\left[\sum_{ij}^n w_{ij}^b \right]^2 - 2Cn^2 \sum_{ij}^n w_{ij}^b \right) \quad (18)$$

$$s.t. \quad (2)-(4),(9)-(11),(12)$$

$$w_{ij}^b = \sum_{kl}^{n} \left[\left(f_{ik}^b - \mathcal{E}(f_{ik}) \right) d_{jl} + C \right] y_{ijkl} \ \forall\ i,j\ \forall\ b \tag{19}$$

$$w_{ij}^b \geq 0\ \forall\ i,j = 1\ \forall\ b \tag{20}$$

Proof. The proof is analogous to the proof of Proposition 1 since the values of w_{ij}^b are determined by (19) and the objective function is identical.

□

Future Research includes the investigation of other layout models, e.g. mixed integer formulations and the maximum planar subgraph problem under multiple criteria and stochastic data as well as the application to real world data.

Acknowledgement

We would like to thank Thomas Stephan from the University of Kaiserslautern for implementing both the BST and the MST algorithm and for conducting the computational experiments as well as Matthias Ehrgott from the University of Auckland for many helpful discussions.

References

1. Adams, W. P. and Johnson, T. A. (1994) Improved Linear Programming-Based Lower Bounds for the Quadratic Assignment Problem. DIMACS **19**, 43–75
2. Adams, W. P. and Sherali, H. D. (1986) A tight Linearization and an Algorithm for Zero-One Quadratic Programming Problems. Management Science **32**, 1274–1290
3. Burkard, R. E. and Çela, E. and Klinz, B. (1994) On the Biquadratic Assignment Problems. DIMACS **16**, 117–146
4. Çela, E. (1998) The Quadratic Assignment Problem, Theory and Algorithms. Kluwer Academic Publishers
5. Ehrgott, M., Hamacher, H. W, Klamroth, K., Nickel, S., and Schoebel, A. (1997) A note on the equivalence of balace points and Pareto solutions in multiple-objective programming. Journal of Optimization Theory and Applications, **92(1)**, 209-212
6. Frieze, A. M. and Yadegar, J. (1983) On the Quadratic Assignment Problem. Discrete Applied Mathematics **5**, 98–98
7. Glover, F. (1972) Improved linear representation of descrete mathematical programs. Management Science Report Series, University of Colorado
8. Hamacher, H. W. and Queyranne, M. (1985) K best Solutions to combinatorial optimization problems. Annals of Operations Research **6**, 123–143
9. Kaufman, L. and Broeckx, F.(1978) An algorithm for the quadratic assignment problem using Benders' decomposition. European Journal of Operational Research **2**, 204–211
10. Li, W.-J. and MacGregor Smith, J. (1994) Stochastic Quadratic Assignment Problems. DIMACS **16**, 221–236
11. Padberg, M. W. and Rijal, M. P. (1996) Location, Scheduling, Design and Integer Programming. Kluwer Academic Publishers, Boston

Section 8

Continuous Optimization

Slice Models in GAMS

Michael C. Ferris, Meta M. Voelker

Oxford University Computing Laboratory, Wolfson Building, Parks Road, Oxford, OX1 3QD (ferris,voelker@cs.wisc.edu)
Permanent address: Computer Sciences Department, University of Wisconsin - Madison, 1210 West Dayton Street, Madison, Wisconsin 53706, USA *

Abstract. We show how to implement the cross-validation technique used in machine learning as a slice model. We describe the formulation in terms of support vector machines and extend the GAMS/DEA interface to allow for efficient solutions of linear, mixed integer and simple quadratic slice models under GAMS.

1 Introduction to Slice Models

Slice models are collections of mathematical programs with the same structure but different data. Mathematically, the k-th program in a slice model is given by

$$\begin{array}{lll} \min_x & f^k(x) & \text{(objective slice)} \\ \text{subject to} & A^k x = b^k & \text{(slice constraints)} \\ & x \in \mathcal{X} & \text{(core constraints)} \end{array} \quad (1)$$

Each problem in the collection differs from the next only in a subset of the constraints, and possibly the objective function. We implicitly assume that much of the difficulty in the problems is hidden in the core constraints; for example, discrete variables, large numbers of, or difficult constraints may all appear as core constraints.

In previous work [3], we described the GAMS/DEA interface that was originally developed to solve large scale Data Envelopment Analysis (DEA) models [2]. In fact, the GAMS/DEA interface provides a way to define and solve more general slice models in GAMS efficiently. Normally, a slice model would be defined and solved in a loop over k, requiring the model to be generated multiple times. However, with the GAMS/DEA interface, the slice constraints are identified by a special alias (`slice`) and are generated all at once. The GAMS/DEA interface then defines the individual problems and passes them to the underlying solver CPLEX [6]. In this way, individual problems are not re-generated, but are instead defined as data modifications of each other, reducing overall model generation time. Furthermore, solutions from one problem can be used as starting points in subsequent problems to speed up overall execution time.

* This material is based on research partially supported by the National Science Foundation Grant CCR-9972372, the Air Force Office of Scientific Research Grant F49620-01-1-0040, Microsoft Corporation and the Guggenheim Foundation

This paper extends the interface to allow efficient testing of support vector machine problems using cross-validation. We have made two enhancements to the interface that allow this technique to be implemented efficiently within a modeling language. Firstly, we provide some simple tools and examples that enable the collection of models arising in cross-validation to be efficiently communicated to the solver. Secondly, we allow quadratic programs (and hence standard support vector machines) to be explicitly formulated as slice models, in addition to linear and mixed integer programs.

2 Support Vector Machines: An Example from Breast Cancer Diagnosis

We focus on an example involving the classification of patients' breast tumors as either malignant or benign. We make use of a support vector machine, trained on a set of data for which the outcome is already known. For the data, we use the Wisconsin Diagnosis Breast Cancer Database (WDBC) [7], consisting of 569 patients and 30 features. In order to test the classifier, the model makes use of cross-validation, leaving a portion of the data out of the training for use as testing data.

We solve the standard linear support vector machine model from [1]:

$$\begin{array}{ll} \min_{w,\gamma,z} & \frac{1}{2}\|w\|_2^2 + Ce^T z \\ \text{subject to} & D(Aw - \gamma) + z \geq 1, \quad z \geq 0 \end{array} \tag{2}$$

Here, A is a matrix containing the training data (patients by features) and D is a diagonal matrix with values ± 1, denoting benign tumors $(+1)$ and malignant tumors (-1) for the corresponding patients. C is a parameter weighting the importance of maximizing the margin between the classes versus minimizing the misclassification error (z). The solution w and γ are used to define a separating hyperplane $\{x | w^T x = \gamma\}$ to distinguish between (unseen) benign and malignant data points.

Prior to solving the quadratic programming (QP) model (2), we first consider a modified version that can be converted into a linear program:

$$\begin{array}{ll} \min_{w,\gamma,z} & \|w\|_1 + Ce^T z \\ \text{subject to} & D(Aw - \gamma) + z \geq 1, \quad z \geq 0 \end{array} \tag{3}$$

Model (3) replaces the Euclidean-norm margin measurement of (2) with the sup-norm measurement (resulting in the minimization of the 1-norm) and can be converted into a linear program by adding additional variables, y:

$$\begin{array}{ll} \min_{w,\gamma,z,y} & e^T y + Ce^T z \\ \text{subject to} & D(Aw - \gamma) + z \geq 1, \quad z \geq 0 \\ & y \geq w \geq -y \end{array} \tag{4}$$

3 GAMS Formulations for the Linear Model

Models (2) and (4) are not slice models per se. They become slice models under cross-validation, where they are solved multiple times on different pieces of data. In these cases, only the data A and D vary between solves, appropriately fitting the definition of slice models.

Figure 1 shows the original GAMS formulation for 10-fold cross-validation applied to (4). Only training is carried out in this model, displaying values that define the separating hyperplane (w, γ) for later testing. The equations are defined over the training sets, generated dynamically from the testing sets inside the solve loop. The testing sets are defined by calling the batch include file `gentestset.inc` (figure 2). A and D are defined in `wdbc.gms`.

Figure 3 shows the GAMS/DEA formulation for the same model (where identical lines above the problem definition have been omitted). The major differences between the original and GAMS/DEA formulations are the lack of the solve loop and training sets in the latter. Also, the GAMS/DEA solver is used with an options file `dea.opt` (figure 4).

The solve loop is eliminated in the GAMS/DEA formulation by the way in which the interface handles the data. All of the data is passed to the interface initially and the key name `slice` is used to determine which equations/data belong to which problems. `slice` is aliased to the set that defines the individual problems (in this case, the fold set). Equations that change between problems (such as `sep_def`) must be declared over `slice`.

The training sets are eliminated from the formulation since they are defined implicitly by deleting in turn each testing set from the complete collection of testing sets. This is an example of a deletion-style slice model (denoted in the options file by `slicetype 0`). DEA-style slice models are addition models: models in which data is added to a core set in order to define a specific problem; addition models are the default slice type in the interface. However, cross-validation models are more appropriately defined by *deleting* slices from the entire (core) set of data.

In the GAMS/DEA formulation, the testing sets are defined prior to the solve statement (since all of the data is passed to the interface at once), and are indexed by `p`. In addition, the equations `sep_def` that depend upon the testing sets are indexed by `p`. By creating a mapping from the slice/fold set `p` to a subset of the patient set `i`, the corresponding (testing) equations are deleted from each of the problems in turn.

Note that equation `obj_def` changes from a sum over the training set to a sum over the entire set in the GAMS/DEA formulation. Variables related to the testing sets in `obj_def` will be eliminated (since they do not appear elsewhere in the model). By default, the GAMS/DEA interface ensures that the number of equations in each problem is the same; most slice models have this property. However, in cross-validation, `card(p)` may not evenly divide `card(i)`. The last problem may have fewer equations than the rest, so we use the option `eqnchk` to turn off the automatic equation check.

```
$title Ten-fold cross-validation example
$eolcom !

$setglobal num_folds 10
set p /1*%num_folds%/;   ! folds to perform
! set i /1*569/;         ! patients (declared in data file)
! set k /1*30/;          ! features (declared in data file)

! Read in data
$include "wdbc.gms"
parameter C /1/;

! Declare testing and training sets
set test(i);
set trai(i);

! Define problem
positive variables z(i), y(k);
variables obj, w(k), gamma;

equations obj_def, sep_def(i), bd1(k), bd2(k);
obj_def..  obj =e= C*sum(trai, z(trai)) + sum(k,y(k));
sep_def(trai)..
           D(trai)*(sum(k, A(trai,k)*w(k)) - gamma) + z(trai) =g= 1;
bd1(k)..   y(k) =g= -w(k);
bd2(k)..   y(k) =g= w(k);

model train /all/;

! Solve
loop(p,

! Generate testing and training sets
$batinclude gentestset.inc i
trai(i) = not test(i);

solve train using lp minimizing obj;
display w.l,gamma.l;

);
```

Fig. 1. The original GAMS formulation for cross-validation applied to model (4).

```
$set i_num floor(card(i) / %num_folds%)

test(%1) = no;

test(%1)$((ord(i) > %i_num%*(ord(p)-1)) and
            (ord(i) <= %i_num%*ord(p))) = yes;
! put leftovers in last set
test(%1)$((ord(p) eq %num_folds%) and
            (ord(i) > %i_num%*%num_folds%)) = yes;
```

Fig. 2. The batch include file **gentestset.inc** for generating the testing sets.

```
! Define problem
alias(p,slice);
positive variables z(i), y(k);
variables obj, w(k), gamma;

equations obj_def, sep_def(i,slice), bd1(k), bd2(k);
obj_def..  obj =e= C*sum(i, z(i)) + sum(k,y(k));
sep_def(i,p)$test(i,p)..
           D(i)*(sum(k, A(i,k)*w(k)) - gamma) + z(i) =g= 1;
bd1(k)..   y(k) =g= -w(k);
bd2(k)..   y(k) =g= w(k);

model train /all/;

! Solve
option lp = dea;  train.optfile = 1;

! Generate testing sets (to be deleted in each problem)
loop(p,

$batinclude gentestset.inc "i,p"

);

solve train using lp minimizing obj;
```

Fig. 3. The GAMS/DEA formulation for cross-validation applied to model (4).

```
slicetype 0
eqnchk 0
primval w,gamma
```

Fig. 4. The options file for the GAMS/DEA formulation.

4 Cross-Validation Testing

Since only one call is made to the solver, the resulting listing file will contain only one set of solutions (for the last problem). Solutions for all of the problems are written to the file `dea.sol`. Inside `dea.sol`, the solutions are given as GAMS parameters, indexed by the slice set. By default, the model status, solver status, and the objective values are written to the solution file.

```
* Ten-fold cross-validation testing file
$eolcom !
$setglobal num_folds 10
set p /1*%num_folds%/;  ! folds to perform

! Read in the data and solution values
$include "wdbc.gms"
$include "dea.sol"

! Declare testing sets
set test(i,p);
loop(p,
$batinclude gentestset.inc "i,p"
);

! Evaluate testing sets
parameter missclass(p);
missclass(p) = sum(i$test(i,p),
  D(i)*(sum(k,A(i,k)*wval('prim',p,k)) - gammaval('prim',p)) le 0);

display missclass;
```

Fig. 5. GAMS code for calculating the number of misclassified test points.

This means that cross-validation testing cannot be done inside of the training program. Since we need to perform testing in the examples above, we write out `w` and `gamma` through the option `primval`. A separate testing file (figure 5) reads in the solution file `dea.sol` and calculates the number of misclassified test points by counting the number of test points i with $D_{ii}(A_{i\cdot}w - \gamma) \leq 0$. Note that the string "val" is appended to the variable name for the solution value.

5 The GAMS Formulation for the QP Model

We extend GAMS/QPWRAP [5] to solve the QP slice model (2). Under GAMS/QPWRAP, a QP model is formulated as a combination of a linear program (specified in the GAMS model) and a Q matrix (specified in a text file). GAMS/QPWRAP passes the Q matrix to the solver.

```
! Define the Q matrix
parameter q(k,k);  q(k,k) = .5;

! Now write the Q matrix to a file named qmatrix.txt
file qp / qmatrix.txt /;
qp.pc=5; qp.nr=2; qp.nd=13; qp.nw=0; ! Some formatting options

put qp 'Q Matrix for svm';
loop(k, put / 'w' k.tl 'w' k.tl  q(k,k))
putclose;

! Define problem
alias(p,slice);
positive variables z(i);
variables obj, w(k), gamma;

equations obj_def, sep_def(i,slice);
obj_def..  obj =e= C*sum(i, z(i));
sep_def(i,p)$test(i,p)..
           D(i)*(sum(k, A(i,k)*w(k)) - gamma) + z(i) =g= 1;

model train /all/;

! Solve
option lp = deaqp;  train.optfile = 1;

! Generate testing sets (to be deleted in each problem)
loop(p,
$batinclude gentestset.inc "i,p"
);

solve train using lp minimizing obj;
```

Fig. 6. The GAMS/DEA formulation for cross-validation applied to model (2).

Figure 6 shows the GAMS formulation for 10-fold cross-validation applied to model (2) (identical lines above the problem definition have been omitted). This formulation is very similar to that of figure 3, with `slice` identifying the slice constraints. It uses the same options file (figure 4). Only the linear portion of model (2) is given explicitly; the quadratic portion is specified by the Q matrix and written to the file `qmatrix.txt` prior to the solver call. In addition, the solver has been changed from `dea` to `deaqp`, that calls GAMS/QPWRAP to read the Q matrix prior to calling GAMS/DEA.

Since the file `qmatrix.txt` must exist prior to the solver call, the ability for GAMS/DEA to solve QP slice models is limited. GAMS/DEA can only solve QP models where the Q matrix does not change between solves, that is, where the Q matrix is part of the core data.

On the WDBC data set, we ran all-but-one testing on both the linear and the quadratic formulations. Testing of model (4) took 541.71 seconds under GAMS/CPLEX and 269.03 seconds under GAMS/DEA; testing accuracy was 95.08%. Testing of model (2) took 7403.03 seconds under GAMS/CPLEXQP and 7221.74 seconds under GAMS/DEAQP; testing accuracy was 95.42%. However, much faster results can be obtained under non-linear solvers. Under GAMS/CONOPT, testing of model (2) took 519.3 seconds; under GAMS/PATH, testing took 691.9 seconds. Eliminating regeneration time from these results, a slice approach would improve these times to 343.18 seconds and 245.45 seconds, respectively.

6 The GAMS/DEA Solver

The linear GAMS/DEA solver is specified with the `option lp=dea` statement; the mixed integer GAMS/DEA solver is specified with the `option mip=dea` statement; and the simple QP GAMS/DEA solver is specified with the `option lp=deaqp` statement. Under any of these, GAMS will call the GAMS/DEA interface, that defines and solves the individual problems (based on the generated slice constraints). CPLEX is used as the underlying solver.

GAMS/DEA requires a properly installed GAMS system (distribution 20.1 or newer). In addition, a valid GAMS license and a GAMS/CPLEX or an ILOG CPLEX 7.1 (or higher) callable library license is needed. The package, including source for all examples, can be downloaded from the GAMS Contributed Software web site [4].

References

1. Bennett, K. P., Campbell, C. (2000): Support Vector Machines: Hype or Hallelujah? SIGKDD Explorations 2(2).
2. Cooper, W. W., Seiford, L. M., Tone, K. (2000): Data Envelopment Analysis: A Comprehensive Text with Models, Applications, References and DEA-Solver Software. Kluwer Academic Publishers, Boston.
3. Ferris, M. C., Voelker, M. M. (2000): Slice Models in General Purpose Modeling Systems. Data Mining Institute Technical Report 00-10, Computer Sciences Department, University of Wisconsin, Madison, Wisconsin. Available from `http://www.cs.wisc.edu/dmi/tech-reports/`.
4. GAMS Development Corporation (2001): Contributed Software. HTML document available from `http://www.gams.com/contrib/contrib.htm`.
5. GAMS Development Corporation (2001): Quadratic Programs in GAMS. HTML document available from `http://www.gams.com/contrib/qpwrap/qpwrap.htm`.
6. ILOG (2001): CPLEX 7.1 Reference Manual. Accessed as HTML document, distributed with ILOG CPLEX 7.1 libraries.
7. Wolberg, W. H., Street, W. N., Mangasarian, O. L. (1996): Wisconsin Diagnosis Breast Cancer Database (WDBC). FTP documents available from `http://www.cs.wisc.edu/~olvi/uwmp/cancer.html`, among other sites.

On Optimal Control of Heating Processes

Stefan W. Pickl[1] and Gerhard-W. Weber[2]

[1] Mathematisches Institut, Zentrum für Angewandte Informatik Köln ZAIK, email: pickl@zpr.uni-koeln.de, Universität zu Köln, D-50931 Köln

[2] Fakultät für Mathematik, email: G.Weber@mathematik.tu-chemnitz.de, Technische Universität Chemnitz, D-09107 Chemnitz

Abstract. This paper is concerned with investigating the nonlinear time-discrete TEM model for environmental protection within the heating problem of global warming.
As the model's steady states are not attractive, the necessary introduction of control parameters into the TEM model's dynamics leads to new results. These give new insights for a Joint-Implementation Program and, herewith, they may improve this important economic tool. The steady states of the TEM model can be considered as limiting values mentioned in Kyoto Protocol.
In a second step, we regard the determined control parameters as candidates for influencing heating of the earth atmosphere (greenhouse effect). We consider the earth as a ball and look for an optimal control such that the temperature in the stratosphere remains within lower and upper bounds. Here, we utilize scientific experience from [KRABS 1987] and [KAPLAN, TICHATSCHKE 1997] and from the authors on time-minimal control of heating by generalized semi-infinite optimization theory. We present a iteration procedure and first numerical results.

Keywords:
TEM-model, CO_2-reduction, dynamical systems, generalized semi-infinite optimization, iteration procedures, time-discrete, control, time-minimal heating, controlled heating

1 A Nonlinear Time-Discrete Model

The conferences of Rio de Janeiro 1992 and Kyoto 1997 demand for new economic instruments which focus on environmental protection in both macro and micro economy. An important economic tool in that area is *Joint-Implementation (JI)* which is mentioned explicitly in Kyoto Protocol. It is an international program which intends to strengthen international cooperations between enterprises on reducing CO_2-emissions. A sustainable development can only be guaranteed if the instrument is embedded in an optimal energy management. In this context, optimal energy management according to *JI* means that it must work on the micro level with minimal costs and it should be protected against misuse on the macro level.

For that reason, the TEM model (Technology-Emissions-Means model) [PICKL 1999] was developed, giving the possibility to simulate such an economic behaviour. The realization of JI is determined subject to technical and financial constraints. Furthermore, the TEM-model can be used to analyze the influence of several cost-allocations on the feasible set of control parameters. In a JI program emissions reduced by technical cooperations are registrated at the *clearing house* whose establishment is also a demand of Kyoto Protocol [KYOTO 1997].

2 The TEM Model

The presented TEM model describes the economic interaction between several actors (players) which intend to minimize their emissions (E_i) caused by technologies (T_i) using expenditures of money (M_i) or financial means, respectively. The index $i \in \{1, \ldots, n\}$ stands for the i-th player. The players are linked by technical cooperations and the market, which expresses itself in the nonlinear time-discrete dynamics of the Technology-Emissions-Means model, in short: TEM model. The TEM model is based on a general model which was refined in [PICKL 1999], namely on the following time-discrete system:

$$\Delta E_i(t) = \sum_{j=1}^{n} em_{ij}(t) M_j(t)$$

$$\Delta M_i(t) = -\lambda_i M_i(t)(M_i^* - M_i(t))(E_i(t) + \varphi_i \Delta E_i(t)) + u_i(t)$$

The first equation describes the time-dependent behaviour of the emissions reduced so far by each player. These levels are influenced by financial investigations which are determined by the second equation. The em_{ij}-parameters determine the effect on the emissions of the i-th actor, if the j-th actor invests money. We can say that it expresses how *effective* technology cooperations are, which is the kernel of a JI program. Besides that, the integration into the TEM model of the memory parameter φ and the growth parameter λ guarantee a realistic economic market behaviour while the $M_i^*(t)$ $(i \in \{1, \ldots, n\},\ t \in N)$ are upper bounds for the financial investigations. The additional parameter $u_i(t)$ is a control parameter which should control and reduce earth warming processes.

3 The Optimal Control Problem of Heating

In [KRABS, PICKL 2000] we control the TEM model and regard an iterative solution applying the gradient method. Now, we insert the determined control parameters into the problem of earth warming which we prepare by a model of time-minimal control.

Based on [KRABS 1987], continued by [WEBER 1999] and [KROPAT, PICKL, RÖSSLER, WEBER 2000], we studied the following problem:
Given a ball consisting of a homogeneous material (see Figure 1).

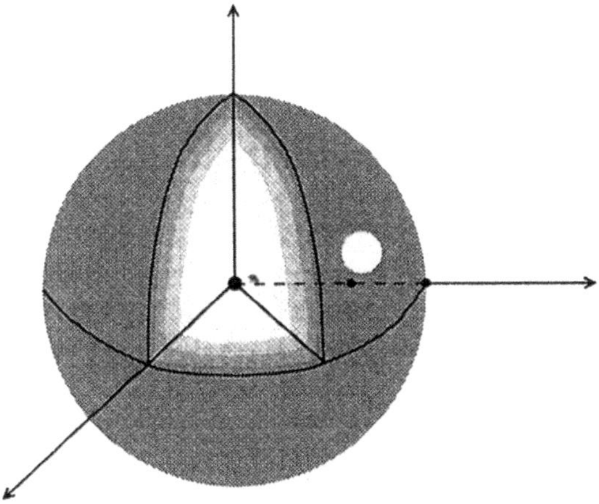

Fig. 1. A ball and its temperature distribution

We want to control the temperature (θ) from a given to a terminal temperature. This process is govered by a heat equation. We consider the temperature at the boundary as out control variable (u) which is equivalent to the control parameter of the TEM model. The emissions described by the TEM model are proportional to (θ).

This leads to the following formulation:

$$\mathcal{P}^{\mathrm{tm}} \left\{ \begin{array}{lr} \text{Min} \quad \mathsf{I}(T, u) := T \quad \text{such that} & (1) \\ \text{there is a bounded function } \theta : [0, R] \times [0, \infty) \rightarrow, \text{ where} & \\ \theta|(0, R] \times (0, \infty) \text{ is partially differentiable,} & \\ u = \theta(R, \cdot)|[0, T] \text{ is continuous, and} & \\ \theta_t(r, t) = a\Delta\theta(r, t) = \frac{a}{r^2}\frac{\partial}{\partial r}(r^2\theta_r(r, t)) \quad ((r, t) \in (0, R] \times (0, \infty)), & (2) \\ \theta(r, 0) = \theta_0 \quad (r \in [0, R]), & (3) \\ \theta(R, T) = \theta_E, & (4) \\ T \geq 0, & (5) \\ |\sigma_u(R, t)| \leq \sigma^* \quad (t \in [0, T]). & (6) \end{array} \right.$$

Here, $\Delta\theta$ represents the Laplacian of θ and R denotes the radius of B. The temperature $\theta(r,t)$ is a function of the radial variable r, where r measures the distance from the center point 0_3 of B, and of the time t. Moreover, we start with an **initial** temperature θ_0 and finish with an intended **target** (end) temperature $\theta_E > \theta_0$ (or $\theta_E < \theta_0$, respectively). Because of this inequality, each time $\widehat{T}$ which is optimal for $\mathcal{P}^{\mathrm{tm}}$, can not be zero (i.e., $\widehat{T} > 0$). The temperature is essentially governed by the heat equation (2), where $a > 0$ describes the coefficient of heat conductivity. This can effectively be realized by the substitution $v(r,t) := r\theta(r,t)$ in (2). We interpret $u_T(\cdot) := u(\cdot) = \theta(R,\cdot)|[0,T]$ as a *control variable* ($T \geq 0$).

As it was show by [KRABS 1987], the *norm-minimal problem* on the termal thress tangential to the boundary has precisely one single solution. We call such an implicit variable a *core function*. (Similar functions appear in optimal control of ordinary differential equations, too; cf. [WEBER 1999].) Below, inserting of this function as our control u will lead to a problem from *generalized semi-infinite optimization*. Here, the set of inequality constraints (enumerated by $y := t$) is usually infinite and depending on the state ($x := T$; [WEBER 1999]).

We are focussing on our partial differential equations with the following unique solution of the (boundary-value) problem:

$$\begin{aligned}\theta(r,t) &= 2r \sum_{k=1}^{\infty} \tfrac{(-1)^{k+1}}{k\pi} \exp(-a(\tfrac{k\pi}{R})^2 t)\theta_0 \tfrac{1}{r} \sin(\tfrac{k\pi}{R} r) + \\ &+ \tfrac{2a}{R} \sum_{k=1}^{\infty} (-1)^{k+1} k\pi \cdot \int_0^t \exp(-a(\tfrac{k\pi}{R})^2 (t-s))\, u(s)\, ds \cdot \tfrac{1}{r} \sin(\tfrac{k\pi}{R} r).\end{aligned} \tag{7}$$

Furthermore, the function $\sigma_u(r,t)$ denotes some thermal stress tangential to the boundary ∂B of B ($r = R$). Finally, σ^* is a given upper bound of the stress.

Under suitable physical assumptions, at the boundary the function $\sigma_u = \sigma_u^{\theta_0}$ has the form

$$(\sigma_u(R,t) =)\ \sigma_u^{\theta_0}(R,t) = \tfrac{E\alpha}{1-\mu} \left(\tfrac{3}{R^3} \int_0^R \theta(r,t) r^2\, dr - u(t)\right). \tag{8}$$

Now, by the preceding functions with the core function inserted, $\mathcal{P}^{\mathrm{tm}}$ becomes the announced problem from generalized semi-infinite optimization. However, based on the solution of the heat equation, this optimization problem need not be smooth. Therefore, we smoothly approximate the defining functions and, then, apply a suitable iteration procedures from generalized semi-infinite optimization. For example, these procedures use, e.g., discretization techniques (see [KROPAT, PICKL, RÖSSLER, WEBER 2000], [WEBER 1999], where several alternative variants and refinements are also presented, e.g., a penalty method) or local linearization within transversally intersecting squares. Detailed information about procedures and underlying theory can be found in [PICKL, WEBER 2000] and [WEBER 1998, 1999, 2001]. From here we derive structural stability results on problem approximation which guarantee the convergence theory for our control problem's numerical treatment.

As we are concerned with the danger of earth warming, we must refine this study by letting the temperature θ of stratosphere also depend from $h :=$ distance from equator, and control it within lower and upper bounds: θ_ℓ and θ_u. Furthermore, we have to replace the ball by the difference of two balls. In fact, we turn from our ball heating problem from production and technique to the warming problem of stratosphere around our earth. (The big size difference of the two balls is a question of so-called *dimension theory*.) Now, we have

$$R_\ell \leq r \leq R_u ,$$

i.e, the radius r varies between a lower bound R_ℓ being half the earth diameter, and an upper bound R_u.

However, one more modification is necessary: We do no longer have a homogeneous material, but the temperature θ also depends on the distance h from equator. In fact, we therefore must *locally* modify our heat equation by an additional h-dependence.

4 Numerical Solutions

We really obtained numerical experience, in particular, by treating our time-minimal problem of heating or cooling as a problem of finding the unique zero T of a strong monotonical decreasing function d defined by the parameters of $\mathcal{P}^{\mathrm{tm}}$. In fact, as it was shown by [KRABS 1987], under certain data constellation, this monotonicity is guaranteed.

We smoothly approximated d by functions d_ν such that their zeroes T_ν tend to the unique solution (T) of our given problem. The following Figure 2 exhibits numerical calculations of functions d_ν and, herewith, of their zeroes.

In our complex optimal control problem, it is hard analytically to verify any constraint qualification, e.g., the extended one of Mangasarian and Fromovitz (EMFCQ). We know that EMFCQ implies that the feasible set is a topological manifold with boundary (see [WEBER 2000]). So, in order to get an imagination how the feasible set (topologically) looks like and, herewith, to give an "indication" of EMFCQ, we make discretization and visualization (cf. [KROPAT, PICKL, RÖSSLER, WEBER 2000], [WEBER 2000]).

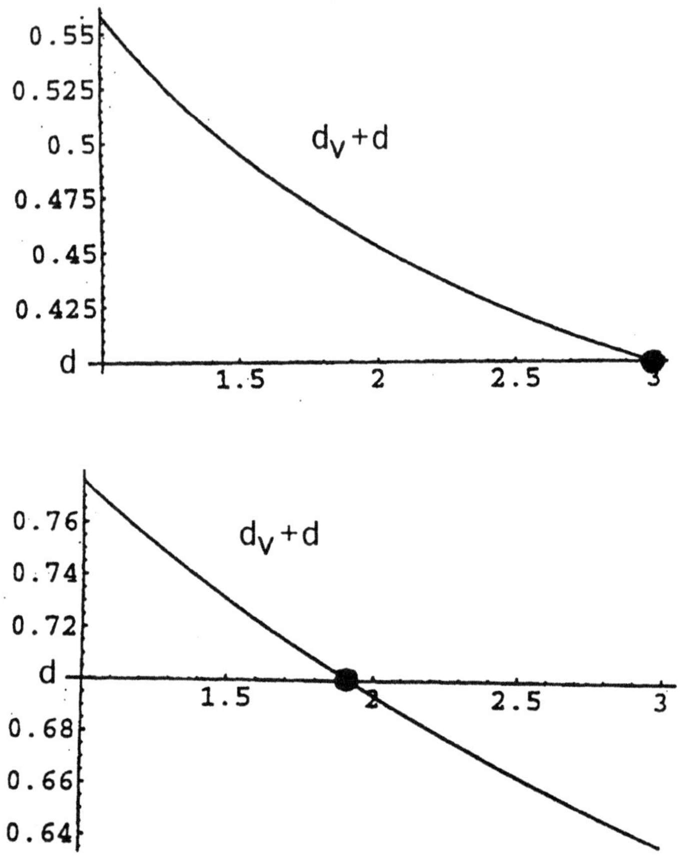

Fig. 2. Finding approximative solutions T_ν under different parameter constellation ($a \in \{0.5, 0.8\}$)

5 Conclusion

We explained how the two approaches from TEM model and time-minimal control come together in a way which is suitable for a control of greenhouse effect. This combination initiates further fruitful research. We also mention the numerical approach presented by the [KAPLAN and TICHATSCHKE 1997].

Let us give the following interpretation of our h-dependence of θ: Such an phenomenon is just the one which we have in the case of a thinning of a uniformly warmed rigid body in the middle. A cooling problem of similar "windmill shaped bodies" is investigated by [EPPLER, TROELTZSCH 2001] in a DFG Forschungsschwerpunkt. Heat processes over diffent bodies are analyzed in [PARKUS 1959a, 1959b]. For future research, we also suggest a transformation of our ball (earth) to such a "hyperboloid".

Moreover, for a realistic refinement, we have to incorporate local and global temperature exchanges in stratospere (e.g., aeodynamical ones) and height (distance from earth). We also will consider a discretization of stratosphere.

Finally, let us mention the alternative approach called *Nordhaus model* studied in [EMMERICH, SCHMITT 2001]. There, an initial value problem of ordinary differential equations is studied in the unknows

- gas concentration,
- temperature,

while a welfare functional has to be maximized.

However, in our approach, we basically integrated the time-discrete TEM model into our optimal control problem of heating (warming) or cooling.

References

1. EMMERICH, E., and SCHMITT, H. (2001): *Numerical simulation of economic control problems: a model of global warming*, Preprint, Berlin University of Technology.
2. EPPLER, K., and TROELTZSCH, F. (2001): *Discrete and continuous optimal control strategies in the selective cooling of steel*, Preprint, in: *Echtzeitoptimierung grosser Systeme* (Schwerpunktprogramm der Deutschen Forschungsgemeinschaft.
3. JONGEN, H. Th., JONKER, P., and TWILT, F. (1986): *Nonlinear Optimization in R^n, II: Transversality, Flows and Parametric Aspects*, Peter Lang Publ. House, Frankfurt a. M., Bern, N.Y..
4. KAPLAN, A., and TICHATSCHKE, R. (1997): *On the numerical treatment of a class of semi-infinite termal problems*, Optimization, 41, 1-36.
5. KRABS, W., and PICKL, S. (2000): *Controllability of a time-discrete dynamical system with the aid of the solution of an approximation problem*, Journal of Control Theory and Cybernetics, submitted for publication.
6. KRABS, W. (1987): *On time-minimal heating or cooling of a ball*, Intern. Ser. Numer. Math., 81, Birkhäuser, Basel, 121-131.
7. KROPAT, E., PICKL, S., RÖSSLER, A., and WEBER, G.-W. (2000): *A new algorithm from semi-infinite optimization for a problem of time-minimal control*, Journal of Computational Technologies, 5, 4, 67-81.
8. KYOTO CONTRACT (1997): see http://www.unfccc.org/resource/convkp.html
9. PARKUS, H. (1959a): *Spannungen beim Abkühlen einer Kugel*, Ing. Arch. 28, 251-254.
10. PARKUS, H. (1959b): *Instationäre Wärmespannungen*, Springer Verlag, Wien.
11. PICKL, S. (1999), *Der τ−value als Kontrollparameter. Modellierung und Analyse eines Joint-Implementation Programmes mithilfe der kooperativen dynamischen Spieltheorie und der diskreten Optimierung*, doctoral thesis, Darmstadt University of Technology, 1998, Shaker Verlag, Aachen.
12. PICKL, S., and WEBER, G.-W. (2000): *An algorithmic approach by linear programming problems in generalized semi-infinite optimization*, Journal of Computational Technologies 5, 3, 62-82.

13. PICKL, S., and WEBER, G.-W. (2001): *Optimization of a time-discrete nonlinear dynamical system from a problem of ecology. An analytical and numerical approach*, Journal of Computational Technologies 6, 1, 43-52.
14. SCHELLNHUBER, H.J. (1998): *Earth System Analysis. The Scope of Challenge.* In: Schellnhuber, H.J. and Wenzel, V. (Eds.), *Earth System Analysis. Integrating Science for Sustainability.* Springer Verlag, Berlin.
15. WEBER, G.-W. (1998): *Generalized semi-infinite optimization: On iteration procedures and topological aspects*, in: *Similarity Methods. International Workshop*, eds.: B. Kröplin, St. Rudolph und St. Brückner, 281-309.
16. WEBER, G.-W. (1999): *Generalized Semi-Infinite Optimization and Related Topics*, TU Darmstadt, habilitation thesis.
17. WEBER, G.-W. (2001): *Structural stability in generalized semi-infinite optimization*, Journal of Computational Technologies 6, 2, 25-46.

On the Relations Between Different Dual Problems in Convex Mathematical Programming

Gert Wanka, Radu-Ioan Boţ

Faculty of Mathematics, University of Technology Chemnitz
e-mail: g.wanka@mathematik.tu-chemnitz.de, bot@mathematik.tu-chemnitz.de

Abstract. Using the Fenchel-Rockafellar approach for the convex mathematical programming problem with inequality constraints different dual optimization problems by means of distinct perturbations of the primal problem are derived and studied. The classical Lagrange dual problem is one of those dual problems obtained by the perturbation of the right hand side of the inequality constraints.
For the various dual problems equality/inequality relations between the optimum values are verified under appropriate assumptions. Moreover, the duality relations to the primal problem are considered, in particular strong duality. Using the dual problems some optimality conditions are established.
The results are illustrated by some examples.
The application and usefulness for the construction of general multiobjective dual problems to the general multiobjective convex optimization problem is mentioned.

1 The constrained optimization problem and its conjugate duals

Consider the mathematical programming problem (called primal problem)

$$(P) \qquad \inf_{x \in G} f(x), \qquad G = \{x \in X : g(x) \leqq 0\},$$

with $X \subseteq \mathbb{R}^n$, $G \neq \emptyset$, $f : \mathbb{R}^n \to \bar{\mathbb{R}} = \mathbb{R} \cup \{\pm\infty\}$, $dom f = X$ (after redefining of an original $f : \mathbb{R}^n \to \mathbb{R}$), $g = (g_1, \ldots, g_k)^T : \mathbb{R}^n \to \mathbb{R}^k$, where $g(x) \leqq 0$ means $g_i(x) \leq 0, i = 1, \ldots, k$.

Applying the well-known so-called Fenchel-Rockafellar approach of defining a dual programming problem to a given primal problem we consider different perturbations of the primal problem (P) by introducing respective perturbation functions Φ_i :

$$(i) \quad \Phi_1(x,q) = \begin{cases} f(x) , \text{ if } x \in X,\ g(x) \leqq q, \\ +\infty \ , \text{ otherwise}, \end{cases}$$

$$(ii) \quad \Phi_2(x,p) = \begin{cases} f(x+p) , \text{ if } x \in X,\ g(x) \leqq 0, \\ +\infty \quad , \text{ otherwise}, \end{cases}$$

$$(iii)\ \Phi_3(x,p,q) = \begin{cases} f(x+p) , \text{ if } x \in X,\ g(x) \leqq q, \\ +\infty \quad , \text{ otherwise}, \end{cases}$$

$p \in \mathbb{R}^n$, $q \in \mathbb{R}^k$ are the respective perturbation variables.
For (i), (ii) and (iii) we consider the corresponding perturbed problems to (P) $\inf\limits_{x\in\mathbb{R}^n} \Phi_1(x,q)$, $\inf\limits_{x\in\mathbb{R}^n} \Phi_2(x,p)$ and $\inf\limits_{x\in\mathbb{R}^n} \Phi_3(x,p,q)$.

Setting the perturbation variables p and q equal to the zero vector in all three cases one gets (P). Now, dual problems may be defined by the conjugates Φ_i^* of the perturbation functions Φ_i as follows

$$(P_1^*) \quad \sup_{q^*\in\mathbb{R}^k} \{-\Phi_1^*(0,q^*)\},$$

$$(P_2^*) \quad \sup_{p^*\in\mathbb{R}^n} \{-\Phi_2^*(0,p^*)\},$$

$$(P_3^*) \quad \sup_{\substack{p^*\in\mathbb{R}^n \\ q^*\in\mathbb{R}^k}} \{-\Phi_3^*(0,p^*,q^*)\}$$

(recall the definition of the conjugate function $f^* : \mathbb{R}^n \to \bar{\mathbb{R}}$ to a function $f : \mathbb{R}^n \to \bar{\mathbb{R}}$ by $f^*(x^*) = \sup\limits_{x\in\mathbb{R}^n} \{x^{*T}x - f(x)\}$).

Calculating the respective conjugate function Φ_i^* we obtain the following three dual problems (P_i^*), $i = 1,2,3$, to (P):

$$(P_1^*) \quad \sup_{q^*\geqq 0} \inf_{x\in X} \{f(x) + q^{*T} g(x)\},$$

$$(P_2^*) \quad \sup_{p^*\in\mathbb{R}^n} \left\{ -f^*(p^*) + \inf_{x\in G} p^{*T} x \right\},$$

$$(P_3^*) \quad \sup_{\substack{p^*\in\mathbb{R}^n \\ q^*\geqq 0}} \left\{ -f^*(p^*) + \inf_{x\in X} [p^{*T} x + q^{*T} g(x)] \right\}.$$

We observe that (P_1^*) indeed is the well-known and classical Lagrange dual problem with the Lagrangian $L(x,q^*) = f(x) + q^{*T} g(x)$, $q^* \geqq 0$. By construction (cf.[1]) there is $\sup(P_i^*) \leq \inf(P)$ (weak duality).

2 The relations between the optimum values of the duals

Proposition 1 *It holds*

$$(i)\ \sup(P_1^*) \geq \sup(P_3^*),$$

$$(ii)\ \sup(P_2^*) \geq \sup(P_3^*).$$

Proof: **(i)** Let $q^* \geqq 0$ and $p^* \in \mathbb{R}^n$ be fixed. By the definition of the conjugate function we have for each $x \in X$ $f(x) \geqq p^{*T}x - f^*(p^*)$ (Young inequality). Adding to both sides $q^{*T}g(x)$ we obtain for each $x \in X$ $f(x) + q^{*T}g(x) \geqq -f^*(p^*) + p^{*T}x + q^{*T}g(x)$. This means that for each $q^* \geqq 0$ and $p^* \in \mathbb{R}^n$ holds

$$\inf_{x\in X}[f(x) + q^{*T}g(x)] \geqq -f^*(p^*) + \inf_{x\in X}[p^{*T}x + q^{*T}g(x)]. \tag{1}$$

We now may calculate the supremum over all $p^* \in \mathbb{R}^n$ and $q^* \geqq 0$ which implies the assertion.

(ii) Let $p^* \in \mathbb{R}^n$ be fixed.

For each $q^* \geqq 0$ the following inequalities are true

$$\inf_{x\in X}[p^{*T}x + q^{*T}g(x)] \leqq \inf_{x\in G}[p^{*T}x + q^{*T}g(x)] \leqq \inf_{x\in G} p^{*T}x.$$

This implies for each $p^* \in \mathbb{R}^n$

$$\sup_{q^*\geqq 0} \inf_{x\in X}[p^{*T}x + q^{*T}g(x)] \leqq \inf_{x\in G} p^{*T}x. \tag{1'}$$

Adding $-f^*(p^*)$ and computing the supremum over $p^* \in \mathbb{R}^n$ yields the wanted result. ■

The following examples show that the inequalities in Proposition 1 may be fulfilled strictly.

Example 1: Let be $X = [0, +\infty]$, $f : \mathbb{R} \to \bar{\mathbb{R}}$, $f(x) = \begin{cases} -x^2 & \text{if}\, x \in X, \\ +\infty & \text{, otherwise,} \end{cases}$

$g : \mathbb{R} \to \mathbb{R}$, $g(x) = x^2 - 1$ and consider (P) for those dates.

A straightforward calculation shows that the supremum of the Lagrange dual is $\sup(P_1^*) = -1$. On the other hand for the dual (P_3^*) we have $\sup(P_3^*) = -\infty$.

Example 2: Let now (P) be defined by $X = [0,+\infty)$, $f : \mathbb{R} \to \bar{\mathbb{R}}$, $g : \mathbb{R} \to \mathbb{R}$, $g(x) = 1-x^2$, $f(x) = \begin{cases} x & \text{if } x \in X, \\ +\infty, & \text{otherwise.} \end{cases}$ Then follows $sup(P_2^*) = 1 > \sup(P_3^*) = 0$.

But, under additional assumptions, e.g. convexity and regularity conditions, one can show that the inequalities in Proposition 1 are fulfilled as equalities.

Theorem 1 *Assume that $X \neq \emptyset$ is a convex set and f and g_i, $i = 1, \ldots, k$, are convex functions. Then it holds* $\sup(P_1^*) = \sup(P_3^*)$.

Theorem 1 results immediately from the following strong duality assertion.

Proposition 2 *Under the assumptions of Theorem 1 there is for all* $q^* \geqq 0$

$$\inf_{x \in X}[f(x) + q^{*T}g(x)] = \sup_{p^* \in \mathbb{R}^n} \left\{ -f^*(p^*) + \inf_{x \in X}[p^{*T}x + q^{*T}g(x)] \right\}.$$

If $\inf_{x\in X}[f(x)+q^{*T}g(x)]$ is finite then the supremum problem on the right hand side has a solution p_0^* such that "sup" may be substituted by "max".

Proof: Let $q^* \in \mathbb{R}^k$, $q^* \geqq 0$, be fixed and define $\alpha := \inf_{x\in X}[f(x)+q^{*T}g(x)]$.

Because of $X \neq \emptyset$ follows $\alpha \in [-\infty, +\infty)$. (1) implies after calculation of the supremum over $p^* \in \mathbb{R}^n$

$$\alpha \geqq \sup_{p^* \in \mathbb{R}^n} \left\{ -f^*(p^*) + \inf_{x \in X}[p^{*T}x + q^{*T}g(x)] \right\}. \tag{2}$$

For $\alpha = -\infty$ the statement is trivial.

For $\alpha > -\infty$ we consider the sets

$$C := \operatorname{epi} f = \{(x,\mu) : x \in X,\ \mu \in \mathbb{R},\ f(x) \leqq \mu\} \subset \mathbb{R}^{n+1},$$

$$D := \left\{(x,\mu) : x \in X,\ \mu \in \mathbb{R},\ \mu + q^{*T}g(x) \leqq \alpha\right\} \subset \mathbb{R}^{n+1}.$$

Obviously, $C, D \neq \emptyset$, are convex sets and $(\operatorname{rint} C) \cap D = \emptyset$ (rint C denotes the relative interior of C) ($(\operatorname{rint} C) \cap D \neq \emptyset$ causes a contradiction to the definition of α).

Therefore also $(\operatorname{rint} C) \cap (\operatorname{rint} D) = \emptyset$ meaning C and D are properly separable (cf. [3]), i.e. there exist $(p^*, \mu^*) \in \mathbb{R}^n \times \mathbb{R}$, $(p^*, \mu^*) \neq (0,0)$, and $\alpha^* \in \mathbb{R}$ such that for the separating hyperplane $H = \{(x,\mu) : p^{*T}x + \mu^*\mu = \alpha^*\}$ holds

$$(a) \qquad \inf\{p^{*T}c + \mu^*\mu : (x,\mu) \in D\} \geqq \alpha^* \geqq \sup\{p^{*T}x + \mu^*\mu : (x,\mu) \in C\}, \tag{3}$$

$$(b) \qquad \sup\{p^{*T}x + \mu^*\mu : (x,\mu) \in D\} > \inf\{p^{*T}x + \mu^*\mu : (x,\mu) \in C\}. \tag{4}$$

Assuming $\mu^* = 0$ allows to conclude a contradiction to (4) because one can see easily that in this case $p^{*T}x = \alpha^*$ for all $x \in X$.
Also $\mu^* > 0$ is not possible since then follows from (3) with letting converge $\mu \to +\infty$ that the supremum on the right hand side of (3) is $+\infty$, therefore $\alpha^* = +\infty$, contradicting $\alpha^* \in \mathbb{R}$ (finite). Therefore must be $\mu^* < 0$.
We may divide (3) by $(-\mu^*)$ having

$$\inf\left\{p_0^{*T}x - \mu : (x,\mu) \in D\right\} \geqq \alpha_0^* \geqq \sup\left\{p_0^{*T}x - \mu : (x,\mu) \in C\right\},$$

where $\alpha_0^* = \alpha^*/(-\mu^*)$ and $p_0^* = p^*/(-\mu^*)$. Hence,

$$p_0^{*T}x - \mu \leqq \alpha_0^* \ \forall (x,\mu) \in C, \tag{5}$$

$$\alpha_0^{*T} \leqq p_0^{*T}x - \mu \ \forall (x,\mu) \in D. \tag{6}$$

(5) yields

$$p_0^{*T}x - f(x) \leqq \alpha_0^* \ \forall x \in X \tag{7}$$

as a consequence of $(x, f(x)) \in C$. Calculating the supremum over $x \in X$ of the left hand side of (7) generates the conjugate $f^*(p_0^*)$, i.e.

$$f^*(p_0^*) \leqq \alpha_0^*. \tag{8}$$

Furthermore, there is $(x, \alpha - q^{*T}g(x)) \in D \ \forall x \in X$ (cf. definition of D). Then by (6) $\alpha_0^* \leqq p_0^{*T}x - (\alpha - q^{*T}g(x)) \ \forall x \in X$, i.e. $\alpha_0^* + \alpha \leqq \inf\limits_{x\in X} [p_0^{*T}x + q^{*T}g(x)]$, which gives with (8)

$$\alpha \leqq -f^*(p_0^*) + \inf_{x\in X} [p_0^{*T}x + q^{*T}g(x)].$$

Together with (2) it results

$$\alpha = \sup_{p^* \in \mathbb{R}^n} \left\{ -f^*(p^*) + \inf_{x \in X} [p^{*T} x + q^{*T} g(x)] \right\}.$$

Finally, by the definition of α this means $\inf_{x \in X} [f(x) + q^{*T} g(x)] = -f^*(p_0^*)$ $+ \inf_{x \in X} [p_0^{*T} x + q^{*T} g(x)] = \max_{p^* \in \mathbb{R}^n} \left\{ -f^*(p^*) + \inf_{x \in X} [p^{*T} x + q^{*T} g(x)] \right\}$. ■

In order to have $\sup(P_2^*) = \sup(P_3^*)$ we need a regularity condition (constraint qualification (CQ)). Therefore consider for $g(x) = (g_1(x), \ldots, g_k(x))^T$ the sets $L = \{i \in \{1, \ldots, k\} : g_i \text{ is an affine function}\}$,
$N = \{i \in \{1, \ldots, k\} : g_i \text{ is not an affine function}\}$.

CQ: There exists an element $x' \in \operatorname{rint} X$ such that $g_i(x') < 0$ for $i \in N$ and $g_i(x') \leqq 0$ for $i \in L$.

Theorem 2 *Assume that X is a convex set, g_i, $i = 1, \ldots, k$, are convex functions such that $G = \{x \in X : g(x) \leqq 0\} \neq \emptyset$ and (CQ) is fulfilled. Then it holds*

$$\sup(P_2^*) = \sup(P_3^*).$$

Proof: For $p^* \in \mathbb{R}^n$ fixed we first show that

$$\sup_{q^* \geqq 0} \inf_{x \in X} [p^{*T} x + q^{*T} g(x)] = \inf_{x \in G} p^{*T} x. \tag{9}$$

Let be $\beta := \inf_{x \in G} p^{*T} x$. Because $G \neq \emptyset$ follows $\beta \in [-\infty, \infty)$.
If $\beta = -\infty$ then by (1') follows

$$\sup_{q^* \geqq 0} \inf_{x \in X} [p^{*T} x + q^{*T} g(x)] = -\infty = \inf_{x \in G} p^{*T} x.$$

Let now be $\beta \in (-\infty, +\infty)$.
If g_L and g_N denote the vector functions with the affine components and the non-affine components of g, respectively, then the system of inequalities $p^{*T} x - \beta < 0$, $g_N(x) \leqq 0$, $g_L(x) \leqq 0$ has no solution in X. Using Theorem 2. 104 in [2] (characterizing the solvability of inequality systems of the above type with (CQ) fulfilled) we observe the existence of $u^* > 0$, $q^* = (v^*, w^*) \in \mathbb{R}_+^k$ such that for each $x \in X$

$$u^*(p^{*T} x - \beta) + q^{*T} g(x) = u^*(p^{*T} x - \beta) + v^{*T} g_N(x) + w^{*T} g_L(x) \geqq 0.$$

Dividing that inequality by u^* we obtain with $q_0^* = q^*/u^*$

$$p^{*T}x - \beta + q_0^{*T}g(x) \geqq 0 \ \forall x \in X$$

and, equivalently,

$$\inf_{x \in X} [p^{*T}x + q_0^{*T}g(x)] \geqq \beta.$$

This last inequality and (1') imply (9). By adding $-f^*(p^*)$ to (9) and calculating the supremum subject to $p^* \in \mathbb{R}^n$ one has $\sup(P_2^*) = \sup(P_3^*)$. ■

3 The strong duality and optimality conditions

Using Theorem 1 and Theorem 2 and verifying that $\inf(P) = \sup(P_1^*)$, which can be done applying the solvability condition of Theorem 2. 104 in [2] to the system $f(x) - \inf(P) < 0$, $g(x) \leqq 0$ in analogous manner as within the proof of Theorem 2, the following strong duality assertion may be derived.

Theorem 3 *Under the assumptions of both Theorem 1, 2 and if* $\inf(P)$ *is finite strong duality holds and* $(P_i^*), i = 1, 2, 3,$ *have solutions*

$$\inf(P) = \max(P_1^*) = \max(P_2^*) = \max(P_3^*).$$

By means of that strong duality one can use each of the different three duals to conclude optimality conditions. We want to restrict ourselves to present the optimality conditions arising from the strong duality between (P) and (P_3^*).

Theorem 4 (a) *Let the assumptions of Theorem 3 be fulfilled and let* x_0 *be a solution of* (P). *Then there exists a tupel* $(p_0^*, q_0^*) \in \mathbb{R}^n \times \mathbb{R}^k, q_0^* \geqq 0$, *such that the following optimality conditions are satisfied*

$$(i) \qquad f(x_0) + f^*(p_0^*) = p_0^{*T}x_0,$$

$$(ii) \qquad q_0^{*T}g(x_0) = 0,$$

$$(iii) \quad p_0^{*T}(x - x_0) \geq q_0^{*T}(g(x) - g(x_0)) \ \forall x \in X.$$

(b) *Let* $x_0 \in G$ *and* $(p_0^*, q_0^*) \in \mathbb{R}^n \times \mathbb{R}^k, q_0^* \geqq 0$, *satisfying (i), (ii) and (iii). Then* x_0 *and* (p_0^*, q_0^*) *turn out to be solutions of* (P) *and* (P_3^*), *respectively, and strong duality holds.*

Proof: (a) Let x_0 be a solution of (P). By Theorem 3 there exists a solution to (P_3^*) $(p_0^*, q_0^*) \in \mathbb{R}^n \times \mathbb{R}^k$, $q_0^* \geqq 0$, such that

$$f(x_0) = -f^*(p_0^*) + \inf_{x \in X} [p_0^{*T} x + q_0^{*T} g(x)].$$

By adding the term $p_0^{*T} x_0 + q_0^{*T} g(x_0)$ we obtain

$$\begin{aligned} & f(x_0) + f^*(p_0^*) - p_0^{*T} x_0 + p_0^{*T} x_0 + q_0^{*T} g(x_0) \\ & - \inf_{x \in X} [p_0^{*T} x + q_0^{*T} g(x)] - q_0^{*T} g(x_0) = 0. \end{aligned} \tag{10}$$

On the other hand the following inequalities hold $f(x_0) + f^*(p_0^*) - p_0^{*T} x_0 \geqq 0$ (Young inequality), $p_0^{*T} x_0 + q_0^{*T} g(x_0) - \inf_{x \in X} [p_0^{*T} x + q_0^{*T} g(x)] \geqq 0$, $-q_0^{*T} g(x_0) \geqq 0$. By (10) all these inequalities have to be equations which means that (i), (ii) and (iii) are fulfilled.

(b) All calculations done within part (a) of the proof may be carried out in the inverse direction starting from (i), (ii) and (iii). Then x_0 solves (P) and (p_0^*, q_0^*) solves (P_3^*) and strong duality holds. ∎

Finally, we would like to remark that, in particular, the strong duality between (P) and (P_3^*) may be applied for the construction of a dual problem in multiobjective convex optimization (cf. [4]).

References

1. Ekeland I., Temam R. (1976) Convex analysis and variational problems. North-Holland Publishing Company, Amsterdam
2. Elster K.-H., Reinhardt R., Schäuble M., Donath G. (1977) Einführung in die nichtlineare Optimierung. Teubner, Leipzig
3. Rockafellar R.T. (1970) Convex Analysis. Princeton University Press, Princeton, NJ
4. Wanka G., Boţ R.-I. (2001) A new duality approach for multiobjective convex optimization problems. Fakultät für Mathematik, TU Chemnitz, Preprint 2001-6

Section 9

Discrete and Combinatorial Optimization

Total Weighted Completion Time Minimization in a Problem of Scheduling Deteriorating Jobs

Aleksander Bachman and Adam Janiak

Institute of Engineering Cybernetics, Wrocław University of Technology, Janiszewskiego 11/17, 50-372 Wrocław, Poland

Abstract. The paper deals with a single machine scheduling problem with start time dependent job processing times. The problem is to find a schedule for which the total weighted completion time is minimized. It is proved that the problem is ordinary NP-hard, if the processing time of one additional job is given by decreasing linear function dependent on its start time. Based on the optimal properties found for some special cases of the general problem, three heuristic algorithms have been constructed and experimentally compared.

1 Introduction

In the classical scheduling theory, all the parameters characterizing jobs are given as constant values. In this paper, we consider problems in which processing times of the jobs are described by start time dependent functions. In the recent literature survey [1] concerning this type of problems, two types of functions describing such a change of job processing time are presented. The first type is related to non-decreasing functions and the second type to non-increasing ones. This paper is devoted to the problem of the total weighted completion time minimization in which jobs with non-increasing processing times are to be scheduled on a single machine. The application example of non-increasing start time dependent job processing time can describe the process by which aerial threats are to be recognized by a radar station [5]. In this case, a radar station has detected some objects approaching it. The time required to recognize the objects decreases as the objects get closer. Thus, the later the objects are detected, the less time for their recognition is.

The remaining part of this paper is organized as follows. In Section 2, a precise formulation of the problem with non-increasing job processing times is given. Section 3 deals with the computational complexity of the considered problem. Section 4 is devoted to some special cases of the problem under study. Three heuristic algorithms with their experimental analysis are given in Section 5. Section 6 contains some conclusions.

2 Problem formulation

There are given a single machine and a set of n independent and non-preemptive jobs immediately available for processing at time 0. The processing time p_i of job i is given by a non-increasing linear function dependent on the starting moment S_i of its execution

$$p_i(S_i) = a_i - b_i S_i, \tag{1}$$

where $a_i > 0$ and b_i denote the normal processing time and the decreasing rate of job i, respectively. It is assumed that the decreasing rates satisfy the following conditions: $0 \leq b_i < 1$ and $b_i(\sum_{j=1}^{n} a_j - a_i) < a_i$. The first condition ensures that the decrease of each job processing time is less than one unit for every unit of delay in its starting moment. The second condition ensures the positive value of the job processing time even if the job i is executed last (see also [5] for detailed explanations). For each job, a positive weight w_i is also given. Define for convenience that $B_i = 1 - b_i$. The objective is to find a schedule π for which the value of the total weighted completion time (TWC) is minimized, i.e.,

$$TWC = \sum_{i=1}^{n} w_{\pi(i)} C_{\pi(i)} = \sum_{i=1}^{n} w_{\pi(i)} \sum_{j=1}^{i} a_{\pi(j)} \prod_{k=j+1}^{i} B_{\pi(k)}, \tag{2}$$

where $C_{\pi(i)}$ denotes the completion time of the job placed on the ith position of the schedule π. The problems considered in the paper are denoted according to the three-field notation $\alpha \mid \beta \mid \gamma$ introduced by Graham *et al.* [4].

3 Computational complexity

In the classical scheduling theory, the problem of minimizing TWC can be solved in $O(n \log n)$ steps by scheduling jobs according to non-decreasing order of the ratio a_i/w_i [6]. In this section, it is proved that the same problem becomes NP-hard, if for one additional job its processing time is given by start time dependent function. In order to show that, we use the NP-complete Partition Problem (PP) [3] in the following form.

PP: There is given a set $X = \{x_1, x_2, \ldots, x_m\}$ of m positive integers for which $\sum_{i=1}^{m} x_i = 2D$. Does there exist a partition of the set X into two disjoint subsets X_1 and X_2 such that $\sum_{x_i \in X_1} x_i = \sum_{x_i \in X_2} x_i = D$?

The decision version of the problem formulated in the previous section, denoted by DV, is given as follows.

DV: There is given a set of n independent and non-preemptive jobs to be scheduled on a single machine. The job processing times are given by $p_i(S_i) =$

$a_i - b_i S_i$, where n, a_i and w_i are positive integers and b_i are non-negative rational numbers. Is there a schedule for which TWC is not greater than y, where y is a positive integer?

Theorem 1. *The problem* $1 \mid p_i(S_i) = a_i - b_i S_i \mid \sum w_i C_i$ *is NP-hard even for the case with one decreasing rate greater than zero.*

Proof. First a reduction from PP to our problem is given. Let DV consist of $n = m + 1$ jobs with the following parameters:

$$\begin{aligned} &a_i = x_i; \ b_i = 0; \qquad w_i = x_i; \quad i = 1, \ldots, m, \\ &a_e = 2D; \ b_e = 1 - D^{-1}; \ w_e = 2D^2. \end{aligned}$$

Note that the parameters of the extra job satisfy the model requirements described in Section 2. The value y is equal to

$$y = \sum_{i=1}^{m} x_i^2 + \sum_{i,j \in X} x_i x_j - D^2 + D(2D+1)^2.$$

We show, that PP has a solution **if and only if** there exists a solution for the constructed instance of $1 \mid p_i(S_i) = a_i - b_i S_i \mid \sum w_i C_i$ with the value $\sum w_j C_j \le y$.

(PP $\Rightarrow$ DV) Assume that there exists a partition of the set X into two disjoint subsets X_1 and X_2, for which $\sum_{x_i \in X_1} x_i = \sum_{x_i \in X_2} x_i = D$. Let J_1 and J_2 denote the subsets containing the jobs constructed on the basis of the elements from the subsets X_1 and X_2, respectively. Assume that the subsets J_1 and J_2 contain k and $m - k$ jobs, respectively. The solution for DV is given as follows. The machine executes at first the jobs from the subset J_1, after then the extra job and finally the jobs from the subset J_2. TWC for the schedule given above is equal to:

$$\begin{aligned} TWC = &\sum_{i=1}^{k} a_{\pi(i)} \sum_{j=i}^{k} w_{\pi(j)} + \sum_{i=k+2}^{n} a_{\pi(i)} \sum_{j=i}^{n} w_{\pi(j)} \\ &+ \left(a_e + B_e \sum_{i=1}^{k} a_{\pi(i)}\right)\left(w_e + \sum_{i=k+2}^{n} w_{\pi(i)}\right). \end{aligned} \tag{3}$$

Since $a_i = x_i$ and $w_i = x_i$, then the expression (3) is given by:

$$\begin{aligned} TWC = &\sum_{i=1}^{n} x_i^2 + \sum_{x_i, x_j \in X_1} x_i x_j + \sum_{x_i, x_j \in X_2} x_i x_j \\ &+ \left(a_e + B_e \sum_{x_i \in X_1} x_i\right)\left(w_e + \sum_{x_i \in X_2} x_i\right) \\ = &\sum_{i=1}^{n} x_i^2 + \sum_{x_i, x_j \in X} x_i x_j - \left(\sum_{x_i \in X_1} x_i\right)\left(\sum_{x_i \in X_2} x_i\right) \end{aligned}$$

$$+\left(a_e + B_e \sum_{x_i \in X_1} x_i\right)\left(w_e + \sum_{x_i \in X_2} x_i\right). \tag{4}$$

Since $\sum_{x_i \in X_1} x_i = \sum_{x_i \in X_2} x_i = D$, then expression (4) is equal to the value y, as it was required

$$TWC = \sum_{i=1}^{m} x_i^2 + \sum_{i,j \in X} x_i x_j - D^2 + D(2D+1)^2 = y.$$

($\neg$PP $\Rightarrow$ $\neg$DV) Assume now that there is no partition of the set X into two disjoint subsets X_1 and X_2, such that $\sum_{x_i \in X_1} x_i = \sum_{x_i \in X_2} x_i$. Let's consider the following case $\sum_{x_i \in X_1} x_i = D - \lambda$ and $\sum_{x_i \in X_2} x_i = D + \lambda$, where λ is a non-zero integer. By (4) we have:

$$\begin{aligned} TWC &= \sum_{i=1}^{n} x_i^2 + \sum_{x_i, x_j \in X} x_i x_j - D^2 + \lambda^2 + [a_e + B_e(D-\lambda)]\,[w_e + D + \lambda] \\ &= \sum_{i=1}^{n} x_i^2 + \sum_{x_i, x_j \in X} x_i x_j - D^2 + \lambda^2 \\ &+ \left[2D + D^{-1}(D-\lambda)\right]\left[2D^2 + D + \lambda\right] = y + \lambda^2(1 - D^{-1}) > y. \end{aligned}$$

Thus, it was shown that DV has a solution, iff PP has also one. □

4 Polynomial time solvable cases

In this section we present the optimal solutions for some special cases of the problem under consideration.

Property 1. The optimal solution for the problem $1 \mid p_i(S_i) = a - bS_i \mid \sum w_i C_i$ can be obtained by scheduling jobs in non-increasing order of their weights.

Proof. Note that each job has the same model of start time dependent processing time. Therefore, it is obvious that the schedule in which the jobs are arranged according to non-increasing order of their weights is optimal. □

Property 2. The non-decreasing order of the ratio $\frac{a_i}{w_i(1-ka_i)}$ gives the optimal solution for the problem $1 \mid p_i(S_i) = a_i - ka_iS_i, k(\sum_{j=1}^{n} a_j - a_{min}) < 1 \mid \sum w_i C_i$, where $a_{min} = \min_{i=1,\ldots,n} a_i$.

Proof. Assume that there are given two permutations π and π'. The permutation π' has been obtained from the permutation π by interchanging the jobs from the ith and the $i+1$st positions. Assume also that the permutation π in which $\frac{a_{\pi(i)}}{w_{\pi(i)}(1-ka_{\pi(i)})} > \frac{a_{\pi(i+1)}}{w_{\pi(i+1)}(1-ka_{\pi(i+1)})}$ is optimal. Note that for the job processing times given above, the completion time is given by the following expression: $C_{\pi(i)} = -\frac{1}{k}\prod_{j=1}^{i}(1 - ka_{\pi(j)}) + \frac{1}{k}$. Therefore, for both permutations only the completion times of the jobs in the ith position are different. We now calculate the difference between the values of the TWC obtained for both permutations.

$$\sum w_{\pi(j)}C_{\pi(j)} - \sum w_{\pi'(j)}C_{\pi'(j)}$$

$$= w_{\pi(i)}\left[-\frac{1}{k}(1 - ka_{\pi(i)})\prod_{j=1}^{i-1}(1 - ka_{\pi(j)}) + \frac{1}{k}\right]$$

$$+ w_{\pi(i+1)}\left[-\frac{1}{k}\prod_{j=1}^{i+1}(1 - ka_{\pi(j)}) + \frac{1}{k}\right]$$

$$- w_{\pi(i+1)}\left[-\frac{1}{k}(1 - ka_{\pi(i+1)})\prod_{j=1}^{i-1}(1 - ka_{\pi(j)}) + \frac{1}{k}\right]$$

$$- w_{\pi(i)}\left[-\frac{1}{k}\prod_{j=1}^{i+1}(1 - ka_{\pi(j)}) + \frac{1}{k}\right]$$

$$= -\left[a_{\pi(i+1)}w_{\pi(i)}(1 - ka_{\pi(i)}) - a_{\pi(i)}w_{\pi(i+1)}(1 - ka_{\pi(i+1)})\right]\prod_{j=1}^{i-1}(1 - ka_{\pi(j)}).$$

Since the expressions $a_{\pi(i+1)}w_{\pi(i)}(1 - ka_{\pi(i)}) - a_{\pi(i)}w_{\pi(i+1)}(1 - ka_{\pi(i+1)}) < 0$ and $\frac{a_{\pi(i)}}{w_{\pi(i)}(1-ka_{\pi(i)})} > \frac{a_{\pi(i+1)}}{w_{\pi(i+1)}(1-ka_{\pi(i+1)})}$ are equivalent, the result obtained above is positive. Thus, the schedule in which the jobs are arranged according to non-decreasing order of the ratio $\frac{a_i}{w_i(1-ka_i)}$ is optimal. □

5 Heuristic algorithms

Since the general problem considered in this paper has been shown to be NP-hard, we do not expect to find an optimal solution of the problem in polynomial time. Therefore, to solve the problem, three heuristic algorithms will be constructed and experimentally compared. In the first algorithm A1, the jobs are scheduled according to non-decreasing order of the ratio $a_i/(w_i(1 - b_i))$, which was proved to be the optimal solution for the problem $1 \mid p_i(S_i) = a_i - ka_iS_i \mid \sum w_iC_i$ (see Property 2). In the second algorithm A2, we apply the Smith's rule [6], which gives the optimal solution for the classical version of the problem under study. In our case, we schedule jobs according to non-decreasing order of the ratio a_i/w_i. Scheduling jobs according to non-decreasing order of the ratio $a_i/(w_ib_i)$ is the main idea applied in the third algorithm A3. The computational complexity of all algorithms formulated above is $O(n \log n)$. The algorithms are tested for three sizes of the problem ($n = 10$, $n = 50$ and $n = 100$). For each problem size, nine different tests are performed. In each test, 100 randomly generated instances are compared. All the parameters are randomly generated according to the uniform distribution. For all the tests, the values of job weights are generated from the interval $(0, 10)$. The values of the normal processing times are generated from the following three intervals: $(0, 1)$, $(0, 10)$ and $(0, 100)$ to check their dependence on job weights. Since the decreasing rates have to satisfy the following conditions: $0 \le b_i < 1$ and $b_i(\sum_{j=1}^{n} a_j - a_i) < a_i$, their values are first generated from the interval $(0, 1)$ and then multiplied by three different values: G, $0.5G$ and $0.1G$, where $G = a_{min}/(\sum_{j=1}^{n} a_j - a_{min})$ and $a_{min} = \min_{i=1,\dots,n} a_i$. For $n = 10$, the solutions constructed by the algorithms A1, A2 and A3 are compared with the optimal one, found by complete enumeration (OPT) (see Table 1). The performance of the algorithms for $n = 50$ and $n = 100$ is verified by the ratios A2/A1 and A3/A1 (see Table 2).

Table 1. The experimental results obtained for $n = 10$

normal proc. time	decreasing rate	A1/OPT	A2/OPT	A3/OPT
	$b_i \in (0, G)$	1.000461	1.000509	1.117639
$a_i \in (0, 1)$	$b_i \in (0, 0.5G)$	1.000178	1.000207	1.117782
	$b_i \in (0, 0.1G)$	1.000034	1.000039	1.118041
	$b_i \in (0, G)$	1.000163	1.000150	1.115803
$a_i \in (0, 10)$	$b_i \in (0, 0.5G)$	1.000019	1.000030	1.115919
	$b_i \in (0, 0.1G)$	1.000003	1.000003	1.116098
	$b_i \in (0, G)$	1.000097	1.000032	1.112727
$a_i \in (0, 100)$	$b_i \in (0, 0.5G)$	1.000024	1.000016	1.112794
	$b_i \in (0, 0.1G)$	1.000002	1.000000	1.112897

The results obtained by algorithms A1 and A2 for the small problem size ($n = 10$) are very close to the optimal. Algorithm A3 gives the worst solutions from all tested algorithms. All the algorithms generate better results for the large values of normal processing times. The same situation occurs for the smaller values of decreasing rates.

For the large problem sizes ($n = 50$ and $n = 100$), the relation between the algorithms remains the same. The algorithms A1 and A2 generate almost the same solutions. However, this remark can be easily explained. Since the values of decreasing rates are very close to zero, the following two expressions $a_i/(w_i(1 - b_i))$ and a_i/w_i are almost the same.

Table 2. The experimental results obtained for $n = 50$ and $n = 100$

		n=50		n=100	
normal proc. time	decreasing rate	A2/A1	A3/A1	A2/A1	A3/A1
	$b_i \in (0, G)$	1.000022	1.140501	1.000021	1.151319
$a_i \in (0, 1)$	$b_i \in (0, 0.5G)$	1.000010	1.141900	1.000010	1.152927
	$b_i \in (0, 0.1G)$	1.000002	1.142927	1.000002	1.154161
	$b_i \in (0, G)$	1.000002	1.143740	1.000000	1.153468
$a_i \in (0, 10)$	$b_i \in (0, 0.5G)$	1.000001	1.144040	1.000000	1.153712
	$b_i \in (0, 0.1G)$	1.000000	1.144288	1.000000	1.153891
	$b_i \in (0, G)$	0.999998	1.149119	1.000000	1.154077
$a_i \in (0, 100)$	$b_i \in (0, 0.5G)$	1.000000	1.149425	1.000000	1.154263
	$b_i \in (0, 0.1G)$	1.000000	1.149669	1.000000	1.154411

6 Conclusions

The computational complexity of the single machine scheduling problem with start time dependent processing times for the total weighted completion time minimization has been proved in this paper. The obtained results deal with the non-increasing linear model of the start time dependent job processing times. The complexity of the problem with the non-decreasing linear model has been cleared up recently in [2]. However, the complexity status of the problems with common weights for the non-decreasing and non-increasing models of job processing times is still an open question.

References

1. Alidaee B., Womer N. K. (1999) Scheduling with time dependent processing times: Review and extensions. Journal of the Operational Research Society 50, 711–720
2. Bachman A., Janiak A., Kovalyov M. Y. (2000) Minimizing the total weighted completion time of deteriorating jobs. (accepted for publication in Information Processing Letters).
3. Garey M. R., Johnson D. S. (1979) Computers and Intractability: A Guide to the Theory of NP-completeness. Freeman, San Francisco
4. Graham R. L., Lawler E. L., Lenstra J. K., Rinnooy Kan A. H. G. (1979) Optimization and approximation in deterministic sequencing and scheduling: a survey. Annals of Discrete Mathematics 3, 287–326
5. Ho K. I-J., Leung J. Y-T., Wei W-D. (1993) Complexity of scheduling tasks with time-dependent execution times. Information Processing Letters 48, 315–320
6. Smith W. E. (1956) Various optimizers for single-stage production. Naval Research Logistics Quarterly 3, 59–66

Sensitive Analysis of a Vector Quadratic Lexicographic Boolean Programming Problem

Yury Nikulin

Belarusian State University, Minsk , Belarus

Abstract. The problem of lexicographic optimization is considered. We study several types of stability understood as preservation of certain invariance by the efficient solution set under "small" changes of input data. Borders of such changes are ascertain. Necessary and sufficient conditions of stability are specified.

A vector optimization problem is usually understood as the problem of finding a set of efficient solutions, i.e. choosing from the set of feasible solutions the alternatives which satisfy to a given optimality principle. In the case where the partial criteria of the problem have an equal importance, the Pareto optimality principle is more often used. If all the partial criteria are ordered by importance in such the manner that each of them is more important than all the subsequent, the principle of lexicographic optimality is used. Investigating stability of a vector optimization problem means usually studying behavior of the set of efficient solutions under perturbations of problem's parameters.

The wide circulation of discrete optimization models in economy, management and design caused a great interest of many specialists to the questions of stability of discrete optimization problems (see, e.g. [1])

The present work adjoins to the cycle of works [2] - [4] devoted to the stability of vector problems of combinatorial optimization.

Let $A = (A_1, A_2, ..., A_m)$, $b = (b_1, b_2, ..., b_m)$, $m \in \mathbf{N}$, where for any index $k \in N_m = \{1, 2, ..., m\}$ matrix $A_k \in \mathbf{R}^{n\times n}$, vector $b_k \in \mathbf{R}^n$, $n \in \mathbf{N}$. We assign the vector criterion

$$f(x) = (f_1(x), f_2(x), ..., f_m(x)) \longrightarrow \min_{x\in X}$$

on the set of feasible solutions $X \subseteq \mathbf{E}^n = \{0,1\}^n$, $|X| > 1$. The partial criteria are the quadratic functions

$$f_k(x, A_k, b_k) = \langle A_k x, x\rangle + \langle b_k, x\rangle, \ k \in N_m,$$

where $\langle\cdot,\cdot\rangle$ is the scalar product of vectors, $x = (x_1, ..., x_n)^T$.

We consider the problem of finding the lexicographic set. It is a subset of the Pareto set and is defined as follows.

Suppose S_n is the set of all $n!$ permutations of the numbers $1, 2, ..., n$. We say that a solution x is a lexicographic optimum if there exists a permutation

$s = \{s_1, s_2, ..., s_n\} \in S_n$ such that one of the following conditions holds for any solution x' :
1) $f(x) = f(x')$;
2) $\exists p \in N_m \ (f_{s_p}(x, A_{s_p}, b_{s_p}) < (f_{s_p}(x', A_{s_p}, b_{s_p}))$ &
&$(\forall k \in N_{p-1} \ f_{s_k}(x, A_{s_k}, b_{s_k}) = f_{s_k}(x', A_{s_k}, b_{s_k}))$.

If $p = 1$ then the last equalities are absent ($N_0 = \emptyset$). Thus s orders the partial criteria with respect to importance such that every previous criterion is more significant than all the consequent ones.

A rigid arrangement of partial criteria with respect to importance is often used for a wide spectrum of important optimization problems, for example, problem of stochastic programming, problems of axiomatic system of utility theory and so on.

We say that the set of all lexicographic optimal solutions defined by all $m!$ permutations is the lexicographic set and write $L^m(A, b)$. The m-criteria quadratic Boolean programming problem of lexicographic optimization is denoted by $Z^m(A, b)$.

For $m = 1$ our problem is the scalar quadratic Boolean programming problem and $L^1(A, b)$ is the set of optimal solutions. Quadratic assignment problem and different optimization problems on graphs are represented in the scheme of this problem. It has many applications in electronics design: partitioning problem, covering problem, packing problem etc.

We define the norm

$$\| z \|_\infty = \max\{|z_i| : i \in N_q\}$$

in the q-dimensional space $\mathbf{R}^q$ and the norm

$$\| z \|_1 = \sum_{i \in N_q} |z_i|$$

in the space conjugate to $\mathbf{R}^q$.

By definition, put

$$\Theta_1 = \{\varepsilon > 0 : \ L^m(A + A', b + b') \subseteq L^m(A, b) \ \forall (A', b') \in \Omega(\varepsilon)\},$$

$$\Theta_2 = \{\varepsilon > 0 : \ L^m(A + A', b + b') \supseteq L^m(A, b) \ \forall (A', b') \in \Omega(\varepsilon)\},$$

$$\Theta_3 = \{\varepsilon > 0 : \ L^m(A + A', b + b') \cap L^m(A, b) \neq \emptyset \ \forall (A', b') \in \Omega(\varepsilon)\},$$

$$\Theta_4 = \{\varepsilon > 0 : \ \exists \, x \in L^m(A, b) \ \forall (A', b') \in \Omega(\varepsilon) \ (x \in L^m(A + A', b + b'))\},$$

where

$$\Omega(\varepsilon) = \{(A', b') \in \mathbf{R}^{n \times n \times m} \times \mathbf{R}^{n \times m} : \| A' \|_\infty < \varepsilon, \ \| b' \|_\infty < \varepsilon\},$$

We introduce the following definitions.

The number

$$\rho_r^m(A, b) = \begin{cases} \sup \Theta_r, & \text{if } \Theta_r \neq \emptyset, \\ 0 & \text{otherwise} \end{cases}$$

is called

$r = 1$ – the stability radius of the vector problem $Z^m(A, b)$;

$r = 2$ – the quasi-stability radius of the vector problem $Z^m(A, b)$;

$r = 3$ – the strong stability radius of the vector problem $Z^m(A, b)$;

$r = 4$ – the strong quasi-stability radius of the vector problem $Z^m(A, b)$.

Obviously, the stability radius (as the strong stability radius) of the vector problem $Z^m(A, b)$ is infinite as the set $L^m(A, b)$ coincides with X. So we exclude the case $L^m(A, b) = X$ from the consideration. If the set $\bar{L}^m(A, b) := X \setminus L^m(A, b)$ is non-empty, then we say that the problem $Z^m(A, b)$ is non-trivial.

Let

$$\varphi_1^m(A,b) = \min_{x \in \bar{L}^m(A,b)} \min_{k \in N_m} \max_{x' \in X \setminus \{x\}} \Gamma_k(x, x', A_k, b_k),$$

$$\varphi_2^m(A,b) = \min_{x' \in L^m(A,b)} \max_{k \in N_m} \min_{x \in X \setminus \{x'\}} \Gamma_k(x, x', A_k, b_k),$$

$$\varphi_3^m(A,b) = \max_{k \in N_m} \min_{x \in \bar{L}^m(A,b)} \max_{x' \in L^m(A,b)} \Gamma_k(x, x', A_k, b_k),$$

$$\varphi_4^m(A,b) = \max_{x' \in L^m(A,b)} \max_{k \in N_m} \min_{x \in X \setminus \{x'\}} \Gamma_k(x, x', A_k, b_k),$$

where

$$\Gamma_k(x, x', A_k, b_k) = \frac{f_k(x, A_k, b_k) - f_k(x', A_k, b_k)}{||x||_1^2 + ||x'||_1^2 + ||x - x'||_1 - 2\langle x, x' \rangle^2}.$$

Theorem 1 *Let the problem $Z^m(A, b), m \geq 1$, be non-trivial. Then the following statements are true*

$$\rho_1^m(A,b) = \varphi_1^m(A,b).$$
$$\rho_3^m(A,b) = \varphi_3^m(A,b).$$

Theorem 2 *The following statements are true for any problem $Z^m(A, b), m \geq 1$,*

$$\rho_2^m(A,b) = \varphi_2^m(A,b),$$
$$\rho_4^m(A,b) = \varphi_4^m(A,b).$$

Definition 1 *The problem $Z^m(A, b)$ is called stable (quasi-stable, strongly stable, strongly quasi-stable) if $\rho_1^m(A, b) > 0$ ($\rho_r^m(A, b) > 0,\ r = 2, 3, 4$).*

We denote by $S_1^m(A, b)$ the set of weak solutions and by $S_2^m(A, b)$ the set of strong solutions:

$$S_1^m(A,b) = \{x \in X :\ \exists k = k(x) \in N_m\ \forall x' \in X \setminus \{x\}\ (f_k(x, A_k, b_k) \leq f_k(x', A_k, b_k))\},$$

$$S_2^m(A,b) = \{x \in X :\ \exists k = k(x) \in N_m\ \forall x' \in X \setminus \{x\}\ (f_k(x, A_k, b_k) < f_k(x', A_k, b_k))\},$$

Corollary 1 *The vector non-trivial problem $Z^m(A,b)$, $m \geq 1$, is stable if and only if the lexicographic set $L^m(A,b)$ coincides with $S_1^m(A,b)$.*

Corollary 2 *The vector problem $Z^m(A,b)$, $m \geq 1$, is quasi-stable if and only if the lexicographic set $L^m(A,b)$ coincides with $S_2^m(A,b)$.*

Let us call the lexicographic set $L^m(A,b)$ of a non-trivial problem $Z^m(A,b)$ regular, if

$$\exists k \in N_m \ \forall x \in \bar{L}^m(A,b) \ \exists x' \in L^m(A,b) \quad (f_k(x,A_k,b_k) > f_k(x',A_k,b_k)).$$

Corollary 3 *The vector non-trivial problem $Z^m(A,b)$, $m \geq 1$, is strongly stable if and only if the lexicographic set $L^m(A,b)$ is regular.*

Corollary 4 *The vector problem $Z^m(A,b)$, $m \geq 1$, is strongly quasi-stable if and only if $S_2^m(A,b) \neq \emptyset$.*

Corollary 5 *The scalar problem $Z^1(A,b)$ is quasi-stable if and only if it has a unique optimal solution.*

References

1. Sotskov Yu.N., Leontev V.K., Gordeev E.N. (1995) . Some concepts of stability analysis in combinatorial optimization. Discrete Applied Mathematics. **58**, 169–190.
2. Emelichev V.A., Berdysheva R.A., et al. (1999) Stability and regularization of vector problems of integer linear programming. Preprint ü 29. Otto-von-Guiricke-Universitat Magdeburg.
3. Emelichev V.A., Nikulin Yu.V. (2000) The stability of efficient solutions of a vector quadratic Boolean programming problem. Izv Akad Nauk Resp. Moldova. **1**, 33 - 39.
4. Emelichev V.A., Nikulin Yu.V. (2000) On the stability of a vector quadratic Boolean programming problem. Proc. of Inter. Conf. on Oper. res. (OR-2000). Dresden, 24 - 25.

Some Relations Between Consecutive Ones and Betweenness Polytopes

Marcus Oswald, Gerhard Reinelt

Institut für Informatik, Universität Heidelberg, Im Neuenheimer Feld 368, D-69120 Heidelberg, Germany,
{Marcus.Oswald,Gerhard.Reinelt}@Informatik.Uni-Heidelberg.De

Abstract. Both the consecutive ones and the betweenness problem deal with the ordering of objects such that certain conditions are satisfied. In the first case certain associated matrices have to have the consecutive ones property and in the second case a set of betweenness conditions has to be satisfied. In this paper we point out some relations between these problems, in particular, some relations between facets of associated polytopes.

Key words: betweenness problem, consecutive ones problem, polytopes, facets

1 Introduction

The input of the *consecutive ones problem* consists of a 0/1-matrix $A \in \{0,1\}^{m \times n}$. The task is to find a matrix A' obtained from A by changing as few entries from 0 to 1 or from 1 to 0 as possible such that A' has the *consecutive ones property (for rows)*. This means that there is an ordering of the columns of A' such that the ones in each row appear consecutively. (For convenience we just say that A' *is C1P*.)

In the *weighted consecutive ones problem (WC1P)* each entry has a specified cost for switching a 0 to a 1 or vice versa. The task is then to find a C1P matrix A' such that the total cost for transforming A into A' is minimal.

The input of the *betweenness problem* is a set of n objects $1, 2, \ldots, n$, a set $\mathcal{B}$ of betweenness conditions, and a set $\overline{\mathcal{B}}$ of non-betweenness conditions. Every element of $\mathcal{B}$ (of $\overline{\mathcal{B}}$) is a triple (i,j,k) (a triple $\overline{(i,j,k)}$) requesting that object j should be placed (should not be placed) between objects i and k. The task is to find an ordering of the objects such that as few betweenness and non-betweenness conditions as possible are violated.

If violations are penalized by weights, we call the problem of finding an ordering which minimizes the sum of penalties the *weighted Betweenness Problem (WBWP)*.

Both the WC1P and a variant of the WBWP occur as models in the so-called physical mapping problem, a central problem in computational biology. In [COR98] and [OR00] branch-and-cut approaches for these two problems are presented.

In this paper we study relations between polytopes that are associated with WC1P and WBWP.

2 The Consecutive Ones Polytope

For a 0/1-matrix A with m rows and n columns let $\chi^A = (a_{11}, \ldots, a_{1n}, \ldots, a_{m1}, \ldots, a_{mn})$ be its characteristic vector. We define the *consecutive ones polytope* as

$$P_{\text{C1}}^{m,n} = \text{conv}\{\chi^A \mid A \text{ is an } (m,n)\text{-matrix with C1P}\}.$$

It is easy to see that $P_{\text{C1}}^{m,n}$ has full dimension $m \cdot n$. Namely, the zero matrix is C1P and, for every $1 \leq i \leq m$ and $1 \leq j \leq n$, the matrix consisting of zeroes only except for a one in position ij is C1P. This gives a set of $n \cdot m + 1$ affinely independent C1P matrices.

Let $a^T x \leq a_0$ be a valid inequality for $P_{\text{C1}}^{m,n}$ and let $m' \geq m$ and $n' \geq n$. We say that the inequality $\overline{a}^T x \leq a_0$ for $P_{\text{C1}}^{m',n'}$ is obtained from $a^T x \leq a_0$ by *trivial lifting* if

$$\overline{a}_{ij} = \begin{cases} a_{ij} & \text{if } i \leq m \text{ and } j \leq n, \\ 0 & \text{otherwise.} \end{cases}$$

Theorem 1 *Let $a^T x \leq a_0$ be a facet-defining inequality for $P_{\text{C1}}^{m,n}$ and let $m' \geq m$ and $n' \geq n$. If $a^T x \leq a_0$ is trivially lifted then the resulting inequality defines a facet of $P_{\text{C1}}^{m',n'}$.*

Inequalities that are obviously valid for $P_{\text{C1}}^{m,n}$ are the *trivial inequalities* $x_{ij} \geq 0$ and $x_{ij} \leq 1$, for all $1 \leq i \leq m$, $1 \leq j \leq n$. It is easily seen that they also define facets.

Theorem 2 *For all $m \geq 1$, $n \geq 1$, $1 \leq i \leq m$, $1 \leq j \leq n$, the inequalities $x_{ij} \geq 0$ and $x_{ij} \leq 1$ define facets of $P_{\text{C1}}^{m,n}$.*

Proofs of these theorems and an integer programming formulation of WC1P that consists only of inequalities which are facet-defining for $P_{\text{C1}}^{m,n}$ are given in [OR00].

3 The Betweenness Polytope

In the following we use indices $i(j)k$ (*betweenness triples*) for pairwise different objects i, j and k, indicating that we consider whether object j is between objects i and k or not. Since the indices $i(j)k$ and $k(j)i$ are equivalent, we only use $i(j)k$ such that $i < k$. In vectors, triples are ordered lexicographically, i.e., we use the order

$$1(2)3, 1(2)4, \ldots, n-1(n-2)n.$$

For each permutation π of $n \geq 3$ elements $1, \ldots, n$ and each betweenness triple $i(j)k$ we define an indicator $\chi^{\pi}_{i(j)k}$ which is 1 if and only if the element

j lies between the elements i and k in the permutation π and 0 otherwise, i.e.,

$$\chi^{\pi}_{i(j)k} = \begin{cases} 1 & \text{if } \pi^{-1}(i) < \pi^{-1}(j) < \pi^{-1}(k) \text{ or } \pi^{-1}(i) > \pi^{-1}(j) > \pi^{-1}(k), \\ 0 & \text{otherwise.} \end{cases}$$

The $3\binom{n}{3}$-dimensional characteristic betweenness vector associated with a permutation π is then

$$\chi^{\pi} = (\chi^{\pi}_{1(2)3}, \chi^{\pi}_{1(2)4}, \ldots, \chi^{\pi}_{n-1(n-2)n}).$$

Now we can define the *betweenness polytope* for $n \geq 3$ as

$$P^n_{\text{BW}} = \text{conv}\{\chi^{\pi} \mid \pi \text{ is a permutation of } \{1, \ldots, n\}\}.$$

The betweenness polytope is not full-dimensional as the following results show.

Lemma 3 *For an arbitrary point* $x = (x_{1(2)3}, \ldots, x_{n-1(n-2)n}) \in P^n_{\text{BW}}$ *and three pairwise different* i, j, k, $1 \leq i, j, k \leq n$, *the equation (betweenness equation)*

$$x_{i(j)k} + x_{i(k)j} + x_{j(i)k} = 1$$

holds.

Proof. For each permutation π and pairwise different i, j and k exactly one of the three indicators $\chi^{\pi}_{i(j)k}$, $\chi^{\pi}_{i(k)j}$ and $\chi^{\pi}_{j(i)k}$ is 1. Thus the equations hold for any vertex of P^n_{BW} and with it for any point of P^n_{BW}.

Theorem 4 P^n_{BW} *has dimension* $2\binom{n}{3}$.

Proof. No variable $x_{i(j)k}$ appears twice in the betweenness equations. Therefore the coefficient matrix of these equations has full row rank $\binom{n}{3}$.

It remains to be shown that all further equations that are valid for P^n_{BW} are linear combinations of betweenness equations implying that the dimension of P^n_{BW} equals $3\binom{n}{3} - \binom{n}{3}$.

We conclude a contradiction from the assumption that there is an equation $a^T x = a_0$ valid for P^n_{BW}, $n \geq 4$, which cannot be written as linear combination of betweenness equations. Then there is at least one betweenness triple $i(j)k$ with $a_{i(j)k} \neq a_{i(k)j}$. W. l. o. g. we set $i = 1$, $j = 2$ and $k = 3$. Now we construct an equation $\tilde{a}^T x = \tilde{a}_0$ with $\tilde{a}_{1(2)3} \neq \tilde{a}_{1(3)2}$ and show that it is valid for P^4_{BW}, which is a contradiction, because P^4_{BW} does not contain such an equation as we have verified by using the software PORTA [CL98]. The coefficients of $\tilde{a}^T x = \tilde{a}_0$ are defined as follows (note that always $i < k$).

$$\tilde{a}_{i(j)k} = \begin{cases} a_{i(j)k} & \text{if } i, j, k \leq 3, \\ \dfrac{1}{n-3} \displaystyle\sum_{l=4}^{n} a_{i(j)l} & \text{if } k = 4, \\ \dfrac{1}{n-3} \displaystyle\sum_{l=4}^{n} a_{i(l)k} & \text{if } j = 4. \end{cases}$$

Now it remains to show that for each permutation $\tilde{\pi} \in S(4)$ the value $\tilde{a}^T\chi^{\tilde{\pi}}$ is a constant value not depending on $\tilde{\pi}$. We construct $(n-3)!$ permutations $\pi_1, \pi_2, \ldots, \pi_{(n-3)!} \in S(n)$ by replacing the element 4 in $\tilde{\pi}$ by all permutations of the $(n-3)$ elements $4, \ldots, n$. According to our assumption $\frac{1}{(n-3)!}\sum_{p=1}^{(n-3)!} a^T\chi^{\pi_p} = a_0$ holds. Therefore it remains to show that

$$\tilde{a}^T\chi^{\tilde{\pi}} + a_0 - \frac{1}{(n-3)!}\sum_{p=1}^{(n-3)!} a^T\chi^{\pi_p}$$

is a constant. And after a short calculation we see that this value equals

$$a_0 - \frac{1}{2}\sum_{i=1}^{3}\sum_{j,k\geq 4} a_{i(j)k} - \frac{1}{3}\sum_{i,j,k\geq 4} a_{i(j)k}.$$

Obviously this term, which is the right hand side $\tilde{a}_0$ of the constructed equation, does not depend on $\chi^{\tilde{\pi}}$ anymore.

In contrast to the consecutive ones polytope, trivial inequalities only define facets in the smallest case.

Theorem 5
(a) The complete linear description of P^3_{BW} is given by the betweenness equation $x_{1(2)3} + x_{1(3)2} + x_{2(3)1} = 1$ and the three trivial inequalities $x_{1(2)3} \geq 0$, $x_{1(3)2} \geq 0$ and $x_{2(1)3} \geq 0$.
(b)For $n \geq 4$ none of the trivial inequalities $x_{i(j)k} \geq 0$ or $x_{i(j)k} \leq 1$ defines a facet.

Proof. The descriptions of P^3_{BW} is easily verified. Now let $n \geq 4$.

For any permutation π the condition $\chi^{\pi}_{i(j)k} = 1$ forces $\chi^{\pi}_{i(k)j} = 0$ and $\chi^{\pi}_{j(i)k} = 0$. Thus the dimension of $\{x \mid x \in P^n_{\text{BW}} \text{ and } x_{i(j)k} = 1\}$ is at most $2\binom{n}{3} - 2$. And consequently the inequality $x_{i(j)k} \leq 1$ cannot be facet-defining for P^n_{BW}.

Similarly, for all permutations of four elements i, j, k, l with $\chi^{\pi}_{i(j)k} = 0$ the additional equation $\chi^{\pi}_{i(j)l} = \chi^{\pi}_{k(j)l}$ holds. This can easily be seen by testing all possibilities. This equation cannot be derived from the betweenness equations. Thus again the dimension of the induced face is at most $2\binom{n}{3} - 2$.

Using PORTA we found, e.g., that the complete description of P^4_{BW} consists of the four betweenness equations and 12 additional facets, all defined by inequalities of the form $x_{i(j)k} + x_{i(k)j} + x_{j(i)l} + x_{k(i)l} \geq 1$.

Since we have equations, the same facet-defining inequality can be stated in various ways. Therefore we define a *normal form* with the property that two facet-defining inequalities define the same facet if and only if their normal forms coincide.

Definition 6 *A facet-defining inequality is stated in normal form if the following properties hold.*
i)The inequality is written as $\sum a_{i(j)k}x_{i(j)k} \geq a_0$.
ii)All coefficients $a_{i(j)k}$ *are nonnegative coprime integers.*
iii)At least one of the three coefficients $a_{i(j)k}$, $a_{i(k)j}$ *and* $a_{j(i)k}$ *is zero for pairwise different elements* i, j *and* k, $1 \leq i,j,k \leq n$.

It is easily seen that the normal form of a facet-defining inequality is unique and that it is easy to convert an inequality to normal form.

4 A Common Polytope

Both the feasible solutions of the WBWP and of the WC1P are based on the permutations of n elements. For examining relations between the two problems we define a master problem combining their constraints. Here we seek for a permutation satisfying betweenness conditions as well as having an associated matrix where the ones appear consecutively. This would model the practical situation where a consecutive ones matrix has to be found that satisfies in addition certain betweenness conditions.

We say that a permutation π of n elements *establishes* the consecutive ones conditions of a 0/1-matrix $A \in \{0,1\}^{m\times n}$ (π *establishes C1C of* A) if in the permuted matrix $A' \in \{0,1\}^{m\times n}$ with $a'_{ri} = a_{r\pi^{-1}(i)}$ the ones occur consecutively in every row. The common betweenness and consecutive ones polytope $P^{m,n}_{\text{BWC1}}$ is then defined as

$$P^{m,n}_{\text{BWC1}} = \text{conv}\{(\chi^{\pi}, \chi^{A}) \mid A \in \{0,1\}^{m\times n} \text{ and } \pi \text{ establishes C1C of } A\}.$$

The single polytopes can be simply obtained from $P^{m,n}_{\text{BWC1}}$.

Remark 7 *The projection of* $P^{m,n}_{\text{BWC1}}$ *on the betweenness variables* $x_{i(j)k}$ *is the betweenness polytope* P^{n}_{BW} *and the projection on the consecutive ones variables* x_{ri} *is the consecutive ones polytope* $P^{m,n}_{\text{C1}}$.

Proof. Clearly the restriction of any vertex of $P^{m,n}_{\text{BWC1}}$ on the betweenness variables, the consecutive ones variables, respectively, is a vertex of P^{n}_{BW}, a vertex of $P^{m,n}_{\text{C1}}$, respectively. It remains to be shown that all vertices of P^{n}_{BW} and $P^{m,n}_{\text{C1}}$ can be reached by projection.

If A is the zero matrix then every permutation π establishes C1C of A. Thus for every vertex v of P^{n}_{BW} (which is associated to a permutation π) there is at least one vertex v' of $P^{m,n}_{\text{BWC1}}$ with $v' = (v, *)$.

Similarly, every vertex w of $P^{m,n}_{\text{C1}}$ is associated with a matrix A that is C1P. Therefore there is at least one permutation establishing C1C of A. And with it there is a vertex w' of $P^{m,n}_{\text{BWC1}}$ such that $w' = (*, w)$.

Of course, all valid inequalities for $P_{\mathrm{C1}}^{m,n}$ or P_{BW}^{n} remain valid for $P_{\mathrm{BWC1}}^{m,n}$. We will now construct valid inequalities for $P_{\mathrm{BWC1}}^{m,n}$ that are formulated both on betweenness and on consecutive ones variables.

Lemma 8 *Let $A = (a_{ri})$ be an (m,n)-matrix ($n \geq 3$) with C1P. Then for all rows r of A, all betweenness triples $i(j)k$ of columns i, j, k of A and all permutations π that establish C1C of A we have*

$$\chi^{\pi}_{i(j)k} \leq 2 - a_{r\pi^{-1}(i)} + a_{r\pi^{-1}(j)} - a_{r\pi^{-1}(k)}.$$

Proof. The only possibility to violate this inequality is to set $\chi^{\pi}_{i(j)k} = 1$, $a_{r\pi^{-1}(i)} = 1$, $a_{r\pi^{-1}(j)} = 0$ and $a_{r\pi^{-1}(k)} = 1$. But according to the definition $\chi^{\pi}_{i(j)k} = 1$ means $\pi^{-1}(i) < \pi^{-1}(j) < \pi^{-1}(k)$ or $\pi^{-1}(i) > \pi^{-1}(j) > \pi^{-1}(k)$. Both cases force the 0-entry $a_{r\pi^{-1}(j)}$ to lie between the 1-entries $a_{r\pi^{-1}(i)}$ and $a_{r\pi^{-1}(k)}$. Thus the ones do not occur consecutively in row r of A in contradiction to the choice of π.

Based on this observation we can define so-called *linking constraints.*

Theorem 9 *The inequalities*

$$x_{i(j)k} \leq 2 - x_{ri} + x_{rj} - x_{rk}$$

are valid for $P_{\mathrm{BWC1}}^{m,n}$ for all $r \in \{1,\ldots,m\}$ and all betweenness triples $i(j)k$.

Proof. According to Lemma 8 the inequalities are fullfilled by all vertices of $P_{\mathrm{BWC1}}^{m,n}$. Therefore they are valid for the polytope.

Intuitively there is a close relationship between the consecutive ones and the betweenness problem. We now establish a connection between the facets of the two polytopes by making use of the linking constraints. These constraints are used to eliminate betweenness variables and replace them by consecutive ones variables.

Since in the consecutive ones problem we actually deal with matrices we will denote a valid inequality for $P_{\mathrm{C1}}^{m,n}$ by $B \circ x \leq b_0$, where B is the coefficient matrix, x is the matrix of the variables and $B \circ x = \sum_{i=1}^{m} \sum_{j=1}^{n} b_{ij} x_{ij}$. Further we will use $B_{i.}^{T} x_{i.}$ instead of $\sum_{j=1}^{n} b_{ij} x_{ij}$.

Theorem 10 *Let $a^T x \geq a_0$ be a facet-defining inequality for P_{BW}^{n}, $n \geq 4$, in normal form. Further let m be the number of nonzero coefficients of a. We assign pairwise different numbers $r_{i(j)k} \in \{1,\ldots,m\}$ to the betweenness triples $i(j)k$ with $a_{i(j)k} > 0$. Let the inequality $B \circ x \leq b_0$ be obtained by summing up $-a^T x \leq -a_0$ and all (scaled) linking constraints $a_{i(j)k} x_{i(j)k} \leq a_{i(j)k}(2 - x_{r_{i(j)k}i} + x_{r_{i(j)k}j} - x_{r_{i(j)k}k})$. Then $B \circ x \leq b_0$ is facet-defining for $P_{\mathrm{C1}}^{m,n}$.*

Proof. See [OR02] for the proof.

Consider for example the inequality

$$x_{1(2)3} + x_{1(3)2} + x_{2(1)4} + x_{3(1)4} \geq 1$$

which is facet-defining for P^4_{BW} *and in normal form. The four linking constraints*

$$\begin{aligned} -x_{1(2)3} &\geq -2 + x_{11} - x_{12} + x_{13} \\ -x_{1(3)2} &\geq -2 + x_{21} - x_{23} + x_{22} \\ -x_{2(1)4} &\geq -2 + x_{32} - x_{31} + x_{34} \\ -x_{3(1)4} &\geq -2 + x_{43} - x_{41} + x_{44} \end{aligned}$$

are valid for $P^{4,4}_{\mathrm{BWC1}}$.

Summing up these five inequalities (thus eliminating the betweenness variables) and multiplying by -1 *yields the inequality*

$$\begin{pmatrix} 1 & -1 & 1 & 0 \\ 1 & 1 & -1 & 0 \\ -1 & 1 & 0 & 1 \\ -1 & 0 & 1 & 1 \end{pmatrix} \circ x \leq 7.$$

As already shown in [OR00]) this inequality is facet-defining for $P^{4,4}_{\mathrm{C1}}$.

If we start with a facet $a^T x \geq a_0$ of P^n_{BW} in normal form, clearly the support of the constructed facet of $P^{m,n}_{\mathrm{C1}}$ has at most n columns. But it is also possible to bound the number of rows. Namely, according to the construction, m is the number of nonzero coefficients of a. Since at least one of three coefficients is zero, we have $m \leq 2\binom{n}{3}$. Taking into account that these facets can be trivially lifted to facets of $P^{m',n}_{\mathrm{C1}}$ with $m' > m$ the total number of constructed facets of $P^{m,n}_{\mathrm{C1}}$ for fixed n and arbitrary m is $O(m^{2\binom{n}{3}})$ and thus polynomial in m.

References

[CL98] T. Christof, A. Loebel (1998) PORTA - A Polyhedron Representation Algorithm. www.informatik.uni-heidelberg.de/groups/comopt/software/PORTA

[COR98] T. Christof, M. Oswald and G. Reinelt (1998) Consecutive Ones and A Betweenness Problem in Computational Biology. *Proceedings of the 6th IPCO Conference, Houston*, 213–228

[OR00] M. Oswald and G. Reinelt (2000) Polyhedral Aspects of the Consecutive Ones Problem, *Proceedings of the 5th Conference on Computing and Combinatorics, Sydney*, 373–382

[OR02] M. Oswald and G. Reinelt (2002) The Weighted Consecutive Ones Polytope for a Fixed Number of Columns *(In preparation)*

Solving One-Dimensional Cutting Stock Problems with Multiple Stock Material Lengths Using Cutting Plane Approach

Guntram Scheithauer, Gleb Belov

Dresden University of Technology, 01062 Dresden, Germany

Abstract. For exactly solving the one-dimensional cutting stock problem (CSP) with several stock material lengths, a cutting plane approach (CPA) is proposed. This work is a continuation of [12] where firstly column generation technique and CPA have been combined for the CSP with identical stock material pieces. In order to ensure efficient implementation of the generation problems in case of no and of additional cutting planes, some modifications of known methods have been performed. The efficiency of the algorithm is demonstrated by extensive numerical results.

1 Introduction

Cutting and packing problems are of wide interest for practical applications and research (cf. [3]). Especially the one-dimensional cutting stock problem (CSP) where smaller lengths (pieces, items) are to be cut from a minimum number of identical stock pieces is considered in a lot of publications. In this work the more general case is studied where different stock types with supply bounds are available. The total material cost minimization is modeled as an integer linear optimization problem.

The Gilmore-Gomory approach [4] solves the continuous relaxation of CSP with multiple stock sizes by means of column generation techniques and applies rounding procedures to the generally non-integral solution. This approach is a heuristic but in case of only one stock type it very often leads to a proved optimum. The reason is that the optimal objective function value of the LP relaxation rounded up very often equals that of the integer problem (cf. [6,10,17]) and is thus an effective lower bound. A series of theoretical investigations is concerned with the *Modified Integer Round-Up Property* (MIRUP) conjecture (cf. [10]) which states that the gap between these values is always smaller than two. For the multiple stock case, [4] states that the possibility to combine different stock sizes helps to achieve better material utilization but it is more difficult to find optimum solutions because of complicated behavior of the objective function: its feasible values are determined by combinations of the stock prices. Our results confirm these conclusions.

In the last years there have been some attempts to solve CSP with one stock type exactly using branch-and-bound and column generation, also denoted as branch-and-price [2,5,11,15,16], using a pure cutting plane algorithm

[12] or others which especially address the bin-packing problem where pieces of each type are to be obtained exactly once [7,13].

This paper addresses an exact cutting plane approach for CSP with different stock sizes. Chvatal-Gomory *(C-G)* cutting planes [8] are used to tighten the continuous relaxation. The investigations are based on [12] where, to our knowledge, the combination of C-G cutting planes and column generation is introduced. This combination is not straightforward. To evaluate a new column during generation, it is necessary to anticipate the values of the cut coefficients which are nonlinear and non-separable in the components of the cutting pattern. The corresponding problem is much more difficult than a usual knapsack problem.

For the practical solvability of the column generation subproblem, the following issues have to be addressed: branching priority list using the dual information on cut constraints; early termination; bounds on cut coefficients. For the overall efficiency of the method, two points are important: the quality of cutting planes and a robust rounding heuristic. The heuristic helps to find an optimal solution long before cutting planes have forced integrality. In this paper two major improvements in comparison to [12] are introduced which concern the upper bound in the subproblem and the rounding heuristic (residual problem extension). Further significant contributions are a piece dominance criterion for bounded knapsack problems and loops elimination in dynamic programming.

In computational results we show characteristics of the proposed approach.

2 Problem Formulation and Solution Approach

CSP with multiple stock sizes is characterized by the following data:

m	Number of piece types	M	Number of stock types
$(l_1,\ldots,l_m)$	Piece lengths	$(L_1,\ldots,L_M)$	Stock lengths
$(b_1,\ldots,b_m)$	Piece order demands	$(B_1,\ldots,B_M)$	Stock supply
		$(\rho_1,\ldots,\rho_M)$	Stock prices

When stock prices are proportional to stock lengths, a trim loss minimization problem arises. Let $n := m + M$. A vector $a = (a_1,\ldots,a_n)^T \in \mathbb{Z}_+^n$ represents a *cutting pattern* if $\sum_{i=1}^m l_i a_i \le \sum_{i=1}^M L_i a_{i+m}$ and $\sum_{i=1}^M a_{i+m} = 1$. The components a_i for $1 \le i \le m$ determine how many pieces of type i will be cut. The component a_{j+m} corresponding to the used stock type j equals one, all other components a_{i+m} for $i \in \{1,\ldots,M\} \setminus j$ equal zero. In the following only *proper* patterns, i. e. $a_i \le b_i$, $1 \le i \le m$, are considered because this does not change the optimal value but the dimension of the solution space becomes smaller and hence, the continuous lower bound is tightened [9].

Let η denote the number of all cutting patterns which is very large in practice. The matrix $A = (a_{ij}) \in \mathbb{Z}_+^{n\times\eta}$ with columns representing all cutting patterns is called a *pattern matrix*. Let the expression '$a \in A$' denote 'a is a column of A' (a is a cutting pattern).

CSP can be formulated as follows. Determine the vector of cutting frequencies $x = (x_1, \ldots, x_\eta)^T$ which minimizes the total material cost so that piece order demands and material supply bounds are fulfilled:

$$z^*_{IP} := \min\{cx : x \in \mathbb{Z}^\eta_+\} \tag{1}$$

$$\text{s. t. } \sum_{j=1}^{\eta} a_{ij}x_j \geq b_i,\ 1 \leq i \leq m, \quad \sum_{j=1}^{\eta} a_{ij}x_j \leq B_{i-m},\ m < i \leq n, \tag{2}$$

where the objective function coefficients $c := (c(a))_{a\in A}$ with $c(a) := \sum_{i=1}^{M} \rho_i a_{i+m}$ are the prices of material used in the columns. They can be chosen equal to one in case of only one stock type.

The *continuous* (or *LP*) *relaxation* of (1)–(2) is obtained by discarding the integrality constraints. Introducing slack variables leads to:

$$\underline{z} := \min\{cx : x \in \mathbb{R}^\eta_+,\ s_i \in \mathbb{R}^1_+,\ 1 \leq i \leq n\} \tag{3}$$

$$\text{s. t. } \sum_{j=1}^{\eta} a_{ij}x_j - s_i = b_i,\ 1{\leq}i{\leq}m, \quad \sum_{j=1}^{\eta} a_{ij}x_j + s_i = B_{i-m},\ m{<}i{\leq}n. \tag{4}$$

The value of $\underline{z}$ is a lower bound *(LP bound)* for z^*_{IP}. A solution of (3)–(4), also called a continuous solution, can be used to construct a feasible integer one by means of rounding heuristics. All columns of the pattern matrix A cannot be obtained for practical instances, so the revised simplex method with column generation has to be applied. Only the basis columns are stored.

In case cutting planes have to be added, the choice of a *cut-generating vector* u as a row of the simplex basis inverse satisfying $u\tilde{b} \notin \mathbb{Z}$ guarantees the infeasibility of the corresponding LP solution when the *basis cut* $\lceil u\tilde{A}\rceil\tilde{x} \geq \lceil u\tilde{b}\rceil$ is added to constraints $\tilde{A}\tilde{x} = \tilde{b}$ (cf. [12]). Some simple cut lifting operations are applied to the vectors u which try to maximize the difference $\lceil u\tilde{b}\rceil - u\tilde{b}$ [12,1]. Then the procedure can be repeated by combining generated inequalities and/or original ones. As we know, any valid inequality for the convex hull of integral solutions can be obtained as a Gomory cut, which provides the exactness and finiteness of the approach. The corresponding properties of C-G cuts can also be proved directly (cf. [8]).

In order to speed up the solution process after adding new cutting planes, the columns of the last optimal simplex basis are used for pivoting at first. It is clear that not all these columns can be used because the last solution becomes infeasible. The rounding direction in C-G procedure was changed (in comparison to [8,12]) in order to improve upper bounds in the column generation procedure.

The cutting plane approach has the following scheme: at first the continuous relaxation (3)–(4) is solved. Then heuristics are applied to construct feasible integer solutions. If the optimality test of the latter fails, some cutting planes are added and the process re-iterates. Let us index this process by κ beginning with $\kappa = 0$. For simplicity of the description, we suppose that only one cut is added in each iteration. Then the valid inequalities present in the formulation in iteration κ are

$$\sum_{j=1}^{\eta} \pi_r(a_j)\, x_j + \sum_{j=1}^{n} \pi_r(a_{j+\eta}) s_j \geq \pi_r(a_0), \quad 1 \leq r \leq \kappa, \tag{5}$$

$$\text{with } \pi_r(a_j) := \left\lceil \sum_{i=1}^{n} a_{ij} u_{ri} + \sum_{s=1}^{r-1} \pi_s(a_j) u_{r,n+s} \right\rceil, \; j=0,\ldots,\eta+n, \tag{6}$$

$$u_r := (u_{r1}, \ldots, u_{r,n+r-1}), \quad 1 \leq r \leq \kappa,$$

where $a_0 := (b_1, \ldots, b_m, B_1, \ldots, B_M)^T$ is the right-hand side of the constraints (4) and the columns a_j, $j > \eta$, correspond to the slack variables (4). The length of the r-th cut-generating vector u_r is the number of constraints $n + r - 1$ when constructing it. The cuts r, $r > 1$, are dependent on those earlier constructed cuts s, $1 \leq s < r$, for which $u_{r,n+s} \neq 0$. Setting $u = u - \lfloor u \rfloor \geq \mathbf{0}$ guarantees that the cut coefficients $\pi(a)$ are non-decreasing in each component of a.

In practice, memory restrictions and column generation complexity necessitate deletion of inactive cuts. Then the finiteness is not guaranteed any longer. But for all tested instances, cycles could be prevented by means of comparison of new cuts with a collection of deleted ones. However, those cuts which are necessary to calculate other ones must not be deleted.

3 Column Generation without Cuts

In the first iteration of the solution approach no cuts are involved. When a primal feasible continuous solution is close to an optimum, the simplex multipliers d_i are nearly proportional to the piece lengths l_i. In problems with very small piece lengths (less than 1/100 of the largest stock length) this makes a branch-and-bound column generation approach inefficient because the quotients d_i/l_i are all nearly equal. In this case dynamic programming is better. Moreover, a single dynamic list can be used to obtain optimal knapsack values for all material types.

The column generation subproblem without cuts is the following:

$$\bar{c}^* := \max \left\{ \bar{c}(a) = da - c(a) \; : \; a \in A \right\},$$

where $d \in \mathbb{R}^n$ are the simplex multipliers. Besides, the slack variables of (4) must be considered but they are only used for pivoting if their reduced cost is best.

In order to formulate dynamic programming recurrences let us define for $1 \leq q \leq m$ and $h \geq 0$

$$f_q(h) := \max \left\{ \sum_{i=1}^{q} d_i a_i \; : \; \sum_{i=1}^{q} l_i a_i \leq h, \; a_i \leq b_i, \; a_i \in \mathbb{Z}_+, \; 1 \leq i \leq q \right\}.$$

Then $\bar{c}^* = \max \{ f_m(L_j) + d_{m+j} - \rho_j \; : \; 1 \leq j \leq M \}$ holds. Using $f_0 \equiv 0$, the following well-known recurrence formula [4,7] is valid in the case when piece

q is *unbounded*, i. e. if $b_q \geq \lfloor h/l_q \rfloor$: $f_q(h) = \max\{f_{q-1}(h), f_q(h-l_q)+d_q\}$; otherwise

$$f_q(h) = \max\{f_{q-1}(h-kl_q)+kd_q : k \in \{0,\ldots,b_q\}\}.$$

In [1] techniques of *overlapping loops elimination* in forward state strategy [14] are presented which improve the applicability of dynamic programming for knapsack problems with upper bounds. Furthermore, the Nicholson principle of dynamic programming is applied. Finally a *piece dominance criterion* for bounded knapsack problems is developed.

4 Column Generation with Cuts

To evaluate a new column when cutting planes are present, besides using the dual information to assess the cutting pattern as done in the standard column generation process, it is necessary to anticipate the values of the coefficients the column will have in the cuts previously introduced, and use the dual information from the cut rows to make a final evaluation of the attractiveness of the column. Selecting the most attractive column involves the solution of an integer programming problem with a non-separable objective function, which does not seem to have any usable structure, and has to be solved with branch-and-bound. It is much more difficult than a usual knapsack problem.

Let κ be the iteration number corresponding to the number of cuts and $d \in I\!R^{n+\kappa}$ be the vector of simplex multipliers corresponding to some feasible solution of the continuous relaxation (3)–(5). The solution is optimal if all columns of the constraint matrix A_κ have non-negative reduced costs. The problem to find a cutting pattern with the minimum reduced cost is the following:

$$\max\{\bar{c}(a) := \textstyle\sum_{i=1}^{n} d_i a_i + \sum_{r=1}^{\kappa} d_{n+r}\pi_r(a) - c(a) : a \in A\}. \qquad (7)$$

The cut coefficients $\pi(a)$ are nonlinear and non-separable in the components of the vector a so that branch-and-bound must be applied. The upper bound on $\bar{c}(a)$ becomes stronger when $\bar{c}(a)$ is non-negative and non-decreasing in each component a_i, $1 \leq i \leq m$. The cut coefficients $\pi_r(a)$ already have this property and hence the condition $d_i \geq 0$, $i \in \{1,\ldots,m\} \cap \{n+1,\ldots,n+\kappa\}$ is sufficient for non-negativity and (weak) monotonicity of $\bar{c}(a)$. To ensure this condition, all slack variables with negative reduced costs are entered into the basis (pivoted) prior to considering the cutting patterns according to (7).

For each stock length $L = L_j$, $1 \leq j \leq M$, the nonlinear knapsack problem

$$\max\{\bar{c}(a) = \textstyle\sum_{i=1}^{m} d_i a_i + \sum_{r=1}^{\kappa} d_{n+r}\pi_r(a) + d_{m+j} - \rho_j : a \in A,\ a_{m+j} = 1\}$$

is solved. The best solution found for each stock length is used as a lower bound for the next type. Problems (1) - (2) with many stock types seem to be easier than problems with very few (but several) types. For details we refer to [1].

5 Computational Results

The algorithm is implemented in GNU C++ under Linux/Unix. Stock prices are equal to stock lengths. All data are supposed to be integer. No external LP solver is used because a division-free simplex basis transformation method has been tested (without significant results).

Each *problem class* for given m and M is characterized by the following parameters: $v_1 L_{max} \le l_i < v_2 L_{max}$ with $0 < v_1 < v_2 \le 1$, $L_{max} = \max\{L_i\}$ and $\underline{b} \le b_i < \overline{b}$ for all pieces i, $1 \le i \le m$; $\underline{L} \le L_i < \overline{L}$ and $\underline{B} \le B_i < \overline{B}$ for all stock types i, $1 \le i \le M$; the greatest common divisor of stock lengths $\gcd(\{L_i\})$. All values are generated according to a uniform distribution.

Table 1. Solution characteristics. Multiple stock types $m = 40$, $M = 5$, time limit 60 sec. after the first LP bound

$v_1 \backslash v_2$	0.2	0.3	0.4	0.5	0.6	0.7	0.8	0.9
	1. Number of instances with cuts (out of 100)							
0.25		87	87	90	85	66	61	55
0.15	80	54	80	92	92	94	76	77
0.01	55	42	56	84	90	95	83	77
0.001	46	46	56	84	89	98	86	71
	2. Average number of iterations							
0.25		37.30	26.70	17.60	5.40	1.20	1.00	0.67
0.15	2.56	2.90	9.38	17.30	12.00	7.52	1.80	1.91
0.01	0.60	0.63	1.67	7.65	12.10	13.40	9.31	4.73
0.001	0.51	0.67	1.40	7.48	12.80	15.20	11.80	6.63
	3. Number of optimally solved instances							
0.25		58	72	91	100	100	100	100
0.15	54	90	71	82	94	94	100	100
0.01	85	94	93	69	72	77	93	97
0.001	84	96	98	74	73	78	90	97
	4. Maximum gap $100 \cdot (z^*_{IP} - \underline{z}_0)/L_{max}$ (solved instances)							
0.25		56.70	6.53	18.90	11.00	20.00	17.50	31.00
0.15	0.98	3.07	1.24	2.85	10.20	12.30	12.50	22.00
0.01	1.60	1.58	3.44	1.00	1.21	7.00	4.69	13.50
0.001	3.19	1.75	1.83	1.57	2.92	8.00	11.00	13.50
	5. Average final gap $100 \cdot (\overline{z} - \underline{z}_{\kappa})/L_{max}$ (all instances)							
0.25		2.86	1.26	0.38	0.00	0.00	0.00	0.00
0.15	1.57	0.11	0.38	0.24	0.08	0.14	0.00	0.00
0.01	0.25	0.08	0.09	0.47	0.56	0.31	0.13	0.04
0.001	0.28	0.05	0.02	0.35	0.39	0.32	0.21	0.06

In order to demonstrate some characteristics of the algorithm, the following series was computed: $m = 40$, $M = 5$, $b_i \in [200, 300)$, $L_1 = 10000$,

$L_i \in \{5000, 5100, \ldots, 9900\}$ for $2 \leq i \leq 5$, i. e. gcd($\{L_i\}$)=100=1/100L_1, and $2000 \leq B_i \leq 2500$ for $1 \leq i \leq 5$. For each problem class (i. e. for each combination v_1/v_2) 100 instances were solved. A maximum running time of 60 seconds after the first LP solution was allowed (Pentium III, 500 MHz). The results are shown in Table 1.

The entries in the first block are the numbers of those instances for which cutting planes were necessary because the integer solution found after the first LP relaxation was not proved to be optimal. In contrast to CSP with one stock type, cuts have to be applied much more often here since the optimality criterion is weaker. Block 2 gives the average numbers of iterations for an instance made within the given time limit. 10 cutting planes were added in the first and 2 in each following iteration. Block 3 contains the numbers of instances solved to an optimum within the given time limit.

For the problem with one stock length the MIRUP-conjecture [10] is known, which states that the gap between the LP lower bound and the optimal integer solution value is always smaller than two stock lengths. Block 4 shows the maximum values of this gap in the optimally solved instances as a fraction of the largest stock length. For many classes the gap is only a few percent.

Because of the weaker continuous lower and heuristic upper bounds in comparison to the case of a single stock type, a significant percentage of instances cannot be solved even with a time limit of hours, as opposed to the case with one stock type. Then efficiency can be measured by the gap between the lower bound $\underline{z}_\kappa$ and the best known integer solution value $\overline{z}$ (upper bound). In block 5 one can see that this gap as a fraction of the largest stock length, called *relative optimality gap*, becomes very small after the given execution time of 1 minute. In all but 3 classes it is less than 1% on average.

6 Conclusions and Outlook

An approach combining a cutting plane algorithm with column generation is applied to the one-dimensional cutting stock problem with multiple stock lengths. The computational difficulties arising in column generation because of the complicated dependence of cut coefficients on the original columns are successfully solved by means of an adapted bound, an appropriate searching strategy and early termination.

As opposed to the standard problem with only one stock type, the optimality criterion in the considered problem is more complicated, which makes the continuous lower and heuristic upper bounds less effective when searching for exact solutions. This necessitates further research concerning the quality of cutting planes and heuristics. However, it seems difficult to use problem structure to derive strong problem-specific valid inequalities. Nevertheless, the efficiency of the implementation both in terms of the number of optimally solved instances and of the material utilization in the solutions encourages to apply this algorithm to cutting stock problems of higher dimensions.

References

1. Belov G., Scheithauer G. (2000) Solving the general one-dimensional cutting stock problem with a cutting plane approach. Technical Report MATH-NM-11-2000, Dresden University of Technology
2. Degraeve Z., Peeters M. (2000) Optimal integer solutions to industrial cutting stock problems: Part 2, Benchmark Results. Technical Report, Katholieke Universiteit Leuven
3. Dyckhoff H., Scheithauer G., Terno J. (1997) Cutting and packing. in: M. Dell'Amico, F. Maffioli, S. Martello (eds), Annotated Bibliographies in Combinatorial Optimization, ch. 22, 393–412, John Wiley & Sons, Chichester
4. Gilmore P.C., Gomory R.E. (1963) A linear programming approach to the cutting-stock problem (Part II). Operetions Research 11:863-887
5. Kupke J. (1998) Lösung von ganzzahligen Verschnittproblemen mit Branch-and-Price. Diplomarbeit, Universität zu Köln
6. Marcotte O. (1986) An instance of the cutting stock problem for which the rounding property does not hold. OR Letters 4(5):239–243
7. Martello S., Toth P. (1990) Knapsack Problems – Algorithms and Computer Implementations. John Wiley & Sons, Chichester et al.
8. Nemhauser G.L., Wolsey L.A. (1988) Integer and Combinatorial Optimization. John Wiley & Sons, New York
9. Nitsche C., Scheithauer G., Terno J. Tighter relaxations for the cutting stock problem. European Journal of Operational Research 112/3:654–663
10. Scheithauer G., Terno J. (1995) The modified integer round-up property of the one-dimensional cutting stock problem. European Journal of Operational Research 84:562–571
11. Scheithauer G., Terno J. (1995) A branch-and-bound algorithm for solving one-dimensional cutting stock problems exactly. Applicationes Mathematicae 23.2:151–167
12. Scheithauer G., Terno J., Müller A., Belov G. (1999) Solving one-dimensional cutting stock problems exactly with a cutting plane algorithm. Technical Report MATH-NM-06-1999, Dresden Univ. of Technology (to appear in JORS)
13. Schwerin P., Wäscher G. (1999) A new lower bound for the bin-packing problem and its integration to MTP. Pesquisa Operacional, Vol.19, No.2, 111–130
14. Terno J., Lindemann R., Scheithauer G. (1987) Zuschnittprobleme und ihre praktische Lösung. Verlag Harri Deutsch Thun und Frankfurt/Main
15. Vance F. (1998) Branch-and-price algorithms for the one-dimensional cutting stock problem. Computational Optimization and Applications 9(3), 212-228
16. Vanderbeck F. (1000) Computational study of a column generation algorithm for bin packing and cutting stock problems. Mathematical Programming 86, Ser.A, 565–594
17. Wäscher G., Gau T. (1996) Heuristics for the one-dimensional cutting stock problem: a computational study. OR Spektrum 18:131–144

Computational Problems that can be Solved Without Computation

Gerhard J. Woeginger

Department of Mathematics, University of Twente, P.O. Box 217
7500 AE Enschede, The Netherlands

Abstract. The paper discusses various optimization problems that look fairly difficult at first sight, but in fact are trivial to solve. Here the word 'trivial' means that there exist primitive computer programs for these problems whose behavior hardly depends on the input data.

Among the discussed problems are: Minimizing the failure probability in a cyclic double-loop system; the travelling salesman problem under specially structured distances; the arrangement of data records in a linear storage array so as to minimize the average access time; balancing the blades of a turbine wheel.

1. Failure probabilities in cyclic double-loop systems. A cyclic double-loop system consists of n items that are ordered into a cycle. Each item has a direct connection to its two immediate predecessors and to its two immediate successors along the cycle. Each item works or fails. The whole system fails if the cycle is broken, i.e., if some two consecutive items both fail. We assume that the n items $I_1, \ldots, I_n$ are stochastically independent, and that item I_i works with probability p_i, $0 < p_i \leq 1$; without loss of generality $p_i \leq p_{i+1}$ for $1 \leq i \leq n-1$. This models the real-world situation where the failure probability of a component only depends on (and grows with) the age of this component. An ordering of the n items along the cycle is called *optimal*, if it maximizes the probability that the system works.

Derman, Lieberman & Ross [1982], Hwang [1982], and Wei, Hwang & Sós [1983] derived some partial results and formulated several conjectures on the combinatorial structure of optimal orderings for cyclic double-loop systems. Later, Du & Hwang [1986] proved all these conjectures by showing that the ordering

$$\langle I_1,\ I_{n-1},\ I_3,\ I_{n-3},\ I_5,\ \ldots, I_{n-4},\ I_4,\ I_{n-2},\ I_2,\ I_n\rangle \tag{1}$$

always minimizes the probability of failure. Rather surprisingly, there exists a *universal optimal ordering* for all instances that only depends on the ranking of the probabilities, but not on their actual values.

2. Some definitions and notations. For a permutation $\phi \in S_n$, we will shortly write $\phi = \langle\phi(1), \phi(2), \ldots, \phi(n)\rangle$. We denote by $\alpha = \langle 1, n-1, 3, n-3, 5, n-5, 7, \ldots, 6, n-4, 4, n-2, 2, n\rangle$ the permutation of indices that corresponds to the universal ordering in (1), and by $\beta =$

$\langle 1,3,5,7,9,11,\ldots,12,10,8,6,4,2\rangle$ the permutation that starts with the odd numbers in increasing order followed by the even numbers in decreasing order.

A symmetric $n\times n$ matrix $A=(a_{i,j})$ is called a *Supnick matrix*, if (i) it satisfies the inequality $a_{i,j}+a_{r,s}\le a_{i,s}+a_{r,j}$ for all $1\le i<r\le n$ and $1\le j<s\le n$ (these inequalities are sometimes called the *Monge inequalities*) and if (ii) $a_{i,j}\le a_{i,j+1}$ and $a_{i,j}\le a_{i+1,j}$ holds for all i,j, i.e., if the entries in every row and in every column are in non-decreasing order. The Supnick matrices form a cone: If we multiply a Supnick matrix by a positive real or if we add two Supnick matrices, we will always end up with another Supnick matrix. Simple examples of Supnick matrices are the *negative product matrices*: A matrix A is a negative product matrix if there exist non-negative real numbers $x_1\le x_2\le\cdots\le x_n$ such that $a_{i,j}=-x_i\cdot x_j$ holds for all $1\le i,j\le n$.

A symmetric $n\times n$ matrix $B=(b_{i,j})$ is called a symmetric *Toeplitz matrix*, if there exists a function $f\colon\{0,\ldots,n-1\}\to\mathbb{R}$ such that $b_{i,j}=f(|i-j|)$ for $1\le i,j\le n$; matrix B is said to be *generated* by this function f. A Toeplitz matrix is constant along every diagonal that is parallel to the main diagonal, and it is completely determined by its first row.

3. The travelling salesman problem on Supnick matrices. In the travelling salesman problem (TSP) the goal is to find a shortest cyclic tour through n cities whose distances are specified by an $n\times n$ distance matrix $C=(c_{i,j})$. In other words, a salesman starts from his home city, visits each city exactly once, in the end returns to his home city, and does all this with the smallest possible amound of gas. We denote by $H=(h_{i,j})$ the adjacency matrix of the undirected Hamiltonian cycle $\langle 1,2,3,4,\ldots,n\rangle$; hence, $h_{i,j}=1$ if $|i-j|=1$ or if $|i-j|=n-1$, and $h_{i,j}=0$ otherwise. Then the travelling salesman problem amounts to finding a permutation $\phi\in S_n$ that minimizes

$$\frac{1}{2}\sum_{i=1}^{n}\sum_{j=1}^{n}c_{\phi(i),\phi(j)}\,h_{i,j} \tag{2}$$

Here the factor $\frac{1}{2}$ accounts for the fact that in the matrix H every edge $[i,i+1]$ of the Hamiltonian cycle is counted twice, once from city i to city $i+1$ and once from city $i+1$ to city i. In general, the TSP is an NP-hard problem. We refer the reader to the book by Lawler, Lenstra, Rinnooy Kan & Shmoys [16] for more information on the TSP.

Supnick [1957] considered the TSP on Supnick distance matrices (in his paper, Supnick just formulated the properties of these matrices, but he did not call them 'Supnick' matrices). He proved by a fairly involved exchange argument that for Supnick distance matrices, the *longest* TSP tour is always given by the permutation α, whereas the *shortest* TSP tour is always given by the permutation β. Hence, in either case the optimal solutions α and β can be written down without looking at the actual distances. No computation! Supnick's results were rediscovered (at least) twice, once by Rubinshtein

[1971] and once by Michalski [1987]. Rudolf & Woeginger [1995] gave very simple proofs for Supnick's results by investigating the underlying geometric cone.

4. The role of the permutation α. The occurence of the permutation α in Supnick's result on the longest TSP tour and in the result of Du & Hwang [1986] on cyclic double-loop systems is not a sheer coincidence. In fact, the result of Du & Hwang is a simple special case of Supnick's result. Let $0 < p_1 \le \cdots \le p_n \le 1$ be the probabilities that the items $I_1, \ldots, I_n$ work. Consider the travelling salesman instance with the distance matrix C where

$$c_{i,j} \doteq \ln(p_i + p_j - p_i p_j). \tag{3}$$

All values $c_{i,j}$ in (3) are non-positive real numbers, and it is easily verified that the resulting matrix C is a Supnick matrix. Then by Supnick's theorem, the permutation α yields a longest TSP tour for C. Since the exponential function is a monotone function, this implies that the expression

$$\max_{\phi \in S_n} \; (p_{\phi(1)} + p_{\phi(n)} - p_{\phi(1)} p_{\phi(n)}) \cdot \prod_{i=1}^{n-1} (p_{\phi(i)} + p_{\phi(i+1)} - p_{\phi(i)} p_{\phi(i+1)}) \tag{4}$$

takes its maximum at $\phi = \alpha$. Now let us interprete this statement for cyclic double-loop systems: For two consecutive items with failure probabilities p and q, the probability that at least one of them works is $1 - (1 - p)(1 - q) = p + q - pq$. The problem of finding an arrangement that maximizes the probability that the whole system works corresponds to finding a permutation $\phi \in S_n$ that maximizes the product in (4). Since this maximum is taken at $\phi = \alpha$, the permutation α yields an optimal ordering of the items in the cyclic double-loop system.

5. Average access time in a linear storage array. Consider a set of n data records $R_1, \ldots, R_n$ which are referenced and accessed repeatedly. Record R_i is referenced with probability q_i $(0 < q_i \le 1)$, and different references are stochastically independent of each other. Without loss of generality the records are numbered such that $q_1 \le q_2 \le \cdots \le q_n$. These n data records are to be stored in a linear array of storage cells, like a magnetic tape. Then they will repeatedly be accessed by a read head that moves along this storage array. The goal is to store the records in such a way that the expected distance between consecutively referenced records is minimized, i. e., one wishes to find a permutation $\phi \in S_n$ that minimizes

$$\sum_{i=1}^{n} \sum_{j=1}^{n} q_{\phi(i)} q_{\phi(j)} \, d_{i,j} \tag{5}$$

where $d_{i,j}$ denotes the distance between the records placed at storage cells i and j respectively.

In the late 1960s and early 1970s, much research has been done on the special case of this problem where the distances $d_{i,j}$ are given by a symmetric Toeplitz matrix, i.e., on the special case where the distance $d_{i,j}$ only depends on the absolute value of the difference between i and j. First, Timofeev & Litvinov [1969] proved that the data arrangement problem is always solved by the permutation β, in case $d_{i,j} = |i-j|$ holds. Just a little bit later, Burkov, Rubinshtein & Sokolov [1969] showed that the data arrangement problem is always solved by the permutation β, in case $d_{i,j} = f(|i-j|)$ holds with a non-decreasing and convex function f. Finally, Metelski [1972] and Pratt [1972] realized that these results are all contained in a much older result due to Hardy, Littlewood & Pólya [1926]: Translated into the language of our paper, this much older result states that the data arrangement problem is always solved by the permutation β, in case $d_{i,j} = f(|i-j|)$ holds with an arbitrary non-decreasing function f. Vickson & Lu [1998] rediscovered the result of Timofeev & Litvinov [1969] for $d_{i,j} = |i-j|$.

Intuitively, it is clear that the records with high access probabilities should be stored around the center of the storage array, whereas the records with low access probabilities should go to the far ends of the array. However, for all the mentioned Toeplitz distances a much stronger statement holds: There exists a universal optimal arrangement for all instances that only depends on the ranking of the probabilities, but not on their actual values. Once again, no computations are needed for finding the optimal solution.

6. Balancing hydraulic turbine runners. Unbalance in a hydraulic turbine runner results in vibrations and in excess stresses on the supports, and thus shortens the life of the runner. Unbalance occurs when the center of gravity of the runner does not coincide with the axis of rotation. A hydraulic turbine runner of the so-called Francis design consists of a cylinder around which a number of blades are welded at regular spacings. A blade may weigh up to 16 tons, but due to manufacturing imperfections the blades are not identical. Mosevich [1986] mentions differences as large as $\pm 5\%$ in the weights of the blades. Hence, one of the main objectives during the welding process is to find a 'good' sequence of the blades around the cylinder which minimizes the deviation of the runner's center of mass from the runner axis. This 'Turbine Problem' has been discussed by Bolotnikov [1978], by Stoyan, Sokolovskii & Yakovlev [1982], by Schlegel [1987], and by Laporte & Mercure [1988].

In a more mathematical formulation of the problem that is due to Laporte & Mercure [1988], we are given n given positive masses (=blades) $0 < m_1 \leq m_2 \leq \cdots \leq m_n$. These masses have to be placed on the vertices $v_i = (\sin(2i\pi/n), \cos(2i\pi/n))$, $1 \leq i \leq n$, of a regular polygon (=cylinder) whose center (=runner axis) lies in the origin. The goal is to assign the masses in such a way to the vertices that the center of gravity of the resulting weighted polygon is as close to the origin as possible, i.e. to find a permutation $\phi \in S_n$ that minimizes the Euclidean norm of the two-dimensional vector

$$\sum_{i=1}^{n} m_{\phi(i)} \begin{pmatrix} \sin \frac{2i\pi}{n} \\ \cos \frac{2i\pi}{n} \end{pmatrix}.$$

An easy calculation reveals that minimizing the Euclidean norm of this vector is equivalent to minimizing the expression

$$\sum_{i=1}^{n} \sum_{j=1}^{n} m_{\phi(i)} m_{\phi(j)} \cos \frac{2(i-j)\pi}{n}. \tag{6}$$

Burkard, Çela, Rote & Woeginger [1998] proved that the minimization version of the turbine problem is NP-hard. This means that the minimization version is very difficult to solve and needs exponential time, unless P=NP. Thus, the turbine problem certainly does not belong to the optimization problems that can be solved without computation, and it is of no further interest for our paper.

In the *maximization* version of the turbine problem the goal is to get the center of gravity of the weighted polygon as far away from the origin as possible, i.e., one wants to find a permutation $\phi \in S_n$ that maximizes the expression in (6). Intuitively, all the heavy masses should go on one side of the polygon and all the light masses should go on the opposite side. Çela & Woeginger [1994] prove that in fact the permutation β yields a universal optimal solution for the maximization version of the turbine problem.

7. The quadratic assignment problem (QAP). Let $A = (a_{i,j})$ and $B = (b_{i,j})$, $1 \leq i, j \leq n$, be two given matrices with real entries. In the QAP in Koopmans-Beckmann form (Koopmans & Beckmann [1957]), the objective is to find a permutation $\phi \in S_n$ that minimizes the cost function

$$\sum_{i=1}^{n} \sum_{j=1}^{n} a_{\phi(i),\phi(j)}\, b_{i,j}. \tag{7}$$

In general, the QAP is an NP-hard problem. We refer the reader to the paper by Lawler [1963] and to the book by Çela [1998] for more information on the QAP.

8. The role of the permutation β. The shortest TSP tour on Supnick matrices, the data arrangement problem in linear storages, and the maximization version of the turbine problem all are special QAPs that all are solved to optimality by the same permutation β. Let us reformulate these three problems such that one of the involved matrices becomes a Supnick matrix and the other matrix becomes a symmetric Toeplitz matrix generated by some $f: \{0, \ldots, n-1\} \to \mathbb{R}$. We will use the fact that the QAP with matrices A and B is equivalent to the QAP with matrices $-A$ and $-B$.

- *The min TSP on Supnick matrices; cf. (2).* Here the first matrix $A = (c_{i,j})$ is a Supnick matrix by definition. The second matrix $B = (h_{i,j})$ is a symmetric Toeplitz matrix that is generated by the function f_1 with $f_1(1) = f_1(n-1) = 1$, and $f_1(i) = 0$ otherwise.
- *The data arrangement problem; cf. (5).* The first matrix $A = (-q_i q_j)$ is a negative product matrix and hence a Supnick matrix. The second matrix $B = (-d_{i,j})$ is a symmetric Toeplitz matrix that is generated by some non-increasing function f_2.
- *The max turbine problem; cf. (6).* The first matrix $A = (-m_i m_j)$ is a negative product matrix and hence a Supnick matrix. The second matrix $B = (\cos \frac{2(i-j)\pi}{n})$ is a symmetric Toeplitz matrix that is generated by the function $f_3(x) = \cos \frac{2x\pi}{n}$.

Now, what are the common properties of these three problems that make them all solvable — without any further computation — by permutation β? In all three problems, the first matrix is a Supnick matrix (and that was our motivation for reformulating the problems). The second matrix is a symmetric Toeplitz matrix. If the reader draws pictures of the generating functions f_1, f_2, and f_3, then he will notice that they all are non-increasing from 0 to $\lfloor n/2 \rfloor$. Moreover, the graphs of f_1 and f_3 are symmetric with respect to the vertical line through $(n-1)/2$. The graph of f_2 is also non-increasing from $\lfloor n/2 \rfloor$ to $n-1$, but apart from this property it may behave quite chaotically.

Let us define another function $g: \{0, \dots, n-1\} \to \mathbb{R}$ by $g(x) = f(n-1-x)$ for $x = 0, \dots, n-1$. Burkard, Çela, Rote & Woeginger [1998] proved that the crucial properties of function f are the following: (i) On the domain $0, \dots, \lfloor n/2 \rfloor$ the function f is non-increasing. (ii) On the domain $\lceil n/2 \rceil, \dots, n-1$ the function f never goes above g, i.e., $f(x) \le g(x)$. If the Toeplitz matrix B is generated by such a function f, and if matrix A is a Supnick matrix, then the QAP is always solved by permutation β. The proof is based on a careful investigation of the interaction between the cone of Supnick matrices and the cone of this special class of Toeplitz matrices.

9. The travelling salesman problem for convex point sets. In one of the first papers on the TSP, Flood [1956] observed that 'in the Euclidean plane the shortest TSP tour does not intersect itself'. This follows quite easily from the fact that in any convex quadrangle, the total length of the diagonals is always greater than the total length of two opposite sides. If some TSP tour contains two intersecting edges, then these edges constitute the diagonals of a convex quadrangle. Then one may swap them with an appropriate pair of opposite sides and thus get a shorter tour. As one consequence, if all cities lie on the boundary of a convex polygon in the Euclidean plane, then the optimal TSP tour is a cyclic walk along the boundary of this polygon in clockwise or counterclockwise direction. Without loss of generality the numbering of the cities goes in clockwise order around the polygon. Then in the language of our paper, we may state that for distance matrices of convex planar point sets the TSP is always solved by the identity permutation $\langle 1, 2, \dots, n \rangle$.

Kalmanson [1975] realized that this statement is not really based on geometric convexity, but on the following much weaker properties of the distance matrix $C = (c_{i,j})$:

$$c_{i,j} + c_{k,\ell} \leq c_{i,k} + c_{j,\ell} \qquad \text{for all } 1 \leq i < j < k < \ell \leq n$$
$$c_{i,\ell} + c_{j,k} \leq c_{i,k} + c_{j,\ell} \qquad \text{for all } 1 \leq i < j < k < \ell \leq n.$$

Any matrix satisfying these conditions is called a *Kalmanson matrix*. It can be seen that the distance matrix of every convex planar point set is a Kalmanson matrix. Moreover, also tree distance matrices are a special case of Kalmanson matrices: In this case all the cities lie somewhere on an edge-weighted undirected tree, and the distance between cities i and j equals their distance in this tree. Kalmanson [1975] proved that for Kalmanson distance matrices, the TSP is always solved by the identity permutation.

10. The role of the identity permutation. Deineko & Woeginger [1998] considered the QAP where one matrix is a Kalmanson matrix and the other matrix is a Toeplitz matrix generated by a function $f: \{0, \ldots, n-1\} \to \mathbb{R}$ with the following properties. First, $f(x) = f(n-x)$ for all $x = 1, \ldots, n-1$, and secondly, $f(x) \geq f(x+1)$ for all $x = 1, \ldots, \lfloor n/2 \rfloor$. They prove that this special case of the QAP is always solved by the identity permutation. This is yet another case of a computational problem that can be solved without any computation.

The result of Deineko & Woeginger [1998] contains the TSP on Kalmanson matrices as a special case: The adjacency matrix $H = (h_{i,j})$ of an undirected Hamiltonian cycle is a symmetric Toeplitz matrix that is generated by the function f with $f(1) = 1$, $f(n-1) = 1$, and $f(i) = 0$ otherwise. Clearly, this generating function is of the type stated above.

References

1. A.A. Bolotnikov [1978]. On the best balance of the disk with masses on its periphery. *Problemi Mashinostroenia 6*, 68–74, (in Russian).
2. R.E. Burkard, E. Çela, G. Rote, and G.J. Woeginger [1998]. The quadratic assignment problem with a monotone anti-Monge and a symmetric Toeplitz matrix: Easy and hard cases. *Mathematical Programming 82*, 125–158.
3. V.N. Burkov, M.I. Rubinshtein, and V.B. Sokolov [1969]. Some problems in optimal allocation of large-volume memories. *Avtomatika i Telemekhanika 9*, 83–91, (in Russian).
4. E. Çela [1994]. *The quadratic assignment problem.* Kluwer, Dordrecht.
5. E. Çela and G.J. Woeginger [1994]. A note on the maximum of a certain bilinear form. Technical Report SFB-8, TU Graz.
6. V.G. Deineko and G.J. Woeginger [1998]. A solvable case of the quadratic assignment problem. *Operations Research Letters 22*, 13–17.
7. C. Derman, G.J. Lieberman, and S.M. Ross [1982]. On the consecutive-k-out-of-n systems. *IEEE Transactions on Reliability 31*, 57–63.

8. D.Z. Du and F.K. Hwang [1986]. Optimal consecutive-2-out-of-n systems. *Mathematics of Operations Research 11*, 187–191.
9. M.M. Flood [1956]. The traveling salesman problem. *Operations Research 4*, 61–75.
10. G.H. Hardy, J.E. Littlewood, and G. Pólya [1926]. The maximum of a certain bilinear form. *Proceedings of the London Mathematical Society 25*, 265–282.
11. F.K. Hwang [1982]. Fast solutions for consecutive-k-out-of-n systems. *IEEE Transactions on Reliability 31*, 447–448.
12. K. Kalmanson [1975]. Edgeconvex circuits and the travelling salesman problem. *Canadian Journal of Mathematics 27*, 1000–1010.
13. T.C. Koopmans and M.J. Beckmann [1957]. Assignment problems and the location of economic activities. *Econometrica 25*, 53–76.
14. G. Laporte and H. Mercure [1988]. Balancing hydraulic turbine runners: A quadratic assignment problem. *European Journal of Operational Research 35*, 378–382.
15. E.L. Lawler [1963]. The quadratic assignment problem. *Management Science 9*, 586–599.
16. E.L. Lawler, J.K. Lenstra, A.H.G. Rinnooy Kan, and D.B. Shmoys (eds.) [1985] *The travelling salesman problem*. John Wiley, Chichester.
17. N.N. Metelski [1972]. On extremal values of quadratic forms on symmetric groups. *Vesti Akad. Navuk BSSR Ser. Fiz.-Mat. Navuk 6*, 107–110, (in Russian).
18. M. Michalski [1987]. On a class of polynomially solvable travelling salesman problems. *Zastosowania Matematyki 19*, 531–539.
19. J. Mosevich [1986]. Balancing hydraulic turbine runners — a discrete combinatorial optimization problem. *European Journal of Operational Research 26*, 202–204.
20. V.R. Pratt [1972]. An $N \log N$ algorithm to distribute N records optimally in a sequential access file. In: *Complexity of Computer Computations*, (R.E. Miller and J.W. Thatcher eds.), Plenum Press, New York, 111–118.
21. M.I. Rubinshtein [1971]. On the symmetric traveling salesman problem. *Automation and Remote Control 32*, 1453–1460.
22. R. Rudolf and G.J. Woeginger [1995]. The cone of Monge matrices: Extremal rays and applications. *ZOR - Mathematical Methods of Operations Research 42*, 161–168.
23. D. Schlegel [1987]. Die Unwucht-optimale Verteilung von Turbinenschaufeln als quadratisches Zuordnungsproblem. Ph. D. Thesis, ETH Zürich.
24. F. Supnick [1957]. Extreme Hamiltonian lines. *Annals of Mathematics 66*, 179–201.
25. Y.G. Stoyan, V.Z. Sokolovskii, and S.V. Yakovlev [1982]. A method for balancing discretely distributed masses under rotation. *Energomashinostroenia 2*, 4–5, (in Russian).
26. B.B. Timofeev and V.A. Litvinov [1969]. On the extremal value of a quadratic form. *Kibernetika 4*, 56–61, (in Russian).
27. R.G. Vickson and X. Lu [1998]. Optimal product and server locations in one-dimensional storage racks. *European Journal of Operational Research 105*, 18–28.
28. V.K. Wei, F.K. Hwang, and V.T. Sós [1983]. Optimal sequencing of items in a consecutive-2-out-of-n system. *IEEE Transactions on Reliability 32*, 952–954.

Section 10

Stochastic Optimization and Simulation

Partitions-requirements-matrices

Regina Hildenbrandt

TU Ilmenau, Institut für Mathematik, Postfach 10 05 65, D - 98684 Ilmenau, Germany

Abstract. Partitions-requirements-matrices (P-R-Ms) are Markov kernels of certain stochastic dynamic programming problems with distance properties (see [5], [7] and [9]). They express optimal decisions or are used as approximate optimal kernels. P-R-Ms have an interesting mathematical structure. For the time being no representation of the elements by a formula is known. (P-R-Ms are computed by means of enumeration.) Beside other things formulas for limits of P-R-Ms in the cases of "sparse" partitions and "non-curtailed heavy" partitions are given.
Further on "kernel-equations" with P-R-Ms are considered. A monotonicity of their solutions (that includes an optimality criterion) is shown in certain cases. The solutions with regard to limits of P-R-Ms have a nice structure in contrast to the formulas of the limits of P-R-Ms.

1 Notation and terminology

- $n \in \mathbf{N}, n \geq 3$ (resp. 2), $k_o \in \mathbf{Z}_+^n, su \in \mathbf{N}$ with $k_o \leq su < nk_o$,
- $S_{n;su;k_o}$ - the set of the restricted (unordered) partitions of su into at most n parts with summands not greater than k_o. (We note down the elements of $S_{n;su;k_o}$ as n-dimensional vectors. That means for $s = (s_1, s_2, \cdots, s_n)$.)
  $r := |S_{n;su;k_o}|$ - the number of the restricted partitions, [1]
- $B_{n;k_o} := \{w \in \mathbf{Z}_+^n | 0 \leq w \leq k_o\}$ (We call this set: set of requirements.)
- We assume that the requirements w are random vectors [2] with a probability function
  $q(x) = \prod_{i=1}^{n} q_o(w_i),$ [3] (1a)
  where the so-called "single probabilities" $q_o(w_i)$ are given:
  $q_o(w_i) > 0$ iff $w_i \in \{0, 1, \cdots, k_o\}$ and $\sum_{w=0}^{k_o} q_o(w) = 1.$ (1b)

---

[1] The theory of (unordered) partitions is complicated and presents a number of interesting problems, cf. [1].

[2] We use the same notations for the random vectors and their realizations.

[3] We assume that the components w_i of requirements are realized independent of each other.

- Let $C[w, su]$ denote the case
$\sum_{i=1}^{n} w_i \leq su$ and $C[su, w]$ the case $\sum_{i=1}^{n} w_i \geq su$.
Let be $s \in S_{n;su;k_o}$ with $s_1 \geq s_2 \geq \cdots \geq s_n$ and $w \in B_{n;k_o}$.
$s^* = s^*(s, w)$ with $\sum_{i=1}^{n} (s_i^*)^2 \to \min$ subject to

$$s^* \in S_{n;su;k_o} \text{ and } \begin{cases} w_i \leq s_i^{*'} \leq max\{s_i; w_i\}, i = 1, \cdots, n \text{ for } C[w, su], \\ \min\{s_i, w_i\} \leq s_i^{*'} \leq w_i, i = 1, \cdots, n \text{ for } C[su, w], \end{cases} \quad (2)$$

for a permutation $s^{*'}$ of s^*
is called **feasible balanced partition with regard to s and w.**

(In the case $C[w, su]$ it is completely possible the requirement w to cover by $s^{*'}$: $w \leq s^{*'}$.
In the case $C[su, w]$ the requirement cannot be covered completely. But it is demanded $s^{*'} \leq w$.)

- $P^* = (p^*_{fl})$ - the **"partitions-requirements-matrix"**.

$$p^*_{fl} = p^*(s^l | s^f) = \sum_{w: s^l = s^*(s^f, w)} q(w) \text{ for } f = 1, \cdots, r, \ l = 1, \cdots, r \quad (3)$$

where $S_{n;su;k_o} = \{s^1, s^2, \cdots, s^r\}$ and $B_{n;k_o}$ be given.

2 A classification of the sets of the restricted partitions

The partitions are arranged in the way of the (partial) dominance order (cf. [10], 1B and 5D, in addition [6]).
Let the **sets** of restricted partitions be arranged in a 3-dimensional lattice in relation to n, su, k_0 with

$$n = 2, 3, \cdots; su = 1, 2, \cdots; k_0 = \lceil \frac{su}{n} \rceil, \lceil \frac{su}{n} \rceil + 1, \cdots, su. \quad (4)$$

Now, we classify the sets of partitions in the following way:

C1.) Sets of **"sparse" partitions**: $S_{n;su;k_0}$ with $n > su$

Many O-s are in the vectors of these sets.
In this case $2 \geq \lceil \frac{2su}{n} \rceil$ follows. Hence $k_0 \geq \lceil \frac{2su}{n} \rceil$ for $2 \leq k_0$ is valid.)

C2.) Sets: $S_{n;su;k_0}$ with $n \leq su$ and $k_0 \geq \lceil \frac{2su}{n} \rceil$

A **finite number** of such sets with any fixed number r of partitions exists only! (Cf. Theorem 3.3,[8].)

C3.) Sets: $S_{n;su;k_0}$ with $n \leq su$ and $k_0 < \lceil \frac{2su}{n} \rceil$
We continue to classify these sets:

C3a.) Sets: $S_{n;su;k_0}$ with $n \leq su, k_0 < \lceil \frac{2su}{n} \rceil$ **and** $\exists\, i : s_i^1 = 0$
(where s^1 is the minimal element of $S_{n;su;k_0}$)
These sets are **equivalent** in relation to the kernel eqation to sets of sparse states (cf.C1.)) or to sets from C2.) (Cf. Definition 2.1.)

C3b.) Sets of "heavy" partitions: $S_{n;su;k_0}$ with $n \leq su, k_0 < \lceil \frac{2su}{n} \rceil$ **and** $s_i^1 > 0$ for $i = 1, \cdots, n$

We continue to classify these sets:

C3b.1) Sets of"**non-curtailed" heavy partitions**:
$S_{n;su;k_0}$ with $n \leq su, k_0 < \lceil \frac{2su}{n} \rceil, s_i^1 > 0$ for $i = 1, \cdots, n$
and $s^r = (k_0, ..., k_0, k_0 - 1, ..., k_0 - 1)$ (that means $\exists i : s_i^r = k_0$)
(where s^r is the maximal element of $S_{n;su;k_0}$)

C3b.2) Sets of **"curtailed" heavy partitions**:
$S_{n;su;k_0}$ with $n \leq su, k_0 < \lceil \frac{2su}{n} \rceil, s_i^1 > 0$ for $i = 1, \cdots, n$ **and**
$s_i^r < k_0$ for $i = 1, \cdots, n$.[1ex] (Example: $S_{2;2k_0-3;k_0}$ with $k_0 > 3$.)

Definition 2.1.: Let $n; su; k_0$ and $\overline{su}, \overline{k}_0$ be given.

(a) If $\mid S_n; su; k_0 \mid = \mid S_n; \overline{su}, \overline{k}_0 \mid$ and $k_c \in \mathbf{Z}_+$ and a one-to-one correspondence of the partitions exist with $s + \bar{s} = (k_c, ..., k_c)^T$ for permutations of s and $\bar{s}$ then $S_{n;su;k_0}$ and $S_{n;\overline{su};\overline{k}_0}$ are called equivalent (in relation to the partial order [4]). s and $\bar{s}$ are called complementary partitions.

(b) If (a) is fulfilled and $k_c = k_0 = \overline{k}_0$ then $S_{n;su;k_0}$ and $S_{n;\overline{su};\overline{k}_0}$ are called equivalent in relation to the kernel equation.

Remark: Sets of
- sparse partitions (cf. C1.)),
- non-curtailed heavy partition (cf. C3b.1)),
- curtailed heavy partitions (cf. C3b.2))

yield an **infinite** number of sets of restricted partitions **with a fixed number r of partitions.**
(That means the corresponding partitions-requirements-matrices are of the same size r by r.)

3 To partitions-requirements-matrices

3.1 Partial results

- We have calculated the P-R-Ms with r = 3 or r = 4.
- We have shown that a dominance property is valid for a small number of P-R-Ms (cf.[7], 3.3.5).

[4] The proof is simple.

3.2 Limits of partitions-requirements-matrices with regard to sets of sparse states

Remark:
Set of sparse partitions with $k_0 = 1$ have only one state. No mathematical problem is to solve. If $k_0 \geq 2$ then $k_0 \geq \lceil \frac{2su}{n} \rceil$ (cf. the classification C1)). An infinite number of sets of sparse partitions with a given fixed number r of partitions is only yielded by $n \to \infty$. (Cf. Theorem 3.3,[8].)
In this section we compute the limits of transition probabilities p^*_{fl} for $n \to \infty$ in general case. We use definitions of monotonous successors, and of principle parts of partitions.

Definition 3.1:

(a) Let a partition $s \in S_{n;su;k_0}$ with $n \underset{(=)}{>} su$ and (w.l.o.g.) $s_1 \geq s_2 \geq \cdots \geq s_n$ be given.
Then $s_H = (s_1, \cdots, s_\eta)$ with $s_i > 1$ for $i = 1, \cdots, \eta$ and $s_i \in \{0, 1\}$ for $i = \eta + 1, \cdots, n$ is called **principal part** of s.
(b) Let s^1_H resp. s^2_H be principal parts of partitions $s^1 \in S_{n_1;su_1;k_{01}}(n_1 \underset{(=)}{>} su_1;$ $s^1_1 \geq s^1_2 \geq \cdots \geq s^1_{n_1})$ resp. $s^2 \in S_{n_2;su_2;k_{02}}(n_2 \underset{(=)}{>} su_2; s^2_1 \geq \cdots \geq s^2_{n_2})$.
Then s^2 is called **monotonous successor** of s^1 if $\eta_1 \geq \eta_2$ and $s^1_i \geq s^2_i$ for $i = 1, \cdots, \eta_2(\eta_1)$.

Theorem 3.1 (Limits of partitions-requirement-matrices)
Let $\{s^1(n), s^2(n)\} \subseteq S_{n;su;k_0}(n \underset{(=)}{>} su)$ be with (w.l.o.g.) $s^\lambda(n)_{su+1} = s^\lambda(n)_{su+2}$ $= \cdots = s^\lambda(n)_n = 0$ for $\lambda = 1, 2$.
(That means $s^\lambda(n) = (s^\lambda(n)_1, \cdots, s^\lambda(n)_{su}, 0, \cdots, 0)$ for $\lambda = 1; 2$ and let be $s^\lambda := (s^\lambda(n)_1, \cdots, s^\lambda(n)_{su}) \in S_{su;su;k_0}$.)
Then $\lim_{n \to \infty} p^*(s^2(n)|s^1(n)) =$

$$\begin{cases} 0, \text{ if } s^2 \text{ is no monotonous successor of } s^1 \\ (q_0(0) + q_0(1))^{\eta_1 - \eta_2} \sum\limits_{s^2_\Pi \in S^2_\Pi} \prod\limits_{i: s^1_i > s^2_{\Pi_i} \geq 2} q_0(s^2_{\Pi_i}) \prod\limits_{i: s^1_i = s^2_{\Pi_i} \geq 2} (q_0(s^1_i) + \cdots + q_0(k_0)), \\ \text{if } s^2 \text{ is a monotonous successor of } s^1 \end{cases}$$

where S^2_Π is the set of permutations s^2_Π of the principal part of s^2 in relation to $i = 1, \cdots, \eta_1$ with $s^2_{\Pi i} \leq s^1(n)_i$ (or i is not occupied) for $i = 1, \cdots, \eta_1$.
(Cf. Def. 3.1(b) for η_1, η_2.)
(Proof: Cf. [8], Theorem 5.2.)

Corollary 3.2: Let $su, k_0 (su \geq k_0)$ be given and let the numerations of the partitions of the sets $S_{n;su;k_0} = \{s^1(n), \cdots, s^r(n)\}$ be so that
$\sum\limits_i (s^f_i)^2 > \sum\limits_i (s^l_i)^2 \Longrightarrow f < l.$
Then the matrix $\lim\limits_{n \to \infty} P^*(n)$ is a triangular matrix.

3.3 Limits of partitions-requirements-matrices with regard to sets of non-curtailed heavy states

Lemma 3.3: Sets of non-curtailed heavy partitions can be represented in the form of $S_{n;nk_0-\bar{su};k_0}$ with $\bar{su} < k_0, \bar{su} < n, (n \geq 2, k_0 \geq 2)$
(that means $su = nk_0 - \bar{su}$). (5)

(The equivalence to the properties of the classification $C3b_1$.) to show is simple.)

Remark: An infinite number of sets of non-curtailed heavy partitions with a given fixed number of partitions is yielded by $k_0 \to \infty$ or $n \to \infty$.
The case $n \to \infty$ can be discussed on the analogy of section 3.2.
Here we consider **the case** $k_0 \to \infty$.
That requires additional properties of the probability functions q:
Let q^{k_0} denote probability functions in relation to $B_{n;k_0}$ (where n is fixed).
Then we assume

$$\lim_{k_0 \to \infty} q_0^{k_0}(w_i) = 0 \text{ for } w_i = 0, 1, \ldots \tag{6}$$

and

$\exists c(k_0)$ (real numbers with $1 > c(k_0) > 0$) :

$$\exists \lim_{k_0 \to \infty} \frac{q_0^{k_0}(w_i)}{c(k_0)} \neq 0 \text{ for } w_i = 0, 1, \ldots . \tag{7}$$

$$\Big(\lim_{k_0 \to \infty} \frac{(q_0^{k_0}(w_i))^l}{c(k_0)} = 0 \text{ for } l \geq 2 \text{ and } w_i = 0, 1, \ldots . \tag{8}$$

follows from (5) and (6).)

(The limits $\lim_{k_0 \to \infty} \frac{q_0^{k_0}(\cdot)}{c(k_0)}$ are unique but a constant multiple (in relation to different $c(k_0)$)).

Definition 3.2:

(a) Let a non-curtailed heavy partition s with (w.l.o.g.) $s_1 \geq s_2 \geq \cdots \geq s_n$ be given. Then $s_H = (s_\eta, \cdots, s_n)$ with $s_i < k_0 - 1$ for $i = \eta, \cdots, n$ and $s_i \in \{k_0, k_0 - 1\}$ for $i = 1, \cdots, \eta - 1$ is called **principal part** of s.

(b) Let s_H^1 resp. s_H^2 be principal parts of non-curtailed heavy partitions $s^1 \in S_{n_1;su_1;k_{0_1}}$ with $s_1^1 \geq s_2^1 \geq \cdots \geq s_{n_1}^1$) resp. $s^2 \in S_{n_2;su_2;k_0}$ with $s_1^2 \geq s_2^2 \geq \cdots \geq s_{n_2}^2$). Then s^2 is called **narrow monotonous successor** of s^1 if $\eta_1 \leq \eta_2$, $s_i^1 \leq s_i^2$ for $i = (\eta_1)\eta_2, \cdots, n$

and a permutation s_π^2 of the principal part of s^2 exists that
$s_i^1 \neq s_{\pi i}^2$ for one $i \in \{\eta_1, \cdots, n\}$ **at most**. (9)

Remark: We can not derive the Definition 3.2 (b) from Definition 3.1 (b) and the equivalence of sets with non-curtailed heavy states and certain sets with sparse states because of condition (9)!

Theorem 3.4: (Limits of partitions-requirements-matrices)
Let $s^1(k_0), s^2(k_0)$ be non-curtailed heavy partition with $\{s^1(k_0), s^2(k_0)\} \subseteq S_{n;su;k_0}$ (where su is representable in the kind $su = nk_0 - \bar{su}$ with fixed $\bar{su}$ and $\bar{su} < n, (\bar{su} < k_0, n \geq 2, k_0 \geq 2)$ and (w.l.o.g.)
$s^\lambda(k_0)_1 = \cdots = s^\lambda(k_0)_{n-\bar{su}} = k_0$ for $\lambda = 1; 2$ and let be
$s^\lambda = (s^\lambda_{n-\bar{su}+1}, \cdots, s^\lambda_n) := (s^\lambda(k_0)_{n-\bar{su}+1}, \cdots, s^\lambda(k_0)_n) \in S_{\bar{su},su,k_0}$.
Further on let given probability functions q^{k_0} fulfill (6) and (7) for certain $c(k_0)$ and $q_0^0(w_i) := \lim\limits_{k_0 \to \infty} \frac{q_0^{k_0}(w_i)}{c(k_0)}$.

Then $\lim\limits_{k_0 \to \infty} \frac{1}{c(k_0)}(p^*(s^2(k_0)|s^1(k_0)) - \delta(s^2(k_0); s^1(k_0))) =$

$$\begin{cases} 0, & \text{if } s^2 \text{ is no narrow monotonous successor of } s^1 \\ -\sum\limits_{i: s_i^1 \leq k_0 - 2}{}^{[13]} (q_0^0(s_i^1 + 1) + \cdots + q_0^0(k_0)), & \text{if } s^1 = s^2 \\ \sum\limits_{s_\pi^2 \in S_\pi^2} |\{j | s_j^1 = s_{i_0}^1 < s_{\pi i_0}^2| \begin{cases} q_0^0(s_{\pi i_0}^2) \text{ for } s_{\pi i_0}^2 < k_0 - 1 \\ (q_0^0(k_0 - 1) + q_0^0(k_0) \text{ for } s_{\pi i_0}^2 = k_0 - 1, \end{cases} \\ \qquad \text{if } s^2 \text{ is a narrow monotonous successor of } s^1 \text{ and} s^2 \neq s^1. \end{cases}$$

There is $\delta(s^2(k_0); s^1(k_0)) = \begin{cases} 1 \text{ if} s^1(k_0) = s^2(k_0) \\ 0 \text{ if} s^1(k_0) \neq s^2(k_0) \end{cases}$.

S_π^2 is the set of permutations s_π^2 of the principal part of s^2 in relation to $i = \eta_1, \cdots, n$ with $s_{\pi_i}^2 \geq s_i^1$ and if $\eta_1 < \eta_2$ and several i are not occupied then $s_{\pi_i}^2 := k_0 - 1$ for the referred i, with the further property that $s_{\pi_i}^2 \neq s_i^1$ for one $i_0 \in \eta_1, \cdots, n\}$ *at most*. (Cf. Def. 3.2(b) for η_1.) (Proof: Cf. [8], Theorem 6.4.)

4 The kernel equation and monotonicity

4.1 The kernel equation

At first we introduce vectors γ: Let a set of restricted partitions
$S_{n;su;k_0} = \{s^1, s^2, \cdots, s^r\}$ (and a probability function q) be given. Then

$$\gamma_l := \sum_{i=1}^{n} \sum_{w_i=0}^{s_i^l} (s_i^l - w_i)\, q_0(w_i), \quad l = 1, ..., r \tag{10}$$

$$\gamma = (\gamma_l, \cdots, \gamma_r)^T \tag{11}$$

($\gamma_l = \frac{1}{2}\frac{1}{k_0+1} \sum\limits_{i=1}^{n} (s_i^l)^2 + R(n; su; k_0)$ follows in the case of discrete uniformly distributed requirements where $R(n, su, k_0)$ is independent of s^l. (Cf. [7],

[13] That is a finite number of such i since $\bar{su}$ is fixed.

3.3.1, Lemma 3.12 and remark b).)

Definition 4.1.
Let a set of restricted partitions $S_{n;su;k_0} = \{s^1, s^2, \cdots, s^r\}$ and a probability function q be given. Further on let be P^* the corresponding P-R-M and γ as in (10). The equation systems (with the variables $(g, \nu) \in \mathbf{R} \times \mathbf{R}^r$)

$$g(-1, ..., -1)^T + (P^* - I)\nu = -\gamma',$$

(where I is the identity matrix and γ' is any affine transformation of γ with $\gamma' = \alpha\,\gamma + \beta(1, ..., 1)^T$ are called kernel eqations [5]

(This equation system does have solutions. We have any choice for one ν_l, the remaining variables are uniquely defined.)

Definition 4.2.
Let the numeration of the partitions of a set $S_{n;su;k_0} = \{s^1, s^2, \cdots, s^r\}$ be so that $\sum_i (s_i^f)^2 > \sum_i (s_i^l)^2 \Rightarrow f < l$. The solution of the corresponding kernel equation is called monotonous (in ν) (in relation to the partial order) if $\sum_i (s_i^f)^2 > \sum_i (s_i^l)^2 \Rightarrow \nu_f < \nu_l$.

Assertion: Solutions of every kernel equation are monotonous!

4.2 Partial results

- The kernel equations have monotonous solutions for sets with 2 or 3 partitions resp. for sets with 4 partitions for discrete uniformly distributed requirements.
- The solutions of a kernel equation are monotonous if the the corresponding partitions-requirements-matrix fulfils the dominance property.

4.3 The kernel equation with regard to sets of sparse partitions of great dimension n

A difficulty is in which way we can include all the problems with $n > su$ in one proof. We use definitions of monotonous successors, and of principle parts of partitions and a certain affine transformation γ':

Theorem 4.1:
Let $S_{n;su;k_0}$ be the sets of sparse partitions ($n \underset{(=)}{>} su$) with a fixed number r of partitions and let q be any probability function.
Then $\nu((1, 1, \cdots, 1, 0, 0, \cdots, 0)^T) := 0$,

$$\nu(s) = \sum_{i: s_i \geq 2} \left(\frac{q_0(1)}{q_0(0)+q_0(1)} + \frac{q_0(1)+q_0(2)}{q_0(0)+q_0(1)+q_0(2)} + \cdots + \frac{q_0(1)+q_0(2)+\cdots+q_0(s_i-1)}{q_0(0)+q_0(1)+\cdots+q_0(s_i-1)} \right),$$

[5] The monotonicity of the solutions of the kernel eqation includes an optimality criterion for certain stochastic dynamic programming problems.

$s \in S_{n;su;k_0}$ are solutions of the equation system

$$g(-1, ..., -1)^T + \left(\lim_{n\to\infty} P^*(n) - I\right) \nu = -\gamma' \tag{12}$$

$$\text{(where } \gamma'(s) = \sum_{i:s_i \geq 2} \sum_{w_i=1}^{s_i} (s_i - w_i) q_0(w_i),\ \gamma' = (\gamma'(s))_{s\in S_{n;su;k_0}}). \tag{13}$$

(Proof: Cf. [8], Theorem 6.5.)

Corollary 4.2: Let $S_{n;su;k_0}$ be the sets of sparse partitions ($n \underset{(=)}{>} su$) with a fixed number r of partitions and let be q any probability function. Then the corresponding kernel equations - but perhaps a finite number of them - have monotonous solutions.

In the case of **non-curtailed heavy partitions** we have analogous results.

Conclusion: We have answered several important questions to P-R-Ms and corresponding kernel equations. Other questions remain, for instance: formulas for the elements of P-R-Ms; the monotonicity of solutions of the kernel equations with regard to sets of curtailed heavy partitions for great k_0; the monotonicity of the solutions of kernel equations for small n and k_0 in the case where the P-R-Ms do not fulfil the dominance property.

References

1. Andrews, G.E. (1976). The Theory of partitions. Addison-Wesley Publishing Company, London, ⋯.
2. Brylawski, T. (1973). The lattice of integer partitions. Discrete Mathematics 6; pp. 201 - 219
3. Girlich, E. und Kowaljew, M.M.(1981). Nichtlineare diskrete Optimierung. Akademie-Verlag, Berlin
4. Hernández-Lerma, O.(1989) Adaptive Markov control Processes. Springer-Verlag, New York-Berlin.
5. Hildenbrandt, R. (1993). A special stochastic decision problem. Optimization, Vol. 28, pp. 95 - 110.
6. Hildenbrandt, R. (1994). Eine Verbandsstruktur für Partitionen .... Preprint No. M 14/94, TU Ilmenau.
7. Hildenbrandt, R. (1995). Methoden aus ganzzahliger Optimierung und Verbandstheorie zur Behandlung eines stoch. dynamischen Transportproblems. Habilitationsschrift, TU Ilmenau. (LibriBoD, 2000)
8. Hildenbrandt, R. (1999). Checking of an optimality criterion in relation to a combinatorial optimization problem - connected with partitions-requirements-matrices. Preprint No. M 30/00, TU Ilmenau.
9. Hildenbrandt, R.(2000). Notes of stoch. dynamic programming, where the random disturbances are observed before the decision is made at each stage .... Preprint No. M 8/2000, TU Ilmenau. (submitted)
10. Marshall, A.W. und Olkin, I. (1979). Theorie of majorization and its appl. Academic Press, New York.

On-Line Simulation of Large Scale Networks

Andreas Pottmeier[1], Roland Chrobok[1], Joachim Wahle[2], Michael Schreckenberg[1]

[1] Gerhard-Mercator-Universität, Physik von Transport und Verkehr, 47048 Duisburg, Germany,
Email: [pottmeier,chrobok,schreckenberg]@traffic.uni-duisburg.de

[2] TraffGo GmbH, Grabenstr. 132, 47057 Duisburg, Germany,
Email: wahle@traffgo.com

Abstract. In this contribution a micro-simulator for freeway traffic in large scale networks is presented. Based on a cellular automaton traffic flow model the simulation, which is coupled with measured data from induction loops etc., can provide network-wide information about the current traffic state. The simulation of the whole network is performed in multiple real-time. The framework is applied to the freeway network of North Rhine-Westphalia (NRW), where traffic data from about 3,500 inductive loops are available and provided on-line minute by minute. Technical aspects of the simulation, e.g., the network structure and the incorporation of the on-line data, are illustrated. This is followed by a discussion of the efficiency of the simulator. Furthermore, the results of the on-line simulation, i.e., daily traffic demands and the traffic state of NRW are given.

1 Introduction

Mobility is a vital good for the economic development of modern societies. Especially in densely populated regions, like the state of North Rhine-Westphalia (NRW), the capacity of the road network is not able to cope with the still growing demand and it is socially untenable to expand this infrastructure. Therefore, new traffic management and information systems to solve these problems are needed.

One vital component of such systems is the detection of the traffic state. Unfortunately, it is not possible to measure the traffic state directly since only local data are available. With regard to this, on-line simulations offer a powerful tool to reproduce traffic states of large scale networks virtually on one machine. Due to the increasing computational power, in recent years a lot of microscopic traffic simulators have been developed [1–3]. To render the simulation as realistic as possible the simulation has to be coupled to real data.

2 Simulation Model

2.1 Principle of On-line Simulation

The starting point for the simulation of a freeway-network should be information about the current traffic state in the whole network. The common sources of information are local measurements provided by inductive loops and infrared or video detection devices which provide data permanently. Other sources are event driven, like messages from police or drivers with cellular phones. However, most information is provided locally. Note that an exception to this rule are floating car data.

Our approach to generates network-wide information is to connect the local data with a micro-simulator. Thus, information can be derived for those regions which are not covered by measurements. The local traffic counts are combined with the network structure (i.e., type of roads, priority regulations at the nodes or on- and off-ramps) under consideration of realistic traffic flow dynamics [4].

2.2 Basic Simulation Model

Since the simulator is based on on-line data the model employed should be efficient. Due to their design cellular automaton models are very effective in large-scale network simulations [4–8].

Models which reproduce the dynamic phases of traffic are still under debate. Thus the object-oriented design of the simulation allows a flexible use of different cellular automata. This includes the original cellular automaton by Nagel and Schreckenberg [9] as well as other approaches which were developed in the recent years (e.g., [10,11]). In a first version of this simulator we used the original cellular automaton by Nagel and Schreckenberg. For completeness, the definition of the model is briefly reviewed. The road is represented by a one dimensional lattice which is subdivided in cells with a length of 7.5m. Each cell can be occupied by one vehicle or is empty. In every time-step $t \rightarrow t+1$ the following update rules are applied to the cars in the lattice:

- Step 1: Acceleration:
  $v_n \rightarrow \min(v_n + 1, v_{\max})$

- Step 2: Braking:
  $v_n \rightarrow \min(v_n, d_n - 1)$

- Step 3: Randomization: with probability p
  $v_n \rightarrow \max(v_n - 1, 0)$

- Step 4: Driving:
  $x_n \rightarrow x_n + v_n$

x_n is the position of the nth car and $v_n = 1, 2, \ldots, v_{max}$ its velocity. v_{max} denotes the maximum velocity and d_n is the number of empty cells (gap) in front of car n. One time-step corresponds to approximately 1s in real time.

The first two rules (step 1 and 2) describe a somehow optimal driving strategy: the driver accelerates if the vehicle has not reached the maximum speed v_{max} and brakes to avoid accidents, which are explicitly excluded. However, drivers do not react in this optimal way: they vary their driving behavior without any obvious reasons, reflected by the *braking noise* p (step 3). It mimics the complex interactions between the vehicles and is also responsible for spontaneous formation of jams.

2.3 Lane Change

In order to handle the more complex topologies in the network the traffic flow models have to be expanded by rules for multilane traffic to ensure realistic lane changing behaviour, a correct implementation of on- and off- ramps as well as lane reduction. For instance, a lane change has to be carried out with regard to safety aspects and legal constraints, which vary between different countries. First, a vehicle checks if it is hindered by the predecessor on its own lane. Then it has to take into account the gap to the successor gap_s and to the predecessor gap_p on the alternative lane. If the gaps allow a safe change the vehicle moves to the other lane.

A systematic approach to multi-lane traffic, i.e., lane-changing rules, can be found in [12]. For a detailed discussion of the different models see [1,2,13] and the references therein.

3 Network Structure

A crucial point in the design of every simulator is the representation of the road network. Like in other simulators (e.g., [4,14]) the network consists of basic elements: links and nodes. A node is either a connection between two links or a sink/source at the boundary or an off/on-ramp. At the sources vehicles are added with regard to input data. On the sinks the cars are simply removed every time step. Links are directional elements representing the roads of the network that connect nodes.

Combining these elements mentioned above one is able to build the complex structures of the freeway networks. Examples for these structures are:

- junctions, where vehicles enter or leave the freeway,
- intersections, at which two freeways are connected, and
- triangular intersections, where two freeways meet, but one ends.

The complexity of an intersection can be derived from Fig. 1. Other geometries can rarely be found in reality. However, they can be constructed easily with the elements used here.

Using the basic elements the freeway network of North Rhine-Westphalia was constructed. It comprises 1,200 nodes, 3,560 links, 830 on- and off-ramps and 67 intersections. The overall length of the lanes is approximately 2,500 km, corresponding to 1,400,000 cells.

The data used for the network were extracted from the *NW*-SIB, a Geographic Information System (GIS) database provided by the state of North Rhine-Westphalia.

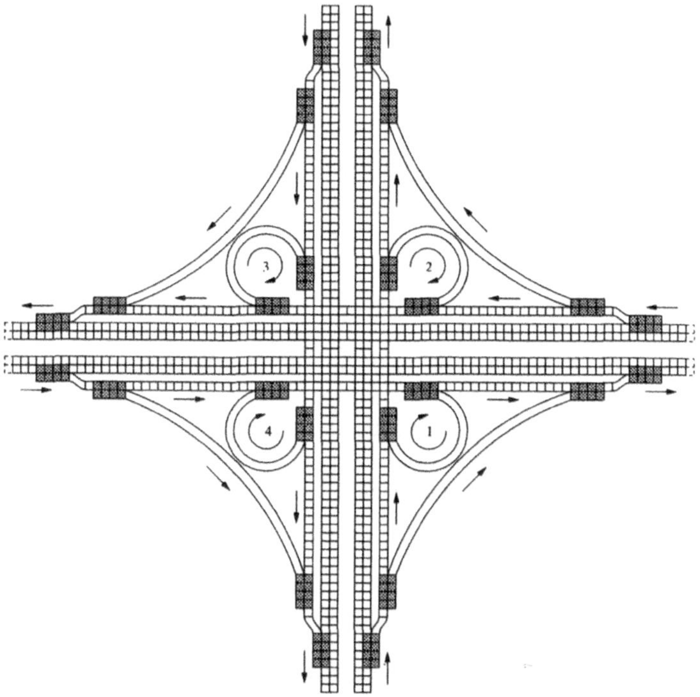

Fig. 1. The complex structure of an intersection.

4 Simulation Technique

Up to now the basic concepts, i.e., the model and the network structure have been introduced. In the following, the guidance of the vehicles in the network and the algorithms used to incorporate the measured data into the simulation are described.

4.1 Guidance of the Vehicles

In principle, there are two different strategies to solve the route choice problem. One can assign an origin and a destination to the road user and they are guided through the network according to this route [5,7]. For our network origin-destination information with a sufficient temporal and spatial resolution is not available. Therefore, the vehicles are guided in the network randomly, according to the tuning percentages calculated on the basis of the measured data.

4.2 Tuning Strategies

To incorporate the real world measurements into the simulation car-inserting and removing algorithms have to be applied. This is done at the so-called checkpoints, which are located at those places in the network, where a complete cross-section is available, i.e., all lanes are covered by an inductive loop. For realistic results it is crucial to incorporate the real world data in the simulation without perturbing the dynamics present in the network. Therefore, we propose a method which follows the idea to add the cars to the network "adiabatically", i.e., without disturbing the system. If the number of cars crossing the checkpoint is lower than measured by the detector, cars are added with regard to the measured mean velocity and mean gap, so that the system is not disturbed, i.e., no car has to brake due to the insertion. This method is therefore called "Tuning of the Mean Gap" [6]. If it is not possible to add the required number of cars some vehicles are left out. Although this is not correct, it is more important to keep the dynamics of the system untouched. It turns out that the strategy is capable of reproducing the traffic state quite well [6].

5 Simulation Run

At the start of every simulation the information of the road network is read from a database. This database comprises the necessary information for the entire network. Thereafter, the on-line connection is established and data is read. These are used to calculate dynamic turning probabilities. Then, the simulation runs for 60 update steps which corresponds to 60 seconds in reality. One update step consists of the update of sinks and sources as well as the tuning of the checkpoints. These steps are followed by the exchange of cars between the links, the lane change and the update of the speed and position according to the cellular automaton model. Afterwards, the system waits for new on-line information from the inductive loops.

Due to its design, i.e., the discrete algorithm, the cellular automaton approach has proved to be efficient in many applications [4–8]. Especially, it has been shown that the whole freeway network of Germany can be simulated in multiple real-time on parallel computers [7]. The simulator presented here, runs on a common personal computer 500 MHZ in multiple real-time.

The on-line simulation also makes it possible to interpolate the traffic state between checkpoints and to extrapolate into areas which are hardly or

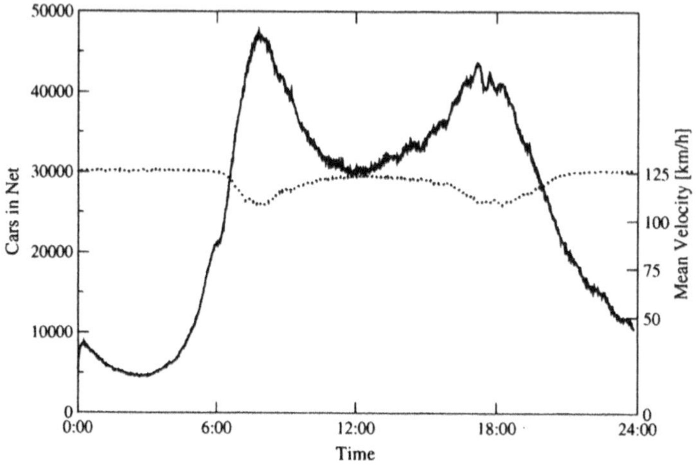

Fig. 2. No. of vehicles and mean velocity vs. time.

not equipped with detection units. Nevertheless, the simulation results allow a detailed examination of the network traffic. In Fig. 2 the number of vehicles and their mean speed during a day is depicted for a Monday.

The dynamic data like link travel times, densities or velocities, are visualised using a Java-applet (Fig. 3).

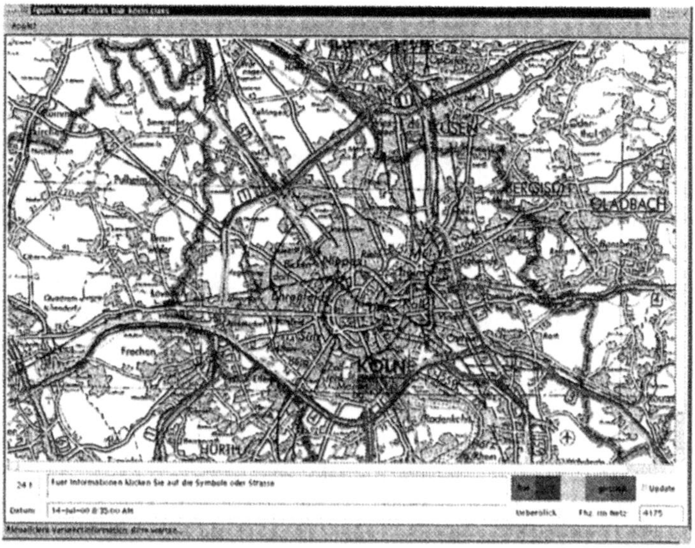

Fig. 3. Screen shot of the visualisation of the current traffic state of the area around Cologne.

6 Summary and Outlook

In this paper an on-line simulation has been presented. The heart of the simulator is a microscopic traffic flow model which is supplemented by the network structure and real-world traffic data provided by about 3,500 detectors of the freeway network of North Rhine-Westphalia.

Besides the model, the network structure has been discussed as well as the simulation techniques, like the vehicle guidance and the tuning strategies, which are used to incorporate real-world data into the simulation.

The results of the simulation show that the typical travel patterns in the freeway network are reproduced by the simulation. Additionally, the simulation can be performed in multiple real-time – a basic requirement for, e.g., a traffic forecast.

Due to the flexibility of the simulation we are able to study the influence of the different cellular automaton models in such a huge network. Also the influence of ramp-metering in parts of the network will be studied. The framework can also be used as a powerful tool for traffic flow control.

Acknowledgement

The authors would like to thank O. Kaumann, K. Froese and L. Neubert for discussion and sharing insights. The authors are also grateful to the "Landesbetrieb Straßenbau NRW" for data support and to the Ministry of Economic Affairs, Technology, and Transport of North Rhine-Westphalia for financial support.

References

1. Helbing D., Herrmann H. J., Schreckenberg M., Wolf D. E. (Eds.) (2000) Traffic and Granular Flow '99. Springer, Heidelberg
2. Schreckenberg M., Wolf D. E. (Eds.) (1998) Traffic and Granular Flow '97. Springer, Singapore
3. Ceder A. (Ed.) (1999) Proceedings of the 14th International Symposium on Transportation and Traffic Theory. Pergamon, Amsterdam
4. Esser J., Schreckenberg M. (1997) Microscopic simulation of urban traffic based on cellular automata. Int. J. of Mod. Phys. C **8**, 1025–1036
5. Nagel K., Esser J., Rickert M. (2000) Large-scale traffic simulations for transport planning. In: Stauffer D. (Ed.), Ann. Rev. of Comp. Phys. **VII**, 151–202, World Scientific, Singapore
6. Kaumann O., Froese K., Chrobok R., Wahle J., Neubert L., Schreckenberg M. (2000) On-line simulation of the freeway network of North Rhine-Westphalia. In: [1], 351–356
7. Rickert M., Wagner P. (1996) Parallel real-time implementation of large-scale, route-plan-driven traffic simulation. Int. J. of Mod. Phys. C **7**, 133–153

8. Schreckenberg M., Neubert L., Wahle J. (2001) Simulation of traffic in large road networks. Future Generation Computer Systems, **17**, 649–657
9. Nagel K., Schreckenberg M. (1992) A cellular automaton model for freeway traffic. J. Physique I **2**, 2221–2229
10. Barlovic R., Santen L., Schadschneider A.,Schreckenberg M. (1998) Metastable states in cellular automata for traffic flow. Eur. Phys. J. B **5**, 793–800
11. Knospe W., Santen L.,Schadschneider A., Schreckenberg M. (2000) Towards a realistic microscopic description of highway traffic. J. Phys. A **33**, L1–L6
12. Nagel K., Wolf D. E., Wagner P., Simon P. (1998) Two-lane traffic rules for cellular automata: A systematic approach. Phys. Rev. E **58**, 1425–1437
13. Chowdhury D., Santen L., Schadschneider A. (2000) Statistical Physics of Vehicular Traffic and Some Related Systems. Physics Reports **329**, 199–329
14. Yang Q., Koutsopoulos H. N. (1996) A microscopic traffic simulator for evaluation of dynamic traffic management systems. Transp. Res. C **4**, 113–129

Section 11

Econometrics, Statistics and Mathematical Economics

Process-Identification and Optimization of Technical Investments with TEMΠ Bubbles, Quelros and Environmental Management

Stefan Pickl

Mathematisches Institut, Zentrum für angewandte Informatik Köln ZAIK,
email: pickl@zpr.uni-koeln.de,
Universität zu Köln, D-50931 Köln

Zusammenfassung The economic analysis of technical investments with the aid of time-discrete dynamical systems and game-theoretical methods can lead to new insights in the area of climate-protection activities and resource-management. Many models in the area of environmental management concentrate themselves either on technical or on economical aspects. There is a lack of dynamical models, which consider both. The TEMΠ approach simulates the technical-economical relationships within joint operations (so called Bubbles) and their influence on quantified CO_2-reduction programs (QUELROS - Quantified Emissions Limitation and Reduction Objectives). Several trend scenarios are compared and analyzed. The TEMΠ tool will be presented, and related theoretical results as well, which exploit the discrete combinatorial structures determining the trajectories of such a time-discrete system.
Keywords: Technical Investment, TEMΠ, Time-Discrete System, Optimization, Quelros, Bubbles, Environmental Management

1 Einleitung

Der Simulation und Modellierung von Technologie-Investitionen mithilfe zeitdiskreter dynamischer Systeme und spieltheoretischen Methoden kommt in der heutigen Zeit angesichts der Resourcenknappheit und Klimaproblematik eine besondere Bedeutung zu. *Da die ganzheitliche Bewertung von Investitionensprojekten* nach Geldermann (2001) *neben der Berücksichtigung technischer und ökonomischer Kriterien resultierende Umwelteinwirkungen miteinschliesst*, kommt der Wechselwirkung zwischen getätigten Investionen einerseits und der möglichen Reduzierung von CO_2- Emissionen andererseits bei einem prozeßorientierten Umweltschutz eine besondere Bedeutung zu.

Mit dem vorzustellenden Instrument TEMΠ (Technologie-Emissionen-Mittel-Prozeß -Identifikation) liegt eine Modellierungsumgebung als Entscheidungsunterstützungssystem vor, die diesen Aspekt aufgreift. Es ist offensichtlich, daß die gleichzeitige Berücksichtigung dieser beiden Elemente (Technik einerseits, ökonomisches Handeln andererseits) für einen praxisorientierten Ansatz steht, der die Betrachtung von verschiedenen Handlungsszenarien mithilfe eines zeit-diskreten Modells ermöglicht. Die angesprochene Technologie-Mittel-Interaktion kann im Rahmen von TEMΠ zusätzlich in den Kontext des Kyoto-Protokolls (1997) eingebettet werden. Dies bedeutet, daß es mit dem TEMΠ zugrundeliegenden TEM-Modell (Technologie-Emissionen-Mittel-Modell) möglich ist, die einzelnen Elemente des Kyoto-Protokolls (kooperatives Vorgehen der einzelnen Akteure, Joint-Implementation Programme, Fonds-Bildungs-Prozesse, Clearing House- Mechanismen) mathematisch zu beschreiben, spieltheoretisch zu modellieren und das dynamische Verhalten zu analysieren. Im Vordergrund steht die Analyse von verschiedenen Allokationen, die im Rahmen eines gemeinsamen Vorgehens (in sogenannten *Bubbles*) auftreten können.

2 Modellierung und Simulation von Klimaschutzaktivitäten mithilfe von TEMΠ

2.1 Die Klimaschutzrahmenkonvention

Seit der Konferenz 1992 in Rio de Janeiro, auf dem die erste Klimaschutzrahmenkonvention verabschiedet wurde, sehen es die unterzeichnenden Staaten als gemeinsame Aufgabe an, die Klimaproblematik zu untersuchen und den möglichen Umweltveränderungen zu begegnen. Obwohl es eine Vielzahl von physikalischen Modellen gibt, die bestimmte Klimaveränderungsprozesse beschreiben, gibt es nur sehr wenige Ansätze, die die Ziel-Mittel Interaktion behandeln. Erschwerend kommt hinzu, daß in dem Vertragstext ein kooperatives Vorgehen innerhalb von internationalen Projekten als mögliche Handlungsoption explizit empfohlen wurde.

Zentraler Gegenstand der Verhandlungen von Kyoto war die Quantifizierung von Emissionsverpflichtungen, sogenannten QUELROS (Quantified Emissions Limitation and Reduction Objectives). Ein zentrales Problem bei derartigen Unternehmungen stellt nun die vertragliche Ausgestaltung der Technologiekooperationen dar. Darüber hinaus kommt der Aufteilung der entstehenden Kosten im Rahmen von gemeinsamen Projekten (siehe hierzu Artikel 6, Kyoto-Protokoll) eine besondere Bedeutung zu. Es scheint offensichtlich, daß eine rein technikorientierte Herangehensweise weder von der Klimaschutzrahmenkonvention gewünscht wird, noch sich für eine konkrete Umsetzung anbietet. Eine Modellierung muß diesen kooperativen Aspekt jedoch berücksichtigen. Im Rahmen des TEMΠ-Ansatzes ist daher eine Fonds-Bildung integriert, die sich mit spieltheoretischen Methoden beschreiben lässt und somit die Allokation in einem sogenannten *Bubble* simulieren kann (Oberthür, 2000): Nach Artikel 4 in Verbindung mit Artikel 3.1 kann jede Gruppe von Vertragsparteien ihre Verpflichtungen aus dem Kyoto-Protokoll gemeinsam erfüllen. Zu diesem Zweck müssen die in Anlage B zugeteilten Mengen jedoch durch interne Vereinbarungen umverteilt und dem Sekretariat mitgeteilt werden.

2.2 Joint Implementation

Aufbauend auf diesen Erfahrungen wurde auf der 1. Vertragsstaatenkonferenz 1995 zum Rahmenübereinkommen über Klimaveränderungen in Berlin beschlossen, Joint-Implementation Programme in einem verstärkten Maße einzusetzen. *Joint-Implementation* meint hierbei die gemeinsame Umsetzung von Klimaschutzzielen im Rahmen von Technologiekooperationen. Im Gegensatz zu anderen Instrumentarien zur CO_2- Minderung trägt Joint-Implementation einen kooperativen Charakter und zeichnet sich durch einen stark kontextberücksichtigenden Lösungsansatz aus.

Das Konzept der gemeinsamen Umsetzung (Joint Implementation) ist der *klassischen Wirtschaftstheorie* entlehnt: Maßnahmen zur Begrenzung von Emissionen können primär dort realisiert werden, wo es am kostengünstigsten bzw. unter Umständen sogar gewinnbringend ist. Dies beruht auf der Tatsache, daß die Kosten aufgrund unterschiedlicher Effizienzmaße von Land zu Land variieren. Im Mittelpunkt des TEM-Modells stehen daher die *em*-Parameter der Effizienzmatrix, die empirisch bestimmbar sind und die Kopplung des nichtlinearen zeit-diskreten System bewirken. In Pickl (1999) wurde neben der Entwicklung des TEM-Modells ein Fonds-Bildungsprozeß eingeführt, der diese gemeinsame Umsetzung von Klimaschutzaktivitäten zum Ziel hatte.

2.3 Die mathematischen Grundgleichungen des TEM-Modells

Im TEM-Modell beschreiben die Parameter E die erzielten Reduktion der Emissionen und M die dafür nötigen finanziellen Mittelaufwendungen der einzelnen Akteure. Das Modell umfaßt $i = 1, \ldots, n$ Akteure. Dies wird durch den Index i angezeigt. Die zweite Gleichung besitzt einen Proportionalitätsfaktor λ, der das Maß der Investitionen beschreibt, sowie einen Gedächtnisparameter φ, der die logistische Differenzengleichung um einen Verzögerungsfaktor erweitert. Die Variable M^* stellt jeweils eine obere Budgetgrenze für die einzelnen Akteure dar.

$$\begin{aligned} \Delta E_i(t) &= \sum_{j=1}^{n} em_{ij} M_j(t) \\ \Delta M_i(t) &= -\lambda_i M_i(t)[M_i^* - M_i(t)][E_i(t) + \varphi_i \Delta E_i(t)] \end{aligned}$$

Das TEM Modell

$E_i(t)$	Reduzierte Emissionen im Rahmen des Kyoto-Prozesses
$M_i(t)$	Finanzielle Mittelaufwendungen
M_i^*	Budgetgrenze
λ_i	Wachstumsparameter
φ_i	Gedächtnisparameter

Im Zentrum des TEM-Modells stehen die *em*-Parameter, die eine Art Effektivitätsmatrix darstellen und die Kopplung der beiden nichtlinearen Gleichungen bewirken.

3 Simulation

Simulationen machen deutlich, daß die nichtlineare Struktur kein gemeinsames Erreichen des gewünschten Endzustands garantiert. Die Akteure starten in dem normalisierten Szenario unterhalb der Koordinatenachse (*Kyoto-Linie*) im negativen Bereich. Die Emissionskurven weisen für das in der angegebenen Dissertation ausführlich beschriebene Beispiel zudem ein starkes Oszillieren auf.

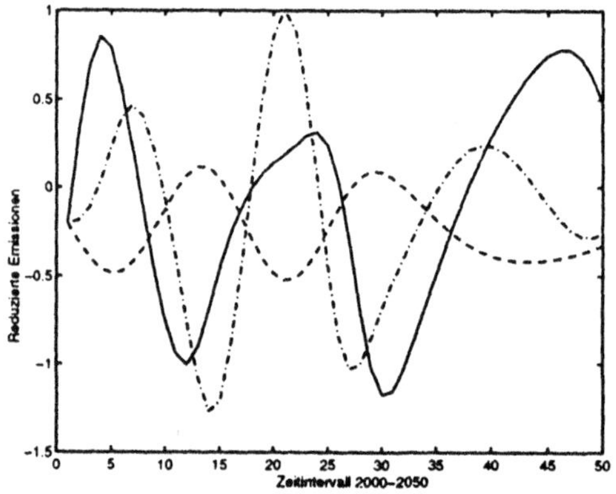

Obwohl kein Akteur den gewünschten Zielzustand dauerhaft erreicht, ist bei allen Akteuren ein Anwachsen der finanziellen Mittelaufwendungen zu beobachten.

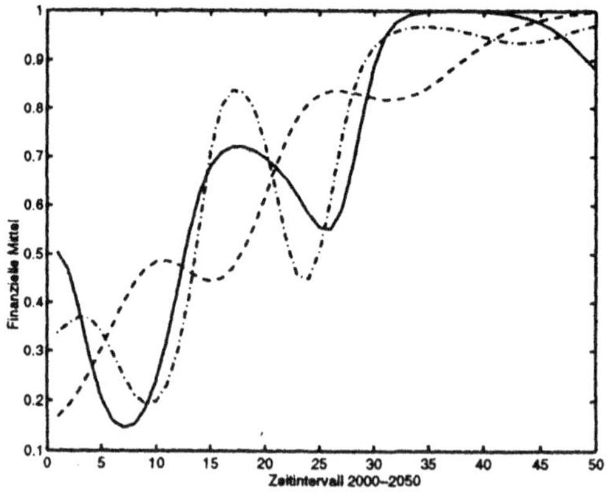

Dies legt die Entwicklung eines steuernd wirkenden Umweltmanagementsystems nahe.

4 Umweltlizenzen

Seit mehr als 30 Jahren werden innerhalb der Umweltökonomie Umweltlizenzen als Alternative zu existierenden ordnungsrechtlichen Verfahren angesehen. In dem Kyoto Protokoll werden handelbare Umweltrechte sogar explizit als marktwirtschaftliche Instrumente empfohlen. Damit kann über die ökonomischen Steuerungsanreize ein effizienter Beitrag zum Umweltschutz geleistet werden. Für die Entwicklung und Etablierung eines solchen Umweltlizenzsystems werden in Hansjürgens (1999) die folgenden vier Phasen als notwendig angesehen:

1. Festlegung einer zulässigen Gesamtemissionsmenge
2. Verteilung der Lizenzen, z.B. durch ein *grandfathering*-Verfahren
3. Vorkehrungen für einen funktionierenden Handel
4. Entwicklung von Kontrollmechanismen

Nach Hansjürgens (1999) kann die Reduktion von CO_2-Emissionen als exemplarisch für die Etablierung eines solchen Umweltlizenzsystems angesehen werden. In diesem Zusammenhang kommt der Bewertung von verschiedenen Technologiepfaden eine zentrale Bedeutung zu.

5 Umweltsystemforschung

Das TEM-Modell bietet die Möglichkeit, die Umsetzung verschiedener Technologiepfade innerhalb eines Energiemanagements über einen längeren Zeitraum zu beschreiben und die damit verbundenen Auswirkungen auf CO_2-Minderungsaktivitäten zu beurteilen. Desweiteren werden durch die Berechnung optimaler zulässiger Trajektorien Handlungskorridore bestimmt, die sich für die konkrete Umsetzung im Rahmen eines verantwortungsvollen und nachhaltigen Umweltmanagements anbieten

5.1 Prozeß-Identifikation - TEMΠ

Da weltweit bisher nur geringe Projekterfahrung mit Joint-Implementation vorliegt, ist man auf eine Simulation von Handlungsszenarien angewiesen. Nach Oberthür (2000), *müsste darüber hinaus geklärt werden, wie sich die Emissionen ohne das Projekt entwickelt hätten* (Berücksichtigung sogenannter Leakage-Effekte), um ein erfolgreiches Gelingen zu garantieren. TEMΠ repräsentiert eine Modellierungsumgebung, die es ermöglicht, die Wirkungsweise von Joint-Implementation und die enthaltenen Möglichkeiten mathematisch zu analysieren. Diese Analyse nimmt Bezug auf die Lösungstrajektorien des Systems ohne steuernden Eingriff sowie die Beschreibung der zulässigen Steuerungsmenge. Diese Prozeß-Identifaktion mithilfe von Referenzszenarien kann eine qualitative und quantitative Beurteilung von Investitionen in diesem innovativen Wirtschaftssektor ermöglichen:

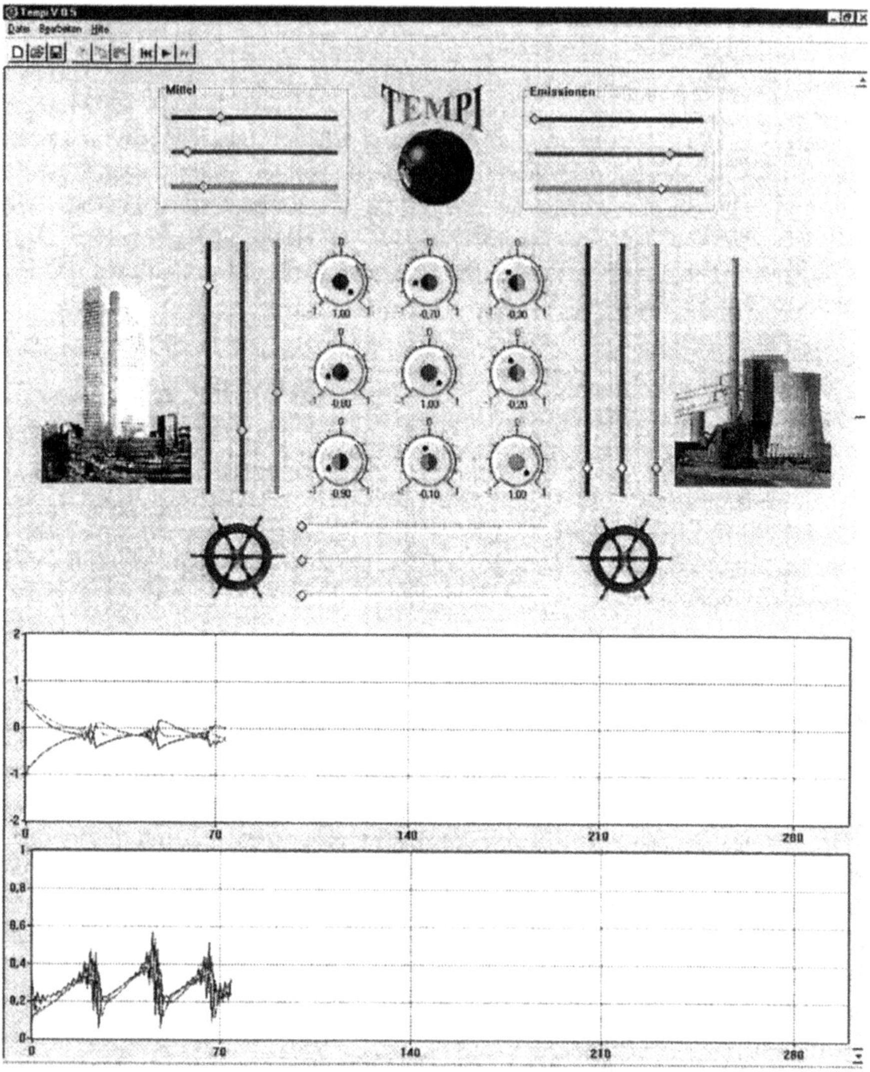

Die Modellierungsumgebung TEMΠ

Im folgenden soll eine kurze Bescchreibung des Prototyps TEMPI, wie er auf der vorhergehenden Seite abgebildet ist, erfolgen. Referenzszenarien erhält man dadurch, indem zunächst die Startparameter für die aktuellen finanziellen Mittelaufwendungen und die bereits erzielten reduzierten Emissionen eingestellt werden (Regler *Mittel* und Regler *Emissionen*). Dies ist für den Fall von drei Akteuren als Beispiel dargestellt. Die Akteure stehen in einer Wirtschaftsbeziehung zueinander, die durch die (normalisierte) Matrix ausgedrückt wird. Die verschiedene Farbgebung ermöglicht eine leichte Identifikation der jeweiligen Wirtschaftsbeziehung. Die Verhältnisse sollen zunächst als statisch angesehen werden.

Eine Simulation liefert dann in zwei getrennten Darstellungen den Verlauf der erzielten Reduktionen (oben) und auch der finanziellen Mittelaufwendungen (unten).

5.2 Investment-Optimierung mit TEMPI

Es ist davon auszugehen, daß eine Ratifizierung des Kyoto-Protokolls mit einem Investitionsanstieg im Technologiebereich verbunden ist, die in lateralen Kooperationen durchgeführt werden. TEMPI kann diese Projekte exemplarisch beschreiben, indem mögliche Handlungsstrategien berechnet und analytisch untersucht werden.

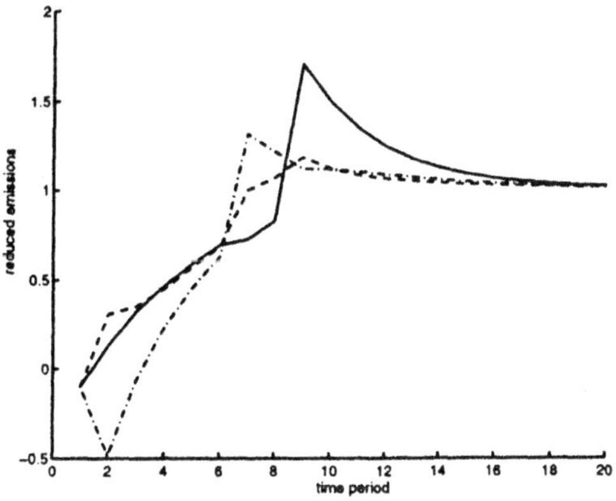

Optimale Bestimmung von Investitionen zur Steuerung des Systems

Ferner stellt es Im Rahmen von Joint-Implementation Projekten ein Modell dar, das in der Lage ist, die Tätigkeit des Clearing House aktiv zu unterstützen.

5.3 Danksagung

Die Weiterentwicklung des TEM-Modells zu TEMΠ wurde durch die Ermutigungen von Frau Prof. Dr. Ulrike Leopold-Wildburger und Herrn Prof. Dr. Wolfgang Eichhorn besonders motiviert.

Literatur

1. EICHHORN, W. et al. (1991): *Economic Theory of Natural Resources*, Würzburg-Wien, Physica-Verlag.
2. EICHHORN, W.; LINDENBERGER, L; and KÜMMEL, R. *Energie, Wirtschaftswachstum und Beschäftigung* in: BRUNE, W. (Hrsg.) (1999) *Die Energiewirtschaft an der Schwelle des neuen Jahrhunderts*, Teubner Verlag.
3. GELDERMANN, J, RENTZ, O. (2001): *Entwicklung eines multikriteriellen Entscheidungsunterstützungssystems zur integrierten Technikbewertung* in: FLEISCHMANN, B. et al. (2001): Operations Research Proceedings 2000, Springer Verlag.
4. KRABS, W. (1997): *Mathematische Modellierung*, B.G. Teubner.
5. KYOTO CONTRACT (1997): see http://www.unfccc.org/resource/convkp.html
6. OBERTHÜHR, S.; OTT, H.E. (2000): *Das Kyoto Protokoll - International Climate Policy for the 21st Century (International and European Environmental Policy Series)*, Springer Verlag Berlin-Heidelberg.
7. PEZZEY, J. (1992): *Analysis of Unilateral CO_2 Control in the European Community and OECD*, Energy Journal, *13*.
8. PICKL, S. (1999): *Der $\tau-$value als Kontrollparameter. Modellierung und Analyse eines Joint-Implementation Programmes mithilfe der kooperativen dynamischen Spieltheorie und der diskreten Optimierung*, Shaker Verlag Aachen.
9. PICKL, S. (2001): *Convex Games and Feasible Sets in Control Theory*, *Mathematical Methods of Operations Research*. Vol. 53, Issue 1, 51-66.
10. RAIFFA, H. (1996): *Lectures on Negotiation Analysis*, Cambridge, MA: The Program on Negotiation.
11. SCHELLNHUBER, H.J. (1998): *Earth System Analysis. The Scope of Challenge. In: Schellnhuber, H.J. and Wenzel, V. (Eds.), Earth System Analysis*, Integrating Science for Sustainability. Springer Verlag Berlin.
12. STROHM, J. (2000): *Internationale Maßnahmen zur nachhaltigen Entwicklung. Auswirkung und Akzeptanz eines ökonomischen, ökologischen und sozial ausgewogenen, weltweiten Ordnungsrahmens*, Hochschulschriften zur *BWL, Bd. 138*.
13. TIJS, S.H. and DRIESSEN, T.S.H. (1986): *Game Theory and Cost Allocation Problems, Management Science, 32*.

Economic Growth, Emission Reduction and the Choice of Energy Technology in a Dynamic-Game Framework

Jürgen Scheffran

Potsdam Institute for Climate Impact Research (PIK), Telegrafenberg C4, P.O. Box 601203, D-14412 Potsdam, Germany, scheffran@pik-potsdam.de

Abstract. To assess options in energy policy, the relationship between energy technology, economic output and emissions is analyzed in a dynamic-game framework which describes the interaction between players aiming at economic output and emission reduction. The players allocate investment to various energy technologies, differing in costs, economic output and emissions per energy unit. Control variables for each player are the amount of investment and its allocation. Game-theoretic analysis is used to discuss conditions under which players would change behavior and select energy technologies to pursue their objectives more cooperatively. Policy instruments are compared with regard to the different energy paths and the compatibility of economic and environmental objectives.

1 Introduction

In economic theory technical progress is a primary source of growth. Its endogenization is an active branch of current research (Barro and Sala-i-Martin 1995, Aghion and Howitt 1998). In neoclassical theory the production of output is a function of capital and labor as input factors while energy and other natural resources are neglected. The energy crises of the 1970s, demonstrated the "limits to growth" and induced an economic crisis, providing a concrete experience of how a rise in energy prices could adversely effect economic growth. The crisis has induced a number of studies on natural resources and energy in economic growth, some of which have early recognized the need for an interdisciplinary perspective (Kümmel 1980, Eichhorn et.al. 1982). Some studies focused on the productive power of energy and used it as a factor in the production function (Renshaw 1981, 36, Hudson and Jorgenson, 1974, 461). Empirical data justified the econometric estimation of the energy elasticity in the Cobb-Douglas production function for the US, Japan and Europe. Being about 0.5 it is roughly equal to the sum of the two other elasticities and ten times the share of the energy cost in total factor cost (Kümmel, Lindenberger and Eichhorn 2000).

The energy-economy debate was revived by the anthropogenic greenhouse effect and projected risks of climate change, analyzed in recent reports (IPCC 2001). As part of the Rio process the Framework Convention on Climate Change (FCCC) agreed on general principles and rules which were specified by the Kyoto Protocol to reduce emissions. Among the policy instruments for emission reduction are administrative concepts (emission taxes), market instruments (tradeable emission permits) and cooperative approaches such as Joint Implementation and the Clean Development Mechanism (CDM).

Economists have estimated the potential damage of rising global temperatures on society and economy and developed models to understand the impacts and costs of avoiding climate change. The Nordhaus model provided a first approach for an integrated assessment (Nordhaus 1994) which induced a number of models to overcome the simplifying assumptions. The macroeconomic perspective, optimizing a global welfare function, has been extended by game-theoretic models on emission reduction (Svirezhev et.al. 1999, Finus 2001, Kemfert 2001). This paper integrates the relationship between economic growth, emission reduction and energy technology into a dynamic-game framework, based on the VCX model developed by the author to analyze dynamic multi-player interaction (the model and some mathematics is explained in Scheffran 2001a; model applications with regard to energy and environment can be found in Scheffran 2000; Scheffran and Pickl 2000; Ipsen, Rösch and Scheffran 2001; Scheffran 2001b).

2 Technical Choice and Conflicting Demands

To model the interaction between energy, evironment and economy, basic variables are energy production E, greenhouse gas emissions G and economic output Q, each for a given time period (e.g. one year) and for particular acting units (players), which can be firms, consumers or nation states. For continuous time, these variables turn into flow variables. The usual measure of economic output is the Gross Domestic Product (GDP). Total emissions can be expressed as a product of the following factors (Edenhofer 2001):

$$G = g^e \times e^q \times q^n \times N$$

where N is the population, $g^e = G/E$ the emission intensity $e^q = E/Q$ the energy intensity and $q^n = Q/N$ the labor productivity. Another relevant ratio is the economic output per energy unit and time unit $q^e = Q/E = 1/e^q$. In general, the factors are not constant and can be functions of time, energy and other variables. For the following analysis they are assumed to be fixed for a particular time period and energy type. Emirical date on the ratios are available and show the decoupling between economic output and emission in industrialized countries.

An increase ΔE of installed power (e.g. more power plants) from one time period to the next requires financial investment $C = c^e \cdot \Delta E$, where c^e is the cost to install one power unit. For fixed g^e and q^e additional power increases both economic output and emissions

$$\Delta G = g^e \cdot \Delta E = g^c \cdot C > 0$$
$$\Delta Q = q^e \cdot \Delta E = q^c \cdot C > 0$$

where $g^c = g^e/c^e$ and $q^c = q^e/c^e$ are emissions and output per cost unit. To simplify notation, index e is ignored, i.e. $g = g^e, q = q^e, c = c^e$.

Each player is assumed to have the choice between different energy paths $k = 1, ..., m$, which correspond to different energy technologies (fossil, solar, nuclear). As a special case, $m = 2$ paths are analysed: the "old" path uses established energy technology ($k = 1$), the "new" path more advanced energy technology ($k = 2$) with less emissions per energy unit, $g_2 < g_1$. $0 \leq p \leq 1$ is the fraction of investment into new energy technology and c_k are the costs per unit of energy technology $k = 1, 2$. Installed power can be reduced by ΔE^-, by shut-down of existing power facilities.

The problem is to find a cost-effective allocation of investments C to energy production to satisfy both economic demands $\Delta Q = V^Q > 0$ (economic growth) and environmental demands $\Delta G = V^G < 0$ (agreed emission reduction), depending on the technical parameters g, q, c (V^Q and V^G are the respective target levels). This leads to two target lines:

$$C^G = \frac{V^G + g_1 \Delta E_1^-}{g_p^c}, \quad C^Q = \frac{V^Q + q_1 \Delta E_1^-}{q_p^c}.$$

where $g_p^c = (1-p)g_1/c_1 + p \cdot g_2/c_2$ is the emission-to-cost ratio for allocation p and $q_p^c = (1-p)q_1/c_1 + p\, q_2/c_2$ is the output-to-cost ratio for allocation p. Both target lines intersect in the point of satisfaction

$$\widehat{C} = \frac{g_1 V^Q - q_1 V^G}{D} \equiv y, \quad \widehat{\Delta E}_1^- = \frac{g_p^c V^Q - q_p^c V^G}{D} \equiv x$$

where $D = g_1 q_p^c - q_1 g_p^c$. Both targets are compatible if they are within the feasible set $(0,0) \leq (\hat{x}, \hat{y}) \leq (x^+, y^+)$ where x^+, y^+ denote the upper limits of action space. If the targets are incompatible, the player can change the allocation p, influence the technical-economic parameters g, q, c or adapt the target levels V^G and V^Q.

With $g_1^q = g_1/q_1$, the equilibrium costs are $\hat{C} = (g_1^q V^Q - V^G)/\bar{D}$ where $\bar{D} = D/q_1$. Since the numerator is positive for emission reduction targets $V^G < 0$, satisfaction can be achieved for positive investments as long as $\bar{D} = q_2^c(g_1^q - g_2^q)\, p > 0$ which is the case for $p > 0$ and $g_1^q = g_1/q_1 > g_2/q_2 = g_2^q$ i.e. the new technology emits less per economic output. As a simple example we use the special case $g_2 = 0.5\, g_1, q_2 = 1.5\, q_1, c_2 = 2\, c_1$, i.e. new technologies are cheaper and more productive but cost twice as much per unit. Then $g_2^q = g_1^q/3 < g_1^q$, $D > 0$ and $\hat{C} > 0$.

Within this framework it is possible to analyze conditions for cooperation between an industrialized country (IC) and a developing country (DC) by shifting to reduced-emission technology in DC with financial support from IC. For the simple example in the previous paragraph, a cooperation channel exists which allows both to achieve their targets at reduced cost by shifting to new technology and joint activities (see Scheffran 2001b).

3 Optimization in Climate Games

3.1 Utility and Damage Associated with Energy

While in the previous analysis economic and environmental targets were treated separately, the usual approach is to integrate the impact of production Q_i and emissions G_i for $i = 1, ..., n$ players Pi into welfare functions $V_i(Q, G) = U_i(Q_i) - D_i(G)$ where $G = \sum_j G_j$ is the sum of emissions of all players Pj and $Q = (Q_1, ..., Q_n)$ is the output vector. Utility function U_i and damage function D_i are defined as $U_i = u_i \cdot Q_i^{\alpha_i}$ and $D_i = d_i \cdot G^{\beta_i}$ where α_i and β_i are the respective elasticities, u_i and d_i are proportionality factors. Various models use $0 < \alpha_i \leq 1$ and $\beta_i \geq 1$. If both production $Q_i = q_i \cdot E_i^{\nu_i}$ and emissions $G_i = g_i \cdot E_i$ depend on energy consumption E_i, welfare of player Pi is a function of energy consumption of all players:

$$V_i(E_1, ..., E_n) = u_i Q_i^{\alpha_i} - d_i G^{\beta_i} = u_i^e E_i^{\eta_i} - d_i (\sum_j g_j E_j)^{\beta_i}$$

where $u_i^e = u_i q_i^{\alpha_i}$ and $\eta_i = \nu_i \alpha_i$. The marginal impact of Pi's energy increase on its welfare $\partial V_i / \partial E_i$ is the difference between marginal utility and damage of energy. Setting this term to zero provides the optimal energy level for Pi, as a function of the energy consumption of all other players Pj.

As an example we treat the case of two players ($n = 2$), each with $\alpha_i = \nu_i = 0.5$ and $\beta_i = 1$ (linear damage function) such that $\eta_i = 1/4$. Then the condition $\partial V_i / \partial E_i = 0$ leads to $E_i^* = (u_i^e / 4 d_i g_i)^{4/3}$. Thus, an increase in u_i^e or a decline in d_i or g_i would increase the optimal energy level which is independent of other players' actions. For $\beta_i = 2$ other players come into the optimization game, and the optimal level can be determined implicitly as $E_j = (u_i^e / 8 d_i g_i g_j) \cdot E_i^{-3/4} - E_i / g_j$ which corresponds to a set of reaction curves in energy space. The special case that Pi is the only emitter ($E_j = 0$) leads to the optimum $E_i^* = (u_i^e / 8 d_i g_i)^{4/7}$.

3.2 Negotiations on Emission Reduction

Now consider $i = 1, ..., n$ players generating economic oputput Q_i, emissions G_i and emission reduction R_i. We assume a linear utility function $U_i = u_i Q_i$, a linear relationship between output and emissions $Q_i = g_i G_i (1 - R_i)$ and a quadratic damage function ($\beta_i = 2$). In addition, the costs are a quadratic function $C_i(R_i) = r_i R_i^2$ of emission reduction. Then as a net value function of utility, damage and costs we define (with $u_i^g = u_i g_i$)

$$V_i = U_i(Q_i) - D_i(G) - C_i(R_i) = u_i^g G_i (1 - R_i) - d_i (\sum_j G_j (1 - R_j))^2 - r_i R_i^2.$$

The optimality condition $\partial V_i / \partial R_i = 0$ yields optimal emission reductions

$$R_i^* = \frac{d_i (\sum_{j \neq i} G_j (1 - R_j)) - u_i^g G_i / 2}{d_i G_i + r_i}.$$

which defines a reaction curve for Pi in emission space. The optimal emission reductions are in the feasible range $0 \leq R_i^* \leq 1$ for

$$G_i^- = \frac{d_i(\sum_j G_j(1-R_j)) - r_i}{u_i^g/2 + d_i} \leq G_i \leq \frac{2d_i(\sum_{j\neq i} G_j(1-R_j))}{u_i^g} = G_i^+$$

If all countries would agree on the same emission reduction $R_i = R$ for $i = 1, ..., n$, then $\partial V_i/\partial R = 0$ gives the optimal emission reductions.

3.3 Position Change and Coalition Formation

If for a player Pi it is individually optimal not to reduce emissions, some incentives are required to change its position $x_i = R_i$ by Δx_i. This can be achieved by sanctions or cooperation and coalition formation which implies that a group of players achieves a "coalition gain". Players are willing to change their positions to jointly concentrate their power resources (critical mass) for achieving a coalition gain which they are unable to do alone. The following terms can influence a position change Δx_i:

1. Position change of player Pi leads to value change $\Delta V_{ii}(\Delta x_i)$.
2. Deviation from Pi's optimum by sharing coalition gain: $\Delta V^i(x)$
3. Impact of other players' power use (reward and punishment) ΔV_{ij}
4. Own costs fractions $\Delta C_{ij} = p_{ij} \cdot C_i$ to influence other players Pj.

Then a condition for a position change Δx_i of player Pi is total value change

$$\Delta V_i = \Delta V_{ii} + \Delta V^i(x) + \sum_j \Delta V_{ij} - \sum_j C_{ij} > 0.$$

4 Technology Choice of Competing Firms

4.1 The Cobb-Douglas Production Function

In the following the analysis is extended to a Cobb Douglas production function, with energy E, capital K and labor L as production factors:

$$Q = \beta \cdot K^{\alpha^K} L^{\alpha^L} E^{\alpha^E}.$$

β measures production efficiency and $\alpha^K, \alpha^L, \alpha^E$ elasticities. Labor $L = p^L C/c^L$ and energy $E = p^E C/c^E$ depend on fractions p^E, p^L of investment C, where c^L, c^E are labor and energy unit costs. Assume capital grows with

$$\dot{K} = p^K e^K C - \delta \cdot K.$$

where δ is the decay rate, e^K is the investment efficiency for capital and p^K the percentage of investment for capital. The steady-state condition $\dot{K} = 0$ leads to $K^* = p^K C/(\delta\, c^K)$ where $c^K = 1/e^K$ is the unit cost to maintain the capital stock K^*. Then produced output can be expressed as a function of costs $Q = \beta \cdot \Pi^\alpha \cdot C^\alpha$ where $\Pi = (p^K/c^K)^{\alpha^K} \cdot (p^L/c^L)^{\alpha^L} \cdot (p^E/c^E)^{\alpha^E}$ and $\alpha = \alpha^K + \alpha^E + \alpha^L$. According to Jensen's inequality, Π and thus Q is maximized for $p^E = \alpha^E/\alpha, p^L = \alpha^L/\alpha, p^K = \alpha^K/\alpha$.

4.2 The Impact of Emission Tax on Competing Firms

We now model firms Pi ($i = 1, ..., n$) with Cobb-Douglas production functions $Q_i = \beta_i K_i^{\alpha_i^K} L_i^{\alpha_i^L} E_i^{\alpha_i^E} = \beta_i \gamma_i \pi_i \cdot C_i$ where $\pi_i = (p_i^K)^{\alpha_i^K} (p_i^L)^{\alpha_i^L} (p_i^E)^{\alpha_i^E}$ is the allocation function and $\gamma_i = (c_i^K)^{-\alpha_i^K} (c_i^L)^{-\alpha_i^L} (c_i^E)^{-\alpha_i^E}$ inverse unit costs. Let $E_i^k = p_i^k C_i / c_i^k$ be energy of type k of firm Pi, where p_i^k and c_i^k are the respective allocation and unit costs. One way to convince firms to switch to more sustainable technologies with less emissions is an emission tax τ. Then the net value for firm Pi with product Q_i can be expressed in terms of costs

$$V_i = q \cdot Q_i - C_i - \tau \sum_k g_i^k E_i^k = (v_i - \sum_j v_{ij} C_j) C_i$$

where $q = a - b \cdot \sum_j Q_j$ is the demand-dependent market price for good Q and v_i, v_{ij} are appropriate factors. For $\partial V_i / \partial p_i > 0$ a change towards new energy technology provides individual value gains. This defines a *lower tax threshold* τ_i^* depending on the model parameters. For $\tau > \tau_i^*$ it would be rational for firm Pi to switch to new energy technologies, if the taxes are not too high to prevent positive value. The condition $V_i > 0$ leads to the *upper tax threshold* $\tau_i^0 > \tau_i$. As long as $\tau_i^0 < \tau_i^*$, environmental and economic demands are compatible. For $\tau_i^0 < \tau_i^*$ however, there is a conflict between both.

4.3 Dynamics and Computer Simulation

Partial derivatives $\partial V_i / \partial C_i = 0$ and $\partial V_i / \partial p_i = 0$ define optimality conditions for costs $\tilde{C}_i$ and allocation $\tilde{p}_i$ which correspond to reaction functions. The adaptation dynamics leads to a system of difference equations

$$\Delta C_i(t) = C_i(t+1) - C_i(t) = \alpha_i^c (\tilde{C}_i(t) - C_i(t))$$
$$\Delta p_i(t) = p_i(t+1) - p_i(t) = \alpha_i^p (\tilde{p}_i(t) - p_i(t))$$

which describe an evolution of firm's behavior over time (see Scheffran 2000a).

For numerical simulation, we assume two firms Pi ($i = 1, 2$) which produce goods Q_i with initial price $a = 1$ and elasticity $b = 0.01$ and two energy technologies E_i^k, where old technology ($k = 1$) initially costs half of new technology ($k = 2$) per energy unit, i.e. $c_i^1 = 1, c_i^2 = 2$, while emissions are twice as much ($g_i^1 = 1, g_i^2 = 0.5$). Capital and labour are assumed to cost the same ($c_i^K = c_i^L = 1$). Exponents and allocation priorities of the production function are highest for energy ($\alpha_i^E = p_i^E = 0.5$), while for capital and labor the estimates $\alpha_i^K = p_i^K = 0.3$ and $\alpha_i^L = p_i^L = 0.2$ are used. Half of initial investment stock $C_i^K = 10$ is used for investment flow C_i. Initially all investment is used for old energy technology $p_i^1 = 1$. The reaction parameters for the adaptation algorithms to approach target costs and allocation are set to $\alpha_i^c = \alpha_i^p = 0.5$. Both firms behave alike, with one exception: production efficiency is slightly higher for firm P1 compared to P2 ($\beta_1 = 5.5, \beta_2 = 4.5$).

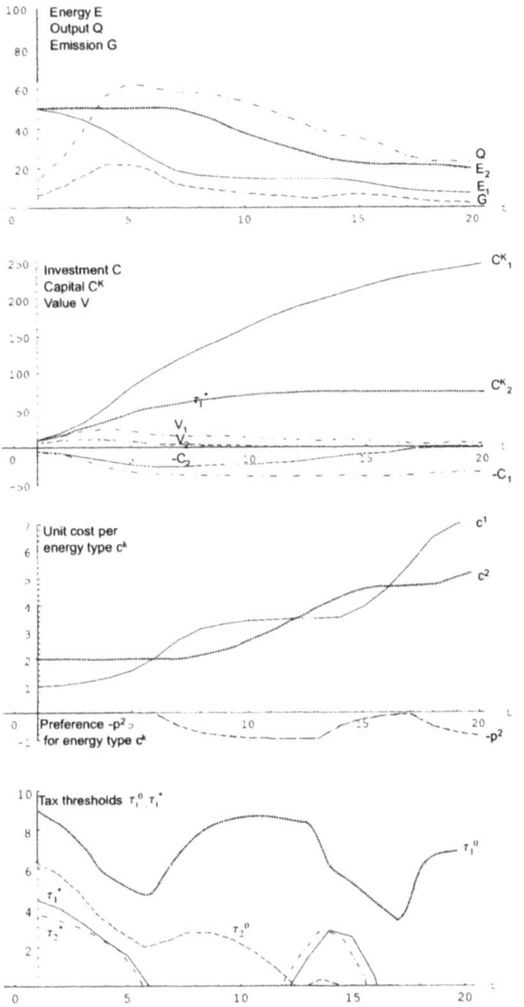

Fig. 1. Simulation of energy, economic output, emissions, investment, value and technology priority for two firms. Tax thresholds are depicted below.

With these initial conditions simulations for 20 iterations are depicted in Figure 1. They suggest that firm P1 with slightly higher production efficiency succeeds in competition with firm P2 by a factor of 5 after 20 time steps. The more energy is depleted, the more the density-dependent unit costs increase (by a factor 3 to 7). Each time when unit costs of one technology become higher than the other, a switch in allocation results. Emissions increase with production but show a decline after unit costs increase due to scarcity. The lower tax thresholds τ_i^* for technology change decline and reach zero after 5 iterations, corresponding to the actual allocation switch. Another peak occurs at around 15 time steps. The upper tax thresholds τ_i^0 reach zero for firm 2 after 13 time steps. Only firm P1 can then profitably exploit the remaining energy and accumulates capital.

References

1. Aghion, P., Howitt, P. (1998) Endogenous Growth Theory. MIT Press, Cambridge
2. Barro, R.J., Sala-i-Martin, X. (1995) Economic Growth. MIT Press, Cambridge
3. Edenhofer O. (2001) The Sustainability Transition and Technological Change. A Research Agenda. Working Paper. Potsdam, PIK
4. Eichhorn, W., Henn, R. et al. (eds.) (1982) Economic Theory of Natural Resources, Physica, Würzburg Wien
5. Finus, M. (2001) Game Theory and International Environmental Cooperation. Edward Elgar, Cheltenham
6. Hudson, E.A., Jorgenson, D.W. (1974) U.S. Energy Policy and Economic Growth, 1975-2000. Bell J. Econ. Managem. Science **5**, 461-514
7. IPCC (2001) Climate Change 2001. Third Assessment Report of the Intergovernmental Panel on Climate Change. Cambridge University Press
8. Ipsen, D., Rösch, R., Scheffran, J. (2001) Cooperation in Global Climate Policy. Potentialities and Limitations. Energy Policy **29/4**, 315-326
9. Kemfert, C., (2001) International Games of Climate Change Policies. The Economic Effectiveness of Partial Coalition Games. Working Paper. University of Oldenburg, Dept. of Economics 1, Febr.
10. Kümmel, R., (1980) Growth Dynamics of the Energy Dependent Economy. Oelgeschlager, Gunn and Hain, Cambridge, Mass.
11. Kümmel, R., Lindenberger, D., Eichhorn, W. (2000) The Productive Power of Energy and Economic Evolution. Indian J. Appl. Econ. **8**, 231-262
12. Nordhaus, W.D. (1994) Managing the global commons. The economics of climate change. The MIT Press
13. Renshaw, E.T. (1981) Energy Efficiency and the Slump in Labor Productivity in the USA. Energy Economies **3**, 36-42
14. Scheffran, J. (2000) The Dynamic Interaction Between Economy and Ecology. Mathematics and Computers in Simulation **53**, 371-380
15. Scheffran, J., Pickl, S. (2000) Control and Game-Theoretic Assessment of Climate Change. Options for Joint Implementation. Annals of Operations Research **97**, 203-212
16. Scheffran, J., (2001a) Stability and Optimal Control of a Multiplayer Dynamic Game, in: Fleischmann, B. et.al., Operations Research Proceedings 2000. Springer, Berlin, Heidelberg, 14-19
17. Scheffran, J., (2001b) Conflict and Cooperation in Energy and Climate Change. The Framework of a Dynamic Game of Power-Value Interaction, to appear in: M. Holler (ed.), Yearbook New Political Economy
18. Svirezhev, Y.M., von Bloh, W., Schellnhuber, H.J. (1999) "Emission game": some applications of the theory of games to the problem of CO2 emission. Environmental Modelling and Assessment **4**, 235-242

Measuring Credit Risk: Can we Benefit from Sequential Nonparametric Control?

Ansgar Steland

Fakultät für Mathematik, Ruhr-Universität Bochum, D-44780 Bochum

Abstract Financial risk measures have to rely on both empirical data and econometric methods to be economically relevant and statistically meaningful. When concerned with credit risk data, credit risk measures are computed and analyzed based on data as spreads, interest rates, ratings, account movements, etc. At least a part of this information, in particular spreads, are available as a sequential stream of data. Consequently, sequential methods to control for credit risk are useful, since they enable a risk manager to react immediately after risk measures have changed significantly.

In this article we briefly review some established approaches to analyze credit risk and explain how nonparametric sequential change–point procedures can be applied to detect credit risk changes as soon as possible. From a practical point of view the method proposed here is attractive, because it is easy to handle if appropriate risk measures are available. Essentially, it is based on a weighted average of past observations. In order to evaluate the quality of such a decision rule, it is reasonable to analyze how the delay of the procedure behaves asymptotically, if the number of observations increases. We report about a mathematical result of the author which analyzes the asymptotic delay for a completely arbitrary local pattern defining the structural change of the mean of the risk process underlying the detection procedure. It is demonstrated how the result can be applied to optimize the procedure when certain a priori knowledge about the structural change is available. We analyze a real data set of credit spreads. It turns out that the empirical findings concerning the choice of the kernel are in perfect agreement with the implications of the theory.

1 Introduction

Measuring and evaluating financial risk is an important task of financial analysts. The evaluation of credit risk is often based on time series of spreads or derived time series providing information about important factors as, e.g., maturity buckets or rating classes. Whereas credit migration, e.g., downgrading, may produce jumps, continuous changes may be attributable to continuous changes of other slowly varying (macroeconomic) factors. Since these effects can overlay and interact, it is reasonable to model a structural change of the expected risk measure by an arbitrary unknown local pattern m_0. Nonparametric sequential credit risk control provides tools to detect the first time point where a credit risk process is affected by such a structural change. Our approach concentrates on accurate detection of these changes.

We do not discuss investment policies to react to changes, since, in general, optimal actions will depend on rather specific characteristics of the concrete decision framework. It is interesting to note that similar methods as discussed here are applied by many analysts using rules-of-thumb. Prominent examples are trading strategies where a buy or sell signal is given when a certain indicator exceeds a threshold or crosses another indicator. From a statistical point of view these ad hoc approaches fit into the framework of sequential econometrics where we aim to detect certain structures in the data (jumps, linear/quadratic trends, changes in the derivative, etc.) as soon as possible. Using statistical theory we are able to study the properties as the average run length or the asymptotic delay of such detection procedures. Nonparametric methods to detect jumps in the mean or its first derivative and the properties of the asymptotic delay of these methods have been studied in Steland (2000, 2001). Here we study the less restrictive case that the time series of credit risk measures is no longer stationary and mean–zero, but is affected by an *arbitrary functional pattern* m_0. The procedure relies on a sequential kernel estimator where the kernel function, K, defines the weighting scheme. The question arises, how the kernel K and the mean function m_0 affect the asymptotic delay of the procedure.

2 Background

Credit risk, which consists of default risk, credit migration risk, exposure uncertainty risk, and recovery risk, can be measured by the difference between the yield of a corporate bond and the yield on a government bond index. Bankers often use the swap curve, in lieu of the corporate curve, as a basic tool for pricing corporate liabilities, assuming that the swap term structure reflects the cost of financing a top–rated corporate issuer for the various maturities. Recall that an interest rate swap is a contract by which a fixed payment stream is replaced against a floating payment stream. The market rate of a swap is then the coupon rate of the fixed leg. The floating leg is usually set at the interbank interest rate, eg., the six–month LIBOR for dollar swaps.

Consequently, time series of spreads or derived time series obtained by estimating more or less complex statistical models for the spread provide a reasonable basis for on–line monitoring procedures. Depending on whether or not the models require knowledge of firm value parameters they are classified as structural–form or reduced–form approaches. Firm value models model the value of the firm's assets relative to its liabilities. This means, they model the firm's capital structure and therefore are also called structural model. The first model for credit-risky bonds is due to Merton (1974) who employs the methodology of Black and Scholes (1973). Here it is assumed that the firm's ability to redeem its debt is determined by the total value V_t of its assets at time t. Suppose the firm has a single liability with promised terminal payoff

K. This can be interpreted as a zero coupon bond. Then the stock price at maturity, T, is given by $\max(0, V_T - K)$, i.e., it can be considered as a call option on the firm. When the bond is small relative to the portfolio of assets of the firm, one can model the default directly as a stochastic point process. Here default is not necessarily related to the firm's value (Duffie and Singleton, 1997). The influence of the rating category and maturity buckets can be analyzed by multiple linear regression models. Such cross–sectional regressions can be performed to produce time series for the respective rating and maturity components.

Spreads are not always available and often the classification in several rating classes is the only observable measure of credit quality. This in particular applies to credits of small companies and consumer credits. In this case, bankers often use score values computed from questionnaires which are then used to rate a credit. Such rating data can be analyzed using generalized linear models (McCullagh and Nelder, 1989), e.g., ordered probit models which are used by JP Morgan's CreditMetrics. Here credit events are driven by underlying latent variables which are affected by external risk factors. More precisely, the model specifies the conditional distribution function given risk factors,

$$P(Z_i \leq r | \widetilde{\mathbf{f}} = \mathbf{f}) = \Phi([\phi_r - \theta_i' \mathbf{f}]/\eta_i), \tag{1}$$

where Z_i denotes the rating category, eg., BBB, $\widetilde{\mathbf{f}}$ are l unobservable risk factors, θ_i is the vector of sensitivities of obligor i with respect to the risk factors, and $\phi_0 < \ldots < \phi_{K-1}$ are cut-off values. The model is motivated as follows. It is assumed that rating category r is observed if and only if $\phi_{r-1} < V_i \leq \phi_r$, where the latent variable V_i can be interpreted as the value of assets of obligor i. For V_i a factor model, $V_i = \theta_i' \widetilde{\mathbf{f}} + \eta_i \epsilon_i$, is assumed, where the weight η_i determines the idiosyncratic risk for the obligor. $\epsilon_i \overset{i.i.d.}{\sim} N(0, 1)$, and $\widetilde{\mathbf{f}} \sim N(\mathbf{0}, \Omega)$ are assumed to be independent. Then, the constraint $V(V_i) = \theta_i' \Omega \theta_i + \eta_i^2 = 1$ yields (1). The model can be estimated by Maximum Likelihood to yield estimates $\widehat{\theta}_{ni}$ for the sensitivities. In this model common dependence on the same underlying risk factors gives rise to correlations (co–movements) in credit events across obligors. Given historical data, one may produce time series of important risk measures which are worth monitoring sequentially. For example, one may define the risk process $Y_n = \widehat{\theta}_{ni}$, where $\widehat{\theta}_{ni}$ is estimated from the data of the nth time period.

3 Nonparametric sequential credit risk control

Nonparametric sequential credit risk control assumes that we are given a credit risk process $\{Y_t\}$ in continuous time (spreads or a derived process) with a mean structure $m(t) = E(Y_t)$. We aim at detecting the first time point where the mean structure is no longer equal to a prespecified one. This means, it is assumed that the process is distributed according to a known distribution as long as the process is in a state of statistical control (*in-control model*). From an economic viewpoint the term statistical control means that risk analyses can be done by relying on the known in-control model. Since then $m(t)$ is known for $t < t_q^*$, we can assume that $m(t) = 0$ as long as $t < t_q^*$ (otherwise we can substract the known mean). This means, the in-control model corresponds to a mean-zero risk process.

The important question arises whether there is a time point such that the process $\{Y_t : t < t_q^*\}$ is in-control but runs out-of-control after the so-called *change-point* t_q^*. In terms of statistical hypothesis testing, the process satisfies a simple null hypothesis for $t < t_q^*$, whereas for $t \geq t_q^*$ the alternative holds true. The case that the risk process remains in a state of statistical control corresponds to $t_q^* = \infty$. In this article we confine ourselves to the case that the change-point is non-stochastic but unknown. This covers the situation that at certain deterministic time points, e.g., when a firm publishes financial reports, investors and analysts get new information which are not contained in market prices and therefore have a well-defined effect. In this case the possible change-points and are known, but the future mean structure is of course unknown. Under certain circumstances random change-points could represent a more realistic model when random events (political events, etc.) affect prices. In this case the change-points are random variables. However, if the underlying event and the risk process are independent, by conditioning on the change-points we can treat them as fixed but unknown times.

In this paper we consider a nonparametric approach where we assume that $m(t)$ vanishes if $t < t_q^*$ and is equal to $m(t;h)$ if $t \geq t_q^*$. $m(t;h)$ is induced by a *generic alternative model* $m_0(t)$ by the rule

$$m(t;h) = m_0([t - t_q^*]/h)$$

if $t \geq t_q^*$, where t_q^* is the change–point. $h > 0$ is a parameter which tends to infinity in our asymptotic analysis. The function m_0 defines the (local) pattern affecting the mean after time t_q^*: jumps, smooth changes, or arbitrary combinations. Observe that

$$\lim_{h \to \infty} m(t;h) = m_0(0).$$

In this sense we say that $m(t;h)$ is a *local alternative*. If we impose the constraint that $m_0(0) = 0$, then $m(t;h)$ converges to the in-control model (null hypothesis), as $h \to \infty$.

It is clear that this flexible local modeling approach can be applied to credit risk data where Y_n denotes a risk measure at time t_n for a certain risky financial instrument with maturity T. The following example shows that it is also possible to handle term structure data.

Example: *Suppose S_t stands for the observed credit spread at time t of a bond of a fixed rating category maturing at time T. Then $m_S(t) = E(S_t)$ is the true credit spread. For any given term structure of credit spreads $TS_0(t)$*

we obtain an in-control model $m_S(t) = TS_0(T-t)$. *The corresponding risk process is defined as* $Y_t = S_t - m_S(t)$. *Then* $E(Y_t) = 0$ *holds true as long as the process is governed by the credit spread curve* $TS_0(t)$. *If* $TS_1(t)$ *stands for an alternative term structure, the corresponding generic alternative is given by*

$$m_0(t) = TS_1(T-t) - TS_0(T-t)$$

and satisfies $m_0(0) = 0$, *if both curves yield the same spread at maturity.*

Now the process is sampled at fixed time points t_n providing observations

$$Y_{nh} = m_{nh} + \epsilon_n, \qquad m_{nh} = m(t_n; h)$$

with mean m_{nh} and an an innovation process $\{\epsilon_n\}$. The change-point is now given by $t_q = [t_q^*] + 1$, where $[x]$ is defined as the integer k satisfying $k \le x < k+1$.

Our theoretical result allows for both non-normality and dependencies. The former is important, since in general spreads are not normal. We also allow for short memory processes. More precisely, we assume that the process $\{\epsilon_n\}$ is stationary and α–mixing.

4 Nonparametric sequential detection based on smoothing

Because we were interested in a theoretical analysis of detection procedures which are commonly applied, we decided to study a simple nonparametric sequential smoother. However, to cover the most important approaches used in practice, we used kernel smoothing which allows for arbitrary weighting schemes. The detection procedure examines the smoothed data (control statistic),

$$\widehat{m}_{nh} = \sum_{i=1}^{n} K_h(t_i - t_n) Y_i, \quad K_h(z) = h^{-1} K(z/h),$$

and provides a signal if $|\widehat{m}_n| > c$, $\widehat{m}_{nh} > c$, or $\widehat{m}_{nh} < c$, depending on the concrete application. Note that $\widehat{m}_{nh}$ is not an estimate for $m(t_n)$, since the weights do not sum up to 1. In this case one should employ the normed version

$$\widetilde{m}_{nh} = \sum_{i=1}^{n} w_h(t_i - t_n) Y_i, \quad w_h(t_i - t_n) = K_h(t_i - t_n) / \sum_{j=1}^{n} K_h(t_j - t_n).$$

However, $\widehat{m}_{nh}$ can be used to detect structural changes in the data and is fast to compute. By specifying the smoothing kernel K we can specify any weighting scheme (exponential smoothing, uniform weights, etc.) In this way important and widely employed smoothers are covered as special cases. The arithmetic mean of the last, say, l, days corresponds to the uniform kernel $K_u(z) = 1/2$, if $z \in [-1, 1]$, and the bandwidth choice $h = l$. This basic tools is commonly applied in technical analysis, e.g., when comparing 200 day- and 30 day averages. Also quite popular are exponentially weighted moving averages (EWMA) defined by $\widehat{m}_{n,e} = (1-\lambda)\widehat{m}_{n-1,e}+\lambda Y_n$, where $\lambda \in (0, 1]$. Considered as a prediction of Y_{n+1}, the exponential smoother is known to be optimal for integrated moving average processes of order 1. Whereas the former approach corresponds to the uniform kernel, the latter one is obtained approximately when using the standardized Laplace kernel, $K_{Lap}(z) = 1/\sqrt{2}\exp(-\sqrt{2}|z|)$ and the equivalent bandwidth $h = h_\lambda = -\frac{\sqrt{2}}{\log(1-\lambda)}$. For details and further discussion see Section 4 of Schmid and Steland (2000).

5 Theory for the normed delay

The first index where a signal is given, the stopping time N_h, is our estimator for the change–point $t_q = [t_q^*] + 1$ in discrete time. In most cases a control procedure is designed to ensure that the average run length,

$$\mathrm{ARL}_h = E(N_h),$$

is not less than some prespecified value, if the process is in control. The theoretical result to be discussed now focuses on another quantity. The quality of the procedure can be defined quite naturally as

$$\rho_h = (N_h - t_q)/h,$$

called *normed delay*. If the kernel vanishes for $|z| > 1$, h is the effective sample size employed by the procedure. In this case ρ_h can be interpreted as the delay expressed as a percentage of the effective sample size. The normed delay is a stochastic quantity. As for the arithmetic mean $\overline{X}$ of n observations $X_1, \ldots, X_n$, it is possible to establish limit theorems. It is known that under certain regularity assumptions $\overline{X}$ converges (in probability) to the mean $\mu = E(X_1)$ of the observations, as the sample size n tends to infinity. For the normed delay the situation is much more involved, but it was possible to establish a similar limit theorem. This result ensures that under mild regularity assumptions, discussed to some extend in the previous sections, the normed delay ρ_h converges in probability to a constant ρ_0, as $h \to \infty$. This means, for large enough h the delay $N_h - t_q$ is roughly equal to $\rho_0 \cdot h$. Further, ρ says in a mathematically rigorous fashion how the performance depends on the smoothing kernel K and the generic alternative model m_0, since the smallest x such that

$$I(x; K; m_0) = \int_0^x K(s - x) m_0(s) ds$$

equals c gives ρ_0. Hence, for a given alternative m_0 we can select an optimal kernel from a finite set of candidate kernels, $\mathbf{K} = \{K_1, \ldots, K_L\}$, by analyzing the corresponding integral functions $I(x; K_1; m_0), \ldots, I(x; K_L; m_0)$.

To illustrate this approach, assume that $m_0(t) = t$ for $t \geq 0$. Figure 1 shows the integral functions for the three different kernels, namely the Gaussian kernel, $K_{Gau}(z) = [2\pi]^{1/2} \exp(-z^2/2)$, the Epanechnikov kernel, for our purposes defined as $K_{Epan}(z) = 0.75 \cdot (1 - z^2)$ if $|z| \leq 1$ and $= 0$ otherwise, and the Laplace kernel $K_{Lap}(z)$ defined above. For a given threshold c, which should be chosen according to some reasonable criterion, these curves allow to compare the asymptotic efficiency in terms of the asymptotic normed delay. Theory suggests to use the kernel with the smallest asymptotic normed delay.

Figure1. *Integral curves $I(x)$ for different kernels: Epanechnikov (bold), Laplace (dashed), and Gauss (thin).*

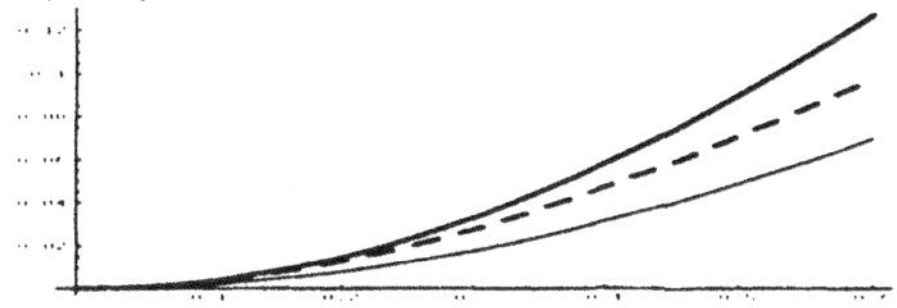

6 Analysis of real credit risk data

To illustrate the nonparametric sequential control of credit risk we analyze credit spread data of Ford Motor Credit Company. The data set consists of $n = 84$ credit spreads $\{S_n : n = 0, \ldots, 83\}$ covering the time period from 11/97 to 11/98. The in-control period was chosen as $\{S_n : 30 \leq n \leq 60\}$. To take account of (positive) autocorrelations, we fitted an AR(2) to the data. That procedure suggests the model

$$S_n - m = 0.601 \cdot (S_{n-1} - m) + 0.298 \cdot (S_{n-2} - m) + 0.0395 \cdot e_n, \quad (2)$$

where $\{e_n\}$ is Gaussian white noise and $m = 0.525$ is the estimated mean. Nonparametric sequential control was based on the sequence of estimators $\widetilde{m}_{nh}$ using three kernels: The Gaussian kernel, the Epanechnikov kernel, and the Laplace kernel. For all kernels we used the bandwidth $h = 5$. Relying on model (2) we simulated time series of credit spreads to estimate the critical value c corresponding to an in-control average run length of $\xi = 60$. We used $50,000$ runs to estimate each average run length for a candidate value c. Figure 2 provides the control chart for all data. Note that there is a considerable influence of the smoothing kernel K. The Gaussian kernel provides smooth estimates but detects the structural change later than the other procedures. This result is in perfect agreement with our theoretical findings for a linear trend as generic alternative.

Conclusions: We may conclude that nonparametric sequential risk control provides an appropriate flexible approach to detect structural changes in risk processes in order to trigger actions. The theoretical result on the normed delay suggests a useful tool to select the smoothing kernel in an optimal fashion.

Figure2. *Control chart for credit spreads of Ford Motor Credit. Estimates* $\widetilde{m}_{nh}$ *are given for three kernels. Horizontal lines mark the critical values. Vertical lines mark the stopping times (signals).*

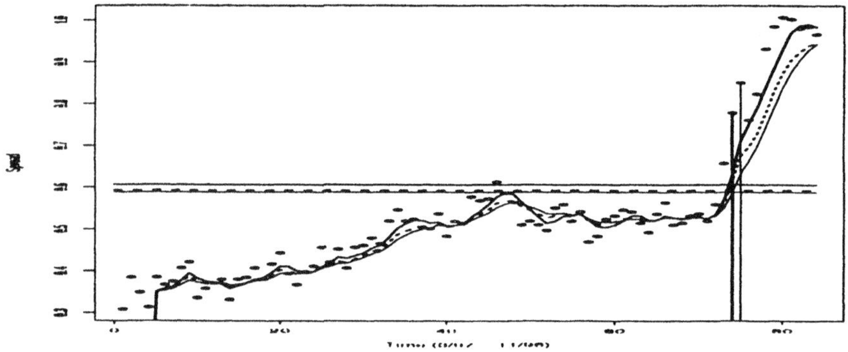

References

1. Black, F and Scholes, M. (1973): The Valuation of Options and Corporate Liabilities, *Journal of Political Economy, 81, 3, 637-654.*
2. Duffie, D. and Singleton K. (1997): An Econometric Model for the Term Structure of Interest–Yield Swap Rates. *Journal of Finance, 52, 1287–321.*
3. McCullagh, P. and Nelder, J.A. (1989): *Generalized Linear Models*, Chapman & Hall, London.
4. Merton, R. C. (1974): On the Pricing of Corporate Debt: The Risk Structure of Interest Rates. *Journal of Finance, 2, 2, 449-470.*
5. Schmid, W. and Steland, A. (2000): Sequential Control of Non-Stationary Processes by Nonparametric Kernel Control Charts, *Journal of the German Statistical Association, 84, 315-336.*
6. Steland, A. (2000): Nonparametric Monitoring of Financial Time Series, to appear: *Statistical Papers.*
7. Steland, A. (2001): On Jump–Preserving Sequential Control for Certain Mixing Processes with Applications in Finance and Econometrics. to appear: *Proceedings of the 4th St. Petersburg Workshop on Computation and Simulation.*
8. Steland, A. (2001): Optimal Sequential Kernel Smoothers under Local Nonparametric Alternatives for Dependent Processes. *Preprint, Faculty of Mathematics, Ruhr-Universität Bochum.*

Section 12

Decision Theory

Die rationale Marketing-Mix Entscheidung bei vager Kenntnis einzelner Instrumentwirkungen

André Ahuja, Quirin Spreiter, Wilhelm Rödder

FernUniversität in Hagen, Profilstraße 8, 58084 Hagen, Germany

andre.ahuja@fernuni-hagen.de, quirin.spreiter@gmx.de,
wilhelm.roedder@fernuni-hagen.de

Zusammenfassung In dieser Arbeit wird ein quantitativer Zugang zur Wirkungsprognose eines Marketing-Mix mittels probabilistischer Konditionallogik vorgestellt. Als Regeln formuliertes vages, d.h. mit Unsicherheit behaftetes Teilwissen über die Absatzwirkung einzelner oder einiger weniger Instrumenteneinsätze bildet die Wissensbasis, welche nach einem Lernprozeß informationstreu eine gemeinsame Verteilung über alle Modellvariable bereitstellt und somit künstlich die vollständige Risikosituation als Grundlage der rationalen Entscheidungsfindung schafft. Nun führen Anfragen an die Wissensbasis zu Prognosen über die Wirkung eines ganzen Maßnahmen-Bündels. Die hier in der probabilistischen Expertensystemshell SPIRIT aufgebaute Wissensbasis besteht aus 20 Variablen und 143 Regeln. Zwei Szenarien führen die Identifikation von Synergie- beziehungsweise Kompensationseffekten sowie die Ableitung des zieladäquaten Instrumenteneinsatzes mit Hilfe von SPIRIT vor.

1 Einleitung

Auf Käufermärkten ist Absatzförderung oder Marketing ein integraler unternehmerischer Aufgabenbereich hoher Komplexität. Dieser verfolgt die operative Umsetzung der aus der Unternehmensstrategie abgeleiteten Marketingstrategie. Die Marketing-Mix (MM) Aufgabe ist die Entscheidung der optimalen Zusammenstellung von Marketingaktivitäten bei (teilweise) gegebenen Produkt- und Marktdaten (situative Variable), unsicheren Marktreaktionen (stochastische Variable) sowie die Unternehmensstrategie unterstützenden verfolgten operativen Zielen (Zielvariable).
Derartige Aktivitäten stellen in der Regel einen betrieblichen Aufwand dar, so dass Effizienzbetrachtungen unerlässlich sind und hier quantitativ durchgeführt werden sollen. Im Falle der Analyse von Absatzwirkungen bedient man sich zumeist statistischer Partialmodelle. Die Bewertung aufgestellter Hypothesen oder geforderte Parameterschätzungen resultiert aus der Evaluation einer entsprechenden Marktstudie. Deren eigener Aufwand bedingt die Begrenzung auf Partialmodelle, wie beispielsweise die Wirkung einer Preisänderung zusammen mit einer Werbeaktion auf Wiederkaufneigung sowie Marktanteil. Alternativ kann zu einer betrachteten Fragestellung eine Expertenmeinung eingeholt werden. Beide Vorgehensweisen liefern in der Regel

nicht sicheres, sondern vages, mit Unsicherheit behaftetes Wissen. Die MM-Aufgabe ist bereits seit den frühen 60er Jahren Untersuchungsgegenstand des OR. Neben der Wirkungsprognose sind hier die Marketingstrategiebildung unter Berücksichtigung des Verhaltens des Wettbewerbs und seit Ende der 80er Jahre die Anwendung von KI-Methoden zur Modellbildung bei unvollständiger Information Forschungsschwerpunkte. In den folgenden Abschnitten wird ein Modell entwickelt, welches den ersten sowie den letztgenannten Aspekt miteinander verbindet.

2 Die probabilistische Expertensystemshell SPIRIT

Die Empirik belegt die weitgehende Akzeptanz insbesondere der computerbasierten MM-Entscheidungsunterstützung [8]. Ein Entscheidungsmodell unter Risiko verlangt die Kenntnis sämtlicher Verteilungen zufallsverteilter Modellvariabler. Wie eingangs beschrieben, sind diese Verteilungen für die MM-Entscheidung typischerweise nur an wenigen Punkten bekannt. Derartige „Wissenshappen" sollen unter Anwendung der SPIRIT-Syntax [RD00] als probabilistische Regeln formuliert werden. SPIRIT ist eine probabilistische Expertensystemshell, welche als Eingabedaten zunächst die Definition diskreter Modellvariabler und auf diesen gebildete Regeln verlangt. Dazu wende man die logischen Operatoren $\neg, \wedge, \vee, |$ auf Elementarausdrücke der Form $< Nominalvariable >=< Wert >$ an und versehe den entstandenen zusammengesetzten logischen Ausdruck mit einer (vom Experten angegebenen oder statistisch ermittelten) Wahrscheinlichkeit für das Zutreffen dieser Aussage, dem Schlusskraftfaktor. Beinhaltet die Regel die bloße W-Zuordnung zu einer einzelenen Merkmalsausprgung, so wird sie Faktum genannt. Die Menge R aller Regeln auf zuvor definierten Variablen bildet die Regelbasis. In einem ersten Verarbeitungsschritt, dem Einlernen der Regelmenge, wird die Wissensbasis erzeugt. Diese wird repräsentiert durch eine gemeinsame Wahrscheinlichkeitsverteilung (W-Verteilung) auf einem endlichen Wahrscheinlichkeitsraum. Die Regelmenge stellt ein lineares Gleichungssystem dar [4], und SPIRIT wendet zur Auswahl einer Verteilung P^* unter allen das Gleichungssystem erfüllenden gemeinsamen W-Verteilungen, der probabilistischen Extension $W(R)$, das informationstreue Prinzip der maximalen Entropie (ME) an.

$$P^* = arg(max H(P), P \in W(R)) \tag{1}$$

Dabei ist die Entropie H einer Verteilung P auf einem diskreten Merkmalsraum M das folgende Funktional.

$$H(P) = -\sum_{v \in M} P(v) \cdot ld\ P(v) \tag{2}$$

Es wird hiermit jene Verteilung bestimmt, welche die vorgegebenen Regelwahrscheinlichkeiten respektiert, sonst jedoch so wenig wie möglich von der

Gleichverteilung abweicht und somit keine nicht intendierten Abhängigkeiten anlegt. Die hochdimensionale nichtlineare Optimierungsaufgabe (1) wird in SPIRIT übrigens mit dem Iterative Proportional Scaling Verfahren (IPS) gelöst, unter Ausnutzung geschickter Überdeckungen der Variablenmenge durch sogenannte Local Event Groups (LEGs) und damit über die Zerlegung in kleinerdimensionale Teilprobleme. Die Konvergenz von IPS ist nachgewiesen [1], und zur Lösung des nichttrivialen Problems der LEG-Bildung existieren verschiedene Heuristiken. Hierzu und zur weitergehenden mathematischen Beschreibung des Einlernprozesses sei aus Platzgründen verwiesen auf [4]. Natürlich birgt die Vorgabe von Regelwahrscheinlichkeiten die Gefahr von Inkonsistenzen, also unvereinbarer Vorgaben für die Gesamtverteilung (das lineare Gleichungssystem der Regelmenge R ist nicht lösbar. Möglichkeiten zur Behebung derartiger Inkonsistenzen werden ausgiebig in [6] diskutiert. Das folgende Beispiel besitzt eine konsistente Regelmenge.

3 Die Marketing-Mix-Aufgabe als SPIRIT-Modell

Das klassische diskrete Entscheidungmodell unter Risiko ist ein gerichtetes Kausalmodell, und die allgemeinste Form der graphischen Abbildung ist das Influenzdiagramm. Prinzipiell lässt sich das gerichtete Modell in der SPIRIT-Shell abbilden [5]. Der Preis dieses Zugangs ist allerdings die unabänderbare Festschreibung der Wirkungsrichtung. SPIRIT nutzt hingegen die Darstellung in einem Markovnetz-ähnlichem Unabhängigkeitsgraphen, welcher bedingten Abhängigkeiten korrespondierende Wirkungen in allen Richtungen zulässt und damit völlig offen ist für die Richtung späterer Anfragen. In vielen Anwendungsfällen sind die Abhängigkeiten der Zustandsvariablen nicht einmal stochastisch bekannt. Um überhaupt eine Entscheidungshilfe geben zu können, *muß* dann eine Verteilung aus der Menge der möglichen gewählt werden. Immerhin geschieht dies durch Anwendung des ME-Prinzips informationstreu.
Das nun auszugsweise vorgestellte SPIRIT-Modell „MARKETING.spi",welches in der Version 2.2 der Shell realisiert wurde, verwendet die folgenden Nominalvariablen.

- Marktvariable: Wettbewerberzahl (`ANZANB`), Lebenszyklus (`MARKTLZY`)
- Kundenvariable: `KAUFKRFT`, `ALTERGRP`, `VERWEND`
- Produktvariable: `OBJQUAL`, `GUTKLASS`, `PRODLZY`
- Instrumentenvariable: `GRATPROB`, `AKTIONEN`, `SONPREIS`, `MASSENWERB`, `ZIELWERB`, `PREIS`
- Zielvariable: `ZUFRIED`, `ERSTKFNG`, `WIEDKFNG`, `ABSATZ`, `IMAGE`, `BEKNTGRD`

Liefert etwa eine Marktstudie eine relative Häufigkeit einer geringen Wiederkaufneigung eines Alltagsgutes in der Käuferschicht mit geringer Kaufkraft von 80%, so lässt sich die korrespondierende Regel

$$P(\mathtt{WIEDKFNG} = \mathit{gering} \mid \mathtt{KAUFKRFT} = \mathit{gering} \wedge \mathtt{VERWEND} = \mathit{alltag}) = 0,8 \quad (3)$$

aufstellen. Das SPIRIT-Modell im vorliegenden Beispiel besteht aus insgesamt 143 derartiger Regeln, einen Ausschnitt zeigt Abbildung 1.

Abbildung 2 repräsentiert die durch das Einlernen innerhalb der Shell erzeugte Wissensbasis in Form sämtlicher einzelner Randverteilungen, jeweils als Knoten dargestellt. Eine zwei Knoten verbindende Kante bedeutet, dass

✓	Nr.	P	Regel
✓	22	0,30	KAUFKRFT=gering
✓	23	0,50	KAUFKRFT=normal
✓	24	0,20	KAUFKRFT=hoch
✓	25	0,70	KAUFKRFT=gering ⇒ ERSTKFNG=gering
✓	26	0,70	KAUFKRFT=normal ⇒ ERSTKFNG=normal
✓	27	0,70	KAUFKRFT=hoch ⇒ ERSTKFNG=hoch
✓	28	0,80	KAUFKRFT=gering ∧ VERWEND=alltag ⇒ WIEDKFNG=gering
✓	29	0,70	KAUFKRFT=normal ∧ VERWEND=alltag ⇒ WIEDKFNG=normal
✓	30	0,60	KAUFKRFT=hoch ∧ VERWEND=alltag ⇒ WIEDKFNG=normal
✓	31	0,70	KAUFKRFT=normal ∧ VERWEND=freizeit ⇒ WIEDKFNG=normal
✓	32	0,70	KAUFKRFT=hoch ∧ VERWEND=freizeit ⇒ WIEDKFNG=hoch
✓	33	0,65	GUTKLASS=convenience ∧ PREIS=gering ⇒ IMAGE=gering
✓	34	0,60	GUTKLASS=convenience ∧ PREIS=normal ⇒ IMAGE=normal

Abbildung1. Ausschnitt des SPIRIT-Regelfensters

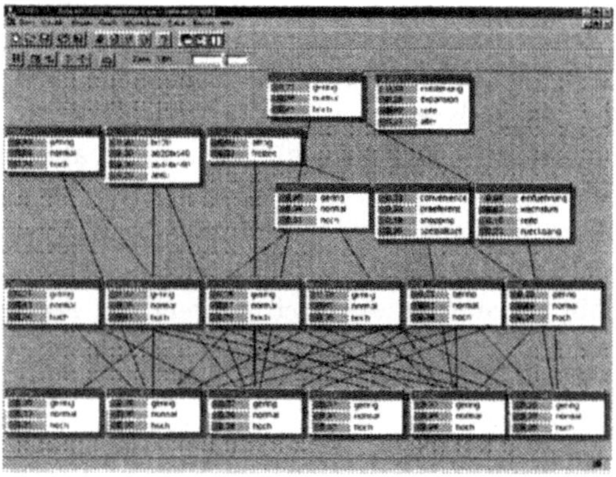

Abbildung2. Inferenzgraph der Wissensbasis „MARKETING.spi“

die entsprechenden Variablen gemeinsam in einer Regel vorkommen, hier also eine kausale Abhängigkeit angegeben wurde. Weiterhin wurden die Knoten derart angeordnet, dass sie zeilenweise von oben nach unten den Markt-, Kunden-, Produkt- sowie Instrumenten- und Zielvariablen korrespondieren. Die Darstellungsweise trägt die Bezeichnung Inferenzgraph, da hier in bequemer Weise die Wissensbasis zum Inferieren herangezogen werden kann. Hierunter versteht man das Hineintragen evidenten, situativen Wissens, etwa die Beobachtung der tatsächlichen Ausprägung einzelner Merkmale, und die Bestimmung der diese Evidenzen berücksichtigenden bedingten gemeinsa-

men Wahrscheinlichkeitverteilung. Die temporäre Fixierung einer Merkmalsausprägung durch Setzen ihrer Wahrscheinlichkeit auf 1 bezeichnet man auch als Evidentiierung. In diesem Sinne stellt eine Wissensbasis ein allgemeines Modell für Expertenwissen dar und die Evidenziierung die Anwendung dieses Expertenwissens auf einen konkreten Fall. So werden weiter unten zwei von der Anwendung her grundverschiedene, aber von der inneren Abhängigkeitsstruktur her sehr ähnliche MM-Szenarien in derselben Wissensbasis, jedoch mittels individueller Evidentiierungen analysiert.
Für die verfolgten Effizienzbetrachtungen eines Marketing-Mix sind nun Nutzenerwartungen der Zielvariablen interessant, etwa zur Vorbereitung einer Entscheidung durch Anwendung des Bernoulli-Prinzips. In SPIRIT lässt sich jeder Ausprägung ausgewählter Merkmale ein Nutzenwert zuordnen, und im Inferenzgraph lässt sich dann für solche Variable der erwartete Nutzen anzeigen. Im vorliegenden Beispiel werden für die Zielvariablen Wertnutzen von 0, 50 und 100 für die Ausprägungen „gering" bzw. „normal" bzw. „hoch" angesetzt.

4 Szenarioanalyse in SPIRIT

Abschließend soll die vorliegende Wissensbasis zur Wirkungsprognose zweier Szenarien herangezogen werden. Angewandt wird hier jeweils die einfachste Form der Wissensauswertung, nämlich die oben genannte Evidentiierung. Im Rahmen des hier bearbeiteten Themas ist diese Einschränkung dadurch gerechtfertigt, dass situatives Markt- sowie Produktwissen im Szenario als sicher angenommen wird, ebenso die situative Gültigkeit einer beliebigen, aber festen Handlungsalternative. Die Behandlung nicht sicheren situativen Wissens gestaltet sich in SPIRIT problemlos über eine sogenannte Anfrage, bestehend aus momentan als gültig angenommenen weiteren Fakten und Regeln. Die Systemantwort erfolgt in Form einer gemeinsamen Verteilung P^{**}, welche diese Fakten und Regeln beachtet und gleichzeitig minimalen informationstheoretischen Abstand zu P^{*}, gemessen mittels relativer Entropie $R(P^{**}, P^{*})$, besitzt. Allerdings bedürfte dieser Ansatz zunächst einer tiefgehenden entscheidungstheoretischen Interpretation, welche den Rahmen dieser Arbeit sprengen würde.

Szenario mit Synergieeffekt
Betrachtet werde auf dem reifen Markt mit vielen Anbietern für Urlaubsreisen das Produkt „Urlaubsreisen für Senioren", ein Shopping-Gut in der Wachstumsphase; angeboten werden soll eine überdurchschnittliche Qualitat. Das betrachtete Kundensegment sind Menschen ab 60 Jahren mit relativ hoher Kaufkraft. Die folgende Abbildung 3 zeigt die entsprechenden sowie zusätzliche Evidentiierungen für einen Marketing-Mix bestehend aus geringem Preis und intensiver Zielgruppenwerbung. Außerdem werden für die Ergebnisvariablen die Nutzenerwartungen eingeblendet. Zu maximieren sei die

Nutzenerwartung von ABSATZ, es liegt also ein eindimensionales Zielsystem vor. Für die Zielvariable ABSATZ wird ein relativ hoher Nutzen von 89,44 aus-

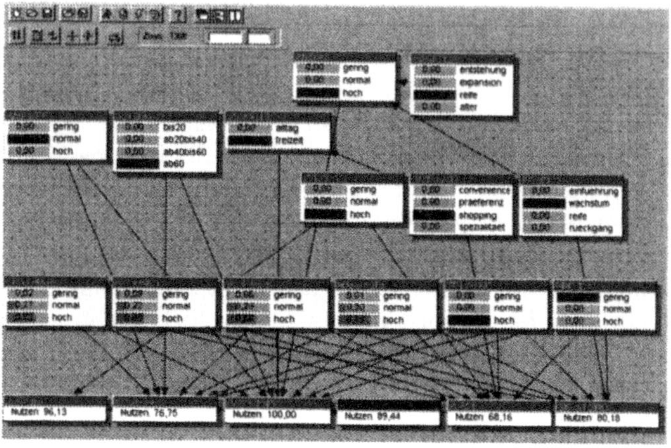

Abbildung3. Absatznutzenerwartung nach Evidentiierung

gegeben. Die folgende Tabelle gibt eine Übersicht über die prognostizierten Absatzwirkungen einiger möglicher Marketing-Mixe.

Interessanterweise ist demnach ein starker Synergieeffekt der eingesetzten

Evidentiierungen	Nutzen von ABSATZ	rel. Änd.
(keine)	70,56	
PREIS=gering	74,79	+6%
MASSENWERB=hoch	71,16	+0,8%
ZIELWERB=hoch	72,41	+2,6%
PREIS=gering ∧ MASSENWERB=hoch	82,13	+16,4%
PREIS=gering ∧ ZIELWERB=hoch	89,44	+26,7%

Tabelle1. prognostizierte Absatzwirkungen verschiedener Marketing-Mixe

Instrumente des letzten Mixes zu beobachten, der aus den isolierten Wirkungsprognosen der beiden beteiligten Instrumente so nicht zu erwarten war.

Szenario mit Kompensationseffekt

Der Markt für PKW der Oberklasse lässt sich beschreiben als reifer Markt mit durchschnittlichem Wettbewerb sowie einem reifem Shopping-Gut hoher Qualität, vgl. Evidentiierungen gemäß Tabelle 2. Ein Anbieter hier möchte sein zweidimensionales Zielsystem ABSATZ und IMAGE optimieren. Tabelle 3 zeigt die Nutzenwerte der sechs Aktionsvariablen des Modells bei unterschiedlichen Evidentiierungen der beiden betrachteten Zielvariablen. Die jeweiligen

Variable	Evidenz
ANZANB	normal
MARKTLZY	reife
GUTKLASS	shopping
PRODLZY	reife
OBJQUALI	hoch

Tabelle2. Evidentiierungen situativer Produkt- und Marktvariabler

Instanziierung	(keine)	ABSATZ=hoch		IMAGE=hoch		ABSATZ=hoch und IMAGE=hoch	
	Nutzen	Nutzen	rel.Ä	Nutzen	rel.Ä	Nutzen	rel.Ä
GRATPROB	73,12	84,56	+16%	76,91	+5%	86,18	+17%
AKTIONEN	81,94	84,60	+3%	92,18	+12%	92,29	+13%
SONPREIS	53,10	64,33	+21%	43,71	-18%	54,57	+3%
MASSWERB	83,01	91,79	+11%	90,30	+9%	95,42	+15%
ZIELWERB	80,28	87,87	+9%	91,60	+14%	94,87	+18%
PREIS	51,93	38,77	-25%	64,61	+24%	50,75	-2%

Tabelle3. Aktionsnutzenwirkung der Zielevidentiierungen

relativen Änderungen gegenüber dem Fall ohne Zielvorgabe sind ebenfalls angegeben. Dieser Fall soll als „Durchschnittswert für Marketingmaßnahmen" verstanden werden, so dass die von den Zielevidentiierungen abhängenden relativen Nutzenänderungen der Aktionsvariablen die Aktion im engeren Sinne beschreiben. Betrachtet man nun zunächst die eindimensionale Zielvorgabe ABSATZ=hoch, so erkennt man, dass für diesen Produkttyp PREIS und SONPREIS die am stärksten anzupassenden Marketing-Instrumente sind. Fordert man dagegen ausschließlich IMAGE=hoch, ergeben sich aus Tabelle 3 gerade die umgekehrten Anforderungen, welche im Rahmen der im Modell formulierten Regeln intuitiv verständlich sind: Mit dem Preis eines Produkts steigt auch dessen Image. Häufige Sonderpreis-Aktionen hinterlassen umgekehrt einen image-senkenden Eindruck. Die Teilaktivitäten sind also nicht unabhängig voneinander. Bei jeweils eindimensionaler Zielvorgabe erweisen sich also PREIS und SONPREIS als Instrumente mit gegenläufigen Ausstrahlungseffekten auf ABSATZ und IMAGE. Das scheinbare Dilemma der gleichzeitigen Verfolgung entgegengesetzter Teilziele mit entgegengesetzt wirkenden, voneinander abhängigen Teilaktivitäten löst SPIRIT auf elegante Weise: Nun wird vorgeschlagen, die beiden ambivalenten Instrumente praktisch nicht zu verändern, dafür aber alle anderen Marketing-Instrumente ungefähr in gleichem Maße zu steigern.

5 Fazit und Ausblick

Fazit

Die Abbildung der Marketing-Mix-Aufgabe in einer SPIRIT-Wissensbasis lässt die Verwertung unvollständiger und unsicherheitsbehafteter Information zu. Gemäß dem Prinzip der maximalen Entropie wird informationstreu künstlich eine Risikosituation geschaffen. Durch Evidentiierung, also Fixieren situativen Wissens über Produkt- und Marktgegebenheiten sowie gewählter Handlungsalternativen oder gewünschter Zielausprägungen, wird jeweils ein Marketingszenario eingestellt und es werden Wirkungsprognosen bzw. Instrumenteneinsätze durch SPIRIT ausgegeben. Hierdurch lassen sich insbesondere Kompensations- sowie Synergieeffekte erkennen.

Ausblick
Das SPIRIT-Modell „Marketing" sollte im nächsten Schritt um eine Optimierungsroutine ergänzt werden, um ein wirkliches Entscheidungsunterstützungssystem darzustellen. Weiter lässt ein Marketing-Mix Modell in der Art des hier vorgestellten wohl auch modelltheoretische Aussagen zu. Die hier noch ausstehende Aufgabe der Zielbeitragsaggregation der einzelnen Ergebnisvariablen wird oftmals mittels hierarchischer Strukturierung und gewichteter Summierung gelöst (analytic hierarchical process AHP, Nutzwertanalyse). Die Erfüllung der hierzu notwendigen Voraussetzung, nämlich der Unabhängigkeit der einzelnen Zielvariablen, lässt sich in SPIRIT anhand sogenannter Kantengewichte im Inferenzgraphen ablesen und gibt einen wichtigen Hinweis auf die rationale Zielhierarchiebildung.

Literatur

1. CSISZÁR, I. (1975) I-Divergence Geometry of Probability Distributions and Minimization Problems, in: The Annals of Probability, Vol. 3, No. 1, S. 146-158.
2. ELIASHBERG, J., LILIEN, G.L. (1993) Mathematical Marketing Models: Some Historical Perspectives and Future Projektions, in: ELIASHBERG, J., LILIEN, G.L. (Hrsg.): Handbooks in Operations Research and Management Science, Vol. 5 Marketing, Elsevier Science, Amsterdam.
3. KOTLER, P., BLIEMEL, F. (1999) Marketing-Management, 9. Aufl., Schäffer-Poeschel, Stuttgart (1999).
4. MEYER, C.-H. (1998) Korrektes Schließen bei unvollständiger Information, Peter Lang, Frankfurt a.M.
5. REUCHER, E., RÖDDER, W. (2001) Modellierung von Entscheidungsproblemen unter Verwendung probabilistischer Konditionale, in FLEISCHMANN, B. ET.AL.: Operations Research Proceedings 2000, Springer, S. 254-259.
6. RÖDDER, W. (2000) Conditional Logic and the Principle of Entropy, Artificial Intelligence, 117, S. 83-106.
7. RÖDDER, W., XU, L. (2001) Behebung von Inkonsistenzen in der probabilistischen Expertensystem-Shell SPIRIT, in FLEISCHMANN, B. ET.AL.: Operations Research Proceedings 2000, Springer, S. 260-265.
8. STONE, R.W., GOOD, D.J. (2001) The assimilation of computer-aided marketing activities, Information & management 38, S. 437-447.

Internet: http://www.xspirit.de

Investitionsentscheidungen bei mehrfachen Zielsetzungen und künstliche Ameisen

Karl Doerner[1], Walter J. Gutjahr[2], Richard F. Hartl[1], Christine Strauss[1], Christian Stummer[1]

[1] Institut für Betriebswirtschaftslehre, Universität Wien
{karl.doerner,richard.hartl,christine.strauss,christian.stummer}@univie.ac.at
[2] Institut für Statistik und Decision Support Systems, Universität Wien
walter.gutjahr@univie.ac.at

Zusammenfassung Die Auswahl des attraktivsten Portfolios von Investitionsprojekten zählt zu den kritischen Managementaufgaben. Angesichts mehrfacher Zielsetzungen und komplexer Projektabhängigkeiten bietet sich dazu ein zweistufiges Vorgehen an, das zunächst effiziente Portfolios identifiziert und den Entscheidungsträger anschließend bei der Suche in diesem Lösungsraum unterstützt. Bei einer großen Zahl an Vorschlägen können die möglichen Projektkombinationen aber nicht mehr in akzeptabler Zeit vollständig enumeriert werden. Adaptierte Meta-Heuristiken bieten hier einen Kompromiß zwischen dem Wunsch nach exakter Bestimmung aller Pareto-optimalen Investitionsprogramme und dem dazu nötigen Rechenaufwand. Dieser Beitrag beschreibt den entsprechenden Einsatz künstlicher Ameisen und diskutiert erste numerische Ergebnisse.

1 Einführung

Mit der Auswahl eines Projektportfolios aus einer Menge von Vorschlägen wird das mittel- und langfristigen Investitionsprogramm eines Unternehmens festgelegt. In Anbetracht der verplanten umfangreichen Ressourcen – dies gilt im besonderen etwa für die Dotierung von Forschungs- und Entwicklungs- (F&E-) Vorhaben – bzw. der drohenden Opportunitätskosten durch gegebenenfalls nicht realisierte Marktchancen kommt dieser Entscheidung hohe praktische Relevanz zu.

Eine zentrale Rolle spielen mehrfache Zielsetzungen wie beispielsweise hoher Cash-Flow, Technologieführerschaft oder Marktdurchdringung. Die dazu häufig vorgeschlagenen Verfahren der „Amerikanischen Schule“ (vgl. [13]) benötigen jedoch eine Nutzenfunktion, welche die Präferenzen des Entscheidungsträgers mit hinreichender Genauigkeit nachbildet und es somit erlaubt, die unterschiedlichen Erfolgsgrößen zu einem aussagekräftigen Gesamtnutzenwert zu aggregieren. Für die Mehrzahl realer Problemstellungen wäre eine solche Funktion allerdings wegen der nichtlinearen Zusammenhänge – etwa bei abnehmendem Grenznutzen – und der vielfältigen Abhängigkeiten zwischen einzelnen Projekten überaus komplex und kann dem Entscheidungsträger nicht abverlangt werden. Eine alternative Vorgehensweise könnte daher auf diese a priori Informationen verzichten und stattdessen in einer ersten

Phase „lediglich“ die effizienten Portfolios ermitteln, damit darin anschließend (interaktiv) nach dem attraktivsten Kandidaten gesucht werden kann.

Die Bestimmung eines solchen Lösungsraums ist jedoch keineswegs trivial. Eine vollständige Enumeration aller Portfolios ist nur für überschaubare Probleme in akzeptabler Zeit möglich, da deren Zahl mit jedem weiteren Projekt exponentiell zunimmt. Heuristische Ansätze bieten hier einen Kompromiß zwischen dem Streben nach einem umfassenden Lösungsraum und dem zu seiner Ermittlung nötigen Rechenaufwand. Beispiele für entsprechende Adaptionen „klassischer“ Meta-Heuristiken finden sich bereits für Genetische Algorithmen, Simulated Annealing sowie Tabu Search (für einen Überblick vgl. [9]). Ein neuerer Ansatz basiert auf der Fähigkeit von Ameisenkolonien, bei der Futtersuche den kürzesten Weg ausfindig zu machen. Für die Suche nach effizienten Portfolios von Investitionsprojekten ergeben sich vor allem angesichts der mehrfachen Zielsetzungen interessante Herausforderungen, die erst durch ein in wesentlichen Teilen neues Konzept für künstliche Ameisen erfolgreich zu bewältigen sind.

Das zweite Kapitel widmet sich zunächst der Modellierung der Investitionsentscheidung. In Kapitel drei wird dann der Einsatz künstlicher Ameisen zur Bestimmung effizienter Projektportfolios diskutiert. Numerische Ergebnisse finden sich schließlich im vierten Kapitel.

2 Zur Modellierung der Investitionsentscheidung

Die Komplexität der Investitionsentscheidung wird insbesondere durch die Anzahl der Ziele mitbestimmt und ist dementsprechend hoch, wenn auch die Entwicklung wesentlicher Kennzahlen im Zeitablauf von Interesse ist (vgl. [11]). Die getrennte Abbildung je Nutzenkategorie (z. B. Cash Flow, Umsatz oder Produktstückzahlen) und Planungsperiode (z. B. Geschäftsjahre) führt dann zum Beispiel bei drei Kategorien und fünf Perioden bereits zu fünfzehn Zielgrößen. Für jede davon können die Beiträge der in dem betreffenden Portfolio enthaltenen Investitionsprojekte addiert und schließlich noch um Effekte durch Projektabhängigkeiten – etwa aufgrund von Synergien – bereinigt werden.

Zulässige Projektkombinationen müssen alle Ressourcenbeschränkungen einhalten sowie gegebenenfalls zusätzliche (strategische) Managementvorgaben erfüllen. Letztere können ausgewählte Projektgruppen positiv oder negativ diskriminieren, weil sie etwa einen wesentlichen Beitrag zur Entwicklung einer zukunftsträchtigen Technologie leisten oder sich mit Produkten für einen Markt beschäftigen, aus dem mittelfristig der Ausstieg angestrebt wird. Unter den verbliebenen Portfolios sind schließlich jene (effizienten, Paretooptimalen) Kandidaten zu ermitteln, für die keine Alternative existiert, welche in allen Zielen zumindest gleich gute und in mindestens einem Ziel bessere Werte aufweist. Für ein detailliertes betriebswirtschaftliches Modell vgl. [12].

3 Zum Einsatz künstlicher Ameisen

Das Anfang der 1990er-Jahre entwickelte Ant System (vgl. [4]) basiert auf einem in der Natur beobachtbaren Phänomen: Einer Gruppe von Ameisen gelingt es – obwohl die Tiere nahezu blind sind – den kürzesten Weg zwischen Nest und Futterstelle zu finden. Jedes Tier hinterläßt dabei auf seinem Weg Pheromon, einen Wirkstoff, der andere Ameisen veranlaßt, diesem Weg zu folgen. Je mehr Ameisen einen bestimmten Weg gewählt haben, desto intensiver die Duft- (Pheromon-) Spur und desto eher wird eine weitere Ameise dieser folgen und selbst wieder zu einer Verstärkung des Pheromons beitragen. Im Zeitablauf werden daher immer mehr Tiere den kürzeren Pfad benützen. Durch diesen autokatalytischen Prozeß wird ein günstiger Weg markiert und schließlich von allen Ameisen als bester Pfad erkannt. Rein zufälliges Verhalten geht auf diese Weise in erlerntes über.

Im nachgebildeten Ant System löst eine Population künstlicher Ameisen als Gemeinschaft kooperierender Individuen eine Optimierungsaufgabe unter Verwendung zusätzlicher globaler Information über das Problem. Künstliche Ameisen unterscheiden sich von ihren natürlichen Vorbildern vor allem durch die Eigenschaft des „Sehvermögens“, mit dem sie ihre Umgebung wahrnehmen und die gewonnenen Informationen auch nutzen können. Es erlaubt den Ameisen, sich bei der Suche zusätzlich auch „greedy“ zu verhalten, und beschleunigt damit den Lernvorgang. Aus dem frühen Ant System entwickelte sich als spezielle Form die „Ant Colony Optimization (ACO)“ (vgl. [8]), die erfolgreich auf Reihenfolgenprobleme (TSP, VRP, SMTTP, etc.; vgl. [7,2,1]) angewendet wurde. Zur Lösung des vorliegenden Portfolioauswahlproblems werden im folgenden ein einfacher heuristischer sowie ein ACO-Ansatz beschrieben.

3.1 Einfache Heuristik

Die einfache Heuristik verzichtet im Gegensatz zur ACO vor allem auf die Auswertung von Lernerfahrungen. Ausgehend von einem anfangs leeren Portfolio werden in jedem Durchlauf so lange Projekte aufgenommen, bis eine Nebenbedingung verletzt wird. Die Wahrscheinlichkeit, daß ein Projekt als nächstes gewählt wird, wird durch einen Attraktivitätswert bestimmt, der widerspiegelt, wie gut es zum aktuellen „Rumpf-Portfolio“ paßt. Er beträgt Null, sofern es zu einer Verletzung einer Ressourcenbeschränkung oder einer Managementvorgabe – derzufolge etwa nicht mehr als eine Maximalzahl an Projekten eines bestimmten Typs in dem Portfolio enthalten sein soll – kommen würde. Andernfalls entspricht er dem Beitrag des Projekts zur Erfüllung von „offenen“ Anforderungen, die beispielsweise eine Mindestanzahl an Vertretern aus einer gegebenen Projektgruppe verlangen. Im günstigsten Fall erfüllt die Hinzunahme alle verbliebenen Vorgaben, der Attraktivitätswert beträgt dann eins.

Nach der „Konstruktion“ des Portfolios wird es mit Hilfe eines Quad-Baums (vgl. [10]) mit anderen bereits identifizierten (potentiell) effizienten Kandidaten verglichen und dabei entweder als dominiert verworfen oder in diese Menge aufgenommen. Nach einer vorgegebenen Zahl an Durchläufen bricht die Heuristik schließlich ab. Die zu diesem Zeitpunkt im Quad-Baum enthaltenen Portfolios bestimmen dann den Lösungsraum.

3.2 Ant Colony Optimization

In der Initialisierungsphase des modifizierten ACO-Algorithmus (vgl. auch den Pseudo-Code in Abbildung 1) werden zunächst Γ künstliche Ameisen mit leeren (Rumpf-)Portfolios $\Psi := \{0, ..., 0\}$ und zufällig bestimmten Lebensdauern Ξ bzw. Zielgewichtungen χ generiert.

Abbildung1. ACO-Algorithmus für Mehrzieloptimierung

```
procedure ACO (...) {
  Initialisierung von ACO;
  for v := 1 to max_iterations {
    for Ant := 1 to Γ {
      bestimme zufällig die Lebensdauer Ξ der Ameise
      aus dem Intervall [1..N];
      /* Die (maximale) Anzahl an Projekten, die eine Ameise wählt,
      entspricht ihrer Lebensdauer Ξ. */
      initialisiere Ψ:={0,...,0};
      bestimme zufällig die Zielgewichtung χ_k für jedes Ziel k;
      ξ := Ξ;
      while ξ > 0 and ∃ η_i(Ψ) > 0 {
        wähle ein Projekt unter Verwendung der Formel (1),
        füge es zu dem Rumpf-Portfolio Ψ;
        verringere Pheromoninformation lokal;
        dekrementiere ξ;
      }
      prüfe Zulässigkeit des Portfolios Ψ;
      if Portfolio Ψ ist zulässig {
        überprüfe die potentielle Effizienz des Portfolios Ψ;
        if Portfolio Ψ ist potentiell effizient {
          speichere Portfolio Ψ;
    } } }
    for each Ziel k {
      bestimme das beste und das zweitbeste Portfolio Ψ_k ;
      verstärke die Pheromoninformation global;
} } }
```

Anschließend versucht jede künstliche Ameise, unter Verwendung von Attraktivitätswerten $\eta_i(\Psi)$ für die Projektkandidaten i ein zulässiges Portfolio Ψ zu „konstruieren". Nach der Überprüfung auf Zulässigkeit sowie Effizienz – letzteres wieder unter Einsatz eines Quad-Baums – werden die Pheromonwerte global aktualisiert, wobei für jedes einzelne Ziel das beste Portfolio der aktuellen Iteration zur Gänze und die zweitbeste Lösung zur Hälfte zum Pheromonupdate beiträgt. Abweichend von der klassischen ACO, in der nur eine Pheromonmatrix verwendet wird, werden hier für alle K Ziele Pheromonmatrizen τ (jeweils mit Zeilen bzw. Spalten für alle N Projekte) eingesetzt. Die Pheromoninformation τ_{ij}^k gibt nun in bezug auf Ziel k an, wie sehr sich bisher eine Kombination der Projekte i und j bewährt hat.

Sowohl die einfache Heuristik als auch die Pheromoninformation werden dann berücksichtigt, um aus der Menge der verbleibenden Projektkandidaten $\Omega = \{i \in N : \eta_i(\Psi) > 0\}$ einen weiteren zum Rumpf-Portfolio hinzuzufügen. Wenn q_0 die vorgegebene Wahrscheinlichkeit der „max-Selektion" und q eine in $[0,1]$ gleichverteilte Zufallszahl bezeichnet, lautet die entsprechende Entscheidungsregel

$$i = \begin{cases} arg\ max \left\{ \left[\sum_{k=1}^{K} \left(\chi_k \cdot \sum_{j \in \Psi} \tau_{ij}^k \right) \right]^\alpha \cdot [\eta_i(\Psi)]^\beta \right\} & \text{if} \quad q \leq q_0 \\ I & sonst, \end{cases} \tag{1}$$

wobei I eine Zufallszahl ist, die nach der diskreten Verteilung

$$\mathcal{P}_i(\Psi) = \begin{cases} \frac{\left[\sum_{k=1}^{K} \left(\chi_k \cdot \sum_{j \in \Psi} \tau_{ij}^k \right) \right]^\alpha \cdot [\eta_i(\Psi)]^\beta}{\sum_{h \in \Omega} \left(\left[\sum_{k=1}^{K} \left(\chi_k \cdot \sum_{j \in \Psi} \tau_{hj}^k \right) \right]^\alpha \cdot [\eta_h(\Psi)]^\beta \right)} & \text{if} \quad i \in \Omega \\ 0 & sonst, \end{cases} \tag{2}$$

gezogen wird.

Sobald eine künstliche Ameise ein Projekt in ihr Portfolio aufgenommen hat, werden die betreffenden Pheromonwerte zunächst aktualisiert. Dabei verringern sich für ein Projekt i alle Werte τ_{ij}^k in Abhängigkeit von den bereits im Rumpf-Portfolio enthaltenen Projekten j für jedes Ziel k, um identische Portfolios zu vermeiden. Nachdem dann die gesamte Ameisenpopulation Lösungen konstruiert hat, verstärkt die ACO-Prozedur schließlich die Pheromonwerte für die besten (vgl. [3]).

4 Numerische Ergebnisse

Das numerische Beispiel basiert auf einer komplexen Entscheidungssituation mit zwanzig Projekten, drei Nutzenkategorien, fünf Planungsperioden und somit fünfzehn Zielen. Die Projekte unterscheiden sich erheblich sowohl in ihren potentiellen Erfolgsgrößen wie auch im jeweiligen Ressourcenbedarf, so daß manche in einzelnen Planungsperioden und/oder Nutzenkategorien besonders hohe und andere weitgehend durchschnittliche Werte aufweisen. Neben Ressourcenbeschränkungen stellen zehn weitere Managementvorgaben sicher, daß z. B. höchstens eines von drei Projekten mit ähnlichem Auftrag oder mindestens zwei Projekte, die zu einer erwünschten Diversifizierung beitragen, in jedem zulässigen Portfolio enthalten sind. Vier weitere Projektabhängigkeiten bilden Synergismen bzw. Kannibalismen ab.

Einige Parametereinstellungen für den ACO-Algorithmus wurden aus erfolgreichen Anwendungen (z. B. [7]: $\alpha = 1$, $\beta = 1$, $\Gamma = 10$) übernommen, andere an die spezifische Aufgabenstellung angepaßt (wie etwa bei der Reduktion von $q_0 = 0,9$ auf $q_0 = 0,4$ zur Förderung von heterogenen Portfolios). Der damit – gemittelt über jeweils zehn Läufe – nach 100, 200, 300, 400 bzw. 500 Iterationen gefundene Anteil an den tatsächlich effizienten Investitionsprogrammen (vgl. Tabelle 1) spiegelt die Entwicklung der Lösungsgüte wider und läßt sich insbesondere auch mit den Ergebnissen bei Einsatz einer Monte Carlo-Simulation bzw. der einfachen Heuristik vergleichen. In umfangreichen Praxisproblemen wird die Identifikation der effizienten Portfolios allerdings zunehmend schwieriger, so daß dann auch die Bereitstellung von hinreichend guten Approximationen hilfreich sein kann (vgl. [5]). Die Tabellenwerte in Klammern beziehen sich dazu auf als effizient ausgewiesene, jedoch – bei Kenntnis des „wahren" Lösungsraums – eigentlich dominierte Portfolios: Beispielsweise wurden von der ACO nach 100 Iterationen durchschnittlich 91,6 Projektportfolios entdeckt; 89,4 finden sich auch unter den 138 mit vollständiger Enumeration ermittelten tatsächlich effizienten Kandidaten (das entspricht den in der Tabelle angegebenen 65%), während 48,6 überhaupt nicht gefunden und nicht zuletzt deshalb im Durchschnitt 2,2 (2% von 138) irrtümlich als Pareto-optimal ausgewiesen werden.

Tabelle1. Numerische Ergebnisse

	100 It.	200 It.	300 It.	400 It.	500 It.
Monte Carlo:	0,00 (0,01)	0,00 (0,03)	0,00 (0,04)	0,00 (0,05)	0,08 (0,05)
Heuristik:	0,55 (0,03)	0,71 (0,02)	0,81 (0,02)	0,85 (0,01)	0,90 (0,00)
ACO:	0,65 (0,02)	0,81 (0,01)	0,88 (0,00)	0,91 (0,00)	0,92 (0,00)

Während das heuristische Verfahren nach 200 Iterationen durchschnittlich nur 71% (d. h. 97,9 der 138 Portfolios) findet, entdeckt der ACO-Algorithmus bereits 81% (112,0 Portfolios). Für die Qualität der ACO-Prozedur spricht auch der relativ kleine Bereich des evaluierten Suchraums, der in unserem Beispiel $2^{20} = 1,048.576$ mögliche Portfolios umfaßt: Die jeweils zehn künstlichen Ameisen haben nämlich selbst nach 500 Iterationen erst weniger als ein halbes Prozent davon untersucht und damit aber im Durchschnitt 92% (126,7

Portfolios) der effizienten Kandidaten entdeckt. Das kann zumindest als ein erstes Indiz für die Eignung des Verfahrens auch für noch wesentlich umfangreichere Probleme gelten. Die ebenfalls getestete Monte Carlo-Simulation erweist sich hier hingegen als offensichtlich völlig unzureichend.

5 Resümee

Die Bestimmung des – unter Berücksichtigung mehrfacher Zielsetzungen – „besten" Investitionsprogramms ist von hoher praktischer Bedeutung. Da häufig ex ante keine geeignete Nutzenfunktion zur Verfügung steht, bietet sich ein zweistufiges Vorgehen an, bei dem zunächst ein Lösungsraum von effizienten Portfolios ermittelt wird. Bei einer größeren Zahl an Projektvorschlägen ist das aber nur mehr durch den Einsatz von Meta-Heuristiken leistbar.

Der hier gewählte Ansatz mit künstlichen Ameisen wurde für diese Aufgabenstellung gegenüber dem ACO-Standardalgorithmus grundlegend modifiziert: (1) Während in der klassischen ACO zur Verwaltung der Pheromonspuren zweidimensionale Pheromonmatrizen ausreichen, wird nunmehr die Spurinformation für jedes einzelne Ziel in einer eigenen Pheromonmatrix abgelegt. (2) Um möglichst viele, effiziente und heterogene Portfolios zu finden, wird außerdem in jeder Iteration für jede einzelne künstliche Ameise ein eigener Gewichtsvektor für die betrachteten Ziele zufällig generiert. (3) Statt der Vorgänger- bzw. Nachfolgebeziehungen als Basis der Pheromondekodierung, wurde schließlich der paarweise Vergleich als Dekodierungsschema gewählt.

Die bisherigen numerischen Ergebnisse belegen das Potential von künstlichen Ameisen bei der Suche nach effizienten Investitionsprogrammen. Dadurch motivierte weitere Forschungsaktivitäten beschäftigen sich in einem nächsten Schritt mit dem Einsatz von ACO bei – in bezug auf die Anzahl der zur Wahl stehenden Projektalternativen – noch wesentlich umfangreicheren Entscheidungsproblemen bzw. mit der Weiterentwicklung von Maßzahlen zur Qualitätsbeurteilung der hierbei generierten Lösungsräume.

Literatur

1. Bauer, A., Bullnheimer, B., Hartl, R.F., Strauss, C. (2000): Minimizing total tardiness on a single machine using Ant Colony Optimization. Central European Journal of Operations Research **8**, 125–141
2. Bullnheimer, B., Hartl, R.F., Strauss, C. (1999): An improved Ant System algorithm for the vehicle routing problem. Annals of Operations Research **89**, 319–328
3. Bullnheimer, B., Hartl, R.F., Strauss, C. (1999): A new rank based version of the Ant System. Central European Journal of Operations Research **7**, 25–38
4. Colorni, A., Dorigo, M., Maniezzo, V. (1992): Distributed optimization by Ant Colonies. In: Varela, F., Bourgine, P. (Hrsg.): Proc. First Europ. Conf. on Artificial Life. Elsevier, Amsterdam, 134–142.

5. Czyzak, P., Jaszkiewicz, C. (1998): Pareto Simulated Annealing. A metaheuristic technique for multiple-objective combinatorial optimization. Journal of Multi-Criteria Decision Analysis **7**, 34–47
6. Dorigo, M. (1992): Optimization, Learning and Natural Algorithms. Doctoral Dissertation, Politecnico di Milano
7. Dorigo, M., Gambardella, L.M. (1997): Ant Colony System. A cooperative learning approach to the Travelling Salesman Problem. IEEE Transactions on Evolutionary Computation **1**, 53–66
8. Dorigo, M., Di Caro, G. (1999): The Ant Colony Optimization meta-heuristic. In: Corne, D., Dorigo, M., Glover, F. (Hrsg.): New Ideas in Optimization. McGraw-Hill, London, 11–32
9. Ehrgott, M., Gandibleux, X. (2000): A survey and annotated bibliography of multiobjective combinatorial optimization. OR Spektrum **22**, 425–460
10. Habenicht, W. (1983): Quad trees. A datastructure for discrete vector optimization problems. In: Hansen, P. (Hrsg.): Essays and Surveys on Multiple Criteria Decision Making. Lecture Notes in Economics and Mathematical Systems 209. Springer, Berlin, 136–145
11. Ringuest, J.L., Graves, S.B. (1990): The linear R&D project selection problem. An alternative to net present value. IEEE Transactions on Engineering Management **37**, 143–146
12. Stummer, C. (2000): Ein interaktives Entscheidungsunterstützungssystem zur Portfolioauswahl bei mehrfachen Zielsetzungen. In: Wanka, G. (Hrsg.): Decision Theory and Optimization in Theory and Practice. Shaker, Aachen, 129-149
13. Vincke, P. (1992): Multicriteria Decision-Aid. Wiley, Chichester

Experience-Based Decision Making and Learning from Examples

Eyke Hüllermeier

University of Dortmund, Computer Science Department,
D-44221 Dortmund, eyke@ls1.cs.uni-dortmund.de

Abstract. We introduce a decision model which takes practically relevant aspects such as limited information and bounded computation time into account. This model, called *satisficing decision trees*, combines the ideas of experience-based decision making and compiled decision models. In experience-based decision making, the decision maker's information consists of experiences in the form of decision problems encountered in the past. Compiled decision models skip the estimation of a utility function and prescribe an action for each problem directly. The combination of these two ideas yields a practically useful decision model which appears to be particularly interesting for real-time decision making.

1 Introduction

The classical approach to decision making in the face of uncertainty relies on the paradigm of *expected utility theory* (EUT) [9,13]. This axiomatic framework has a great theoretical appeal but appears rather restrictive from a practical point of view. Particularly, it assumes the decision maker to have very detailed information at its disposal: a list of the *states of nature*, a probability distribution over these states, a list of potential acts, and a numerical utility value for all act–state pairs. Apart from that, some well-known paradoxes [1,4] as well as psychological studies [15] show that EUT can be challenged as a *descriptive* theory of (human) decision making.

A more faithful assumption is that the information available to a (not necessarily human!) decision maker (agent) simply consists of *experiences* in the form of *cases* (p, a, u), where p is a decision problem, a is an action chosen by the agent to solve that problem, and u is the resulting utility. In fact, *reinforcement learning* is based on this kind of experience, as well as *case-based decision theory* that has recently been introduced [5]. Proceeding from the same type of information, a general framework of *experience-based decision making* (EBDM) has been proposed in [6].

Clearly, the agent's information in EBDM is limited and does not allow for deriving a decision which is provably optimal, e.g. in the sense of expected utility maximization. In this connection, the concept of "bounded rationality" turns out to be useful. The latter is closely related to SIMON's idea of "satisficing" [7,14]: Roughly speaking, a satisficing agent tries to choose actions which are satisfactory, but it does not necessarily try to maximize utility.

Another practically relevant idea in decision making is that of *compiled decision models.* A compiled decision model skips the utility analysis, which involves the computation of an expected or estimated utility for all of the potential acts. Rather, it assigns an (optimal) act to each problem directly. In [12], *compilation* is understood as a method for omitting intermediate computations in some input–output relation. Thus, a compiled model is an execution architecture computing the original mapping, but doing so in a more efficient way.

In this paper, we introduce the *satisficing decision tree* model that combines the ideas of bounded rationality and compiled decision models. Section 2 describes the setting of the problem in more detail. Section 3 gives a very brief introduction to decision tree learning. The satisficing decision tree model is introduced in Section 4. The paper concludes with a summary in Section 5.

2 Problem Setting

Let $\mathcal{P}$ and $\mathcal{A}$ be sets of problems and acts, respectively, and denote by $\mathcal{R}$ a set of results or outcomes. Choosing act $a \in \mathcal{A}$ for solving problem $p \in \mathcal{P}$ leads to the outcome $r = r(p, a) \in \mathcal{R}$. A utility function $u : \mathcal{R} \rightarrow U$ (resp. $u : \mathcal{P} \times \mathcal{A} \rightarrow U$) assigns utility values to such outcomes; the utility scale U is taken as the set of real numbers. Suppose the decision making agent to have a (finite) memory

$$\mathcal{M} = \{(p_1, a_1, r_1), \ldots, (p_n, a_n, r_n)\}$$

of cases at its disposal, where $(p_k, a_k) \in \mathcal{P} \times \mathcal{A}$, $r_k = r(p_k, a_k)$ $(1 \leq k \leq n)$, and suppose furthermore that it has to choose an act for a new problem $p_0 \in \mathcal{P}$.

The *optimal decision function* $\Delta^* : \mathcal{P} \rightarrow \mathcal{A}$ maps problems to optimal decisions, i.e. $\Delta^*(p) = \arg\max_{a \in \mathcal{A}} u(p, a)$. A compiled decision model can be seen as an approximation (estimation) of this function.

Standard (supervised) learning methods assume data in the form of examples (x, y), where $y = f(x)$ and f is the function to be learned. In the context of EBDM, examples are cases of the form (p, a, u). Such a case can be interpreted as a *valued example*, that is $(x, y) = (p, a)$ is an example and u is an evaluation thereof. As can be seen, in the context of EBDM, the compiled decision model approach necessitates an extension of standard learning methods which takes the valuation u of an example (p, a) into account. To this end, we shall fall back on the idea of "satisficing" as mentioned above: A "satisficing" decision maker discriminates between only two types of decisions, namely acceptable and non-acceptable ones. In the context of machine learning, this boils down to a simple classification problem, where acceptable and non-acceptable decisions play the role of positive and negative examples, respectively.

In subsequent sections, we propose a concrete approach to learning compiled decision models, namely a modified version of decision tree induction. Thus, the idea is to implement the decision function $\Delta : \mathcal{P} \to \mathcal{A}$, an estimation of the optimal decision function $\Delta^* : \mathcal{P} \to \mathcal{A}$, as a decision tree.[1] This approach is adequate if the set $\mathcal{A}$ of available acts is relatively small. Moreover, it assumes that problems are represented by attribute–value pairs with discrete-valued (or discretized) attributes. There are no special restrictions concerning the utility scale U.

3 Learning Decision Trees

Decision trees are one of the most successful approaches in machine learning. The basic principle underlying most decision tree learners, well-known examples of which include the ID3 algorithm [10] and its successor C4.5 [11] as well as the CART system [2], is that of partitioning the set of given examples, S, in a recursive manner. Each inner node η of the decision tree defines a partition of a subset $S_\eta \subset S$ of examples assigned to that node. This is done by classifying elements $s \in S_\eta$ according to the value of a specific attribute T. The attribute is selected according to a measure of effectiveness in classifying the examples, thereby supporting the overall objective of constructing a small tree (that is a simple model).

Several selection measures have been devised, such as e.g. a measure based on the so-called GINI function:

$$G(S,T) = I(S) - \sum_t \frac{|S_t|}{|S|} I(S_t), \tag{1}$$

where the "impurity" of an attribute T with respect to the classes $\mathcal{C}$ is defined by the GINI function [2]

$$I(S_t) = \sum_{c \neq c' \in \mathcal{C}} q_c q_{c'} = 1 - \sum_{c \in \mathcal{C}} (q_c)^2 \tag{2}$$

In (2), q_c denotes the proportion of elements $s \in S_t$ with classification c, where S_t is the set of examples whose value for attribute T is t. See [8] for an empirical comparison of splitting measures.

Suppose a set $\mathcal{X}$ of objects to be given, where each object is characterized by several attribute values. Moreover, each $x \in \mathcal{X}$ belongs to one class $c = \text{class}(x) \in \mathcal{C}$. Given a set of training samples $S = \{(x_1, c_1), \ldots, (x_n, c_n)\} \subset \mathcal{X} \times \mathcal{C}$, the basic ID3 algorithm derives a decision tree as follows:

[1] Decision tree learning is actually a classification method. Even though classifications can be considered as decisions, it is not specifically used for decision making in the proper sense. Therefore, one might prefer the alternative term discrimination tree.

- The complete set of training samples, S, is assigned to the root of the decision tree.
- A node η becomes a leaf (answer node) of the tree if all associated samples S_η belong to the same class or if all attributes have already been used along the path from the root of the tree to η.
- Otherwise, node η becomes a decision node: It is split by partitioning the associated set S_η of examples. This is done by selecting an attribute as described above and by classifying the samples $s \in S_\eta$ according to the value for that attribute. Each element of the resulting partition defines one successor node.

Once the decision tree has been constructed, new examples can be classified on the basis of the associated classification rules, namely by looking at the class label of the leaf node whose attribute values match the description of the example. According to the inductive learning hypothesis underlying machine learning, it is assumed that a model consistent with the given examples will also classify the new examples correctly. The induction of as small as possible decision trees is a version of *Occam's razor*. Roughly speaking, the latter suggests that simple models generalize better than complex models. See [16] for an incremental version of decision tree learning which is especially interesting in the context of EBDM.

4 Satisficing Decision Trees

Let $\mathcal{P} = T_1 \times T_2 \times \ldots \times T_m$ be a set of potential problems, where $T_\imath$ denotes the (finite) domain of the $\imath$th attribute. Thus, each problem $p \in \mathcal{P}$ is represented as a vector $p = (t_1, \ldots, t_m)$ of attribute values. Moreover, let $\mathcal{A} = \{\alpha_1, \ldots, \alpha_k\}$ be a set of available actions. U denotes a completely ordered utility scale. Assume a memory of cases

$$\mathcal{M} = \{(p_1, a_1, u_1), \ldots, (p_n, a_n, u_n)\} \in \mathcal{P} \times \mathcal{A} \times U$$

to be given, where $u_\imath = u(p_\imath, a_\imath)$ is the utility that has resulted from applying act $a_\imath$ to problem $p_\imath$. Let $\mathcal{M}_\mathcal{P}$, $\mathcal{M}_\mathcal{A}$, and $\mathcal{M}_{\mathcal{P}\times\mathcal{A}}$ denote the projection of $\mathcal{M}$ to $\mathcal{P}$, $\mathcal{A}$, and $\mathcal{P} \times \mathcal{A}$, respectively. The idea pursued here is to compile this case base into a decision tree which is then used for solving future decision problems.

Let $u^* \in U$ be a utility threshold defined by the decision maker (this threshold corresponds to the so-called *aspiration level* in case-based decision theory [5]): An action a is *acceptable* for a problem p if $u(p, a) \geq u^*$, and it is not acceptable if $u(p, a) < u^*$.

In a first step, each case $(p_\imath, a_\imath, u_\imath)$ is changed into an *example* $(p_\imath, a_\imath)$. The latter is called a *positive example* if $u_\imath \geq u^*$ and a *negative example* if $u_\imath < u^*$. In a second step, a *modified memory*, $\mathcal{M}^*$, is derived from $\mathcal{M}$. For each problem $p \in \mathcal{M}_{\mathcal{P}}$ it contains a *generalized example* (p, A_p), where A_p denotes the set of *feasible acts* for problem p. This set is defined as follows:

$$\forall a \in \mathcal{A} : a \in A_p \Leftrightarrow \begin{cases} u(p,a) = u_{max}(p,\mathcal{M}) & \text{if} \quad u_{max}(p,\mathcal{M}) \geq u^* \\ (p,a) \notin \mathcal{M}_{\mathcal{P}\times\mathcal{A}} & \text{if} \quad u_{max}(p,\mathcal{M}) < u^* \end{cases},$$

where

$$u_{max}(p,\mathcal{M}) \doteq \max_{(p,b)\in\mathcal{M}_{\mathcal{P}\times\mathcal{A}}} u(p,b).$$

In plain words, an action a is feasible for p if it belongs to the best among the actions known to be acceptable for p, or if no acceptable action is known and a has not been tried as yet. Notice that $A_p = \emptyset$ if all available acts have been applied to p but none of them was acceptable, that is if an acceptable act for p does actually not exist. In this case, the decision maker should reduce the utility threshold u^*.[2] Subsequently, we assume $A_p \neq \emptyset$ for all $p \in \mathcal{P}$.

We are now ready to formulate a generalized version of the decision tree learning problem: Given a set of generalized examples

$$S^* = \{(p_1, A_{p_1}), (p_2, A_{p_2}), \ldots, (p_n, A_{p_n})\} \subset \mathcal{P} \times 2^{\mathcal{A}}, \tag{3}$$

induce a decision tree which implements a decision function $\Delta : \mathcal{P} \to \mathcal{A}$ such that

$$\forall p \in \mathcal{M}_{\mathcal{P}} : \Delta(p) \in A_p.$$

Now, the splitting of a set of examples (3) is not necessary if

$$A(S^*) = \bigcap_{i=1}^{n} A_{p_i} \neq \emptyset, \tag{4}$$

hence, (4) defines a natural stopping condition. The corresponding node η in the decision tree then becomes a leaf, and any action $a \in A(S^*)$ can be chosen as the prescribed decision a_η associated with that node.

The main modification concerns the "goodness of split" measure. Let m denote a measure that can be applied for classical decision tree learning. That is, $m(T, S)$ quantifies the quality of the split of the sample S induced by the attribute T. Now let the class of *selections*, $\mathcal{F}(S^*)$, of the generalized set of examples (3) be given by the class of samples

$$\{(p_1, a_1), (p_2, a_2), \ldots, (p_n, a_n)\} \subset \mathcal{P} \times \mathcal{A}$$

such that $a_i \in A_{p_i}$ for all $1 \leq i \leq n$. We extend the measure m to generalized samples S^* as follows:

$$m(T, S^*) \doteq \max_{S\in\mathcal{F}(S^*)} m(T, S). \tag{5}$$

[2] It would also be possible to maintain individual thresholds for the problems.

As can be seen, the extended measure (5) is the ordinary measure obtained for the most favorable instantiation of the generalized examples (p, A_p) and hence defines a "potential" goodness of split. It corresponds to the "true" measure that would have been derived for the attribute T if this instantiation is compatible with the ultimate decision tree. Taking this optimistic attitude is clearly justified since the tree is indeed constructed in an (apparently) optimal manner (hence averaging would hardly make sense).

Computing (5) comes down to solving a combinatorial optimization problem and might hence be regarded as critical from a time complexity point of view, especially in the context of real-time decision making. One should keep in mind, however, that not the construction (or revision) of the decision tree is time critical but rather its application. In fact, real-time decision making must not be confused with real-time learning; as in other decision support systems, learning will rather be realized as an "off-line" procedure [3]. Besides, several heuristic approaches can be devised which allow for a quite efficient computation or approximation of (5). See Appendix A for one such approach.

We have carried out extensive experimental studies in order to test the practical performance of satisficing decision trees. Particularly, we were interested in the influence of the complexity of the decision environment and of the aspiration level on the induced decision model. The results, which are omitted here due to reasons of space, can be found in an extended version of this paper [6]. In summary, the experiments show that the satisficing decision tree model performs excellent in practical applications. Apart from that, the results confirm the following expectations: The more complex the decision environment, the larger the decision tree and the smaller the average utility over time. Moreover, higher aspiration levels entail more complex decision models and higher expected utilities. However, the more complex the decision environment is, the smaller is the influence of the aspiration level on expected utility.

5 Summary

We have considered two ideas which are of considerable relevance for practical decision making. Firstly, the idea of experience-based decision making, where the decision maker has only limited information at its disposal, namely experience in the form of cases. Secondly, the idea of compiled decision models which skip the estimation of a utility function and prescribe an action for each problem directly. Such models are computationally efficient and appear especially adequate for routine decisions.

In order to combine these ideas, we have introduced a decision model called *satisficing decision trees*. This approach draws on concepts from decision theory, bounded rationality (satisficing), and machine learning. The basic idea is to learn an efficient representation of an acceptable decision model, that is an as simple as possible model that prescribes actions which lead to satisfactory outcomes. This type of model appears to be particularly useful in real-time decision making, where deliberation time is bounded.

Currently, the decision model introduced here is extended in several directions, e.g. by methods for adapting the utility threshold u^* is an optimal way. Moreover, we develop a simulation environment that allows for comparing different approaches to experience-based decision making in a systematic way. Here, the objective is to figure out the performance of different methods for certain types of decision environments, thereby providing some guidelines that support the choice of a method. A further topic of ongoing work is the application of our decision model in a real-world decision environment.

A Computation of Extended Splitting Measures

A possible approach to computing the extended splitting measure (5) is to search the space of selections of S^* in a systematic way. This can be realized very efficiently by means of a branch & bound procedure if the basic measure m is given by (1). In fact, a good bounding function can be derived on the basis of the following result, which is proved in [6].

Proposition: Consider subsets $A_1, \ldots, A_n$ of a finite set $\{1, \ldots, N\}$. For each selection $a = (a_1, \ldots, a_n) \in A_1 \times \ldots \times A_n$ define the probability vector $p(a) = (p_1, \ldots, p_n)$ such that $p_\imath = N_\imath / N$, where $N_\imath = \text{card}\{1 \leq \jmath \leq N \mid a_\jmath = \imath\}$. Now, let $(a_1, \ldots, a_m)$ be a partial selection of the first $m < n$ values and denote by $\mathcal{C}$ the class of possible completions $a = (a_1, \ldots, a_n)$. Then the following holds ($p \cdot p$ denotes the dot product $p_1 p_1 + \ldots + p_N p_N$ of a vector p): $\max_{a \in \mathcal{C}} p(a) \cdot p(a) \leq p(a^c) \cdot p(a^c)$, where the vector a^c (which is not necessarily a feasible selection!) is defined as

$$a_\imath^c = \begin{cases} a_\imath & \text{if} \quad 1 \leq \imath \leq m \\ a_{max} & \text{if} \quad m+1 \leq \imath \leq n \end{cases}$$

and $1 \leq a_{max} \leq N$ is any number (e.g. the smallest) such that

$$\text{card}\{1 \leq \imath \leq m \mid a_\imath = a_{max}\} \geq \text{card}\{1 \leq \imath \leq m \mid a_\imath = k\}$$

for all $1 \leq k \leq N$.

This proposition suggests an effective lower bound to the GINI function (2) and, hence, an upper bound to the information gain (1). In fact, this upper bound can be used as a bounding function in a branch & bound algorithm which works as follows: Suppose a set S^* of generalized examples (3) and an attribute T to be given. In order to compute $m(T, S^*)$, a selection of the set A_{p_m} of actions is chosen at the mth level of the algorithm. That is, each node of the branch & bound tree at level m has $\text{card}(A_{p_m})$ successors, each of which corresponds to the choice of one particular act $a_m \in A_{p_m}$. At each node, the aforementioned bounding function can be derived for the associated partial selection $(a_1, \ldots, a_m)$. A branch (partial selection) is continued only if the value of the bounding function is above the current best solution (information gain). Moreover, the following heuristic strategies turned out to be useful in practice: The sets $A_{p_\imath}$ in the extended sample (3) should be re-arranged according to their size (smaller sets before larger ones). Moreover, the order in which actions $a_\imath \in A_{p_\imath}$ are chosen (i.e. the order of successors of an inner node of the search tree) should reflect the following preference: An act a is preferred to (chosen before) an act a' if it is more frequent among the sets $A_{p_\imath}$, that is, $\text{card}\{\imath \,|\, a \in A_{p_\imath}\} \geq \text{card}\{\imath \,|\, a' \in A_{p_\imath}\}$. The resulting algorithm for computing the extended splitting measure proved to be very efficient in practical applications [6].

References

1. M. Allais. Le comportement de l'homme rationnel devant le risque: Critique des postulates et axioms de l'école americaine. *Econometrica*, 21:503–546, 1953.
2. L. Breiman, J. Friedman, R. Olshen, and C. Stone. *Classification and Regression Trees.* Wadsworth International Group, Belmont, CA, 1984.
3. A.R. Chaturvedi and G.K. Hutchinson AndD.L. Nazareth. Supporting complex real-time decision making through machine learning. *Decision Support Systems*, 10:213–233, 1993.
4. D. Ellsberg. Risk, ambiguity, and the Savage axioms. *Quarterly Journal of Economics*, 75:643–669, 1961.
5. I. Gilboa and D. Schmeidler. Case-based decision theory. *Quarterly Journal of Economics*, 110(4):605–639, 1995.
6. E. Hüllermeier. Experience-based decision making. Technical Report 72, Department of Economics, University of Paderborn, September 2001.
7. J.G. March and H.A. Simon. *Organizations.* Wiley, New York, 1958.
8. J. Mingers. An empirical comparison of selection measures for decision-tree induction. *Machine Learning*, 3:319–342, 1989.
9. J. Von Neumann and O. Morgenstern. *Theory of Games and Economic Behavior.* John Wiley and Sons, 1953.
10. J.R. Quinlan. J.r. quinlan. discovering rules by induction from large collections of examples. In D. Michie, editor, *Expert Systems in the Micro Electronic Age.* Edinburgh University Press, 1979.
11. J.R. Quinlan. *C4.5: Programs for Machine Learning.* Morgan Kaufmann, 1993.
12. S.J. Russell and E.H. Wefald. *Do the Right Thing: Studies in Limited Rationality.* MIT Press, Cambridge, Massachusetts, 1991.
13. L.J. Savage. *The Foundations of Statistics.* John Wiley and Sons, 1954.
14. H.A. Simon. *Models of Man.* John Wiley and Sons, New York, 1957.
15. A. Tversky and D. Kahneman. Rational choice and the framing of decisions. *Journal of Business*, 59:251–278, 1986.
16. P.E. Utgoff. Incremental induction of decision trees. *Machine Learning*, 4:161–186, 1989.

Vektoroptimierung bei korrelierten Zielen

Andreas Kleine

Universität des Saarlandes, Rechts- und Wirtschaftswissenschaftliche Fakultät, Postfach 15 11 50, D-66041 Saarbrücken, E-mail: ak@wiwi.uni-sb.de

Zusammenfassung Ansätze der Vektoroptimierung zur Lösung von Entscheidungsproblemen bei mehrfacher Zielsetzung - wie etwa die Zielgewichtung oder das Goal Programming - unterstellen in der Regel mehrere unabhängige Ziele. Häufig bestehen zwischen Zielen jedoch Abhängigkeiten, die nicht im Prozess der Entscheidungsfindung berücksichtigt werden. Wie Abhängigkeiten zwischen korrelierten, aber nicht redundanten Zielen auf Grundlage einer nichtadditiven Gewichtung in ein Kompromissmodell einfließen können, ist Gegenstand dieses Beitrags. Durch die Berücksichtigung von zusätzlichen Interaktionsgewichten, d.h. durch eine nichtadditive Gewichtung, lässt sich eine kompromissoptimale Lösung auf Basis eines Choquet Integrals ermitteln. Die Formulierung des entsprechenden Kompromissmodells als gemischt ganzzahliges Programms verdeutlicht, dass dieser Ansatz für zahlreiche betriebswirtschaftliche Anwendungen - wie etwa für eine Investitionsprogrammplanung mit konfliktären Zielen - unmittelbar anwendbar ist.

1 Einführung

Bei betrieblichen Entscheidungsproblemen verfolgt das Management eines Unternehmens häufig nicht nur ein Ziel, sondern mehrere Ziele simultan. Ansätze der Vektoroptimierung bzw. des Multiobjective Decision Making unterstützen Entscheidungsträger bei der Lösung derartiger Problemstellungen auf vielfältige Weise. Aufbauend auf generellen Überlegungen in Bezug auf die Effizienz der zur Auswahl stehenden Alternativen existieren unterschiedliche Kompromissmodelle zur Lösung von Zielkonflikten, wie z.B. Zielgewichtung, Goal Programming, Abstandsminimierung, Referenzpunktverfahren oder entsprechende interaktive Verfahren [2,14,16]. Diese Ansätze basieren in der Regel auf der Annahme, dass es sich bei den unterschiedlichen Zielen um unabhängige Kriterien handelt [7, S. 51ff].

Vielfach bestehen zwischen Zielen jedoch bestimmte generelle Abhängigkeiten, deren Einbeziehung in die Analyse eines vektoriellen Entscheidungsproblems als wünschenswert erscheint. Im Unterschied zu den multilinearen Verfahren der Multiattributiven Nutzentheorie [7,12] stellt dieser Beitrag einen Ansatz vor, der generell - und damit unabhängig vom Niveau eines Zieles - die Möglichkeit bietet, Abhängigkeiten zwischen Zielen zu erfassen. Derartige Abhängigkeiten können etwa bei zwei Zielen auftreten, die „ähnliche" Kriterien erfassen, die sich allerdings derart unterscheiden, dass eine Vernachlässigung des einen oder anderen Zieles nicht in Frage kommt. Ausgehend von einer Zielgewichtung wird ein modifiziertes Kompromissmodell zur

Lösung eines vektoriellen Entscheidungsmodells bei korrelierten Zielen eingeführt. Ergänzend wird eine entscheidungstheoretische Einordnung vorgenommen und gezeigt, wie sich dieses Kompromissmodell als gemischt ganzzahliges Programm formulieren lässt.

2 Vektorielle Entscheidungsmodelle, Effizienz und Zielgewichtung

Zur Analyse von Entscheidungsproblemen, bei denen ein Entscheidungsträger mehrere Ziele simultan verfolgt, bieten sich vektorielle Entscheidungsmodelle (VEM) an [2, S. 33ff]. Die Konsequenzen, die mit einer Alternative $\boldsymbol{x}$ aus der zugrunde liegenden Alternativenmenge X verbunden sind, erfassen K zu maximierende Zielfunktionen z_k ($z_k : X \to \mathbb{R}$ für alle $k \in \mathsf{K} := \{1, \ldots, K\}$):

$$\text{(VEM)} \quad max \left\{ \begin{pmatrix} z_1(\boldsymbol{x}) \\ \vdots \\ z_K(\boldsymbol{x}) \end{pmatrix} \middle| \, \boldsymbol{x} \in X \right\} .$$

Da in der Regel keine perfekte Alternative existiert, die bei allen K Zielen gleichzeitig zu einer individuell optimalen Lösung führt, interessieren insbesondere solche Alternativen, die nicht dominiert werden. Eine Alternative $\boldsymbol{x}' \in X$ dominiert eine Alternative $\boldsymbol{x}'' \in X$ genau dann, wenn $\boldsymbol{x}'$ im Vergleich zu $\boldsymbol{x}''$ bei mindestens einer Zielfunktion einen höheren Wert aufweist und bei keiner der K Zielfunktionen einen schlechteren Wert. Alle nicht dominierten Alternativen heißen effizient bezüglich (VEM).

Die Ermittlung aller bezüglich (VEM) effizienten Alternativen ist in der Regel kein triviales Problem, so dass häufig direkt Kompromissmodelle zur Anwendungen kommen. Ein resultierender Kompromissvorschlag, d.h. eine kompromissoptimale Lösung, sollte nach Möglichkeit der Menge der effizienten Alternativen angehören. Eines der bekanntesten Kompromissmodelle stellt die Zielgewichtung (ZG) dar. Bei diesem Kompromissmodell wird jede Zielfunktion mit einem positiven Zielgewicht t_k gewichtet und die Summe über alle gewichteten Zielfunktionen gebildet:

$$\begin{aligned} &\text{(ZG)} \quad \max \left\{ \sum_{k \in \mathsf{K}} t_k \cdot z_k(\boldsymbol{x}) \, \middle| \, \boldsymbol{x} \in X \right\} , && \\ &\textit{wobei} \sum_{k \in \mathsf{K}} t_k = 1 && (\textit{Normierung}) \\ &\qquad t_k > 0 \text{ für alle } k \in \mathsf{K} && (\textit{Monotonie}) . \end{aligned}$$

Die – nicht notwendige – Normierung der Gewichte t_k auf eine Summe von genau eins findet hierbei meist Verwendung, um direkt die relative Bedeutung der einzelnen Ziele ausdrücken zu können. Durch die Wahl ausschließlich positiver Zielgewichte t_k ist zudem sichergestellt, dass jede kompromissoptimale Lösung auch effizient bezüglich (VEM) ist [9]. Allerdings beschränken sich kompromissoptimale Lösungen bei der Zielgewichtung auf eine Teilmenge der effizienten Alternativen, die so genannten wesentlich effizienten Alternativen. Diese erweisen sich in Bezug auf die konvexe Hülle des zur Alternativenmenge korrespondierenden Zielraums von (VEM) als effizient [1,3].

Die möglichen Zielfunktionswerte können zudem auf unterschiedlichen Skalen variieren, so dass sich eine Skalierung der Zielfunktionswerte anbietet [14, S. 200ff]. Eine Skalierung kann z.B. die Bandbreite der effizienten Zielfunktionswerte für jede der K Ziele auf einer gemeinsamen Skala abbilden. Hier wird etwa für jedes Ziel eine Bandbreite zwischen den Werten Null und Eins vorgeschlagen ($\forall\, k \in \mathsf{K}$):

$$\tilde{z}_k(\boldsymbol{x}) := \frac{z_k(\boldsymbol{x}) - \underline{z}_k}{\overline{z}_k - \underline{z}_k} \quad (\textit{Skalierung})\,,$$

wobei $\underline{z}_k$ bzw. $\overline{z}_k$ das Minimum bzw. Maximum über alle effizienten Zielfunktionswerte eines k-ten Zieles symbolisiert (mit $-\infty < \underline{z}_k < \overline{z}_k < \infty$). Diese Skalierung hat auf die Effizienz und wesentliche Effizienz von Alternativen keinen Einfluss, d.h., jede Alternative die bezüglich einer Zielgewichtung (ZG) optimal sein kann, ist auch bei skalierten Zielen mit entsprechend modifizierten Zielfunktionen bezüglich (ZG) optimal. Durch dieses Vorgehen lassen sich die Funktionswerte unterschiedlicher Ziele einer Alternative anschaulich vergleichen. Dieser Aspekt erweist sich insbesondere im Hinblick auf die folgenden Darstellungen bei korrelierten Zielen als hilfreich.

3 Zielgewichtung bei korrelierten Zielen

Die im vorangehenden Abschnitt kurz skizzierte Zielgewichtung (ZG) unterstellt stets unabhängige Ziele. Dennoch sind Situationen denkbar, in denen zwischen Zielen Abhängigkeiten bestehen (vgl. u.a. das Beispiel von Grabisch in [4]). So können etwa auch bei der Beurteilung anzuschaffender Investitionsobjekte zum einen ökonomische Ziele (wie z.B. Kapitalwert und Amortisationsdauer) und zum anderen ein ökologisches Ziel (wie z.B. Schadstoffemissionen) in ein entsprechendes vektorielles Entscheidungsmodell (VEM) einfließen (vgl. Abschnitt 5). Bei einer Zielgewichtung kommt den beiden ökonomischen Zielen tendenziell eine höhere Bedeutung zu als dem einen ökologischen Ziel, denn die ökonomischen Aspekte kommen in zwei nicht unabhängigen, aber unterschiedlichen „Facetten“ (Zielen) zum Ausdruck, während sich der Umweltaspekt „nur“ in einem Ziel niederschlägt. Zwar könnten die in Betracht gezogenen Zielgewichte für die beiden ökonomischen Ziele einfach reduziert werden, allerdings soll hier ein Ansatz vorgestellt werden, der speziell das Gewicht für die Summe dieser beiden Ziele reduziert, aber nicht die jeweils einzelne Gewichtung.

Um entsprechende Effekte zwischen zwei Zielen z_j und z_k berücksichtigen zu können, bietet sich die Einführung von zusätzlichen (reellwertigen)

Interaktionsgewichten t_{jk} an. Durch die Wahl eines negativen Interaktionsgewichtes t_{jk} kann ein Entscheidungsträger beispielsweise eine Überschneidung (Overlap) zweier Ziele abschwächen. Umgekehrt lässt sich eine in den einzelnen Zielgewichten zum Ausdruck kommende geringe relative Bedeutung zweier Ziele durch positive Interaktionsgewichte verstärken (Synergie). In diesem Fall würde zwei Zielen z_j und z_k in der gemeinsamen Kombination eine höhere Bedeutung zugeordnet als es die Summe der beiden Zielgewichte aus t_j und t_k zum Ausdruck bringt.

Ein Kompromissmodell, das derartige Abhängigkeiten berücksichtigt, lässt sich in Anlehnung an die Zielgewichtung für skalierte Ziele formulieren:

$$\text{(ZG-I)} \quad \max \left\{ \sum_{k \in \mathsf{K}} t_k \cdot \tilde{z}_k(\boldsymbol{x}) + \sum_{\{j,k\} \subseteq \mathsf{K}} t_{jk} \cdot \min\left\{\tilde{z}_j(\boldsymbol{x}), \tilde{z}_k(\boldsymbol{x})\right\} \;\middle|\; \boldsymbol{x} \in X \right\},$$

$$\textit{wobei} \sum_{k \in \mathsf{K}} t_k + \sum_{\{j,k\} \subseteq \mathsf{K}} t_{jk} = 1 \qquad (\textit{Normierung})$$

$$t_k + \sum_{j \in \mathsf{J}} t_{jk} > 0 \quad \text{für alle } k \in \mathsf{K} \text{ und } \mathsf{J} \subseteq \mathsf{K} \setminus \{k\} \qquad (\textit{Monotonie}).$$

Die Kompromisszielfunktion dieses Kompromissmodells setzt sich wie bei der Zielgewichtung aus der gewichteten Summe über alle Ziele zusammen, ergänzt um einen Interaktionsterm. Dieser Term korrigiert die relative Bedeutung zweier Ziele gemäß des unterstellten Interaktionsgewichtes t_{jk} durch Multiplikation mit dem jeweiligen minimalen Zielfunktionswert. Für Interaktionsgewichte mit einem Wert von null ergibt sich gerade die Zielgewichtung mit den im Abschnitt 2 erläuterten Charakteristika.

Wie bei der Zielgewichtung genügen die Ziel- und Interaktionsgewichte einer Normierungs- und einer Monotoniebedingung, die im Vergleich zu (ZG) jeweils anzupassen sind. Die Normierung erfolgt aufgrund der reellwertigen Interaktionsgewichte nun über alle Gewichte, so dass die Summe aller Zielgewichte nicht mehr zwangsläufig eins beträgt. Neben positiven Zielgewichten ist ebenfalls darauf zu achten, dass auch jeweils die Summe aus Zielgewicht und entsprechender Summe von Interaktionsgewichten positiv ist.

Diese beiden Bedingung stehen unter formalen Aspekten unmittelbar mit dem hier implizit verwendeten Fuzzy Measure in Verbindung, wie es aus Ansätzen bei partiellen Informationen bekannt ist [8,13,15]. In Bezug auf die eingeführte Notation ist ein Fuzzy Measure eine auf der Menge $\mathcal{K}$, d.h. der Potenzmenge von K, definierte reellwertige Funktion τ ($\tau : \mathcal{K} \to [0,1]$) mit folgenden beiden Eigenschaften:

a) $\tau(\emptyset) = 0$ und $\tau(\mathsf{K}) = 1$ (*Normierung*)

b) $\mathsf{J}_1 \subset \mathsf{J}_2 \implies \tau(\mathsf{J}_1) < \tau(\mathsf{J}_2)$ für alle $\mathsf{J}_1, \mathsf{J}_2 \in \mathcal{K}$ (*Monotonie*).

Diese generelle Formulierung erlaubt die Vorgabe von Interaktionsgewichten nicht nur für Paare von jeweils zwei Zielen, sondern auch für entsprechende

Gruppen mit drei, vier etc. Zielen. Da diese zusätzliche Freiheit in der Vorgabe von Interaktionsgewichten mit einem erheblich größeren Aufwand für einen Entscheidungsträger verbunden ist, beschränken sich Betrachtung des Multiobjective Decision Making in der Regel auf so genannte 2-Order Fuzzy Measures [5], die zu den hier eingeführten Ziel- und Interaktionsgewichten korrespondieren:

$$\begin{aligned} \tau(\{k\}) &= t_k && \text{für alle } k \in \mathsf{K} \\ \tau(\{j,k\}) &= t_j + t_k + t_{jk} && \text{für alle } \{j,k\} \subseteq \mathsf{K}\,, \end{aligned}$$

woraus sich für alle restlichen Mengen folgende Werte für τ ergeben:

$$\tau(\mathsf{J}) = \sum_{k\in \mathsf{J}} t_k + \sum_{\{j,k\}\subseteq \mathsf{J}} t_{jk} \quad \text{für alle } \mathsf{J} \subseteq \mathsf{K} \text{ und } |\mathsf{J}| > 2\,.$$

Aus dieser letzten Gleichung resultiert speziell für die Menge aller Ziele K mit gemäß a) $\tau(\mathsf{K}) = 1$ die im obigen Kompromissmodell (ZG-I) zu berücksichtigende Normierungsvorschrift. Diese Normierung leitet sich somit aus den Eigenschaften eines 2-Order Fuzzy Measures ab. Die im Modell (ZG-I) zudem formulierte Bedingung der Monotonie ergibt sich analog aus der Eigenschaft b). Des Weiteren sind gemäß dieser Formulierung zum einen die Zielgewichte t_k ($t_k > 0$) positiv und schwanken zum anderen die Interaktionsgewichte zwischen den Werten minus eins und plus eins ($t_{jk} \in\,]-1,+1[\,$). Da diese Eigenschaft der Interaktionsgewichte mit denen der Korrelationskoeffizienten vergleichbar ist, spricht man analog auch von positiv bzw. negativ korrelierten Zielen. Die Interaktionsgewichte lassen sich außerdem als Shapley-Werte interpretieren, wie sie für kooperative Spiele bekannt sind [4].

Auf Basis dieser Erläuterungen lässt sich ergänzend unmittelbar ableiten, dass die Kompromisszielfunktion der Formulierung (ZG-I) für die speziellen Gewichte (Fuzzy-Measure) genau einem Choquet-Integral entspricht [11]. Unter der Voraussetzung, dass die Zielfunktionswerte einer Alternative $\boldsymbol{x} \in X$ – ggf. nach entsprechender Sortierung o.B.d.A. – absteigend geordnet seien ($\tilde{z}_1(\boldsymbol{x}) \geqq \tilde{z}_2(\boldsymbol{x}) \geqq \ldots \geqq \tilde{z}_K(\boldsymbol{x})$), ist das Choquet-Integral von $\boldsymbol{x}$ definiert als:

$$\begin{aligned} &\tau(\{1\}) \cdot \tilde{z}_1(\boldsymbol{x}) + \sum_{k=2}^{K} \big(\tau(\{1,\ldots,k\}) - \tau(\{1,\ldots,k-1\})\big) \cdot \tilde{z}_k(\boldsymbol{x}) \\ &= t_1 \cdot \tilde{z}_1(\boldsymbol{x}) + \sum_{k=2}^{K} \Big(t_k + \sum_{j=1}^{k-1} t_{jk}\Big) \cdot \tilde{z}_k(\boldsymbol{x}) \\ &= \sum_{k=1}^{K} t_k \cdot \tilde{z}_k(\boldsymbol{x}) + \sum_{k=2}^{K} \sum_{j=1}^{k-1} t_{jk} \cdot \tilde{z}_k(\boldsymbol{x}), \end{aligned}$$

was sich bei nicht sortierten Zielfunktionswerten schreiben lässt als:

$$\sum_{k\in \mathsf{K}} t_k \cdot \tilde{z}_k(\boldsymbol{x}) + \sum_{\{j,k\}\subseteq \mathsf{K}} t_{jk} \cdot \min\big\{\tilde{z}_j(\boldsymbol{x}), \tilde{z}_k(\boldsymbol{x})\big\}\,.$$

Aufgrund des zugrunde liegenden Choquet-Integrals im Kompromissmodell (ZG-I) ist folglich auch sichergestellt, dass ausschließlich bezüglich (VEM) effiziente Alternativen als kompromissoptimale Lösungen in Frage kommen.

Damit können wie bei der klassischen Zielgewichtung dominierte Alternativen nicht bezüglich (ZG-I) kompromissoptimal sein. Im Unterschied zur Zielgewichtung ist eine kompromissoptimale Lösung nicht auf die Menge der wesentlich effizienten Alternativen beschränkt, wie unter anderem das Beispiel im Abschnitt 5 belegt.

4 Gemischt ganzzahlige Modellformulierung

Formulierungen unter Berücksichtigung korrelierter Ziele finden bisher insbesondere Verwendung, um bereits getroffene Entscheidungen (deskriptiv) zu erklären [6,10]. Sollen auf der Basis dieses Kompromissmodells Entscheidungsträger im Prozess der Entscheidungsfindung unterstützt werden, so bietet sich eine Formulierung als gemischt ganzzahliges Programm an, die bei entsprechenden Strukturen einen Rückgriff auf klassische Algorithmen bzw. Standardsoftware des Operations Research ermöglicht.

In der folgenden Modellformulierung wird daher die in der Kompromisszielfunktion von (ZG-I) verwendete Minimumfunktion in zusätzlichen Restriktionen berücksichtigt. Dabei ist bei positiven Interaktionsgewichten eine Formulierung verwendbar, wie sie für Maximin-Probleme bekannt ist (Restriktionen 2a und 2b). Bei negativen Interaktionsgewichten sind zusätzlich Binärvariablen y_{jk} einzuführen, die in Verbindung mit einer hinreichend großen Zahl M (Big-M-Methode) die Berücksichtigung der entsprechenden unteren Schranken garantieren (Maximax-Modell, Restriktionen 3a bis 3c). Zusammenfassend führt dies zu folgendem Kompromissmodell:

$$\begin{array}{lll}
\max & \sum\limits_{k \in \mathsf{K}} t_k \cdot \tilde{z}_k(\boldsymbol{x}) + \sum\limits_{\{j,k\} \subseteq \mathsf{K}} t_{jk} \cdot \tilde{z}_{jk} & \\
\text{u.d.N.} & (1) \quad \boldsymbol{x} \in X & \\
& \left.\begin{array}{l} (2a) \quad \tilde{z}_k(\boldsymbol{x}) \geqq \tilde{z}_{jk} \\ (2b) \quad \tilde{z}_j(\boldsymbol{x}) \geqq \tilde{z}_{jk} \end{array}\right\} & \text{für alle } \{j,k\} \subseteq \mathsf{K} \text{ mit } t_{jk} > 0 \\
& \left.\begin{array}{l} (3a) \quad \tilde{z}_k(\boldsymbol{x}) \leqq \tilde{z}_{jk} + M \cdot y_{jk} \\ (3b) \quad \tilde{z}_j(\boldsymbol{x}) \leqq \tilde{z}_{jk} + M \cdot (1 - y_{jk}) \\ (3c) \quad y_{jk} \in \{0, 1\} \end{array}\right\} & \text{für alle } \{j,k\} \subseteq \mathsf{K} \text{ mit } t_{jk} < 0 \\
& (4) \quad \tilde{z}_{jk} \in \mathbb{R} & \text{für alle } \{j,k\} \subseteq \mathsf{K}\,.
\end{array}$$

Handelt es sich beim zugrunde liegenden vektoriellen Entscheidungsmodell um ein gemischt ganzzahliges lineares vektorielles Programm, so ist auch das zugehörige Kompromissmodell als gemischt ganzzahliges Programm formulierbar.

Während kompromissoptimale Lösungen bei linearen vektoriellen Entscheidungsmodellen – mit Ausnahme von Mehrfachlösungen – stets auf die

Ecken des zugehörigen Polyeders im Alternativenraum beschränkt sind, können bei (ZG-I) aufgrund der zusätzlich zu berücksichtigenden Interaktionsgewichte auch Nicht-Basislösungen kompromissoptimal sein.

5 Beispiel

Abschließend soll ein kleines numerisches Beispiel die Darstellungen zum Kompromissmodell (ZG-I) illustrieren. Angenommen, einem Investor stehen - drei voneinander unabhängige und nicht teilbare - Investitionsobjekte zur Auswahl. Das Budget erlaubt die Anschaffung von genau drei Objekten. Alle drei Objekte erfordern die gleiche Anschaffungsauszahlung, so dass ein Objekt auch mehrfach beschaffbar ist. Die drei Ziele erfassen die mit den Investitionsobjekten erwirtschaftbaren zu maximierenden Kapitalwerte (Ziel z_1 in Mio Euro), die zu minimierenden Amortisationsdauern (Ziel z_2 in Perioden) und die zu minimierenden Schadstoffemissionen (Ziel z_3 in Schadstoffeinheiten). Es ergibt sich das folgende vektorielle Entscheidungsmodell:

$$\text{(VEM)} \quad max \left\{ \begin{pmatrix} +3{,}6x_1 + 2{,}2x_2 + 2{,}8x_3 \\ -3x_1 - 11x_2 - 7x_3 \\ -700x_1 - 350x_2 - 525x_3 \end{pmatrix} \middle| \begin{array}{l} x_1 + x_2 + x_3 = 3 \\ x_1, x_2, x_3 \in \mathbb{N}_0 \end{array} \right\}$$

Nach einer entsprechenden Skalierung der Zielfunktionen gemäß der vorgeschlagenen Bandbreiten lässt sich zunächst eine optimale Lösung bezüglich einer Zielgewichtung (ZG) bestimmen. Bei einer gleichen Gewichtung aller drei Ziele ($t_1 = t_2 = t_3$) würde sich ein Investor für die Anschaffung des ausschließlich ersten Investitionsobjektes entscheiden ($x_1^* = 3$, $x_2^* = x_3^* = 0$). Erfolgt zusätzlich eine Berücksichtigung der bestehenden Abhängigkeit zwischen den Zielen, so dass die Interaktion zwischen den beiden ökonomischen Zielen ($\tilde{z}_1$ und $\tilde{z}_2$) durch $t_{12}{=}-0{,}4$ abgeschwächt und der ökologische Aspekt durch $t_{13}{=}t_{23}{=}0{,}1$ verstärkt wird - bei gleicher Gewichtung der einzelnen Ziele $t_1{=}t_2{=}t_3{=}0{,}4$ -, dann ergibt sich als kompromissoptimale Lösung bezüglich (ZG-I) die Anschaffung aller drei Investitionsobjekte ($x_1^*{=}x_2^*{=}x_3^*{=}1$). Diese optimale Lösung ist effizient, aber nichtwesentlich effizient ist, d.h. diese effiziente Alternative kann nie bezüglich einer Zielgewichtung (ZG) optimal sein.

Würde es sich bei den Investitionsobjekten etwa um beliebig teilbare Objekte handeln, d.h. x_1, x_2, $x_3 \in \mathbb{R}_+$, dann ergäbe sich als bezüglich (ZG-I) optimale Lösung die Anschaffung von jeweils $x_1{=}x_2{=}1{,}5$. Diese optimale Lösung ist aufgrund der unterstellten konvexen Alternativenmenge eine wesentlich effiziente Lösung, aber keine Basislösung.

6 Ausblick

Die Ausführungen der vorangehenden Abschnitte zeigen, wie sich Abhängigkeiten zwischen Zielen bei Zielkonflikten auf der Grundlage eines modifizierten Zielgewichtungsmodells berücksichtigen lassen. Die hier exemplarischen Ausführungen lassen sich auf weitere Kompromissmodelle wie z.B. das Goal Programming übertragen. Diese „klassischen“ Ansätze berücksichtigen bisher in der Regel nur Gewichtungen unabhängiger Ziele. Allerdings müssen Entscheidungsträger bei diesen erweiterten Ansätzen im Hinblick auf die Vorgabe von Ziel- und Interaktionsgewichten ausreichend unterstützt werden, damit sie in der Lage sind, entsprechende Abhängigkeiten geeignet zum Ausdruck zu bringen.

Literatur

1. Brucker P. (1972) Diskrete parametrische Optimierungsprobleme und wesentliche effiziente Punkte. ZOR, **16**, 189–197
2. Dinkelbach W., Kleine A. (1996) Elemente einer betriebswirtschaftlichen Entscheidungslehre. Springer, Berlin Heidelberg New York
3. Dinkelbach W., Kleine A. (2001) Produktionscontrolling – Effiziente Produktionen in diskreten Technologien. ZfB-Ergänzungsheft **2**, 51–80
4. Grabisch M. (1996) The application of fuzzy integrals in multicriteria decision making. EJOR **89**, 445–456
5. Grabisch M. (1997) K-order additive discrete fuzzy measures and their representation. Fuzzy Sets and Systems **92**, 167–189
6. Grabisch M., Roubens M. (2000) Application of the Choquet Integral in Multicriteria Decision Making. In Grabisch M., Murofushi T., Sugeno M. (Eds.): Fuzzy Measures and Integrals. Physica, Heidelberg New York, 348–374
7. Keeney R. L., Raiffa H. (1993) Decisions with Multiple Objectives. Cambridge Uni. Press, Cambridge
8. Kleine A. (1999) Unvollkommene Informationen in Leontief-Technologien, ZfB-Ergänzungsheft **4**, 21–45
9. Kuhn H. W., Tucker A. W. (1951) Nonlinear Programming. In Neyman J.(Ed.): Proceedings of the Second Berkeley Symposium on Mathematical Statistics and Probability. California Press, Berkeley, 481–492
10. Marchial J.-L., Roubens M. (2000) Determination of weights of interacting criteria from a reference set. EJOR **124**, 641–650
11. Murofushi T., Sugeno M. (1989) An interpretation of fuzzy measures and the Choquet integral as an integral with respect to a fuzzy measure. Fuzzy Sets and Systems **29**, 201–227
12. Nitzsch R. von (1992) Entscheidung bei Zielkonflikten. Gabler, Wiesbaden.
13. Schmeidler D. (1989) Subjective probability and expected utility without additivity. Econometrica **57**, 571–587
14. Steuer R. E. (1986) Multiple Criteria Optimization: Theory, Computation, and Application. Wiley, New York
15. Wakker P. (1990) A behavioral foundation for fuzzy feasures. Fuzzy Sets and Systems **37**, 327–350
16. Zeleny M. (1973). Compromise Programming. In Cochrane J. L., Zeleny M. (Eds.): Multiple Criteria Decision Making. Columbia: Uni. of South Carolina Press, 262–301

Relevanz von Information in konditionalen Entscheidungsmodellen

Elmar Reucher, Wilhelm Rödder

Lehrstuhl für Betriebswirtschaftslehre, insb. Operations Research, FernUniversität in Hagen, elmar.reucher@fernuni-hagen.de, wilhelm.roedder@fernuni-hagen.de

Zusammenfassung Die Lösung von Entscheidungsproblemen unter unvollständiger Information ist aufgrund der praktischen Relevanz ein in der Literatur oft diskutierter Forschungsgegenstand. Dem Informationsbeschaffungsprozeß, der den Handlungsentscheidungen zeitlich vorgelagert ist, kommt dabei eine äußerst wichtige Bedeutung zu. Ziel ist es hierbei, unvollständiges Wissen durch Zusatzinformation weiter zu verdichten, um somit Entscheidungssituationen verbessern zu können. Aufgrund zeitlicher und finanzieller Restriktionen ist man dabei jedoch gezwungen, eine effiziente Auswahl aus einer Menge verfügbarer Informationsangebote zu treffen. Basierend auf der entropieoptimalen Verarbeitung unvollständigen Wissens wird in diesem Beitrag ein Bewertungskriterium für alternative Informationsangebote mit dem Ziel der Reduktion verbleibender Ungewißheit hinsichtlich des Erwartungsnutzens vorgestellt. Anhand eines abschließenden Beispiels, das unter Anwendung der Expertensystem-Shell SPIRIT verarbeitet wird, findet das entwickelte Bewertungskonzept seine praktische Anwendung.

1 Einleitung

In der Realität sind Entscheidungen unter Unsicherheit zu treffen, die gemeinhin durch eine Wahrscheinlichkeitsverteilung repräsentiert wird; in ihr sind alle bekannten Informationen über eine Wissensdomäne abgelegt. Bei vollständigem Informationsstand ist die Verteilung eindeutig bestimmt, womit insbesondere auch der Erwartungsnutzen festgelegt ist. Dabei handelt es sich um ein Entscheidungsproblem unter Risiko, das unter Anwendung des Bernoulli-Prinzips gelöst werden kann [6]. Diese Situation ist aber realitätsfremd, da der Entscheidungsträger im allgemeinen nur einen unvollständigen Informationsstand über eine Domäne besitzt. In diesem Fall existiert eine Familie überabzählbar vieler zulässiger Verteilungen; es liegt ein Entscheidungsproblem unter partieller Information vor [4]. Insbesondere ist noch ungewiß, welcher der für jede Verteilung berechenbare Erwartungsnutzen der richtige ist, da sich die Menge aller Wahrscheinlichkeitsverteilungen nach dem Erwerb zusätzlicher Information noch verändern kann. Im Zuge eines Informationsbeschaffungsprozesses wird man natürlich daran interessiert sein, diese Ungewißheit durch weitere Information soweit zu reduzieren, bis Informationsvollständigkeit besteht. Aufgrund zeitlicher und finanzieller Restriktionen kann dieses Ziel aber ohne Verwendung einer effizienten Methode zur Informationsbeschaffung und -bewertung kaum erreicht werden [8]. Ob nun weitere

Information *relevant* bezüglich der Entscheidungssituation ist, hängt dabei von dem schon bekannten Wissen ab. Die Relevanz von Information wird dahingehend gemessen, ob und wie stark durch ihren Erwerb die verbleibende Ungewißheit in Bezug auf den Erwartungsnutzen reduziert wird. Da die Beschaffung von Information zumeist mit (monetären) Kosten verbunden ist, ist schließlich durch einen Kosten-Nutzen-Vergleich zu entscheiden, ob die Beschaffung weiterer Information ökonomisch sinnvoll ist oder nicht.

2 Mathematische Grundlagen

2.1 Probabilistische Konditionale

Auf einer Menge $\mathcal{V} = \{V_1, ..., V_n\}$, wobei jede Variable V_i eine endliche Anzahl diskreter Ausprägungen v_i besitzt, sei eine Verteilung P definiert. p bezeichne dabei die Wahrscheinlichkeitsfunktion mit $P(\mathcal{V} = \mathbf{v}) = p(\mathbf{v})$ für alle Vollkonjunkte $\mathbf{v} = v_1, ..., v_n$. Faßt man die Variablenrealisationen v_i als Literale auf, so läßt sich eine Sprache $\mathcal{L}$ bestehend aus syntaktischen Ausdrücken formulieren, indem die Literale durch Disjunktion $\vee$, Negation $\neg$ und Konjunktion $\wedge$ miteinander verknüpft werden. Solche aussagelogischen Formeln werden als Fakten bezeichnet und die Elemente aus $\mathcal{L}$ werden im folgenden durch Großbuchstaben $A, B, C, ...$ symbolisiert. Regeln sind nun Ausdrücke der Form $B|A(v)$, wobei das Konditional entsprechend der dreiwertigen Logik von CALABRESE wohldefiniert ist. Zu lesen ist dieser Ausdruck dabei in der Form *Wenn A dann B.* $B|A(v)$ ist wahr, falls die Realisation BA unter v wahr, falsch, für $\overline{B}A$ unter v wahr und nicht definiert für $\overline{B}$ unter v [1]. Bei dem im folgenden vorgestellten Bewertungsansatz werden Sachverhalte über Variablenabhängigkeiten in Form probabilistischer Konditionale formuliert.

2.2 Informationstreue Wissensverarbeitung

Sachverhalte in Form einer Regelmenge $\mathcal{R} = \{B_i|A_i\ [x_i]\ i = 1, ..., I\}$ werden in diesem Beitrag entropieoptimal durch Lösen der Aufgabe

$$P^* = \arg\min R(Q, P^0) = \arg\min \sum_{\mathbf{v}} q(\mathbf{v}) \cdot \log(\frac{q(\mathbf{v})}{p^0(\mathbf{v})}) \qquad (1)$$

$$\text{u. d. N.:} \qquad (1 - x_i) \cdot Q(B_i A_i) - x_i \cdot Q(\overline{B}_i A_i) = 0 \ \forall i = 1, ..., I$$

zu Wissen verarbeitet; P^0 bezeichnet dabei die Gleichverteilung [11]. Das in (1) formulierte Prinzip der *Minimierung Relativer Entropie* haben verschiedene Autoren unabhängig voneinander axiomatisch fundiert; es garantiert eine informationstreue Wissensverarbeitung [3], [12]. Die auf diese Weise erzeugte Verteilung P^* ist gerade jene unter allen Verteilungen $Q \in \mathcal{P}$, in der alle in $\mathcal{R}$ explizit formulierten Abhängigkeiten repräsentiert sind - und keine weiteren. Gelöst wird die Aufgabe (1) mit einem verallgemeinerten *Iterative Proprotional Scaling* IPS-Verfahren, dessen Konvergenz von CSISZÀR bewiesen wurde [2], [11].

3 Informationsbewertung in konditionalen Entscheidungsmodellen

3.1 Problembeschreibung

Im Rahmen einer konditionalen Umgebung lassen sich beliebige Entscheidungsmodelle abbilden und die Abhängigkeiten der Variablen untereinander mit dem entropieoptimalen Ansatz (1) informationstreu verarbeiten [10]. Als Grundlage der weiteren Ausführungen wird ein Entscheidungsmodell betrachtet, dessen Variablenmenge $\mathcal{V}$ aus einer Aktionsvariablen A mit Ausprägungen a, einer endlichen Menge von Zustandsvariablen $\{Z_1, ..., Z_n\}$ mit jeweils endlichen Ausprägungen z_i und einer Nutzenvariablen U mit einer endlichen Anzahl von reellwertigen Nutzenausprägungen u besteht. Eine Konfiguration auf dem Gesamtraum der Dimension $|A| \cdot |Z_1| \cdot .... \cdot |Z_n| \cdot |U|$ ist dann durch $\dot{\mathbf{v}} = \dot{a}, \dot{z}_1, ..., \dot{z}_n, \dot{u}$ mit $u(\dot{\mathbf{z}}, \dot{a}) = \dot{u}$ gegeben.
Es sei partielles Wissen in Form einer Regelmenge $\mathcal{R}$ über die Domäne gegeben, das nach Lösung der Aufgabe (1) durch die entropieoptimale Verteilung P^* repräsentiert wird. Das in P^* vorhandene Wissen soll nun weiter ergänzt werden. Dazu steht eine endliche Menge alternativer Informationspakete $\{\mathcal{I}_1, ..., \mathcal{I}_K\}$ zur Verfügung, wobei ein Informationspaket $\mathcal{I}_k$ aus einer endlichen Menge von probabilistischen Konditionalen $\mathcal{I}_k = \{B_{ik}|A_{ik}\ [x_{ik}],\ i = 1, ..., I_k\}$ besteht, die sowohl untereinander als auch mit dem a-priori Wissen $\mathcal{R}$ konsistent sind.

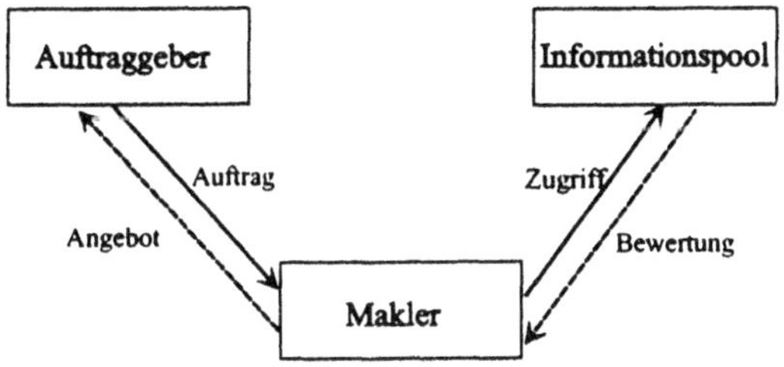

Abbildung 1. Modell für einen Informationsmarkt.

Man betrachte das Informationsbeschaffungsmodell aus Abbildung 1. Ein Auftraggeber (Entscheider) ist an der Beschaffung neuer Information zwecks Verbesserung seiner momentanen Entscheidungssituation interessiert. Er beauftragt einen Makler, der Zugriff auf eine Menge verschiedener Informationspakete hat, die in einem Informationspool zusammengefaßt sind. Der Auftraggeber stellt dem Makler sein a-priori Wissen $\mathcal{R}$ zur Verfügung. Der Makler bewertet daraufhin die verfügbaren Pakete und nennt dem Auftraggeber Preise, zu dem die einzelnen Pakete erworben werden können. Jedes

Informationspaket $\mathcal{I}_k$ wird durch Lösung der Aufgabe

$$P_k^* = arg \min_{\mathcal{I}_k \cup \mathcal{R} \models Q} R(Q, P^0) \tag{2}$$

entropieoptimal verarbeitet, womit sich ein Erwartungsnutzen für eine Aktion $A = a$ von

$$eu_k(a) = \sum_{\mathbf{z}} p_k^*(\mathbf{z}) \cdot u(a, \mathbf{z}) \tag{3}$$

ergibt. Der Wert in (3) kann aber im Zuge weiterer Informationsbeschaffungen noch variieren; er ist noch ungewiß.

3.2 Relevanz von Information

Die noch verbleibende Ungewißheit bezüglich des Erwartungsnutzens wird durch ein Intervall repräsentiert, das, abgesehen von der enormen Größe des dabei zu lösenden Optimierungsproblems bei einem hochdimensionalen Zustandsraum, durch Lösung der Aufgaben

$$eu_k^-(a) = \min_{\mathcal{R} \cup \mathcal{I}_k \models P_k} \sum_{\mathbf{z}} p_k(\mathbf{z}) \cdot u(a, \mathbf{z}) \tag{4}$$

$$eu_k^+(a) = \max_{\mathcal{R} \cup \mathcal{I}_k \models P_k} \sum_{\mathbf{z}} p_k(\mathbf{z}) \cdot u(a, \mathbf{z}) \tag{5}$$

prinzipiell mittels kommerziell verfügbarer Software bestimmt werden kann [7]. $\Delta eu_k(a) := [eu_k^-(a); eu_k^+(a)]$ ist dann das Erwartungsnutzen-Ungewißheitsintervall zum Informationspaket $\mathcal{I}_k$ für eine Aktion $A = a$. Dieses Intervall gibt somit, ausgehend von dem Wissen $\mathcal{R} \cup \mathcal{I}_k$, die im Zuge noch weiterer Informationsbeschaffungen möglichen Erwartungsnutzen an, von denen jeder letzendlich der unter vollständiger Information "wahre" sein kann; insbesondere auch der minimale Erwartungsnutzen $eu_k^-(a)$. Das Informationsentscheidungsproblem läßt sich nun folgendermaßen formulieren:
Aus einer Menge von Informationspaketen ist dasjenige zu bestimmen, welches die noch verbleibende Ungewißheit bezüglich des Erwartungsnutzens am stärksten reduziert.
Die Bewertung der Informationspakete mittels des minimalen Erwartungsnutzens erscheint nach folgender Überlegung gerechtfertigt. Der Wert einer Information ergibt sich aus der Differenz der Erwartungsnutzen, die aus der besten Aktion *vor* und *nach* der Informationsbeschaffung resultieren [6]. Durch die Hinzunahme zusätzlicher Information wird schon vorhandenes Wissen weiter verdichtet, der Informationswert kann nicht daher nicht negativ sein. Aus dem Nutzenungewißheitsintervall ist aber $eu_k^-(a)$ der einzige Wert, der im Zuge weiterer Informationsbeschaffungen auf keinen Fall mehr unterschritten wird, so daß der Wert eines Informationspakets $W(\mathcal{I}_k)$ durch

$$W(\mathcal{I}_k) := \max_a eu_k^-(a) - \max_a eu^-(a) \geq 0 \tag{6}$$

definiert werden kann. $eu^-(a)$ bezeichnet dabei den minimalen Erwartungsnutzen bei Aktion $A = a$ *vor* der Informationsbeschaffung. Diese Überlegungen führen schließlich zur Definition der Relevanz von Information in konditionalen Entscheidungsmodellen.

Ein Informationspaket $\mathcal{I}_k$ heißt <u>*relevant*</u>*, wenn sein Wert $W(\mathcal{I}_k)$ aus (6) echt positiv ist.*

Aus der Menge relevanter Informationsangebote läßt sich schließlich das optimale Paket bestimmen [6], [8], [9]. Bezeichnen $K(\mathcal{I}_k)$ die Kosten für den Erwerb des Informationspaketes $\mathcal{I}_k$, so ist dasjenige $\mathcal{I}_{opt}$ optimal, für das gilt:

$$\mathcal{I}_{opt} = \arg\max_k (W(\mathcal{I}_k) - K(\mathcal{I}_k)). \tag{7}$$

Im folgenden wird ein Verfahren vorgestellt, mit dem die Aufgaben (4) und (5) und somit die Beurteilung über die Relevanz von Information auch mit einem entropieoptimalen Ansatz in der Expertensystem-Shell SPIRIT gelöst werden können. Dem Benutzer steht damit eine systemimmanente Methode zur Auslotung der verbleibenden Nutzenungewißheit unter einem aktuellen Informationsstand zur Verfügung.

4 Ein Näherungsverfahren zur Messung der Relevanz von Information

Gegeben sei schon vorhandenes Wissen $\mathcal{R}$ in Form probabilistischer Abhängigkeiten über Zustandsvariable eines Entscheidungsmodells und eine Menge alternativer Informationsangebote $\mathcal{I}_1, ..., \mathcal{I}_K$, um die $\mathcal{R}$ ergänzt werden kann. Es bezeichne $u_{min}(a) = min_{\mathbf{z}} u(a, \mathbf{z})$ den minimal realisierbaren Nutzen bei einer Aktion $A = a$.

1. Schritt: Man bestimme die entropieoptimale Verteilung mit einem Erwartungswert $u_{min}(a) + \epsilon$ zum Niveau $\epsilon > 0$ durch Lösen der Aufgabe:

$$\begin{aligned} \overline{P}^0_{u_{min},\epsilon} &= \arg\min R(Q, P^0) \\ \text{u. d. N. :} \quad & \sum_{\mathbf{z}} q(\mathbf{z}) \cdot u(a, \mathbf{z}) = u_{min}(a) + \epsilon. \end{aligned} \tag{8}$$

Da die lineare Restriktion eine andere Form als die der Konditionale in (1) hat, bedarf es zur Lösung von (8) anderer Verfahren. Hier wird das Newton-Verfahren genommen [13].

2. Schritt: Man adaptiere das in 1. erzeugte a-priori Wissen an $\mathcal{I}_k$.

$$P^{k,0}_{u_{min},\epsilon} = \arg \min_{\mathcal{I}_k \cup \mathcal{R} \models Q} R(Q, \overline{P}^0_{u_{min},\epsilon}).$$

In der Verteilung $P^{k,0}_{u_{min},\epsilon}$ wird der minimale Erwartungswert zum Niveau ϵ im informationstheoretischen Sinne soweit wie möglich beibehalten, wie es das Wissen aus $\mathcal{R} \cup \mathcal{I}_k$ zuläßt. Ein erste Näherung für den minimalen Erwartungsnutzen unter der Information $\mathcal{R} \cup \mathcal{I}_k$ ist dann durch

$$eu^-_{k,\epsilon}(a) = \sum_{\mathbf{z}} p^{k,0}_{u_{min},\epsilon}(\mathbf{z}) \cdot u(a, \mathbf{z})$$

gegeben. Ob er noch weiter reduziert werden kann, wird mit den beiden folgenden Schritten überprüft.

3. Schritt: Lösen der Aufgabe

$$\overline{P}^1_{u_{min},\epsilon} = \arg\min R(Q, P^{k,0}_{u_{min},\epsilon})$$
$$\text{u. d. N.}: \quad \sum_{\mathbf{z}} q(\mathbf{z}) \cdot u(a, \mathbf{z}) = u_{min}(a) + \epsilon.$$

4. Schritt: Lösen der Aufgabe

$$P^{k,1}_{u_{min},\epsilon} = \arg \min_{\mathcal{I}_k \cup \mathcal{R} \models Q} R(Q, \overline{P}^1_{u_{min},\epsilon}).$$

5. Schritt: Überprüfung, ob die erzeugten Verteilungen in Schritt 2 und 4 noch voneinander differieren.
   Gilt

$$R(P^{k,1}_{u_{min},\epsilon}, P^{k,0}_{u_{min},\epsilon}) = 0: \text{ Abbruch des Verfahrens.}$$
$$R(P^{k,1}_{u_{min},\epsilon}, P^{k,0}_{u_{min},\epsilon}) > 0: \text{ Man setze } P^{k,1}_{u_{min},\epsilon} = P^{k,0}_{u_{min},\epsilon} \text{ und gehe zu 3.}$$

Entsprechend kann auch die Obergrenze des Erwartungsnutzen-Ungewißheitsintervall berechnet werden, wobei dann die Restriktion in Schritt 1 durch $u_{max} - \epsilon$, den maximalen Erwartungsnutzen zum Niveau ϵ, ersetzt wird. Das Ergebnis ist dann ein Erwartungsnutzen-Ungewißheitsintervall zum Niveau ϵ: $\Delta eu_{k,\epsilon}(a) = [eu^-_{k,\epsilon}(a); eu^+_{k,\epsilon}(a)]$ bezüglich der Aktion $A = a$ und dem Informationspaket $\mathcal{I}_k$.

5 Ein Beispiel

Anhand des folgenden Beispiels sollen die theoretischen Aussagen näher illustriert werden. Dabei wurden Teile des Modells aus einem unveröffentlichten Arbeitspapier entnommen [5]. Eine Unternehmung steht vor einer Investitionsentscheidung im kommenden Jahr. Es stehen zwei alternative Projekte $P1$ und $P2$ zur Verfügung, die je nach Projektverlauf mit den Gewinnen in [Tausend Geldeinheiten], kurz [TGE], in der Entscheidungsmatrix in Tabelle 1 bewertet werden; a_i bezeichnet dabei die Aktion *Investition in Projekt Pi*. Die Unternehmung besitzt das a-priori Wissen $\mathcal{R} = \{PV = g \vee PV = m\ [0.7]\}$. Der Algorithmus wird zur Berechnung der minimalen Gewinnerwartung bei Aktion a_1 mit $u_{min}(a_1) = -550$ gestartet, das Ergebnis ist $eu^-(a_1) = 45$ [TGE]. Entsprechend erhält man die übrigen Werte aus Tabelle 2 [1]. Un-

[1] Die Werte wurden mit dem Algorithmus für $\epsilon = 10^{-3}$ berechnet und gerundet.

Tabelle 1. Entscheidungsmatrix.

	Projektverlauf PV		
Aktion	*gut*	*mittel*	*schlecht*
a_1	780	300	-550
a_2	420	270	-180

Tabelle 2. Erwartungsnutzen-Ungewißheit, Wert von Information [TGE].

$\mathcal{R}$	$eu^-(a_1) = 45;\ eu^+(a_1) = 381$	$eu^-(a_2) = 135;\ eu^+(a_2) = 240$	
$\mathcal{I}_1$	$eu^-(a_1) = 45;\ eu^+(a_1) = 381$	$eu^-(a_2) = 135;\ eu^+(a_2) = 240$	$W(\mathcal{I}_1) = 0$
$\mathcal{I}_2$	$eu^-(a_1) = 253;\ eu^+(a_1) = 353$	$eu^-(a_2) = 200;\ eu^+(a_2) = 232$	$W(\mathcal{I}_2) = 118$

ter dem Wissen aus $\mathcal{R}$ ist dann a_2 mit einer minimalen Gewinnerwartung von 135 [TGE] optimal. Zur Reduktion der noch verbleibenden Ungewißheit bezüglich der Erwartungswerte beauftragt die Unternehmung einen Makler, der von zwei unabhängigen Experten *E1*, *E2* entsprechend ihrer Verlaufsprognosen das Zusatzwissen in Form von Informationspaketen erhält:
$\mathcal{I}_1 = \{PV = g|E1 = g\ [0.95], PV = m|E1 = m\ [0.85], PV = s|E1 = s\ [0.99]\}$,
$\mathcal{I}_2 = \{PV = g|E2 = g \vee E2 = m\ [0.65], PV = s|E2 = m \vee E2 = s\ [0.90]\}$.
Aus den Werten in der Tabelle 2 erkennt man, daß das Informationspaket $\mathcal{I}_1$ in Ergänzung zu dem schon bekannten Wissen aus $\mathcal{R}$ mit einem Wert von $W(\mathcal{I}_1) = 0$ nicht relevant ist, wohingegen das Wissen des Experten $E2$ aus Paket $\mathcal{I}_2$ einen positiven Wert von $W(\mathcal{I}_2) = 253 - 135 = 118$ [TGE] besitzt und daher sehr relevant ist. Aus ökonomischer Sicht sollte es von der Unternehmung erworben werden, solange die Kosten für den Informationserwerb unterhalb dieses Wertes liegen. Es ist dann Aktion a_1 optimal mit einer minimalen Gewinnerwartung von 253 [TGE].

6 Fazit und Ausblick

In dem vorliegenden Beitrag wurde auf der Grundlage eines informationstreuen Wissensverarbeitungsprozesses die Relevanz zusätzlicher Information in Form probabilistischer Konditionale unter ökonomischen Aspekten gemessen. Durch die Orientierung an dem minimalen Erwartungsnutzen wurde die Philosophie der vorsichtigen Vorwegnahme zukünftiger Informationsentwicklungen verfolgt. Die Berechnungen der Erwartungsnutzen-Ungewißheitsintervalle sind mit einem Näherungsverfahren unter Verwendung der Expertensystem-Shell SPIRIT durchführbar, womit die Grundlagen weiterer Forschungsaktivitäten auf dem Gebiet der Benutzerunterstützung bei einer effizienten Informationsbeschaffung mit der Shell geschaffen sind.

Literatur

1. Calabrese, P. M. (1991) Deduction and Inference Using Conditional Logic and Probability, Conditional Logic in Expert Systems. I. R. Goodman, M. M. Gupta, H. T. Nguyen, G. S. Rogers (editors), Elsevier Science Publishers B. V.
2. Csiszàr, I. (1975) I-divergence Geometry of Probability Distributions and Minimization Problems. The Annals of Probability, Vol.**3**.
3. Kern-Isberner, G. (1998) Characterizing the principle of minimum cross-entropy within a conditional-logical framework. Artificial Intelligence, Vol. **98**, 169-208.
4. Kofler, E., Menges, G. (1976) Entscheidungen bei unvollständiger Information. Springer, Berlin.
5. Kopittke, B. (1997) Investitionsentscheidung mit stochastischer dynamischer Programmierung. Interner Arbeitsbericht des Lehrstuhls für Operations Research, FernUniversität in Hagen.
6. Laux, H. (1998) Entscheidungstheorie. Springer, Berlin, Heidelberg.
7. Lingo (2000) Optimization Modelling with LINGO. Lindo Systems Inc. http://www.lindo.com.
8. Mag, W. (1977) Entscheidung und Information. Vahlen, München.
9. Müller, A. (1992) Informationsbeschaffung in Entscheidungssituationen. Wissenschaft und Praxis, Ludwigsburg, Berlin.
10. Reucher, R., Rödder, W. (2000) Modellierung von Entscheidungsproblemen unter Verwendung probabilistischer Konditionale. OR Proceedings, SOR2000 in Dresden, Springer, Berlin, Heidelberg, New York, 254–259.
11. Rödder, W., Meyer, C.-H. (1996) Coherent knowledge processing at maximum entropy by SPIRIT. Proceedings 12th Conference on Uncertainty in Artificial Intelligence, Morgan Kaufmann, San Francisco, 470–476.
12. Shore, J.-E., Johnson R.-W. (1980) Axiomatic Derivation of the Principle of Maximum Entropy and the Principle of Minimum Cross-Entropy, IEEE Transact Inf Theory IT-**26**(1): 26–37.
13. Xu, L. (2001) Interner Arbeitsbericht des Lehrstuhl für Operations Research, FernUniversität in Hagen.

Section 13

Artificial Intelligence and Fuzzy Logic

Bestimmung der Gewichte bei Mehrzielentscheidungen. Eine vergleichende Analyse ausgewählter Verfahren

Susanne Eickemeier

Institut für Statistik und Mathematik, Fachbereich Wirtschaftswissenschaften, Johann Wolfgang Goethe-Universität, Mertonstr. 17-23, 60054 Frankfurt, Eickemeier@wiwi.uni-frankfurt.de

1. Einleitung

Sowohl in der betrieblichen Praxis als auch bei privaten Entscheidungen hat man es zumeist mit Problemen zu tun, die sich wegen ihrer Komplexität nur durch eine Vielzahl unterschiedlichster Kriterien in ihrer ökonomischen Bedeutung voll erfassen lassen. Um dem Wunsch der Praktiker nach realistischeren und anwendbaren Entscheidungsunterstützungsmethoden zu entsprechen und die beschränkte Informationsverarbeitungskapazität bzw. die eingeschränkte Rationalität eines Entscheiders zu berücksichtigen, wird häufig vorgeschlagen, bei Vorliegen sehr vieler Ziele diese in einem hierarchisch aufgebauten Zielsystem zu strukturieren.

Die hierarchische Aufbereitung der Ziele dient einer möglichst treffenden Zuordnung der Zielgewichte, da in einem derart aufgebauten Zielsystem in den Teilsystemen in der Regel nur zwei oder drei Teilziele gleichzeitig zu betrachten und mit Stufen-Gewichten zu versehen sind. Das Gewicht für jedes Basisziel errechnet sich als Produkt der Stufengewichte über die Kanten, die vom Basisziel zum Gesamtziel führen. Ist auf diese Weise für jedes Basisziel das Gewicht ermittelt, wird die Summe über die derart gewichteten Nutzenwerte der Zielgrößen dieser Alternative gebildet. Die Alternative mit dem höchsten Wert ist auszuwählen.

Die Zielgewichte sollten dabei so gewählt werden, daß das Verhältnis zweier Gewichte die Grenzrate der Substitution dieser Zielkriterien wiedergibt. Bezeichnen wir die Gewichte zu den Zielen k und r mit g_k und g_r, so ist die Austausch- oder Substitutionsrate $\alpha_{kr} = g_r / g_k$ definiert als der Faktor, um den der Nutzenwert bezüglich des Ziels k erhöht werden muß, wenn der Nutzenwert des Ziels r um den absoluten Wert Δ reduziert wird, d. h. $\alpha_{kr} \cdot \Delta \cdot g_k = \Delta \cdot g_r$.
Zur Berechnung der Austauschraten ist eine kardinale Nutzenmessung notwendig und die α_{kr} sind dann eindeutig bestimmt, wenn die Summe der Gewichte auf 1 normiert wird. "Widerspruchsfreie Präferenzen" liegen bei einer linearen additiven Entscheidungsregel dann vor, wenn die ermittelten Austauschraten der Konsis-

tenzbedingung $a_{kr} \cdot a_{rs} = a_{ks}$ genügen. Sind alle Austauschraten a_{rs} positiv, so folgt aus der Konsistenzbedingung und $a_{kk} = 1$ die Formel $a_{rk} = \frac{1}{a_{kr}}$.

Das zentrale Problem besteht in diesem Zusammenhang in der Formulierung der Austauschraten durch den Entscheider. Da die präzise Angabe, um wieviel einem Entscheider ein Kriterium wichtiger ist als ein anderes, in der Praxis sehr große Schwierigkeiten bereitet, wird häufig vorgeschlagen, die Paarvergleiche anhand von Punkteskalen vorzunehmen, deren Bedeutung verbal beschrieben wird.

Die bekannteste Punkteskala wurde von Thoma L. SAATY [1980] vorgeschlagen, die er in seinem Analytic Hierarchy Process (AHP) verwendet und die in Tabelle 1 dargestellt ist. Andere Vorschläge stammen z.B. von J. BARZILAI UND F.A. LOOTSMA [1997] oder von KAPRCYK und NURMI [1998]. In dieser Arbeit soll vergleichend untersucht werden, welchen Einfluß die gewählte Punkteskala auf den aus der Paarvergleichsmatrix abzuleitenden Gewichtungsvektor hat.

2. Vergleich der Punktbewertungsskalen von SAATY und BARZILAI/LOOTSMA

In seinem Analytic Hierarchy Process geht SAATY von einem hierarchischen Zielsystem aus. Er läßt zur Bestimmung der Austauschraten eine 9-Punkte-Skala zu. Für die Paarvergleichsmatrix **A** kommen damit nur die Koeffizienten 1, 2, 3, 4, 5, 6, 7, 8, 9 und die Kehrwerte $1, \frac{1}{2}, \frac{1}{3}, \frac{1}{4}, \frac{1}{5}, \frac{1}{6}, \frac{1}{7}, \frac{1}{8}, \frac{1}{9}$ in Betracht, die inhaltlich entsprechend der Tab. 1 zu interpretieren sind.

1	gleiche Bedeutung	Beide verglichenen Elemente haben die gleiche Bedeutung für das nächst-höhere Element (Ziel).
3	etwas größere Bedeutung	Erfahrung und Einschätzung sprechen für eine etwas größere Bedeutung
5	erheblich größere Bedeutung	Erfahrung und Einschätzung sprechen für eine erheblich größere Bedeutung
7	sehr viel größere Bedeutung	Die sehr viel größere Bedeutung eines Elements hat sich in der Vergangenheit klar gezeigt
9	absolut dominierend	Es handelt sich um den größtmöglichen Bedeutungsunterschied zwischen Elementen
2,4, 6,8	Zwischenwerte	Zwischen zwei benachbarten Urteilen muß eine Übereinkunft getroffen werden, ein Kompromiß

Tab. 1: Punkteskala von Saaty für die Paarvergleiche

LOOTSMA/BARZILAI [1997] verwenden in ihrem multiplikativen AHP eine vergleichbar aufgebaute Punkteskala, vgl. Tab. 2. Die Entscheider stellen die Paarvergleichsmatrizen gemäß ihrer Präferenzen auf, wobei die Präferenzaussagen δ_{ck} erst in einem zweiten Schritt in Abhängigkeit von dem Skalierungsparameter γ in Substitutionsraten transformiert werden:

$$r_{ck} = \exp(\gamma \cdot \delta_{ck}) \tag{2.1}$$

Als plausibler Wert für den Skalierungsparameter wird von BARZILAI/LOOTSMA je nach Hierarchiestufe $\gamma = \ln 2$ bzw. $\gamma = \ln\sqrt{2}$ vorgeschlagen.

VERBALE BEWERTUNGEN	δ_{ck}	r_{ck} $\gamma = \ln 2$	r_{ck} $\gamma = \ln\sqrt{2}$	SAATY	μ_R
Sehr starke Präferenz zugunsten von S_k	-8	$\frac{1}{256}$	$\frac{1}{16}$	$\frac{1}{9}$	0
Starke Präferenz zugunsten von S_k	-6	$\frac{1}{64}$	$\frac{1}{8}$	$\frac{1}{7}$	$\frac{1}{8}$
Deutliche Präferenz zugunsten von S_k	-4	$\frac{1}{16}$	$\frac{1}{4}$	$\frac{1}{5}$	$\frac{1}{4}$
Schwache Präferenz zugunsten von S_k	-2	$\frac{1}{4}$	$\frac{1}{2}$	$\frac{1}{3}$	$\frac{3}{8}$
Indifferenz zwischen S_c und S_k	0	1	1	1	$\frac{1}{2}$
Schwache Präferenz zugunsten von S_c	+2	4	2	3	$\frac{5}{8}$
Deutliche Präferenz zugunsten von S_c	+4	16	4	5	$\frac{3}{4}$
Starke Präferenz zugunsten von S_c	+6	64	8	7	$\frac{7}{8}$
Sehr starke Präferenz zugunsten von S_c	+8	256	16	9	1

Tab. 2: Punkteskalen im Vergleich

Auch wenn BARZILAI und LOOTSMA die Wahl des Skalierungsparameters γ nicht ausführlicher diskutieren, verdeutlicht Tab. 2 sehr anschaulich, daß sich je nach zugrunde gelegter Skala die Austauschraten unterscheiden und damit auch die daraus abzuleitenden Gewichte verschieden ausfallen werden.

Sowohl bei SAATY als auch bei BARZILAI/LOOTSMA wird durch den reziproken Aufbau der Paarvergleichsmatrix eine Ungleichbehandlung der Präferenzen gefördert. Präferenzaussagen zugunsten eines Kriteriums werden per se sehr viel stärker berücksichtigt als Abneigungen, die durch die Kehrwertbildung immer relativ nahe beieinander liegen.

Mit einem $\gamma = \ln\sqrt{2}$ unterstellen BARZILAI/LOOTSMA im Vergleich zu SAATY's Skala eine relative Präferenz, die konkaver verläuft. Die gesamte Skala ist stärker gestreckt, wobei größere Werte eine relativ höhere und kleinere Werte eine relativ geringere Bedeutung beigemessen bekommen. Viel deutlicher wird diese Tendenz bei einer Wahl von $\gamma = \ln 2$: Das Spektrum der Skala wird noch erweitert und die Substitutionsraten werden offensichtlich extremer, obwohl weiterhin die gleiche Interpretation dahinter steht.

Gerade die Skalierung von BARZILAI/LOOTSMA scheint schwer begründbar: Für einen Entscheider sind die Konsequenzen seiner Punktbewertung nicht erkennbar, da er seine Präferenzaussagen in Bezug auf die Skala –8 bis 8 ausrichtet. Die nachträglich vorgenommene Transformation in Substitutionsraten unterstellt einen allgemeingültigen Präferenzverlauf. Es scheint zweifelhaft, ob damit den subjektiven Vorstellungen der Entscheider in allen Fällen angemessen entsprochen werden kann.

3. Austauschraten in Form von Zugehörigkeitswerten

Eine andere Variante bezüglich der Angabe von Austauschraten repräsentiert der Ansatz von KAPRCYK und NURMI [1998]. Die Paarvergleiche werden anhand von Präferenzaussagen der Entscheider in Form von Fuzzy-Präferenzrelationen formuliert. Damit soll – wie in den anderen Ansätzen auch – dem Problem Rechnung getragen werden, daß Entscheider häufig ihre individuellen Substitutionsraten nicht präzise ausdrücken können.

Eine individuelle Fuzzy-Präferenzrelation eines Entscheiders, R, läßt sich beschreiben anhand der Zugehörigkeitsfunktion $\mu_R : S \times S \rightarrow [0,1]$ für die gilt:

$$\mu_R = \begin{cases} 1 & \text{wenn } S_c \text{ sehr stark gegenüber } S_k \text{ präferiert wird} \\ c \in \,]0{,}5,1[& \text{wenn } S_c \text{ gegenüber } S_k \text{ präferiert wird} \\ 0{,}5 & \text{Indifferenz} \\ d \in \,]0, 0{,}5[& \text{wenn } S_k \text{ gegenüber } S_c \text{ präferiert wird} \\ 0 & \text{wenn } S_k \text{ sehr stark gegenüber } S_c \text{ präferiert wird} \end{cases}$$

Eine Fuzzy-Präferenzrelation eines Individuums wird dargestellt durch eine n×m-Matrix $R = [r_{ck}]$ mit $r_{ck} = \mu_R(S_c, S_k)$, wobei R so aufgebaut ist, daß $r_{ck} + r_{kc} = 1$.

Betrachtet man diese Art der Skalierung, vgl. Tab. 2, so fällt auf, daß die Skala – im Vergleich zu den bereits vorgestellten - nun gestaucht wurde. Zu beachten ist, daß positive und negative Präferenzaussagen einen gleich großen Bewertungsspielraum haben, nämlich [0,5, 1] und [0, 0,5[. Dadurch sind – im Gegensatz zu den oben betrachteten Skalen – die Abstände zwischen den positiven und den negativen Präferenzaussagewerten nicht mehr so unterschiedlich groß. Die bisher zu beobachtende extreme Gewichtung hoher Präferenzaussagen wird vermindert und der Trend geht zu mittleren Gewichten.

4. Vergleich verschiedener Punkteskalen zur Bestimmung der Zielgewichte

Bisher wurden die den unterschiedlichen Mehrzielverfahren zugrunde gelegten Bewertungssskalen diskutiert. Abgesehen von den Skalen differieren die Ansätze darüber hinaus in ihrer Vorgehensweise zur Ermittlung der benötigten Gewichte.

Gemäß der Definition der Austauschraten weist eine konsistente Paarvergleichsmatrix A eine spezielle Form auf, bei der alle Spaltenvektoren Vielfache von einander sind und jede Spalte somit einen äquivalenten Gewichtevektor darstellt. Durch Normierung der Komponenten des r-ten Spaltenvektors durch die Komponentensumme $\sum_{k=1}^{K} a_{kr}$, $r \in \{1,...,K\}$, erhält man den auf die Summe 1 normierten Gewichtungsvektor. Dieser Vektor ergibt sich auch, wenn das arithmetische oder das geometrische Mittel der Spaltenvektoren berechnet wird und diese Größen dann normiert werden. Darüber hinaus entspricht bei einer konsistenten Matrix **A** der Gewichtevektor dem normierten Eigenvektor von **A** zum größten Eigenwert von A, so daß sich auch mit Hilfe dieses Ansatzes der identische Gewichtungsvektor einstellt.

Für den Fall des Vorliegens einer inkonsistenten Matrix geht man davon aus, daß alle drei Verfahren weiterhin ihre Berechtigung zur Bestimmung des normierten Gewichtevektors behalten, solange die Inkonsistenz in einem gewissen Rahmen verläuft, vgl. z.B. die Berechnung des Konsistenzwertes bei SAATY [1980].

Betrachten wir im folgenden ein Beispiel zur Verdeutlichung der Konsequenzen der unterschiedlichen Skalen und Vorgehensweisen. Gegenübergestellt werden

- die Eigenvektormethode in SAATY's AHP,
- das multiplikative AHP von BARZILAI/LOOTSMA basierend auf dem geometrischen Mittel der Spaltenvektoren und
- Zugehörigkeitswerte nach KAPRCYK/NURMI unter Verwendung sowohl des arithmetischen als auch des geometrischen Spaltenmittels.

Als Beispiel ziehen wir eine Autokaufentscheidung heran, bei der zur Vereinfachung nur die 4 Ziele "Technik", "Werkstatt", "Design" und "Kaufpreis" als Beurteilungskriterien für die Alternativen betrachtet werden. Zugrunde gelegt wird die folgende Paarvergleichsmatrix eines Entscheiders, der seine Präferenzaussagen in Hinblick auf SAATY's Punktbewertung formuliert hat, vgl. Tab. 3.

	Technik	Werkstatt	Design	Kaufpreis
Technik	1	5	3	$\frac{1}{3}$
Werkstatt	$\frac{1}{5}$	1	$\frac{1}{2}$	$\frac{1}{9}$
Design	$\frac{1}{3}$	2	1	$\frac{1}{6}$
Kaufpreis	3	9	6	1

Tab. 3: Paarvergleichsmatrix nach SAATY

Als Gewichtevektoren dieser drei Varianten ergeben sich dann:

	SAATY	BARZILAI/LOOTSMA $\gamma = \ln\sqrt{2}$	BARZILAI/LOOTSMA $\gamma = \ln 2$	KAPRCYK Arithm.	KAPRCYK Geom.
T	0,250	0,236	0,126	0,281	0,334
W	0,054	0,054	0,007	0,148	0,000
D	0,096	0,099	0,022	0,203	0,230
KP	0,600	0,611	0,845	0,367	0,436

Tab. 4: Vergleich der Gewichtevektoren

Im Rahmen dieser vergleichenden Betrachtung der Gewichtevektoren bestätigen sich die bisherigen Ausführungen. Zieht man den Gewichtevektor von SAATY als Referenzgröße heran, so ist erkennbar, daß selbst die logarithmische Skala mit $\gamma = \ln\sqrt{2}$ von BARZILAI/LOOTSMA dazu tendiert, die Gewichteordnung deutlicher zu betonen. Insbesondere bei $\gamma = \ln 2$ wird die Verteilung sehr extrem, die schwächeren Gewichte werden verringert zugunsten des höchsten Gewichts, das sehr viel stärker angehoben wird.
Bei den beiden Varianten von KAPRCYK/NURMI liegen dagegen die Gewichte relativ nah beieinander; die Ordnung bleibt zwar erhalten, die Abstände sind aber erkennbar verringert. Offensichtlich ist das geometrische Mittel bei der Verwendung von Zugehörigkeitswerten nicht unbedingt opportun, da bei Verwendung der Skalenendpunkte das entsprechende Gewicht gleich 0 wird.

Angesichts der doch recht gravierenden Unterschiede in den Gewichtevektoren, die sich in Abhängigkeit der Skalen und Verfahren herauskristallisieren, wird offenkundig, daß mit der Wahl der Skala der Gewichtevektor und damit gegebenenfalls die Alternativenauswahl manipuliert werden kann. Es drängt sich die Frage nach der "wahren" Skala auf, die nämlich Gewichte produziert, die den Vorstellungen eines Entscheiders tatsächlich auch entsprechen. Die Beantwortung dieser Frage bleibt sicher in großen Teilen den Individuen selbst vorbehalten, nichtsdestotrotz wäre es hilfreich, mit Hilfe empirischer Untersuchungen Argumente für oder wider einzelne Skalen zusammenzutragen.

Grundsätzlich ist festzuhalten, daß jegliche Punkteskala immer nur einen Ersatz darstellen kann, da sie den Entscheider auf eine vorgegebene Bewertung beschränkt und diesen seine Einschätzung nicht völlig frei ausdrücken läßt. Auch läßt die verbale Erläuterung der Punkteskalen einen relativ großen individuellen Spielraum, da z.B. Begriffe wie "deutliche" und "starke" Präferenz nicht eindeutig gegeneinander abgegrenzt sind. Ganz entscheidend ist desweiteren, daß die Inputdaten bedingt durch die Klassifizierung ursprünglich nur als ordinal skalierte Daten vorliegen Die im Rahmen der dargestellten Ansätze vorgenommenen Operationen erfordern jedoch das anspruchsvolle Verhältnisskalenniveau, ohne das die Kehrwertbildung oder die Bestimung des arithmetischen bzw. geometrischen Mittels aus theoretischer Sicht nicht akzeptabel sind.

Berücksichtigt man, daß mit der Klassifizierung der Präferenzaussagen methodisch erhebliche Unwägbarkeiten verbunden sind, so ist es letztendlich doch sinnvoller, dem Entscheider abzuverlangen, Austauschraten zu formulieren. Ein derartiger Ansatz wird im folgenden Abschnitt ausführlicher diskutiert.

5. Fuzzy-AHP

In praktischen Anwendungen ist ein Entscheider oft nicht in der Lage ist, beim Paarvergleich alle Ausgleichsraten exakt anzugeben. Trotzdem besitzt er meist eine ungefähre Vorstellung, um wieviel wichtiger er das eine Ziel erachtet als das andere. Daher stellt sich die Frage, ob es überhaupt notwendig ist, daß jede Substitutionsrate auf eine eindeutige reelle Zahl verschärft wird, oder ob man nicht besser das vorhandene Wissen ausnutzt und im Gegensatz zu Saaty diese Interpretation bei der Bewertung der Paarvergleiche zugrunde legt. Die nur größenordnungsmäßig bekannten Werte lassen sich dann durch Fuzzy-Mengen mathematisch beschreiben. Dabei bietet es sich an, Fuzzy-Intervalle des ε-λ-Typs zu verwenden, die auch den Vorteil besitzen, daß arithmetische Rechnungen leicht ausführbar sind. Fuzzy-Intervalle $\tilde{a}_{ij}$ des ε-λ-Typs haben stückweise lineare Zugehörigkeitsfunktionen und lassen sich durch 6 Werte hinreichend genau beschreiben: $\tilde{a}_{ij} = (\underline{a}_{ij}^{\varepsilon}; \underline{a}_{ij}^{\lambda}; \underline{a}_{ij}; \overline{a}_{ij}; \overline{a}_{ij}^{\lambda}; \overline{a}_{ij}^{\varepsilon})^{\varepsilon, \lambda}$.

Die gewählte Vorgehensweise basiert auf der Ermittlung des Gewichtevektors aus den normierten Spaltenvektoren. Als Operator bietet sich unserer Ansicht nach vor allem das Arithmetische Mittel an, da die so ermittelten Gewichte zur gewichteten Addition der Teilnutzen Verwendung finden.

Um dem in diesem Zusammenhang unabdingbaren Begriff "Normierung auf 1" adäquat genügen zu können, schlagen EICKEMEIER und ROMMELFANGER [2001] ein neues Verfahren zur Normierung der Spalten der Paarvergleichsmatrix vor, nach dem alle Parameter eines Fuzzy-Intervalls vom ε-λ-Typ durch die gleiche reelle Zahl dividiert werden. Um zu erreichen, daß die zum 1-Niveau gehörenden Gewichtungssummen nahe bei 1 liegen, empfehlen sie zur Normierung der j-ten Spalte das arithmetische Mittel

$$\sigma_j^* = \frac{1}{2}(\underline{\sigma}_j + \overline{\sigma}_j) \qquad (4.1)$$

zu verwenden. Die Normierung erfolgt dann nach der Formel

$$\tilde{a}_{ij}^* = \frac{\tilde{a}_{ij}}{\sigma_j^*} = \left(\frac{\underline{a}_{ij}^{\varepsilon}}{\sigma_j^*};\frac{\underline{a}_{ij}^{\lambda}}{\sigma_j^*};\frac{\underline{a}_{ij}}{\sigma_j^*};\frac{\overline{a}_{ij}}{\sigma_j^*};\frac{\overline{a}_{ij}^{\lambda}}{\sigma_j^*};\frac{\overline{a}_{ij}^{\varepsilon}}{\sigma_j^*}\right)^{\varepsilon,\lambda} . \qquad (4.2)$$

Mit den so normierten Ausgleichsraten berechnet man dann die Gewichte

$$\tilde{g}_i^* = \frac{1}{n}\cdot(\tilde{a}_{i1}^* \oplus \tilde{a}_{i2}^* \oplus \cdots \oplus \tilde{a}_{in}^*), \qquad (4.3)$$

die den Gewichtevektor $\tilde{g}\,'^* = (\tilde{g}_1^*, \tilde{g}_2^*, \ldots, \tilde{g}_n^*)$ bilden.

Mit Hilfe dieses Normierungsverfahren bleibt die in den Austauschraten vorgegebene Unschärfe bei der Normierung erhalten und wird nicht vergrößert, auch wird die Gestalt der Fuzzy-Mengen bewahrt. Es läßt sich zeigen, daß auch bei Verwendung der aus diesem Ansatz resultierenden Fuzzy-Gewichte meist eine Alternativenauswahl möglich ist. Eine nicht der Informationslage entsprechende künstliche Verschärfung der Daten ist daher nicht notwendig.

Eine weitere Konsequenz dieser Normierung ist, daß die Gesamtnutzenbewertungen weniger fuzzy sind und damit die Präferenzordnung der Alternativen deutlicher wird. Ist in Bezug auf die Alternativen die Trennschärfe der Präferenzrelationen nicht ausreichend, lassen sich zumindest nicht in Frage kommende Alternativen aussondern und zusätzliche Informationen sind dann nur zielgerichtet unter Kosten- und Nutzenabwägungen für die restlichen Alternativen zu beschaffen.

Literatur

BARZILAI, J; LOOTSMA, FA (1997) Power Relations and Group Aggregation in the Multiplicative AHP and SMART. Journal of Multi-Criteria Decision Analysis 6, 155-165

EICKEMEIER, S; ROMMELFANGER, H (2001) Fuzzy utility value analysis and fuzzy analytic hierarchy process. Proceeding of EUROFUSE Workshop on Preference Modelling and Applications. Granada, 139-146

KACPRZYK, J; NURMI H (1998) Group decision making under Fuzziness. In: SLOWINSKI, R.: Fuzzy Sets in Decision Analysis, Operations Research and Statistic. Volume of „The Handbook of Fuzzy Sets", Kluwer Academic Publishers, Boston, 1998

SAATY, TL (1980) The Analytic Hierarchy Process. McGraw-Hill, New York

SAATY, TL (1995) Decision Making for Leaders. RWS Publications, Pittsburgh

Wissen ist meßbar

Wilhelm Rödder

Lehrstuhl für Betriebswirtschaftslehre, insb. Operations Research
FernUniversität in Hagen, wilhelm.roedder@fernuni-hagen.de

Zusammenfassung Der Abgleich von a priori-Wissen auf neue relevante Sachverhalte und die Verwendung des so erworbenen Wissens zur Beantwortung einer Anfrage sind das zentrale Thema dieses Aufsatzes. Hierbei wird der gesamte Wissensverarbeitungsprozeß in einem probabilistischen Modell konditionaler Struktur informationstreu umgesetzt. Er ist meßbar: Wissen(sumfänge), Schlußstärken und Relevanz sind ebenso quantifizierbar wie die verbleibende Unsicherheit in erlernten Zusammenhängen und die Ungewißheit über die Korrektheit gegebener Antworten. Die Expertensystemshell SPIRIT unterstützt diese Form der Wissensverarbeitung.

1 Wissensverarbeitung auf Konditionalstrukturen

1.1 Syntax, Semantik, Konditionallogik und Wahrscheinlichkeit

Auf semantischer Ebene beschäftigen wir uns mit einer Population von Individuen oder Objekten und deren Eigenschaften. Ziel der Betrachtungen ist der Erwerb von Wissen über diese Eigenschaften, d.h. die Häufigkeit ihres Auftretens und ihre wechselseitige Bedingtheit.

Auf der syntaktischen Ebene werden die Eigenschaften durch eine endliche Menge von endlichwertigen Variablen $V = \{V_1, ..., V_L\}$ mit Werten oder Ausprägungen v_l von V_l beschrieben. Als Variablenbezeichner werden oft Zeichenketten von mnemotechnischen Großbuchstaben verwendet, als Ausprägungsbezeichner häufig Kleinbuchstaben oder ggf. Zahlen. GESCHLECHT = m oder ALTER = 60 sind typische Wertzuweisungen zu Variablen. Ausdrücke der Form $V_l = v_l$ werden durch die Verknüpfungszeichen $^-$ (nicht), $\wedge$ (und), $\vee$ (oder) und Klammern bei Geltung entsprechender Syntaxregeln zusammengefaßt. Allgemeine Ausdrücke dieser Art bezeichnen wir in der vorliegenden Arbeit mit $A, B, C...$, ggf. indiziert: A_i, E_j, G_k. Einfache Konjunkte von Elementarausdrücken $V_l = v_l$ seien oft abgekürzt als ungeordnete Werte-Tupel dargestellt, wie z. B. das Vollkonjunkt $v = v_1...v_L$.

Auf der logischen Ebene heißen Ausdrücke der Form $V_l = v_l$ Literale. Sie sind aussagenlogische Formeln, die für einzelne Individuen wahr (t) oder falsch (f) sein können. Gleiches gilt für die Ausdrücke A, B, C ... .
Bei dieser Interpretation und den syntaktischen Vereinbarungen heißt die Menge aller logischen Formeln die Sprache $\mathcal{L}$. Sie ist mit $^-, \wedge, \vee$ und den neutralen Elementen t, f auf der Menge der Klassen logisch äquivalenter Formeln eine Algebra.

Will man auf semantischer Ebene ausdrücken, daß die Eigenschaft A die Eigenschaft B bedingt, konditioniert man „B unter der Bedingung, daß A“ oder sagt umgekehrt „aus A folgt B“.

Die hier verwendete Syntax für solche Sachverhalte ist die des Konditionals $B \mid A$. Hierbei ist $\mid$ der Konditionaloperator. Durch Klammerung und eine geeignete Festlegung syntaktischer Regeln für die Verknüpfung solcher Konditionale mit $^-, \wedge, \vee$ entsteht eine komplexe Hierarchie erlaubter syntaktischer Ausdrücke, die wir als die Sprache $\mathcal{L} \mid \mathcal{L}$ bezeichnen wollen. Zur Rückführbarkeit solcher Ausdrücke auf einfache Konditionale $B \mid A$ vgl. [1].

Auf logischer Ebene definiert man, siehe wiederum [1]:

$$B|A(v) = \begin{cases} \text{t} & \text{für} \quad v \subset BA \qquad (B \wedge A) \\ \text{f} & \text{für} \quad v \subset \overline{B}A \qquad (\overline{B} \wedge A) \\ \text{u} & \text{für} \quad v \subset \overline{A}. \end{cases} \tag{1}$$

Ein Konditional kann wahr oder falsch sein, falls der Bedingungsteil wahr ist, andernfalls ist diese Frage nicht entscheidbar und der Wahrheitswert undefiniert.

Ist P ein Wahrscheinlichkeitsmaß auf $\mathcal{L}$, so kann es in natürlicher Weise auf $\mathcal{L} \mid \mathcal{L}$ erweitert werden.

Definition 1 (Bedingte Wahrscheinlichkeit)

Ist $B \mid A \in \mathcal{L} \mid \mathcal{L}$ und P eine Wahrscheinlichkeitsverteilung auf $\mathcal{L}$, so heißt jede Zahl $P(B \mid A)$, die die Gleichung $P(BA) = P(B \mid A) \cdot P(A)$ erfüllt, eine bedingte Wahrscheinlichkeit des Konditionals.

Die Wahrscheinlichkeit des Konditionals $B \mid A$ ist natürlich gleich der bedingten Wahrscheinlichkeit B unter A für positive $P(A)$, wie es auch sein sollte.

Damit sind die Präliminarien bereitgestellt, um im Folgeabschnitt den Prozeß informationstreuer Wissensverarbeitung zu entwickeln.

1.2 Wissenserwerb, Anfrage und Antwort

Informationstreue Wissensverarbeitung vollzieht sich in drei Schritten: Festlegung der Wissensdomäne, Wissenserwerb und Anfrage. Im einzelnen:

a) **Festlegung der Wissensdomäne**: Die Wissensdomäne wird durch Wahl der Variablen V_l und ihre Ausprägungen festgelegt. Damit ist die Menge aller v das feinste verfügbare Eigenschaftsmuster des Erfahrungsbereichs. Im Sinne des vorigen Abschnitts verfügt man somit über die Sprache $\mathcal{L} \mid \mathcal{L}$ als Kommunikationsmittel.

b) **Wissenserwerb**: Der Experte kennt Sachverhalte in der Wissensdomäne.

Sachverhalte sind eine Menge probabilistischer Konditionale $\mathcal{R} = \{B_i \mid A_i[x_i], i = 1...I\}$. Probabilistische Konditionale sind Konditionale $B_i \mid A_i$ und je eine Wahrscheinlichkeit x_i, mit der sie in der Population gelten sollen. Gesucht wird nun nach einer Wahrscheinlichkeitsverteilung P^*, die eine gegebene a priori-Verteilung P über der Domäne möglichst informationstreu an die neuen Sachverhalte anpaßt. P, P^* sind Wissen(szustände) auf der Domäne – siehe auch das nächste Kapitel.

Bei einem erstmaligen Wissenserwerb ist das a priori -Wissen oft die Gleichverteilung P^0, sie repräsentiert Unwissen oder Nicht-Wissen. Die Adaption erfolgt dann durch die Lösung der mathematischen Optimierungsaufgabe

$$P^* = \arg\min R(Q, P^0), \text{ s.d. } Q(B_i \mid A_i) = x_i, i = 1...I. \quad (2)$$

R ist eine informationstheoretische Abstandsfunktion, die relative Entropie zwischen der Verteilung Q und P^0. (5) sucht also den Wissensstand P^*, der alle Sachverhalte respektiert und so wenig wie möglich vom früheren Wissen(szustand) abweicht. P^* heißt Domänenwissen. (2) wird in den Arbeiten [7], [3] ausführlich begründet.

c) **Anfrage**: ca) **Annahmen und Frage**: Die Annahmen sind wieder als Menge probabilistischer Konditionale $\mathcal{E} = \{F_j \mid E_j[y_j], j = 1...J\}$ formulierbar. Hypothetische Sachverhalte stellen in der Anfrage die Prämisse dar, die jedoch sprachlich ebenso reich sein kann wie die Sachverhalte zum Wissenserwerb in (2). Sie sind allerdings temporär oder situativ. Nach Beantwortung der Frage wird die „gedankliche Situation“ wieder verlassen.

Eine Frage hat die konditionale Form $H \mid G =?$. Die Anfrage als Annahme und Frage ist die Aufforderung, die Frage $H \mid G =?$ in der angenommen gedanklichen Situation vor dem Hintergrund des früher erworbenen Domänenwissens zu reflektieren und die Wahrscheinlichkeit des Konditionals $H \mid G$ auszurechnen. Dazu bedarf es eines Schlusses.

c) **Anfrage**: cb) **Schluß**:

Die Adaption des Domänenwissens an die Annahmen erfolgt mit folgender mathematischen Optimierungsaufgabe:

$$P^{**} = \arg\min R(Q, P^*), \text{ s.d. } Q(F_j \mid E_j) = y_j, j = 1...J. \quad (3)$$

c) **Anfrage**: cc) **Beantwortung der Frage**: Es bleibt nun noch die Berechnung von

$$P^{**}(H \mid G) = z. \quad (4)$$

z ist die Antwort auf die Frage $H \mid G =?$

2 Maße informationstreuer Wissensverarbeitung

2.1 Bedingtheit, Entropie und Information

Wissen ist die (wechselseitige) Bedingtheit von Aussagen durch Aussagen. Je sicherer man aus der Richtigkeit von Aussagen auf die Richtigkeit anderer Aussagen schließen kann – da sie einander bedingen – umso mehr weiß man! Im Umkehrschluß ist Nicht-Wissen die (wechselseitige) Un-Bedingtheit von Aussagen durch Aussagen.

Da unser Modell probabilistisch ist, sagen wir ein wenig genauer: Wissen in P ist die mittlere Abweichung bedingter Wahrscheinlichkeiten auf $\mathcal{L} \mid \mathcal{L}$ gegenüber denen in der Gleichverteilung P^0. In P^0 ist die Vorhersagbarkeit von Aussagen bei Kenntnis anderer Aussagen minimal, die (bedingte) Unsicherheit aufgrund der probabilistischen Unabhängigkeit maximal. Mit zunehmender wahrscheinlichkeitstheoretischer Bedingtheit wächst das Wissen.

Nun ist seit JAGLOM, JAGLOM [2] bekannt, daß die in einem P auf $\{v\}$ inhärente Unsicherheit (über den Ausgang bedingter Experimente) durch die Entropie

$$H(P) = -\sum_{v \subset t} P(v) \cdot ld\ P(v) \tag{5}$$

gemessen wird. H hat die Dimension [bit].

Rödder und Reucher übertragen in [6] auf Seite 8 diese Überlegungen auf die probabilistische Konditionalstruktur in $\mathcal{L} \mid \mathcal{L}$. Das Funktional (5) mißt auch die in P steckende mittlere Un-Bedingtheit oder Unabhängigkeit von Aussagen durch Aussagen. In der gleichen Arbeit wird nachgewiesen, daß die relative Entropie

$$R(Q,P) = \sum_{v \subset t} Q(v) \cdot ld \frac{Q(v)}{P(v)} \tag{6}$$

die Abweichung der mittleren Un-Bedingtheit oder Unabhängigkeit in Q relativ zu der in P mißt. R hat die Dimension [bit].

Man zeigt sofort, daß die relative Entropie bzgl. der Gleichverteilung die Gleichung $R(Q,P^0) = ld \mid \{v\} \mid -H(Q)$ erfüllt. Die Änderung des relativen Unabhängigkeitsmaßes bei Übergang von P^0 zu Q ist stets gleich dem Wert maximal möglicher Entropie $ld \mid \{v\} \mid$ abzüglich der Entropie.

Unabhängigkeitsstruktur und *Unsicherheitsstruktur* sind zwei verschiedene Sichtweisen des gleichen Tatbestands. Ist $P(A)$ die Wahrscheinlichkeit von A, wächst $-ld\ P(A)$ [bit] Information zu, falls A eintritt. Jedes Lehrbuch über Informationstheorie zeigt diesen Zusammenhang auf. Ist $P(B \mid A)$ die Wahrscheinlichkeit des Konditionals $B \mid A$, wächst die Information $-ld\ P(BA) - (-ld\ P(A))$ zu, falls es eintritt. Und das ist der Fall, wenn ein Individuum der Population die Eigenschaft $v \subset BA$ hat! Information ist in beiden Fällen Reduktion von Unsicherheit. Vor diesem Hintergrund erfahren Entropie und relative Entropie neue Interpretationen. In [5] wird nämlich gezeigt, daß die in P über die gesamte Konditionalstruktur $\mathcal{L} \mid \mathcal{L}$ erwartete

Unsicherheitsreduktion ebenfalls gleich der Entropie (5) und daß die Unsicherheitsänderung beim Übergang von P zu Q gleich der relativen Entropie (6) ist.

Wissen ist die Bedingtheitsstruktur in einem zur Verfügung stehenden P über alle Welten v. Wissen(sumfang) ist ein Maß für diese wechselseitige Bedingtheit, nämlich $R(P, P^0)$. Wie so oft verwenden wir den gleichen Begriff für das *Wesen des Wissens* und das *Maß des Wissens.*

Wissenserwerb ist die Verbesserung der Bedingtheitsstruktur in P beim Übergang zu Q. Wissenserwerb(sumfang) ist ein Maß für diese Verbesserung beim Erfahren neuer, mit bereits bekannten kompatibler, Sachverhalte.

$R(Q, P^0) - R(P, P^0) = H(P) - H(Q)$ ist das Maß dieser Wissensverbesserung. Wie so oft verwenden wir den gleichen Begriff für das *Wesen des Erwerbs* und sein *Maß.*

Wissen und Wissenserwerb werden in [bit] gemessen.

2.2 Wissen, Schlußstärke, Einfluß

In diesem Abschnitt erfolgt die Messung der Wissensverarbeitungsschritte aus 1.2. Nach Festlegung der Wissensdomäne wird in b) das a priori-Wissen P^0 auf die Sachverhalte $\mathcal{R} = \{B_i \mid A_i[x_i], i = 1...I\}$ durch Lösung der Aufgabe (2) abgeglichen. Lesen Sie dort nochmals nach! Solange nur die Sachverhalte in $\mathcal{R}$ formuliert sind, kann eine Messung des in ihnen steckenden Wissens noch nicht erfolgen. Der Grund ist die Unüberschaubarkeit der wechselseitigen Beeinflussung zwischen den einzelnen Gleichungen $Q(B_i \mid A_i) = x_i$. Erst durch Lösung von (2) wird P^* erzeugt, das alle Forderungen erfüllt und eine geringstmögliche Änderung der Bedingtheitsstruktur gegenüber P^0 erfährt.

$$P^* \text{ ist Wissen und } R(P^*, P^0) \text{ ist sein Maß.} \qquad (7)$$

Durch seinen Erwerb ist *Information* zugeflossen, nämlich $R(P^*, P^0)$ [bit]. Die mittlere Unsicherheit wurde von $H(P^0)$ zu $H(P^*) = H(P^0) - R(P^*, P^0)$ reduziert. Der Informationsbegriff ist relativ, er bezieht sich hier auf die a priori-Verteilung P^0. Wäre $\bar{P} \neq P^0$ a priori-Verteilung, wären sowohl das aus den Sachverhalten $\mathcal{R}$ durch Minimierung von $R(Q, \bar{P})$ zu erwerbende Wissen als auch die beim Erwerb zufließende Information anders. Ist im Extremfall $\bar{P} = P^*$, enthält $\mathcal{R}$ keine Information für ein mit P^* ausgestattetes a priori-Wissen!

In Punkt c) werden in dem Unterpunkt ca) die Annahmen und die Frage formuliert und in Punkt cb) der Schluß durch Lösen der Aufgabe (3) vollzogen.

$$P^{**} \text{ ist Fallwissen und } R(P^{**}, P^0) \text{ ist sein Maß} \qquad (8)$$

in der angenommenen Situation; es wird in [bit] gemessen.

$R(P^{**}, P^*)$ ist lt. Abschnitt 1.2 die Wissensveränderung durch Evidenziierung von $\mathcal{E}$; sie wird in [bit] gemessen.

$$R(P^{**}, P^*) \text{ mißt die } \textit{Schlußstärke}. \tag{9}$$

Im Punkt cc) wird $P^*(H \mid G) = z$ ausgewertet, womit die Frage beantwortet ist; siehe (4). Die informationstheoretische Abweichung zwischen $(P^*(H \mid G), P^*(\bar{H} \mid G))$ und $(P^{**}(H \mid G), P^{**}(\bar{H} \mid G))$ berechnet sich mittels

$$R((P^{**}(H \mid G), P^{**}(\bar{H} \mid G)), (P^*(H \mid G), (P^*(\bar{H} \mid G))) \tag{10}$$

und ist ein Maß für den Einfluß der Evidenz auf die Antwort, er wird in [bit] gemessen.

3 Vermutung, Ungewißheit und Relevanz

Die Nebenbedingungen $\mathcal{R}$ in (2) stellen ein konvexes Polyeder in Q dar (Umformung!), das i.a. nicht einpunktig ist. Aus ihm wird P^* aus hinlänglich diskutierten Gründen ausgewählt. Berechnet man nun P^{**} unter der Evidenz $\mathcal{E}$ gemäß (3), so ist $P^{**}(H \mid G)$ natürlich nur eine *Schätzung* für die bedingte Wahrscheinlichkeit, die wir ab jetzt *Vermutung* nennen wollen. Werden nämlich weitere Sachverhalte zur Wissensbasis $\mathcal{R} = \mathcal{R}^1$ in Form weiterer Nebenbedingungen $\mathcal{R}^2$ bekannt, kann sich P^* und ggf. auch P^{**} ändern – mit der Folge einer möglicherweise anderen Vermutung: Das konvexe Polyeder birgt *Ungewißheit.* Diese Ungewißheit ist für Spezialfälle meßbar; einen solchen Spezialfall behandeln wir hier.
Dazu vereinbaren wir folgende Symbolik:

$Q \models \mathcal{R}, Q \models \mathcal{E}$ bedeuten $Q(B_i \mid A_i) = x_i, i = 1....I$ bzw. $Q(F_j \mid E_j) = y_j, j = 1...J$. Wir sagen auch: In Q gilt $\mathcal{R}$ bzw. $\mathcal{E}$.

Gegeben sei $\mathcal{R}, \mathcal{E} = \{F \mid \mathrm{t}[1.0]\}$ und $H \mid G$ bei beliebigem G. Zu beantworten ist also eine Anfrage bei sicherer Evidenz und bedingter ($G \neq \mathrm{t}$) oder unbedingter ($G = t$) Frage. Folgt man obiger Überlegung, so ist prinzipiell für jedes $Q \models \mathcal{R}$ F [1.0] zu evidenziieren, mit dem Ergebnis $P^{**}(Q, F)$, und dann $P^{**}(H \mid G)$ auszuwerten. Nun ist aber $P^{**}(H \mid G)$ berechenbar, indem in P^{**} G [1.] evidenziiert und in der Ergebnisverteilung nur H ausgewertet wird. Diese „doppelte Bedingung“$(H \mid G) \mid F$ formt man um zu $(H \mid G) \mid F = H \mid G\, F$. Durch Lösung der Aufgaben

$$\bar{u} = \min Q(H \mid GF) = \frac{Q(HGF)}{Q(GF)} \text{ s.d. } Q \models \mathcal{R} \tag{11}$$

und

$$\bar{\bar{u}} = \max Q(H \mid GF) = \frac{Q(HGF)}{Q(GF)} \text{ s.d. } Q \models \mathcal{R} \tag{12}$$

bestimmt sich hiermit das Ungewißheitsintervall $[\bar{u}, \bar{\bar{u}}]$. In diesem Intervall kann die Antwort zur Frage $H \mid G = ?$ bei sicherer Evidenz variieren.

$m = -ld\bar{u} - (-ld\bar{\bar{u}})$ [bit] ist das zu $[\bar{u}, \bar{\bar{u}}]$ gehörige Ungewißheitsmaß. m mißt die maximal mögliche Informationsdifferenz, wenn man erfährt, daß $H \mid G$ wahr wird. m ist ein Maß für die Ungewißheit im konvexen Polyeder $\mathcal{R}$ hinsichtlich $H \mid G$.

Beispiel (Ungewißheit über Lea Sombe, vgl. [8], [4])

In einer fiktiven Gesellschaft betrachten wir vier Eigenschaften: Student sein oder nicht $S = s/\bar{s}$; jung sein oder nicht $Y = y/\bar{y}$; Familienstand ledig, verheiratet, zusammenlebend $M = s/m/c$; Eltern sein oder nicht $P = p/\bar{p}$.

Folgende Sachverhalte sind bekannt:

- Etwa 90 % aller Studenten sind jung, $Y = y \mid S = s$ [0.9].
- Etwa 80 % aller jungen Leute sind ledig, $M = s \mid Y = y$ [0.8].
- 70 % der Ledigen sind jung, $Y = y \mid M = s$ [0.7].
- 30 % junger Menschen studieren, $S = s \mid Y = y$ [0.3].
- Lebt jemand in einer eheähnlichen Gemeinschaft, so ist er zu 80 % jung, $Y = y \mid M = c$ [0.8].
- Studenten mit Kindern sind nur zu 10 % ledig, $M = s \mid S = s \wedge P = p$ [0.1].

Diese Sachverhalte bilden die Menge probabilistischer Konditionale $\mathcal{R}$; sie wurden durch die Expertensystemshell SPIRIT erlernt, und es wurde über dem Merkmalsraum die Verteilung P^* erzeugt.

Eine erste Anfrage wird gestellt: Lea Sombe ist Studentin und hat ein Kind; ist sie jung? In der Terminologie dieses Aufsatzes: $\mathcal{E} \equiv \{S = s \wedge P = p\ [1.0]\}, H \mid G \equiv Y = y$. Die Berechnung von P^{**} und die Auswertung $P^{**}(Y = y) = 0.81$ zeigt die Vermutung: Mit 81 % Wahrscheinlichkeit ist Lea Sombe jung. Diese Antwort ist jedoch äußert ungewiß! Gemäß (11), (12) und $\mathcal{E} \equiv \{S = s \wedge P = p\ [1.0]\}$ errechnet sich das Ungewißheitsintervall zu $[\bar{u}, \bar{\bar{u}}] = [0, 1]$ mit $m = \infty$ [bit].

Eine zweite Anfrage lautet: Über Lea Sombe wissen wir, daß sie Studentin und ledig ist; ist sie jung? Diesmal lautet die Antwort $P^{**}(Y = y) = 0.95$. Der Familienstand *ledig* ist also bei Studenten ein stärkeres Indiz für Jugend als es die Mutterschaft ist. Und diese Erkenntnis ist auch noch wesentlich gewisser: $[\bar{u}, \bar{\bar{u}}] = [0.75, 1]$ mit $m = 0.42$ [bit].

Mit den bisherigen Überlegungen in diesem Abschnitt, untermauert durch Beispiele, ist der Ungewißheitsbegriff meßbar geworden. Nun wenden wir uns nochmals der Frage der Bedeutung oder *Relevanz* zu. Nach Erlernen von $\mathcal{R}^1 = \mathcal{R}$ und dem Erhalt von Zusatzinformation in Form von weiteren Sachverhalten $\mathcal{R}^2$ erhöht die gemeinsame Verarbeitung von $\mathcal{R}^1 \cup \mathcal{R}^2$ im allgemeinen das Wissen. Dieser Wissenszuwachs, so sagt unsere Intuition, muß nicht immer auch *relevant* für eine bestimmte Anfrage sein. Wir wollen daher definieren, daß die Zusatzinformation $\mathcal{R}^2$ relevant für die Anfrage $\mathcal{E}, H \mid G$ ist, wenn das Unbestimmtheitsmaß m unter $\mathcal{R}^1 \cup \mathcal{R}^2$ kleiner als das unter $\mathcal{R}^1$ ist.

In Fortführung des Beispiels Lea Sombe kann man mittels der Shell SPIRIT zeigen, daß der zusätzlich bekannt gewordene Sachverhalt $\mathcal{R}^2 = \{P = p \mid S = s\ [0.15]\}$ die Ungewißheitssituation grundlegend bessert. Zwar bleiben die Vermutungen, eine studierende Mutter bzw. eine ledige Studentin sei jung, mit 0.81 und mit 0.95 unverändert, jedoch reduziert sich die Ungewißheit zu $[\bar{u}, \bar{\bar{u}}] = [1/3, 1]$ mit $m = 1.58$ [bit] bzw. zu $[\bar{u}, \bar{\bar{u}}] = [0.79, 1]$ mit $m = 0.35$ [bit].

Sind Sie interessiert an der Expertensystemshell SPIRIT, mit der der hier vorgestellte Wissensverarbeitungsprozeß auch für große betriebswirtschaftliche Modelle unterstützt wird, so besuchen Sie unsere Internetseite http://www.xspirit.de.

Literatur

1. Calabrese, P. M.(1991): Deduction and Inference Using Conditional Logic and Probability, Conditional Logic in Expert Systems, I. R. Goodman, M. M. Gupta, H. T. Nguyen, G. S. Rogers (editors), Elsevier Science Publishers B. V.
2. Jaglom, A. M., Jaglom, I. M. (1967) Wahrscheinlichkeit und Information, VEB Deutscher Verlag der Wissenschaften, Berlin.
3. Kern-Isberner, G. (1998) Characterizing the principle of minimum cross-entropy within a conditional-logical framework, Artificial Intelligence, Vol. 98, 169-208.
4. Rödder, W., Kern-Isberner, G. (1997) Lea Sombe und entropie-optimale Informationsverarbeitung mit der Expertensystem-Shell SPIRIT, OR Spektrum, 19, 41-46.
5. Rödder, W. (2001) Knowledge Processing under Information Fidelity, Proc. IJCAI 2001, Seattle, Vol. 1, 749-754.
6. Rödder, W., Reucher, E. (2001) Wissen und Folgern aus relevanter Information Diskussionsbeitrag Nr. 301 des FB WiWi, FernUniversität in Hagen.
7. Shore, J.-E., Johnson R.-W. (1980), Axiomatic Derivation of the Principle of Maximum Entropy and the Principle of Minimum Cross-Entropy, IEEE Transact Inf Theory IT-26(1): 26-37.
8. Sombe, L. (1992) Schließen bei unsicherem Wissen in der Künstlichen Intelligenz, Vieweg, Braunschweig, Wiesbaden.

Constraint Satisfaction by Means of Dynamic Polyhedra

Miguel A. Salido, Adriana Giret, Federico Barber

Universidad Politécnica de Valencia, 46071, Valencia, Spain

Abstract. *Nowadays, many real problems in the Artificial Intelligence environments, can be efficiently modelled as Constraint Satisfaction Problems (CSP's) and can be solved by means of Operational Research techniques. It is well known that any non-binary CSP can be transformed into an equivalent binary one, using some of the current techniques. However, this transformation may not be practical in problems with certain properties. Therefore, it is necessary to manage these non-binary constraints directly. In this paper, we propose an algorithm called "HSA≠" that solves non-binary constraint satisfaction problems in a natural way as an incremental and non-binary CSP solver. This non-binary CSP solver carries out the search through a polyhedron that maintains in its vertices those solutions that satisfy all non-binary constraints.*

1 Introduction

Nowadays, many researchers are working on combined methods of Artificial Intelligence and Operational Research (AI/OR) because many real problems can be modelled as constraint satisfaction problems (CSP's) and can be solved using linear programming techniques. These include problems from fields such as artificial intelligence, operational research, expert systems, databases, etc. Most of these problems can be naturally modelled using non-binary (or n-ary) constraints. In the constraint satisfaction literature, the need to address issues regarding non-binary constraints has only recently started to be widely recognised. Researchers have traditionally focused on binary constraints. The basic reasons were the simplicity of dealing with binary constraints compared to non-binary ones and the fact that any non-binary constraint satisfaction problem can be transformed into an equivalent binary one [5]. However, this transformation has several drawbacks:

- It should be mentioned that transforming a non-binary CSP into a binary one produces a significant increase in the problems size, so the transformation may not be practical [2][4].

- The translation process generates new variables, which may have very large domains, causing extra memory requirements for algorithms. So, in some problems, solving the binary formulation can be very inefficient [1].

In any case, this forced binarization generates unnatural formulations, which cause extra difficulties for constraint solver interfaces with human users [3].

In this paper, we propose an algorithm called "*The Hyperpolyhedron Search Algorithm*" (HSA≠) that handles non-binary CSPs by means of OR techniques where the non-binary constraints (inequalities) are hyperplanes that are intersected in order to obtain the polyhedron vertices. However, in Artificial Intelligent framework, traditional CSP techniques obtain the solution by searching systematically through the possible assignments of values to variables. These AI techniques increase their complexity exponentially with the domain length and the number of disequational constraints, so it is necessary to use methods which are guided by heuristics. By combining AI/OR techniques, we can obtain methods, like HSA≠, that do not depend on the variable domain length and on the number of disequational constraints.

Thus, HSA≠ overcomes some of the weaknesses of other techniques. Moreover, it can manage constraints that can be inserted incrementally into the problem without needing to solve the entire problem again.

2 Preliminaries

Briefly, the constraint satisfaction problem (CSP) that HSA≠ manages consists of:

- a set of variables $X = \{x_1, ..., x_n\}$;
- each variable $x_i \in X$ has a continuous set D_i of possible values (its domain);
- and a finite collection of constraints $C = \{c_1, ..., c_p\}$ restricting the values that the variables can simultaneously take.

A solution to a CSP is an assignment of a value from its domain to every variable, such that all constraints are satisfied.

The objective in a CSP may be:

- to determine whether a solution exists,
- to find one solution, with no preference as to which one or to find all solutions,
- to find the variable domains,
- to find an optimal, or a good solution by means of an objective or multi-objective function defined in terms of some variables.

2.1 Notation and definitions

In this section, we summarize the notation that is used in this paper.

Generic: The number of variables in a CSP will be denoted by n. The domain of the variable x_i will be denoted by D_i. The constraints will be denoted by c, and all the constraint have the maximum arity n. We will always use this notation when analysing the behaviour of the algorithms.

Variables: To represent variables, we will use x with an index, for example x_1, x_i, x_n.

Domains: The domain of the variable x_i will be noted by $D_i = [l_i, u_i]$, so the domain length of the variable x_i is $u_i - l_i$. It is important to realise that the domain is continuous.

Constraints: Let $X = \{x_1, ..., x_n\}$ be a set of real-valued variables. Let α be a polynomial of degree n (i.e., $\alpha = \sum_{i=1}^{n} p_i x_i$) over X and b an integer. A linear relation over X is an expression of the form $\alpha r b$ where $r \in \{<, \leq, =, \neq, \geq, >\}$. A linear disequation over X is an expression of the form $\alpha \neq b$. A linear equality over X is an expression of the form $\alpha = b$. The constraints that we are going to manage are linear relations and linear disequations of the form:

$$\text{Inequalities: } \sum_{i=1}^{n} p_i x_i \leq b \tag{1}$$

$$\text{Disequations: } \sum_{i=1}^{n} p_i x_i \neq b \wedge F(x) \neq b : F \text{ non-linear} \tag{2}$$

where x_i are variables ranging over continuous intervals $x_i \in [l_i, u_i]$, b is a real constant, and $n \geq 1$. Using the above constraints, equalities can be written as conjunctions of two inequalities. Similarly, strict inequalities can be written as a conjunction of an inequality and an disequation. Thus, we can manage all possible combinations: $\{<, \leq, =, \neq, \geq, >\}$.

3 Specification of the Hyperpolyhedron Search Algorithm

Initially, HSA≠ can be considered as a classic CSP solver, where there is a static set of constraints that the solutions must satisfy. The graph and pseudo-code specification of HSA≠ is presented in Figures 1 and 2, respectively. HSA≠ generates an initial polyhedron (step 1) with 2^n vertices created by means of the Cartesian Product of the variable domain bounds $([l_1,u_1]\times[l_2,u_2]\times\cdots\times[l_n,u_n])$.

For each (≤) constraint, HSA≠ carries out the consistency check (step 2). If the (≤) constraint is not consistent, HSA≠ returns *not consistent problem*, else, HSA≠ determines if the (≤) constraint is not redundant, updating the polyhedron (step 3). This updating is carried out by Linear Programming techniques. The hyperplane generated by this (≤) constraint intersects with the polyhedron and thus, this hyperplane is the new face of the updated polyhedron. Finally, when all (≤) constraint are analysed and the problem is consistent with these (≤) constraints, HSA≠ checks the consistency with the (≠) constraints (step 4). The consistency study with the (≠) constraints does not update the polyhedron. This study only checks the consistency of the (≠) constraints with the polyhedron vertices.

Thus, solutions to CSP are the all vertices and all the convex combinations between any two vertices which satisfy all disequational constraints.

Finally, HSA≠ can obtain important results such as: (1) the problem consistency; (2) one or all solutions; (3) the new variable domains; and (4) the vertex of the polyhedron that minimises or maximises an objective or multi-objective function.

Furthermore, when HSA≠ finishes its static behaviour (classic CSP solver), new constraints can be dynamically inserted into the problem, and HSA≠ studies the consistency check such as an incremental CSP solver.

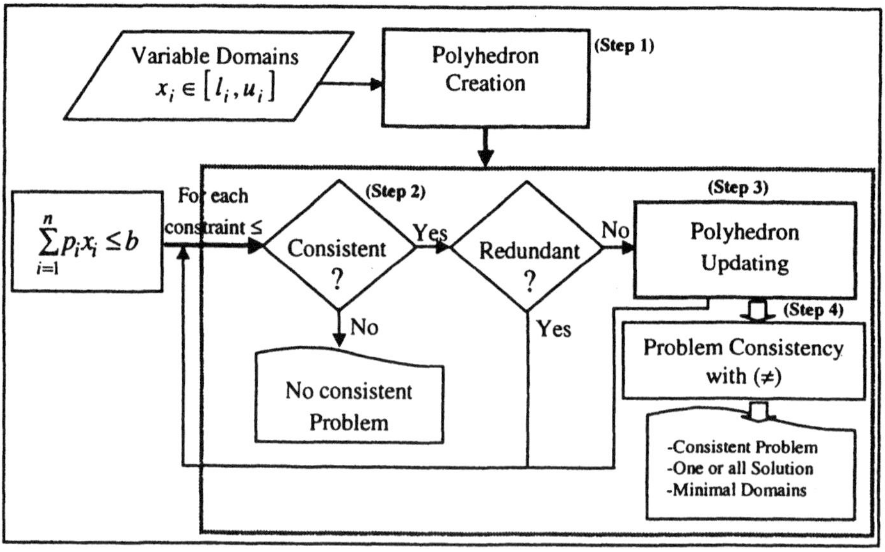

Fig. 1. General Scheme of the Hyperpolyhedron Search Algorithm.

HSA≠(Dimension, Domains, Constraints≤ , Constraints≠, Solutions)

Step 1 $ListV \leftarrow$ *Polyhedron_creation (Dimension, Domains, Constraints≤);*

$L_{yes} \leftarrow \phi$; $L_{no} \leftarrow \phi$;

Step 2 For each $C_i \in$ *Constraints≤ do:*

{

$\forall v_i \in ListV$ do:

{

If *Satisfy*(C_i, v_i) then: *// checks if the v_i satisfies the constraint C_i*

$L_{yes} \leftarrow L_{yes} \cup \{v_i\}$; *// C_i is consistent with the system*

else $L_{no} \leftarrow L_{no} \cup \{v_i\}$;

}

If $L_{yes} = \phi \Rightarrow STOP$; *// C_i is not consistent with the system*

If $L_{no} = \phi \Rightarrow$ " *C_i is consistent and redundant* ";

else

Step 3 $\forall v' \in L_{no}$ *Polyhedron_Updation* (v' , L_{yes}, L_{no});

}

Step 4 *Satisfaction*(*Constraints≠*, L_{yes} , *Solutions*);

return output; // HSA≠ returns the consistency check, and the extreme solutions

Polyhedron_Updation (v' , L_{yes} , L_{no})

For each adjacent vertex v of v' do:

HSA obtains the straight line $\bar{l}$ that unites both v' and v points.

HSA intersects $\bar{l}$ with the polyhedron obtaining the new point v''

$L_{yes} \leftarrow L_{yes} \cup v''$

return L_{yes};

Fig. 2. Hyperpolyhedron Search Algorithm.

4 Analysis of Hyperpolyhedron Search Algorithm

The HSA≠ spatial cost is determined by the number of vertices generated. Initially, the HSA≠ generates 2^n vertices, where n is the number of problem variables. For each constraint (step 2), HSA≠ might generate n new vertices and eliminate only one. Thus, the number of polyhedron vertices is $2^n+c_{\leq}(n-1)$ where $c_{\leq}$ is the number of (≤) constraints. Therefore, the spatial cost is $O(c_{\leq}2^n)$.

The temporal cost is divided into three steps: initialisation, consistency check and actualisation. The initialisation cost (step 1) is $O(2^n)$ because the algorithm only generates 2^n vertices. For each (≤) constraint (step 2), the consistency check cost depends linearly on the number of polyhedron vertices, but not on the variable domains. Therefore the temporal cost is $O(2^n)$. Finally, the actualisation cost (step 3) depends on the number of variables $O(2^n)$. Also, the algorithm checks the consistency with each (≠) constraint in O(n).

Thus, the temporal cost is: $O(2^n)+c_{\leq}*\left(O(2^n)+O(2^n)\right)+c_{\neq}*O(n)\Rightarrow O(c_{\leq}2^n)$.

5 Evaluation of Hyperpolyhedron Search Algorithm

In this section, we compare the performance of HSA≠ with some of the more current CSP solvers. We have selected Forward-checking (FC) and Real Full Look-ahead (RFLA) because of they are the most appropriate techniques that are able to handle this problem typology . To evaluate this performance, the computer used for the tests was a PIII-800 with 128 Mb. of memory and Windows NT operating system.

The problems generated to evaluate the performance depended on four parameters $<v,c_{\leq},c_{\neq},d>$, where v was the number of variables, $c_{\leq}$ the number of inequational constraints, $c_{\neq}$ the number of disequational constraints and d the length of the variable domains. The problems were randomly generated by modifying the parameters. The constraints $c_{\leq}$ and $c_{\neq}$ are constraints type (1) and (2) respectively, with the coefficients $p_i\neq 0$.

Thus, each of the graphs shown, set three parameters and varied the other one in order to evaluate the algorithm performance when this last parameter increased. We tested 100 test cases for each problem and each value of the variable parameter, and we present the mean CPU time for each technique.

Four graphs are shown in Fig. 3 which correspond to the four significant parameters. The *unsolved problems* were assigned a 200-second run time. Thus, these graphs contain a horizontal asymptote in time=200.

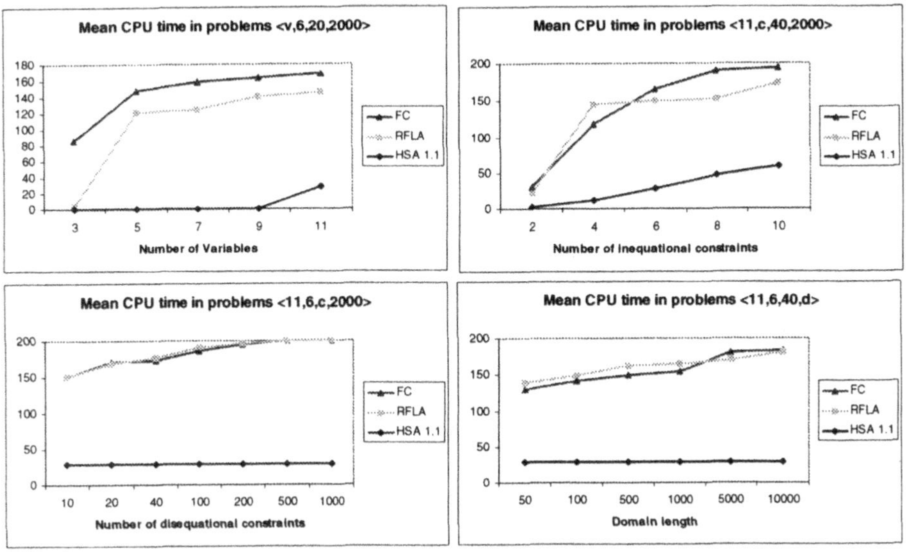

Fig. 3. Temporal Cost in problems with different parameters

Each graph in Fig. 3 shows that HSA≠ had better behaviour than FC and RFLA. The left upper graphic shows that HSA≠ increased it temporal cost later than the other approaches. The right upper graph summarises that our algorithm had a lower temporal cost than the others, when the number of inequational constraints increased. The lower graphs show that HSA≠ did not increase its temporal complexity when the number of disequational constraints and the domain length increased. However, FC and RFLA were unable to solve many of these problems.

C≤ = 6	V	3	5	7	9	11
C≠ = 20	RFLA	0	52	54	63	65
D=2000	FC	34	64	72	77	78
V = 11	C≤	2	4	6	8	10
C≠ = 40	RFLA	10	69	71	77	85
D = 2000	FC	14	52	80	94	96
V = 11	C≠	10	20	40	100	200	500
C≤ = 6	RFLA	75	82	86	92	95	100
D = 2000	FC	75	85	86	92	96	100
V = 11	D	± 50	± 100	± 500	± 1000	± 5000	± 10000
C≠ = 6	RFLA	66	70	78	82	84	90
D = 2000	FC	62	70	74	76	88	90

Table 1. Number of unsolved problems

The number of unsolved problems is presented in table 1. HSA≠ was able to satisfactorily solve all the problems (2200 problems) while FC was unable to satisfactorily solve 1655 problems and RFLA 1546.

6 Conclusion and Future Works

In this paper, we have proposed an algorithm called HSA≠ as an incremental and non-binary CSP solver. This proposal carries out the consistency study through a polyhedron that is updated by means of LP techniques, and maintains (in its vertices) those values that satisfy all non-binary constraints. We are currently, working on disjunctive and non-binary constraint satisfaction problems to be modelled as distributed and incremental CSP solvers. These proposals would work on a polyhedron whose vertices are also polyhedra, and the disjunctive and non-binary CSP's are solved by means of metaheuristics techniques. Also, it is appropriate to generate algorithms to be configured dynamically depending on the user's inputs. This proposal will be composed by HSA≠ as a complete algorithm, helped by heuristics like OFHH[6] and NFHH[7].

References

[1] Bacchus, F., van Beek, P. (1998) On the conversion between non-binary and binary constraint satisfaction problems. In proceeding of AAAI-98, 311-318

[2] Bessière, C., Meseguer, P., Freuder, E.C., Larrosa, J. (1999) On Forward Checking for Non-binary Constraint Satisfaction. In Proc. Principles and Practice of Constraint Programming (CP-99), 88-102

[3] Bessiere, C. (1999) Non-Binary Constraints. In Proc. Principles and Practice of Constraint Programming (CP-99), 24-27

[4] Larrosa J. (1998) Algorithms and Heuristics for total and partial Constraint Satisfaction, Phd Disertation, Barcelona, Spain,

[5] Rossi, F., Petrie, C., Dhar, V. (1990) On the equivalence of constraint satisfaction problems. In proceeding of European Conference of Artificial Intelligence, ECAI-90, 550-556

[6] Salido, M.A., Giret, A., Barber, F. (2001) A Non-binary Constraint Satisfaction Solver: The One-face Hyperpolyhedron Heuristic. In book: Research and Development in Intelligent Systems XVIII. (Ed. Springer Verlag).

[7] Salido, M.A., Giret, A., Barber, F. (2001) A combination of AI and OR methods for solving Non-binary Constraint Satisfaction problems. In Proceedings of Workshop on New Results in Planning, Scheduling and Design (PUK2001), 78-88

Multi-Agent FX-Market Modeling by Neural Networks

H.-G. Zimmermann, R. Grothmann, R. Neuneier

Siemens AG, Corporate Technology, Munich, Germany

Abstract. A market is basically driven by a superposition of agents decisions. The price dynamic results of the excess demand / supply created on the micro level. The behavior of a few agents is well understood by game theory. In case of a large number of agents one may refer to the assumption of an atomic market structure, which allows the aggregation of agents by statistics. We can omit the latter assumption, if we model the market by a multi-agent approach.

This paper deals with the *explicit* and *implicit* dynamics of the market price, and their implementation into neural networks. Neural networks allow the fitting of high dimensional nonlinear models, which is often utilized in econometrics. Furthermore, there is a close relationship to economics because a neuron can be interpreted as a decision making model. Thus, a neural network models the interaction of many decisions and, hence, can be interpreted as a market mechanism.

1 Introduction

A market can be interpreted as a network consisting of a large number of interacting agents. Agents buying / selling decisions can be summed up, resulting in the aggregated demand / supply of the market. These aggregates give a rise to prospective market price changes. Hence, the natural way to explain or even to predict prospective market prices is to give a detailed causal analysis of all agents decisions and their interaction.

Surveying recent literature, one can find several publications attempting to model a (financial) market in the latter manner. For an overview see [2].

An alternative way of describing market prices is provided by the field of statistics: Here, economic relationships are modeled by referring to quantitative variables and inter-variable dependencies. From an econometrical point of view, the contribution of neural networks to market modeling can be described as follows: Because of their functional approximation abilities, neural networks allow the fitting of high dimensional nonlinear economic systems. As a further advantage, neural networks are able to handle high dimensional and nonlinear interactions at the same time. This is often required in economics.

On the other hand, there is a close relationship between neural networks and economic theory: A neuron can be seen as an elementary economic decision making process. From this point of view, a neural network reflects the interaction of a large number of decision makers resp. a market process [6].

The purpose of this paper is to establish a crosslink between econometric and economic modeling by the application of neural networks. In chapter 2, we explain the basics of a single agent and the elementary market mechanism. Thereafter two market paradigms are concerned. The first one (sec. 3, 4) considers the impact of the microeconomic structure on the price formation, whereas the second one (sec. 5, 6) discusses the influence of the market on the micro level. Both paradigms will be embedded into neural networks, which are evaluated in an empirical study of the DEM / USD FX-market (sec. 7).

2 From Agents Behavior to Market Dynamics

Looking at the micro-structure of a single market, one surely can identify various market participants interacting with one another. Since these agents decide independently from each other in accordance with their objective function (e. g. profit maximization), it is *not* quite clear, if their trading activities entirely match: if so, the market is in an equilibrium state. Else, the amount of planed purchases is bigger than the quantity of desired sales, or vice versa.

Both situations of a market disequilibrium have an impact on the market price: In case of an excess demand, a price increase will re-balance demand and supply, because more agents will come to the market willing to sell respective goods at the higher price, while other agents who planed to purchase are withdrawn from the market. Likewise, an excess supply is compensated by lowering the market price, since agents scheduled to sell are withdrawn from the market and some agents add themselves to the list of buyers.

Here, we apply the market mechanism sightly mentioned above to a FX-market, where a country's currency is expressed in another country's currency (e. g. on the DEM / USD FX-market, US-Dollars are priced in German Mark). Agents participating in a FX-market try to maximize their profit by anticipating changes in the FX-rate to expand their net capital balance. Referring to the example of the DEM / USD FX-market, agents believing in a devaluation of the DEM would allocate their funds in USD, while agents predicting a devaluation of the USD place their funds in DEM.

Modeling the trading behavior of the agents, we assume, that each agent has two accounts reflecting his dispositions. Let x_t^a(in USD) reflect the balance of the USD account of agent a in time period t. Analogous, y_t^a(in DEM) is the current balance of account y expressed in DEM.

If agent a decides to deal z_t^a units of USD, this will automatically have an effect on the account balances. For example, if he buys $z_t^a = 100$ USD, an equivalent value of DEM will be subtracted from y_t^a(in DEM), while the 100 USD are credited to x_t^a(in USD). Regarding to this example, the trading behavior of agent a can be modeled by two simple equations, reflecting the upcoming cash balances x_{t+1}^a(in USD) and y_{t+1}^a(in DEM). Recognizing that p_t is the DEM / USD FX-rate, we obtain

$$\begin{aligned} x_{t+1}^a(\text{in USD}) &= x_t^a(\text{in USD}) + z_t^a \\ y_{t+1}^a(\text{in DEM}) &= y_t^a(\text{in DEM}) - p_t z_t^a \end{aligned} \tag{1}$$

Eq. 1 describes the elementary trading mechanism of each agent a in a very compact form. If $z_t^a < 0$, the agent decides to sell USD. For simplicity, transaction costs are equal to zero. Furthermore, we assume that each agent is doing one trading decision ($z_t^a \gtrless 0$) per time period t.

In order to solve the decision problem of buying / selling USD, we refer to an elementary decision making process, which can be characterized in *three* stages [6]. These stages are (*i.*) information filtering, market evaluation and (*iii.*) performance of a trading activity.

For an explanation of this decision making process, let us focus on an agent dealing in a FX-market. Nowadays, due to the progress in the field of information technology, this agent is overwhelmed by a huge flood of information. In order to handle the latter, the agent has to focus his attention on a couple of news, which, in his opinion, seem to have the deepest impact on the related FX-rate. This procedure of sorting out relevant items of information is the *first* stage of the decision making process. In the *second* stage, the agent has to analyze the collected pieces of information. This evaluation of the data leads directly to the *third* stage of the decision making process, where the agent carries out an action, which refers to his market prediction.

The elementary decision making scheme can be modeled by a single neuron, which precisely reflects the *three* stages of decision making.

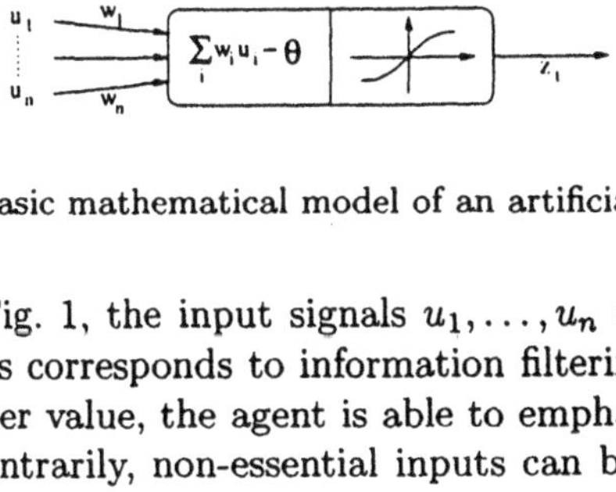

Fig. 1. Basic mathematical model of an artificial neuron.

As depicted in Fig. 1, the input signals $u_1, \ldots, u_n$ are weighted by factors $w_1, \ldots, w_n$. This corresponds to information filtering. By adjusting the weights w_i to a higher value, the agent is able to emphasize the importance of the inputs u_i. Contrarily, non-essential inputs can be ignored by setting their weights equal to zero. Adding up the input signals and comparing it to a threshold value, the agent forms his opinion of the further development of the FX-market. Afterwards the result is transferred through a non-linearity (e. g. tanh(·)) in order to generate the neurons output z_t^i. The output reflects the trading activity of the agent. Regarding to this interpretation, a neural network can be seen 'as an interaction model between a large number of decision makers (model of a market process)' [6].

Besides the decision process, an objective function is needed which guides the agents behavior. Typically, an agent is driven by a profit maximization:

$$\frac{1}{T}\sum_t \ln\left(\frac{p_{t+1}}{p_t}\right) z_t^a \rightarrow \max_{w_i,\theta} \tag{2}$$

Following Eq. 2, each agent tries to utilize upcoming FX-rate shifts to expand his net capital balance. This is done by maximizing the average profit gained from the trading over all time periods.

Up to now, the agents are unable to interact with each other. In order to enable the agents to interact, we include a market clearance condition:

$$\sum_a z_t^a = 0 \quad \text{for} \quad t = 1, \ldots, T \tag{3}$$

The market clearance is the macro economical side of the FX-market. As stated in Eq. 3, the sum of all buying and selling decisions have to be equal to zero. Hence, agents plans are indirectly associated with each other.

3 Explicit Dynamics of the FX-Rate

The market clearance of Eq. 3 is an economic equilibrium condition for the FX-market: If the market is in an equilibrium, we have market clearance, i. e. demand and supply of the traded currency are balanced. Contrarily, if we denote a violation of the market clearance condition, the FX-market is in a disequilibrium. We either observe an excess demand or an excess supply.

This motivates us to define the *explicit market price paradigm*, which claims that the excess demand / supply is the cause of market price changes. The price level on the market is influenced by agents decisions.

In case of a market equilibrium, no price adaptations are required:

$$\sum_a z_t^a = 0 \quad \Rightarrow \quad \ln\left(\frac{p_{t+1}}{p_t}\right) = 0 \tag{4}$$

In contrast, if the aggregated decisions are not zero, there is either an excess demand or supply. If so, price shifts return the market to an equilibrium. Recognizing, that $c > 0$ is a constant, the required price shift is

$$\sum_a z_t^a \neq 0 \quad \Rightarrow \quad \ln\left(\frac{p_{t+1}}{p_t}\right) = c \sum_a z_t^a \tag{5}$$

The equilibrium price shift π_t can be estimated by scaling the excess demand / supply by a positive constant $c > 0$. For example, if we observe an *excess supply*, the total of agents planed sales $\sum z_t^a < 0$ exceeds the number of desired purchases $\sum z_t^a > 0$. Since the supply of the concerned currency is bigger than its demand, the particular currency is closed to a devaluation. This leads to a declining exchange rate p_t, i. e. a negative price shift $(\pi_t < 0)$. This price shift is technically computed by scaling the *negative* sum of the z_t^a by $c > 0$. In case of an excess demand, a *positive* price shift $\pi_t^a > 0$, i. e. a revaluation of the concerned currency will balance demand and supply. As before, the price shift is estimated by Eq. 5: Multiplying the *positive* sum of the z_t^a with a constant $c > 0$ always results in a *positive* price shift π_t.

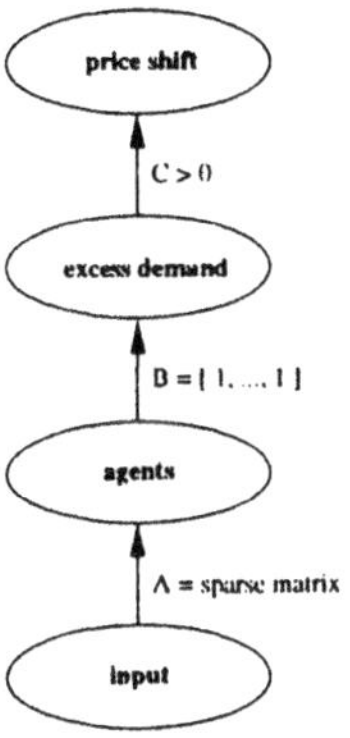

Fig. 2. Multi-agent modeling of the explicit dynamics of the FX-rate with a *four* layer feed-forward neural network architecture. Due to visual clarity, the bias weights θ to the agents cluster are suppressed.

4 Modeling the Explicit Market Dynamics

Now, we integrate the explicit price dynamics into a neural network (Fig. 2).

The input cluster contains N input signals. The agents cluster models each agent by a single neuron with a tanh(.) squashing. Therefore the agents cluster consists of n neurons representing n agents participating in the FX-market. The behavior of a each agent is guided by profit maximization. Here, we use the (*prof-max*) error function: $-\sum(\textit{target} \cdot \textit{output}) \rightarrow \min$ [5].

Matrix A is a sparse connector, i. e. most of the weights are equal to zero. This construction is based on the following two arguments: (1.) due to the information filtering of the decision making process, to ignore useless information; (2.) we have to guarantee a wide spectrum of different trading decisions, i. e. heterogeneous decision behavior. Thus, the available information is strewn among our agents. Note, that we initialize matrix A at random.

The computation of the excess demand / supply, is done by the fixed matrix ($B = (1, ..., 1)$), which aggregates agents decisions. Then, the price shift is estimated by scaling the excess demand.

5 Implicit Dynamics of the FX-Rate

In contrast to the *explicit* price dynamics, the *implicit* market paradigm argues, that the market treats the price shift π_t^i as a control parameter to avoid situations of a market disequilibria. Thus, the market directly influences agents decisions by adjusting the market price. Being so, agents decisions z_t^a are a function of both, the inputs x_t and the price shift $\pi_t = \ln(\frac{p_{t+1}}{p_t})$:

$$ex(x_t, \pi_t) = \sum_a z_t^a(x_t, \pi_t) = 0 \quad \text{with} \quad \frac{\delta ex(.)}{\delta \pi_t} < 0 \tag{6}$$

Eq. 6 reflects the implicit market dynamics. The market clearance condition depends not only on the external inputs but also on the price shift π_t. Thus, π_t acts as a control parameter of the market to avoid disequilibria.

To compute the implicit market price, we have to find a price shift π_t such that the excess demand $ex(\pi_t)$ is equal to zero. This can be done by calculating a fix-point π iterating a series $\{\pi^n\}$ which converges to π:

$$\pi^{n+1} = \pi^n + \epsilon \cdot ex(\pi^n) \tag{7}$$

The series of Eq. 7 converges to a fix-point π, if ϵ is positive and sufficiently small. The convergence properties of Eq. 7 can be proofed by the contraction inequality of Eq. 8. Note, that the partial derivative of $ex(.)$ is negative [6].

$$\|\pi^{n+1} - \pi\| \leq \|1 + \epsilon \cdot ex'(\tilde{\pi})\| \cdot \|\pi^n - \pi\| \tag{8}$$

6 Modeling the Explicit Market Dynamics

In Fig. 3 and 4 we introduce a fix-point neural network to model the implicit market dynamics. Here, we distinguish between a training and testing phase.

Fig. 3 depicts the training phase of the implicit price dynamics, which is done by a feedforward neural network. Depending on the inputs x_t and price shifts π_t the agents optimize their expected return, while the market maker adjusts the price to vanish excess demands. Influencing the agents decisions making, matrix D provides the agents with information about the price shifts π_t. Since we have to guarantee that the derivative of the excess demand is negative (Eq. 6), we only allow negative weights $d_i < 0$ in D.

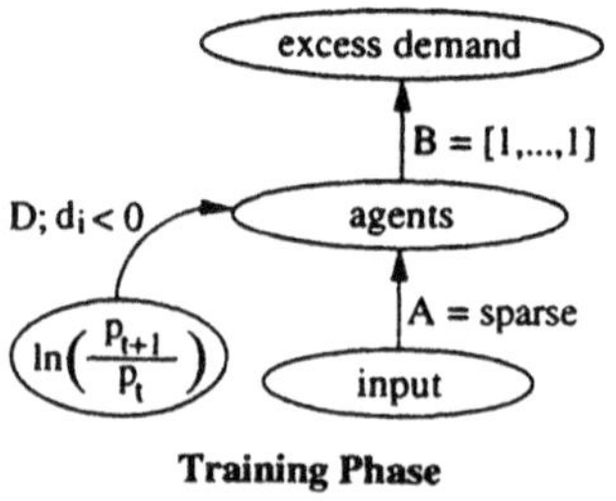

Fig. 3. Neural Network capturing the implicit price dynamics: Training.

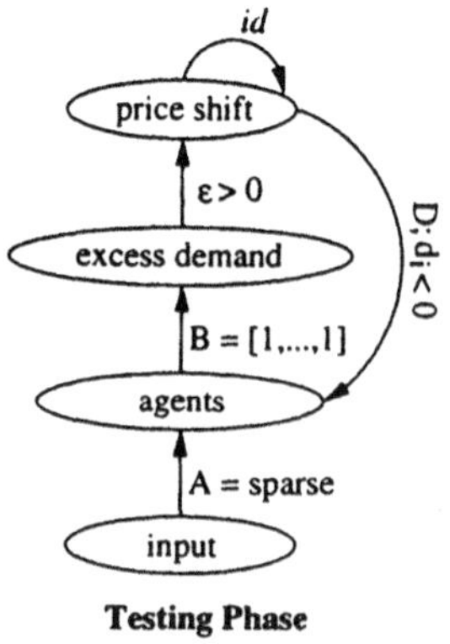

Fig. 4. Neural Network capturing the implicit price dynamics: Testing.

During recall we omit the future information π_t. Instead, we determine the price shift by the implicit condition of Eq. 6. The fix-point neural network in Fig. 4 solves the implicit condition by iterating the series π^n (Eq. 7). Note, that connectors A and D are recaptured from the training phase, whereas $\epsilon > 0$ is initialized sufficiently small to guarantee convergence (Eq. 8).

7 Empirical Study

Now, we present results of an empirical study concerning the application of our models to a real world FX-market (DEM / USD). We demonstrate our approaches on the basis of daily data from Jan. 91 to Dec. 97 to forecast daily price shifts. We divided the data set into two parts (Jan. 91 to Sep. 96 as training set and Oct. 96 to Dec. 97 as test set).

The selection of raw input data to support the agents decision making processes was guided by the idea, that every major currency is at least driven by *four* influences. These are the development of *i.* other major currencies, *ii.* the national and international bond, *iii.* stock and *iv.* commodity markets [3]. With respect to these influences, we composed a data set of 11 indicators, based on technical and fundamental data. The preprocessing of the raw input data is done by calculating the momentum and force of each signal [5].

In our experiments we found that 400 agents are a suitable size to model the FX-market. We trained the networks by *pattern-by-pattern* learning [5].

First, we compare the average number of inputs, which is used by the agents during their decision making, with the resulting excess demand (see Fig. 5). For *both* market paradigms it turns out, that for the case of higher average number of inputs analyzed by an agent, the FX-market is driven by huge excess demands. Thus, if the agents are too homogeneous in their behavior the market tends to be very volatile. Such huge imbalances of demand and supply are improper to adapt the FX-market dynamics. On the other hand, if the average size of the agents data sets to small, their decisions are too heterogeneous. Therefore, one can only observe small market movements, which likewise do not reflect real world behavior. For both paradigms, we found, that an average data base of 9 inputs allows is suitable.

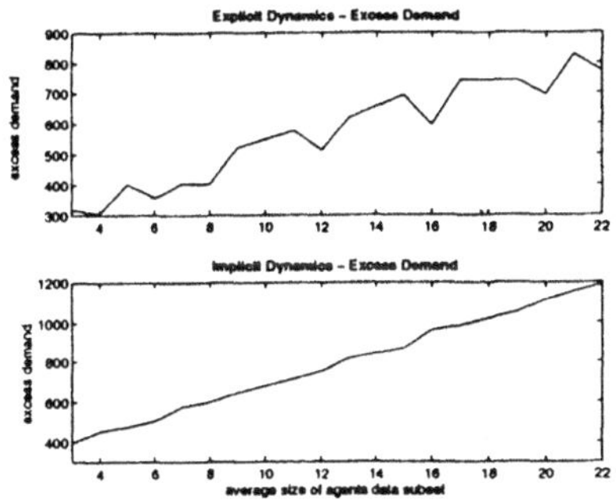

Fig. 5. Observed excess demand / supply of the FX-market depending on the average size of the agents data bases.

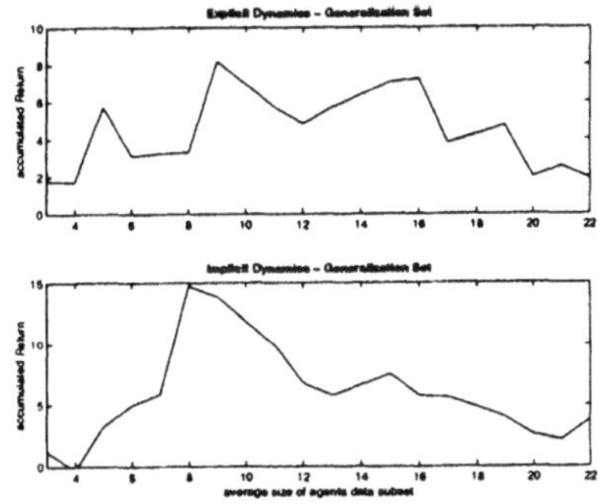

Fig. 6. Accumulated model return in dependence of the average size of the agents data base.

Second, we compare the accumulated return of the multi-agent models to the average size of agents data bases (see Fig. 6). The upper panel of Fig. 6

refers to the accumulated return of a trading strategy using the explicit price dynamics, while the lower panel of Fig. 6 is related to the implicit market paradigm. Both accumulated returns are calculated on the generalization set. Analyzing the realized returns of both models, it turns out, that the highest return can be realized by using 9 inputs on the average for each agent.

Examining the matrix A in both models, we found that several weights are pushed to zero during the learning. Thus, the true size of the average database used by an agent is 7 instead of 9 inputs. Remarkably, an attention span of 7 input signals is also known in psychology. Other data base sizes lead to unsatisfactory results, because the market would be either driven by huge or by small excess demands, which do not reflect the FX-rate dynamics.

Comparing both market paradigms, the explicit price dynamics represents a slow price adjustment, while the implicit approach proceeds fast price adjustments. The empirical results suggest that both price regimes can be found in the DEM / USD FX-market with similar importance.

8 Conclusion

We integrated three views: (*1.*) explanatory multi-agent market models, (*3.*) an econometric framework, and (*2.*) the embedding of both tasks in a neural network. We showed that such an integration is natural: (*i.*) a single neuron represents an elementary decision model and (*ii.*) the neural network allows the modeling of the market price dynamic. Our models are based on semantic specifications instead of assuming ad-hoc functional relationships. Based on local specifications and global learning we pointed out that our integrated models fit real world behavior of FX-markets.

Further work will concern the modeling of agents decision behavior by application of cognitive systems. As a structural representation of cognitive system we suggest recurrent error correction neural networks [7].

References

1. Beltratti, A., Margarita, S. and Terna, P.: *Neural Networks for economic and financial modeling*, International Thomson Computer Press, London, UK. 1996.
2. LeBaron, B. *Agent-based computational finance: Suggested readings and early research*, in: Journal of Economic Dynamics & Control, Vol. 24, 2000.
3. Murphy, J.J., *Intermarket Technical Analysis*, New York.
4. Tesfatsion, L.:*Agent-Based Computational Economics: A Brief Guide to the Literature*, Reader's Guide to the Social Sciences, Fitzroy-Dearborn, London, 2000.
5. Neuneier, R., Zimmermann, H.G., *How to Train Neural Networks*, in: Tricks of the Trade: How to make algorithms really to work, Springer, Berlin 1998.
6. Zimmermann, Hans-Georg, Neuneier, Ralph and Grothmann, Ralph *Multi-Agent Modeling of Multiple FX-Markets by Neural Networks.* IEEE Transactions on Neural Networks Special Issue 12(4):1-9, 2001.
7. Zimmermann, Georg, Neuneier, Ralph and Grothmann, Ralph *An Approach of Multi-Agent FX-Market Modeling based on Cognitive Systems.* In Proceedings of the Int. Conference of Neural Networks, ICANN 2001, Vienna 2001.

Section 14

Informations and Decision Support Systems / OR in Internet

Zeitoptimierte Assoziationsanalyse durch Stichprobenauswahl dargestellt am Beispiel aus der Telekommunikationsbranche

Frank Beekmann[1], Andreas Rudolph[2]

[1] Gerhard Mercator Universität Duisburg, Fachgebiet Wirtschaftsinformatik und Operations Research, Lotharstr. 65, 47057 Duisburg

[2] Wissenschaftliches Zentrum IBM, Vangerowerstr. 18, 69115 Heidelberg

Der vorliegende Artikel zeigt die Anwendbarkeit der Stichprobentheorie zur schnelleren Berechnung von Assoziationsregeln in großen Datenbeständen. Assoziationsregeln beschreiben Zusammenhänge zwischen Attributwerten von Datensätzen einer Datei. Statt diese Regeln durch Analyse des gesamten Datenbestandes zu generieren reicht es aus, eine Stichprobe aus diesem Datenbestand zu analysieren und die hier erhaltenen Maßzahlen für die Regeln als Schätzer für die im Datenbestand tatsächlich vorliegenden Maßzahlen anzusehen. Es werden Formeln angegeben, mit denen Stichprobengrößen bestimmt werden können, so dass die in der Stichprobe gefundenen Maßzahlen der Regeln hinreichend genau die tatsächlich vorliegenden Maßzahlen der Regeln darstellen. Durch dieses Vorgehen lässt sich die Gesamtdauer des zugehörigen Prozesses des „Knowledge Discovery in Databases" erheblich verringern. Dazu werden die Ergebnisse einiger Tests angegeben. Die praktische Anwendbarkeit dieses Ansatzes der Assoziationsanalyse wird durch eine Analyse von Telekommunikationsdaten zur Ermittlung von Fehlroutings sowie Bündelvertauschungen dargestellt.

Keywords: *Assoziationsregeln, Stichproben, KDD, Telekommunikation*

1 Einleitung

Bei der Durchführung eines KDD (Knowledge Discovery in Databases) Prozesses fällt der größte Zeitaufwand in der Phase der Datenaufbereitung an. Je weniger Datensätze in dieser Phase für die Analyse bereitgestellt werden müssen, desto größer ist die Zeitersparnis in dieser sowie in der nachfolgenden Analysephase. Eine Möglichkeit zur Reduktion der Anzahl der Datensätze besteht in der Durchführung einer Stichprobenauswahl. Hierbei ist allerdings zu berücksichtigen, dass dadurch Ergebnisse ungenau bzw. Muster nicht erkannt werden können, weshalb damit die Analyse der Genauigkeit der Ergebnisse notwendig ist. Ein Verfahren, das sich für eine solche Betrachtung eignet, ist die Assoziationsanalyse [AgIS93].

Im ersten Teil dieses Artikels werden Genauigkeitsschranken für die durch eine Assoziationsanalyse auf Basis einer Stichprobe generierten Maßzahlen von Regeln angegeben. Diese Schranken basieren auf univariaten Konfidenzintervallen und liefern einen Zusammenhang mit dem notwendigen Stichprobenumfang, worin eine Verbesserung bereits bekannter Ansätze liegt, in denen Chernoff-Schranken benutzt wurden [Toiv96], [ZPLO96]. Anschließend wird mittels einer Testreihe die Größenordnung der Zeitersparnis und des Genauigkeitsverlustes aufzeigt. Dies geschieht anhand von anonymisierten Daten, die während eines Projektes mit der Deutschen Telekom AG anfielen. Das Projekt zeigt eine wichtige Anwendungsmöglichkeit der Assoziationsanalyse in der Praxis.

Um geeignete statistische Verfahren anzuwenden, wird die Assoziationsanalyse zunächst vom statistischen Standpunkt aus genauer untersucht.

2 Assoziationsanalyse

Durch eine Assoziationsanalyse werden Regeln generiert, welche die Beziehungen zwischen den in den Datensätzen eines Datenbestandes vorkommenden Elementen (Items) beschreiben. Eine Regel könnte lauten: Wenn Item A vorkommt, tritt auch Item B auf. Dabei sind die Items als Ausprägungen von Attributwerten des Datensatzes zu verstehen. Eine Regel wird durch einen Datensatz unterstützt, falls der Datensatz die Items der Regel beinhaltet. Der Support einer Regel ist definiert als das Verhältnis zwischen der Anzahl der Datensätze, die diese Regel unterstützen und der Gesamtanzahl aller Datensätze in der Analysedatei, so dass der Support einer Regel einen Anteilswert darstellt. Je größer der Supportwert ist, desto relevanter ist die Regel. Ein weiteres Maß für die Stärke einer Assoziationsregel ist die Konfidenz, welche das Verhältnis der Anzahl der Datensätze angibt, die eine bestimmte Regel unterstützen, zu der Anzahl der Regeln, die den Prämissenteil der Regel unterstützen. Diese lässt sich aus dem Quotienten zweier Supportwerte bestimmen.

Assoziationsregeln werden durch Support und Konfidenz bewertet. Als Beispiel möge folgende Regel dienen: In 80% der Fälle, in denen Item A in einem Datensatz enthalten ist, tritt auch Item B auf. Insgesamt erscheint diese Regel in 5% aller Datensätze im Gesamtbestand.

Zur Generierung von Assoziationsregeln müssen zunächst ein Mindestsupport und eine Mindestkonfidenz der gesuchten Regeln definiert werden. Zugehörige Algorithmen laufen grundsätzlich in zwei Phasen ab. In der ersten Phase werden alle Kombinationen von Items bestimmt, deren Support größer als die vorgegebene Mindestsupportschranke ist. Anschließend werden für jede dieser Kombinationen alle Regeln gebildet, deren Konfidenz größer ist als der vorgegebene Mindestkonfidenzwert.

3 Schätzer für Supportwerte

Die Supportwerte werden nun mit Hilfe einer Stichprobe geschätzt. Da es sich um einen Anteilswert handelt, wird im Folgenden nur auf Anteilsschätzer eingegangen. Dabei wird hier nur die Möglichkeit aufgezeigt, mit univariaten Schätzern zu arbeiten, obwohl auch ein multivariater Ansatz überprüft wurde. Für den Supportwert p als Anteilswert wird ein erwartungstreuer und konsistenter Schätzer $\hat{p}$ auf einer Stichprobe vom Umfang n definiert durch

$$\hat{p} = \frac{1}{n}\sum_{i=1}^{n} B_i$$

wobei die Zufallsvariable B_i den Wert 1 annimmt, wenn beim i-ten Element in der Stichprobe das gesuchte Ereignis „Datensatz i unterstützt eine Regel R" eintritt, anderenfalls gilt $B_i=0$.

Für eine Genauigkeitsbetrachtung für den Schätzwert kann folgendes (approximatives) Konfidenzintervall herangezogen werden, welches einen Bereich angibt, der mit einer Wahrscheinlichkeit von 1-α den in der Grundgesamtheit tatsächlich vorliegenden Anteilswert p überdeckt [LeLe99, S. 63]

$$\left(\hat{p} - u_{1-\frac{\alpha}{2}}\sqrt{\frac{N-n}{N}}\sqrt{\frac{\hat{p}(1-\hat{p})}{n-1}}\ ;\hat{p} + u_{1-\frac{\alpha}{2}}\sqrt{\frac{N-n}{N}}\sqrt{\frac{\hat{p}(1-\hat{p})}{n-1}}\right),$$

wobei $u_{1-\frac{\alpha}{2}}$ das $(1-\frac{\alpha}{2})$-Fraktil der Standard-Normal Verteilung ist.

Hieraus ist ersichtlich, dass die Größe des Konfidenzintervalls von der Größe der Fehlerwahrscheinlichkeit α abhängt, aber auch vom Stichprobenumfang n und der Größe der Grundgesamtheit N. Soll die Größe der Abweichung bei vorgegebener Fehlerwahrscheinlichkeit α beschränkt werden, so kann dazu als Maß ein Faktor d vom wirklich vorliegenden Wert gewählt werden. Damit handelt es sich um eine relative Abweichung beispielsweise als Prozentsatz vom wahren Supportwert (der Ansatz, einen absoluten Fehler vorzugeben, beispielsweise durch 0,1 Prozentpunkte, ist ebenfalls möglich).

Bei gegebenem Faktor d kann folgende Formel zur Bestimmung des notwendigen Stichprobenumfangs herangezogen werden [LeLe99, S. 74].

$$n \geq \frac{u^2_{1-\frac{\alpha}{2}} Np(1-p)}{(N-1)d^2p^2 + u^2_{1-\frac{\alpha}{2}} p(1-p)}$$

Damit ist nun eine Bestimmung des Stichprobenumfangs für den Anteilswert in einem Intervall mit vorgegebener Sicherheitswahrscheinlichkeit $1-\alpha$ und vorgegebener maximaler relativer Abweichung dp möglich. Der Stichprobenumfang hängt allerdings vom zu schätzenden Anteilswert p ab. Da die Funktion der Stichprobengröße bezüglich p monoton fallend ist, kann der kleinste zu schätzende Anteilswert benutzt werden, der gerade durch die Größe Mindestsupport vorbestimmt wird.

Dieser Ansatz kann auf die Supportschätzung von Assoziationsregeln angewendet werden. Zum Beispiel wird ausgehend von einem Datenbestand mit 7,6 Mio Datensätzen für die Supportschätzung von Regeln, deren Supportwerte größer als ein Mindestsupport von 0,01% sind und mit einer Wahrscheinlichkeit von 99% nicht mehr als 20% von den im Gesamtdatenbestand vorliegenden Supportwerten abweichen, eine Stichprobe vom Umfang von 528570 Datensätzen benötigt.

4 Testergebnisse

Um die Effizienz des Ansatzes zu testen, wurden Analysen auf einem realen Datenbestand durchgeführt. Das Projekt und die Bedeutung der Daten wird in Kapitel 6 erläutert.

Untersucht wurde ein Datenbestand mit rund 7,6 Millionen Datensätzen, die jeweils aus 8 Attributen bestanden. Es wurden Assoziationsanalysen durchgeführt, um Abhängigkeiten zwischen Ausprägungen der Attributwerte zu entdecken. Dabei wurden drei Auswertungen mit unterschiedlichen Vorgaben der Parameter MinSup für den Mindestsupportwert durchgeführt. Für MinSup wurden Werte von 1%, von 0,1% und von 0,01% gewählt, die Mindestkonfidenz-Parameter waren bei allen Analysen gleich und betrugen 75%. Diese Analysen wurden anschließend analog auf Stichproben des Gesamtdatenbestandes durchgeführt.

4.1 Zeitersparnis

Für jede Analyse wurde sowohl die Zeit für die eigentliche Assoziationsanalyse wie auch die Zeit für das gesamte Preprocessing gemessen. Letztere setzt sich zusammen aus der Zeit für das Laden der Datensätze in eine DB2 Datenbank für weitere Vorbereitungsschritte und der Zeit für die sogenannte Pivotisierung des Datenbestandes, bei der jeder aus einer Identifikationsnummer und k Attributen bestehende Datensatz in k sukzessive, einzelne Datensätze bestehend aus Identifikationsnummer und jeweiligem Attribut aufgespaltet wird.

Tabelle 1 umfasst die Verabeitungszeiten auf den jeweiligen Datenbeständen mit MinSup 0,01%. Die Bearbeitung fand auf einer IBM_RS/6000 mit 1 GB Hauptspeicher und mit dem Analysewerkzeug „Intelligent Miner“ statt. Zusätzlich sind in der Tabelle die Zeiten für die Stichprobenbildung sowie die Gesamtprozessdauer angegeben.

	Stichproben-bildung	Ladezeit	Pivoti-sierung	Analyse	Gesamt
Gesamt		24m 20s	35m 20s	1h 32m 38s	2h 32m 18s
10%Stichprobe	8m 36s	2m 25s	3m 33s	10m 17s	24m 51s
1% Stichprobe	7m 37s	15s	21s	1m 40s	9m 53s

Tabelle 1: Zeitaufstellung beim Analyseprozess

Es zeigt sich eine erhebliche Verringerung der Analysezeit, wenn die Analyse auf dem Stichprobenbestand durchgeführt wird. Die Zeiten für die Analyse verhalten sich annähernd linear zum Stichprobenumfang. Im nächsten Abschnitt wird gezeigt, dass trotz dieser Zeitersparnis die Ergebnisqualität nur wenig beeinträchtigt wird.

4.2 Genauigkeit

Zur Analyse der Genauigkeit der mit dem Stichprobenansatz erzielten Ergebnisse wurden die Regeln, die bei den Auswertungen auf den verschiedenen Datenbeständen mit verschiedenen MinSup-Parametern generiert wurden, verglichen. Dabei zeigt sich zunächst, dass die Anzahl der generierten Regeln auf dem Gesamtdatenbestand und den Stichprobendatenbeständen bezüglich gleicher Analyseparameter nicht mehr als maximal 2% voneinander abweichen.

Die nächste Frage ist damit die nach der Anzahl der gemeinsamen Regeln. In Tabelle 2 ist angegeben, wie viele Regeln des Gesamtdatenbestandes auch unter den Regeln des Stichprobenbestandes vorkommen. Zum Vergleich ist in der ersten Zeile die Anzahl der im Gesamtdatenbestand auftretenden Regeln angegeben.

	MinSup 1%	MinSup 0,1%	MinSup 0,01%
Gesamt	10148	120116	206544
Gesamt und 10%-Stichprobe	9876	118685	204365
Gesamt und 1% Stichprobe	9048	111008	190296

Tabelle 2: Anzahl gemeinsamer Regeln

Offenbar ist die Anzahl der gemeinsamen Regeln sehr hoch, weshalb durch die Stichproben kaum Informationen verloren gehen. Bei der 10% Stichprobe beträgt der Anteil der gemeinsamen Regeln bei einem MinSup von 0,01 beispielsweise 98,95 %, im Fall des 1% Stichprobenbestandes ist dieser Wert immerhin noch 92,12%. Durch Vorgabe der Stichprobenanzahl lässt sich dieser Fehler, wie oben beschrieben, eingrenzen.

5 Ergebnis

Durch den vorgestellten Ansatz kann eine erhebliche Zeitersparnis bei der Assoziationsanalyse erreicht werden. Die durch die Stichproben generierten Regeln spiegeln sehr gut die tatsächlich vorliegenden Regeln wider. Im theoretischen Teil wurden Genauigkeitsparameter angegeben, die näherungsweise auf die gezogenen Stichprobengrößen zutreffen. Die Regeln, die durch die Stichprobe nicht erkannt wurden, weisen einen Support auf, der in der Nähe der Mindestsupportschranke liegt. Soll ein Datenbestand mit vorgegebenem Mindestsupport analysiert werden, so bietet sich an, die Analyse auf einem Stichprobendatenbestand mit einer etwas kleineren Mindestsupportschranke durchzuführen. Damit steigt die Sicherheit, alle relevanten Regeln zu erfassen. Eine Analyse der zugehörigen Konfidenzwerte zeigte eine maximale relative Abweichung von 1,7 %.

6 Anwendungsmöglichkeit

In diesem Kapitel wird die Anwendbarkeit der Assoziationsanalyse und des vorgeschlagenen Ansatzes auf ein Praxisproblem vorgeführt.

6.1 Problembeschreibung

Bei der Deutschen Telekom AG treten bei der Vermittlung von Gesprächen, die über private Mitbewerber, auch Carrier genannt, durchgeführt werden, zwei Arten von Problemen auf. Da die meisten Carrier noch keine eigenen Anschlüsse zu ihren Endkunden besitzen, muss ein Gespräch am Startort aus dem Netz der Telekom in das Netz des Carriers überführt werden und anschließend am Zielort wieder in das Netz der Telekom zurückgeleitet werden. Derartige Übergabepunkte werden „Ort der Zusammenschaltung“ (ODZ) genannt und sind für jeden Teilnehmeranschluss vorgegeben.

Gelegentlich wird ein Gespräch beispielsweise wegen Überbelastung nicht über den vorgesehenen Übergabepunkt geleitet, was als Fehlrouting bezeichnet wird. Ein zweites Problem besteht in der sogenannten Bündelvertauschung, wobei ein Bündel ein Kabelstrang an einem Vermittlungscomputer ist. Ist dieser Kabelstrang fehlerhaft geschaltet, so dass das Gespräch z. B. de facto in das Netz der Telekom übergeben wird, eigentlich aber aus dem Telekomnetz in ein Carriernetz übergeben werden sollte, so wird der zugehörige Vermittlungscomputer das Gespräch an einen anderen Computer im Telekomnetz weitervermitteln, das Abrechnungssystem jedoch falsch fakturieren.

Eine wichtige Aufgabe ist es, Fehlroutings und Bündelvertauschungen aufzufinden, die nicht gelegentlich infolge einer Netzüberlastung auftreten, sondern mit einer Systematik. An dieser Stelle setzen die Assoziationsregeln an, die das Aufdecken derartiger Regelmäßigkeiten ermöglichen.

6.2 Analyse

Übergabepunkte werden durch Vermittlungscomputer realisiert, die Telekommunikationsdatensätze speichern. Diese beinhalten für jedes Gespräch eine Menge von Merkmalen wie „anrufender Anschluss“, „Endteilnehmer“, „Datum“, „Uhrzeit“, „Dauer des Gesprächs“, „Bündel, über das ein Gespräch weitergeleitet wird“, usw. Darüber hinaus liegen für jeden Teilnehmeranschluss Einzugsbereichstabellen vor, die einem Anschluss einen ODZ zuordnen, über den ein Gespräch vermittelt werden muss, ein sogenannter Soll-ODZ. Liegt eine Abweichung des tatsächlich vorliegenden ODZ und des Soll-ODZ vor, so indiziert dies entweder ein Fehlrouting oder eine Bündelvertauschung.

Anhand der Kommunikationsdatensätze ist es möglich, Assoziationsregeln zu finden, die konkret angeben, wann oben angegebene Probleme auftreten. Dazu wurden die Telekommunikationsdatensätze zunächst mit dem Feld ODZ_ABW, das die Abweichung des Ist-ODZ zum Soll-ODZ anzeigt, angereichert und in die Datenbank geladen. Anschließend wurde der Pivotisierungsschritt auf einem Datenbestand mit ausgewählten Attributen durchgeführt. Danach wurde mit dem „Intelligent Miner“ eine Assoziationsanalyse angesetzt und eine Regeldatei erzeugt, die genauer untersucht wurde. Insbesondere konnten Regeln herausgefiltert werden, deren Konsequenz die Ausprägung [ODZ_ABW=J] ist, was eine ODZ-Abweichung bedeutet. Anhand dieser Regeln konnte auf die Ursache der ODZ-Abweichungen geschlossen werden.

Bei der Telekom ist dieses Analyse-System implementiert und soll wöchentlich Regeln finden, die Aufschluss über Fehlroutings bzw. Bündelvertauschungen geben. Da die Anzahl der zu analysierenden Datensätze mit über 10 Millionen sehr hoch ist, zeigt die Anwendung des vorgestellten Ansatzes erhebliche zeitliche Verbesserungen.

Literatur

[AgIS93]

Agrawal, R.; Imielinski T.; Swami, A.: Mining association rules between sets of items in large databases. In: Proceedings of ACM SIGMOD conference on management of data, Washington D. C. 1993, S. 207-216.

[Toiv96]

Toivonen, H.: Sampling large databases for association rules. In: Proceedings of the 22nd VLDB conference, Mumbai 1996, S. 134-145.

[LeLe99]

Levy, P.; Lemeshow, S.: Sampling of populations. New York et al. 1999.

[ZPLO96]

Zaki, M.; Parthasarathy, S.; Li, W.; Ogihara, M.: Evaluation of sampling for data mining of association rules. Technical report 617, University of Rochester, New York 1996.

A Metaheuristic-based DSS for Portfolio Optimization

Ulrich Derigs, Nils-H. Nickel

Seminar for Information Systems and Operations Research (WINFORS)
University of Cologne, Pohligstr. 1, 50969 Cologne, Germany
{derigs, nickel}@winfors.uni-koeln.de

Abstract. In this paper we present a framework, i.e. a concept and design as well as results with a prototypical implementation of a metaheuristic-based decision support system PM-DSS© for portfolio optimization and managing investment guidelines. PM-DSS© can be used for active as well as passive fund management.

1 The general problem

In this paper we present a framework, i.e. a concept and design as well as results with a prototypical implementation of a metaheuristic-based decision support system PM-DSS© for portfolio optimization and managing investment guidelines. PM-DSS© is a DSS-generator, i.e. specific decision support systems for active as well as passive fund management can be customized using the models and methods which are packaged in PM-DSS©.
For an introduction into modern portfolio theory see [3] and [4]. In [1] a concept of optimization in passive portfolio management is discussed.

The decision situation that we assume is the following:

After the initial design of a portfolio the fund manager is constantly confronted with the following decision problem of Portfolio Re-Optimization:
Due to changes on the market the given portfolio has become infeasible because of violations to the investment guidelines and/or sub-optimal due to under-performance with respect to yield, risk etc. In this daily situation the fund manager has to restructure the portfolio through a series of sell and buy transactions. Here the transactions themself may be under some restrictions as for instance minimal/maximal amounts etc.

There are highly sophisticated software systems on the market which routinely check portfolios as well as transactions for violations of constraints from investment guidelines. Yet, still the fund manager has no support in identifying promising transactions which will result in a feasible and profitable portfolio.

2 The DSS framework

The main components of our DSS for re-optimizing portfolios are

- a **constraint checker** representing the problem-specific constraints stemming from legal regulations, contractual fund constraints and internal investment guidelines,
- an **evaluation module** modeling the assets in the market and the fund managers objectives, and,
- a **metaheuristic search procedure** for controlling an efficient and intelligent construction of "good" portfolios.

The architecture of our DSS is depicted in figure 1.

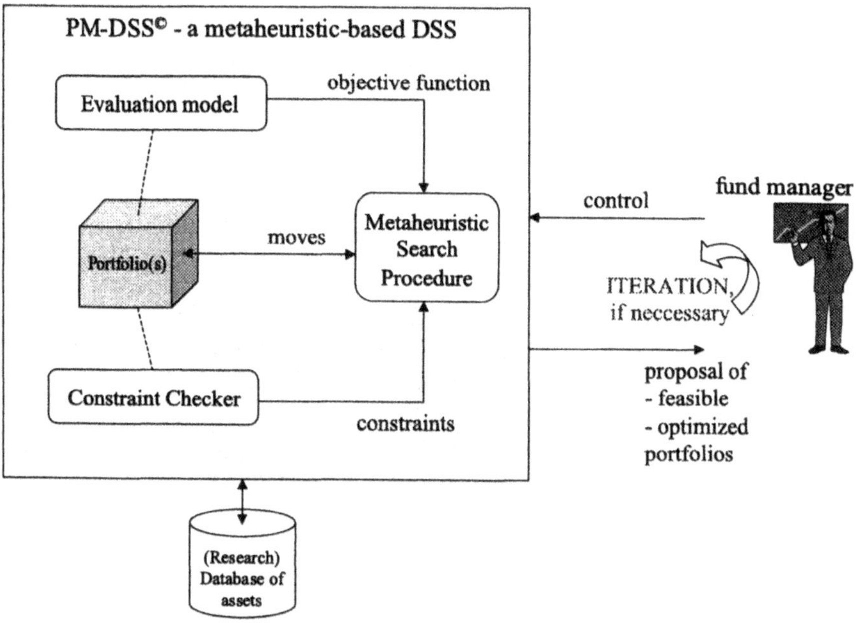

Fig. 1. The architecture of our DSS

The concept and architecture of PM-DSS© is open with respect to different implementations of these basic modules:

- We are able to integrate commercial checkers or we may encode specific rule sets individually. For the specific DSS's sets of constraints are packaged in a called rule-base card which is attached to portfolio data-sets and may be activated in a very flexible manner. The use of a professional checker reliefs the trust company from designing and controlling the complicated system of rules.
- Evaluation of portfolios and transactions may be based on factor models as well as on a Monte Carlo Simulation of a stochastic process. Using Monte Carlo Simulation all relevant asset types including derivatives as

well as different risk measures as for instance shortfall probability, value at risk etc. can be represented.

- In the heuristic search component we have implemented different local search strategies. Due to the fact that checking and evaluating a portfolio or a transaction is a rather complicated and costly task the neighborhood structure and the corresponding moves of the local search has to be appropriately designed to result in acceptable performance. An overview of local search concepts is given in [5] and the application of meta-heuristics to portfolio optimization has been treated in [2].

3 An application and implementation

In this section we describe the application of PM-DSS© to a (real-world) portfolio management problem of an investment trust company, for confidentiality called ABC in the sequel. With this description the general functionality of the DSS as well as the optimization model and heuristic search procedure implemented for the concrete application should become clear.

3.1 The Model

ABC offers an equity fund, lets call it ABC-GS, which should reproduce the German Stock Index DAX, i.e. ABC-GS is passively managed with the DAX as benchmark portfolio. Several constraints of the German Law on investment trust companies (KAGG) like "no asset is allowed to have a share of more than 10%" or "the share of the set of those assets with an individual share of more than 5% should not exceed 40%" prohibit that the benchmark DAX30 can be identically reproduced. Thus the portfolio manager has the possibility to introduce into the GS-portfolio a limited number of the 180 non-DAX assets contained in the STOXX200-Index. Additionally, a cash position is contained in the portfolio and a position in a future on the DAX in order to invest temporary cash flows to the fund. These two assets, the DAX assets and the STOXX200 assets form the so-called investment universe. The problem is now to construct a portfolio of assets from the universe fulfilling all fund-constraints, which is "as close as possible" to the DAX-benchmark. We will formalize this problem in the following:

Representation of a portfolio

In general, we represent a portfolio of assets from the investment universe by its share-vector $x = (x_1, \ldots, x_N)$ with N the number of stocks and $\sum_{i=1}^{N} x_i + x_{Fut} + x_{Cash} = 1$ (here $N = 210$). The components x_{Cash} and x_{Fut} represent the share of cash or DAX-futures, respectively (in order to use the vector formulation, $x_{N+1} := x_{Fut}$ and $x_{N+2} := x_{Cash}$). In the GS-management

problem the actual DAX-portfolio
$x^{(D)} = (x_1^{(D)}, \ldots, x_{30}^{(D)}, \underbrace{x_{31}^{(D)}, \ldots, x_{210}^{(D)}, x_{Fut}^{(D)}, x_{Cash}^{(D)}}_{=0})$ is given data.

Similarity measure, evaluation of portfolios and objective function

In the application the return of a portfolio is estimated by a factor model. Here it is assumed that the return r_i of asset i ($i = 1, \ldots, N+2$) is linearly dependent on K (=43) economical factors F_j ($j = 1, \ldots, K$):

$$r_i = \alpha_i + \sum_{j=1}^{K} \beta_{ij} F_j + \epsilon_i$$

The return r_x of a portfolio x is linearly dependent on these factors, too, and given by :

$$r_x = \sum_{i=1}^{N+2} x_i(\alpha_i + \sum_{j=1}^{K} \beta_{ij} F_j + \epsilon_i) = \underbrace{\sum_{i=1}^{N+2} x_i \alpha_i}_{=:\alpha_x} + \sum_{j=1}^{K} \underbrace{(\sum_{i=1}^{N} x_i \beta_{ij})}_{=:\beta_{xj}} F_j + \underbrace{\sum_{i=1}^{N+2} x_i \epsilon_i}_{=:\epsilon_x}$$

Now, the objective is to minimize the so-called "tracking error" (TE), which is the variance of the difference $r_x - r_{x^{(D)}}$:

$$\begin{aligned} TE(x) &= Var(r_x - r_{x^{(D)}}) \\ &= (x - x^{(D)})^t \beta^t \cdot C \cdot \beta\,(x - x^{(D)}) + \sum_{i=1}^{N+2} (x_i - x_i^{(D)})^2\, Var(\epsilon_i) \end{aligned}$$

with specified, i.e constant parameter matrices $\beta = (\beta_{ij})_{i=1,\ldots,N+2\,;j=1,\ldots,K}$ and $C := (Cov(F_i, F_j))_{i,j=1,\ldots,K}$.

Constraints

There are two sets of constraints: the general legal constraints from the KAGG and GS-specific constraints, i.e. internal guidelines for the portfolio management.
In this rather simple application (no complex derivatives etc.) the set of meaningful KAGG-constraints reduces to:

(1) $x_i \geq 0 \qquad (i = 1, \ldots, 212)$
(2) $x_{Fut} \leq 0.2$
(3) $x_i \leq 0.1$
(4) $\sum_{i=1}^{210} (\max\{x_i - 0.05, 0\} \cdot \frac{x_i}{x_i - 0.05}) \leq 0.4$
(i.e. the share of the set of those assets with an individual share of more than 5% should not exceed 40%)

(5) $\sum_{i=1}^{210} x_i \geq 0.51$ and obviously

(6) $\sum_{i=1}^{212} x_i = 1$

There are three different GS-specific constraints which have to be introduced into the model:

(7) For every DAX asset the share in the portfolio should differ from the share in the actual DAX or from the upper bound 10%, respectively, by at most Δ_i, with Δ_i prespecified and constant. Since the actual DAX shares $x_i^{(D)}$ and Δ are given as parameters these constraints can be modeled as individual simple bounds:
$L_i \leq x_i \leq U_i \qquad i = 1, \ldots, N$

(8) The number of non-DAX assets should be at most N_P (with N_P predefined and constant):
$|\{x_i \neq 0 \mid i = 31, \ldots, 210\}| \leq N_P$

(9) The share of the cash asset should range between 3% and 10%:
$0.03 \leq x_{Cash} \leq 0.1$

Thus the portfolio optimization problem for ABC-GS is given by

$$\text{(ABC-GS-PMP)} \qquad \begin{array}{ll} \text{minimize} & TE(x) \\ \text{subject to} & (1) - (9) \\ & x \in \mathbb{R}^{N+2} \end{array}$$

ABC-GS-PMP is a complex nonlinear discrete optimization problem.

3.2 The Heuristic

For solving ABC-GS-PMP we have implemented a simulated annealing heuristic. In the following we describe the basic specifications:

Every solution (portfolio) is represented by its share-vector. There are several options for incorporating the feasibility-aspect into the search heuristic and there is a trade-off with respect to the neighborhood-structure and the move-operators, respectively:

- the penalty approach with relaxing sets of constraints and penalyzing violations in the objective function,
- the use of a repair function for non-feasible solutions (portfolios),
- the use of moves generating feasible solutions only.

Experiments with this application and on similar instances have shown that when relaxing cardinality constraints as for instance (8) the search procedure in spite of penalties does not converge to a feasible solution, since very low shares are maintained for too many assets. Since feasibilty of a portfolio with respect to such simple cardinality constraints is easy to check for the manager and at the same time the fulfillment of these constraints is generally guaranteed in the portfolio managers portfolio redesign process, such solutions are not acceptable and the fulfillment of these constraints is essential for the acceptance the Decision Support System as a whole. Therefore we have relaxed the constraints (2),(4), and (5) and implemented a move-operator ensuring feasibility with respect to (1),(3), and (6)-(9). Let

$H_2(x) := x_{Fut}$,

$H_4(x) := \sum_{i=1}^{210} (\max\{x_i - 0.05, 0\} \cdot \frac{x_i}{x_i - 0.05})$,

$H_5(x) := \sum_{i=1}^{210} x_i$ and

γ_i sufficiently large ($i = 2, 4, 5$).

Thus we are solving the following auxiliary problem

$$\text{minimize} \quad TE(x) + \gamma_2 \cdot \underbrace{\max\{H_2(x) - 0.2, 0\}}_{\text{Violation of (2)}} + \gamma_4 \cdot \underbrace{\max\{H_4(x) - 0.4, 0\}}_{\text{Violation of (4)}} + \gamma_5 \cdot \underbrace{\max\{H_5(x) - 0.51, 0\}}_{\text{Violation of (5)}}$$

$$\text{subject to} \quad (1),(3),(6)\text{-}(9), \quad x \in \mathbb{R}^{N+2}$$

4 Results and comments

When applying PM-DSS© for restructuring a specific portfolio instance, i.e. the ABS-GS fund on a specific day, the user may vary several (model) parameters to analyze the stability and sensitivity of the solutions with respect to certain constraints. This is an important feature since (in certain scenarios) some constraints in portfolio management are "soft" in the sense that they may be violated for a short period of time and recovered later and due to the fact that every portfolio manager has some set of individual strategic objectives and tacit knowledge which is not explicit in the "standard" model.

Of special interest for the portfolio manager is for instance the influence of N_P on the solution quality. Constructing a (feasible) portfolio for $N_P = 0$ gives a certain kind of reference value for the tracking error and the tracking error of every other solution may be evaluated against this base value. Increasing N_P from 0 to the actual bound given in (8) the solution quality increases. Yet, depending on the gradient or the absolute improvement the manager may prefer a non-optimal solution with a smaller number of non-DAX assets or modify the optimal solution since the transformation from the actual portfolio to the solution portfolio requires fewer transactions.

Thus PM-DSS© and the model given above are used not only to optimize but to assist the manager in generating and evaluating a variety of different alternative solutions. For that purpose the heuristic search procedure has to be fast enough.

The overall reaction of the fond managers has shown that the solutions constructed by the model are consistent, that the resulting values for the tracking error are convincing, and that the solution times are acceptable such that PM-DSS© is a valuable assistant in the process of portfolio management.

References

1. Rudd, A. (1980) Optimal Selection of Passive Portfolios. Financial Management **9/10**, Spring, 57–66
2. Chang, T.-J., Meade, N., Beasley, J.E. and Sharaiha, Y.M. (2000) Heuristics for cardinality constrained portfolio optimisation. Computers & Operations Research **27**, 1271–1302
3. Reilly, F.K. and Brown, K.C. (1997) Investment Analysis and Portfolio Management. The Dryden Press, Fort Worth
4. Elton, J.E. and Gruber, M.J. (1994) Modern Portfolio Theory and Investment Analysis. John Wiley & Sons, New York - London
5. Aarts, E. and Lenstra, J.K. (1997) Local Search in Combinatorial Optimization. John Wiley & Sons

Section 15

Didactics of OR / Public Understanding

Kritische Erfolgsfaktoren für die Erstellung universitärer Multimediakurse
– mit einem Beispiel aus dem Bereich des OR –

André Ahuja, Wilhelm Rödder

FernUniversität in Hagen, Profilstraße 8, 58084 Hagen, Germany

andre.ahuja@fernuni-hagen.de, wilhelm.roedder@fernuni-hagen.de

Zusammenfassung Die Ziele des Einsatzes multimedialer Lehre an Hochschulen sind erweiterte Möglichkeiten beim Transport abstrakter Inhalte, Zukunftsorientierung, langfristige Personalkostensenkung in der Kursberatung sowie die Antizipation möglicher Umstrukturierungen in der Hochschullandschaft. Damit verbietet sich eine strikte Orientierung an Multimediaprodukten für den industriellen Einsatz und deren Erstellungsprozessen. Vielmehr sind die kritischen Erfolgsfaktoren für die Erstellung universitärer Multimediakurse zum einen die Anwendung von Projektmanagement, und hier insbesondere die Aufstellung des relevanten Zielsystems, welches hier auch als verbales Optimierungsmodell angegeben wird, die Zeitplanung sowie die geeignete Personalwahl, und zum anderen die geeignete Wahl der technischen Plattform, des Front-Ends sowie der einzusetzenden Medien. Gerade die Wahl der einzusetzenden Medien ist entscheidend für an Studierende der Wirtschaftswissenschaften gerichtete Inhalte des Operations Research. Weiter lassen sich wichtige Hinweise zwecks Projektrisikoverkleinerung aussprechen.
Jene kritischen Erfolgsfaktoren dienen als Maßstab zur Beurteilung der Durchführungsqualität eines Multimedia-Projekts des Lehrstuhls für Operations Research der FernUniversität in Hagen.

1 Einleitung

Multimediagestütztes Lernen ist in Lehre und Praxis auf dem Vormarsch, befördert durch Kostenvorteile, alternative Zugänge zu Inhalten sowie Freiheit in der individuellen Zeitplanung gegenüber traditionellen Medien [5]. Auch stellen technische Voraussetzungen heutzutage meist kein Hindernis für einen erfolgreichen Zugang des Anwenders mehr dar. Im Bereich Bildung stehen Industrie sowie Hochschulen gleichermaßen vor einem weitreichenden Umbruch. Doch die derzeit oft anzutreffende Orientierung an sehr wohl für die Industrie geltenden Maßstäben für die Vorgehensweise der Multimediaprodukterstellung sowie der Produktgestaltung selbst ist auf Grund anderer Gegebenheiten an Hochschulen - wie organisatorische und personelle Strukturen, Planungshorizonte und -zyklen, Arbeitsweisen, Adressatenkreis und auch innere sowie äußere Marktsituation - nicht zu rechtfertigen. Zumindest für die genannte Phase des Umbruchs erscheint ein Bewußtsein für kritische Erfolgsfaktoren ein positives Risikomanagement zu ermöglichen.

Dabei bietet die multimediale Lehre besondere Vorzüge für das Studium des Operations Research (OR), geprägt durch die Verbindung mathematischer sowie betriebswirtschaftlicher Inhalte. Die Studierenden sind ausgestattet mit einer über viele Jahre erworbenen, mehr oder weniger tiefen mathematischen Ausbildung. Einerseits kann es nicht originäre Aufgabe des OR sein, hier aufzufüllen; andererseits ist ein tieferes Verständnis mathematischer Modelle und Verfahren unbedingte Voraussetzung für die richtige Anwendung von OR-Methoden in der betrieblichen Praxis. Hier kann die multimediale Lehre durch alternative Zugänge zu abstrakten Inhalten und eine individuelle Lerngeschwindigkeit sehr wohl eine positive Wirkung haben. Somit ist es ein Gebot, Möglichkeiten zur Entwicklung von Multimediakursen unter für die Universität typischen organisatorischen, personellen, zeitlichen, finanziellen sowie technischen Restriktionen zu nutzen. Wichtige Hinweise zur Vorgehensweise bei der Kursentwicklung werden durch die Unterschiede in Zielen sowie Restriktionen zur traditionellen universitären Lehre einerseits sowie zur multimediagestützen Weiterbildung auf dem Industriesektor andererseits offenbar. Allerdings können im Rahmen dieser Arbeit nicht sämtliche Unterschiede explizit genannt werden.

2 Projektzielsystem

Neuartigkeit, Abgrenzung von der Routine, Komplexität, Fehlschlagrisiko, spezifische Organisation und zeitliche sowie andere Zielvorgaben sind die wesentlichen Merkmale eines Projekts. Damit ist die Entwicklung eines Multimediakurses ein Projekt, und in erster Konsequenz empfiehlt sich Projektmanagement mit seinen Führungs-, Strukturierungs-, Planungs- und Kontrollwerkzeugen zur Umsetzung, vgl. auch [1]. Im folgenden wird unter dem Begriff Projekt stets die Multimediakurserstellung verstanden.
Bei gegebener Projektidee beginnt Projektmanagement mit der Zielsystembildung, wobei ein Zielsystem aus Zielfunktionen sowie (Budget-, Zeit- und anderen) Restriktionen im klassischen Verständnis besteht. Das Projektzielsystem verfolgt operativ ein strategisches Ziel, hier wohl meist Verbesserung der Lehre und Außendarstellung der Innovationsorientierung. Typische Projektteilziele sind neben dem unten eingegrenzten Sachziel die über Personalkostenvorteile und Erhebung von Serviceaufwand mittelbar messbare Senkung der Kursbetreuungsleistung und weiter die direkt messbaren Termin- und Kostenziele. Die Zeitrestriktion ist bei Einbindung ins Curriculum absolut fix, nämlich der Termin des Ersteinsatzes abzüglich der Dauer für technische und lizenzrechtliche Prüfung, Vervielfältigung und Distribution. Planungsspielraum lässt sich hier durch die mögliche Aufteilung auf stufenweise multimedial angereicherte Kursversionen mit Einsatz zu jeweils einem Semesterbeginn gewinnen.
Für die Zuteilung eines Projektbudgets gibt es im wesentlichen zwei Möglichkeiten, abhängig davon, ob es sich um ein gefördertes Projekt handelt oder

nicht. Im ersten Fall wird ein monetär ausgedrückter zweckbezogener Mittelbedarf meist weit vor der Durchführungsplanung angefordert und von der Zuteilerseite oftmals noch gekürzt. Die Chance der Deckung von Budgetbedarf und -zuteilung ist damit eher gering, insbesondere ist ein Nachtragsbudget ohne Ausweitung der Sachziele meist nicht möglich. Im anderen Fall könnte das Budget aus Lehrstuhlmitteln finanziert werden, wäre damit jedoch eher marginal. Stets empfiehlt sich somit die Verwendung von zugänglichem bereits Erstelltem, Zurückhaltung in der Produktgestaltung bei Konzentration auf den Wissenstransport sowie weitgehende Anwendungserstellung im eigenen Haus. Insgesamt ist damit, und dies deckt sich durchaus mit den Erfahrungen der Autoren, die konkrete Ausgestaltung des Sachziels im engeren Sinne, nämlich welche Multimedia-Elemente in welchem Umfang und zu welchen Themen innerhalb des Produkts zwecks Verbesserung des Wissenstransports zu realisieren sind, die zu optimierende Zielgröße unter Einhaltung der harten Zeit- sowie Budgetrestriktionen.

3 Ressourcen und technische Plattform

Erstellung im eigenen Haus zieht zwei wesentliche Konsequenzen nach sich. Zunächst gehen die Anforderungen an technisches Wissen und die Bereitschaft zur Einarbeitung in Entwicklungssoftware direkt in ein Anforderungsprofil an den Kursentwickler und damit Projektleiter ein. Durch die erwünschte Anwendung von Projektmanagement wird das Profil ergänzt um entsprechende Managementqualifikationen. Des weiteren sehr wichtig, und dies soll hier zum Anforderungsprofil genügen, ist die fachlich-inhaltliche Kompetenz. Die wissenschaftliche Sprache ist unter anderem gekennzeichnet durch eine hohe Informationsdichte. Dies ist zu übertragen auf die Ausgestaltung multimedialer Elemente. So mag etwa ein Software-Designer allenfalls zuarbeiten. Vielmehr ist die Nähe von Autor und Kursentwickler für den Realisierungserfolg entscheidend, insbesondere mit Wirkung auf den Wissenstransport und die Projektdauer. Es empfiehlt sich somit die Besetzung der Projektleitung durch Lehrstuhlpersonal, allerdings auf Grund der zu erwartenden zeitlichen Belastung nicht durch den Lehrstuhlinhaber.

Die zweite Konsequenz betrifft die Wahl einer geeigneten Plattform. Hierunter versteht man den technischen Zugang, die unterstützten Betriebssysteme, die Entwicklungsumgebung für Benutzersoftware und Multimediamodule sowie das vom Benutzer zu verwendende Front-End.

Neben Lizenzkostengesichtspunkten bestimmt bei weitestgehender Entwicklung im eigenen Haus das vorhandene und kurzfristig zu erwerbende Know-how die Wahl der Software, und die Richtung dieser Abhängigkeit zeigt ein weiteres Mal die besondere Situation der Multimedia-Entwicklung für Hochschulen. Der technische Zugang beinhaltet zwei Aspekte, nämlich den physikalischen Zugang selbst sowie die an den Anwender zu stellenden Hardwarevoraussetzungen. Letztere müssen gerade bei Einbindung des Kurses ins

Curriculum eher niedrig ausfallen und beeinflussen damit auch die Produktausgestaltung. Der physikalische Zugang kann ein fester Datenträger wie die CD-ROM oder aber das Internet, also Online sein. Die Wahl hier hängt wesentlich von der Gültigkeitsdauer der Inhalte, den zu übermittelnden Datenmengen und vom Bedarf eines angeschlossenen Online-Auswertungssystems etwa für Selbsttests oder Prüfungsvorleistungen ab. Derzeit sind auf jeden Fall Onlinekosten, Datendurchsatz und Netzverfügbarkeit mit entscheidungsrelevant, so dass insbesondere auch eine Hybridlösung in Betracht zu ziehen ist.

4 Interdependenz von Produktgestaltung und Projektorganisation

Gängige Multimediaelemente sind Text, Bild, Ton, Video, Animation und Applikation. Für die Verwendungsauswahl stellt das Projektzielsystem ein natürliches Maß dar, insbesondere Erstellungsaufwand und erwarteter zusätzlicher Lernnutzen, aber auch Homogenität der Erscheinung des gesamten Produkts in Stil und Qualität sowie Möglichkeit der Realisierung durch verfügbare Ressourcen. Andererseits genießt ein universitäres Multimedia-Produkt bei Einbindung ins Curriculum im Vergleich zu einem kommerziellen regelmäßig einen besonderen Vorteil: Wegen der weitgehend festen Nachfrage kann sich die Produktgestaltung von nachfragewirksamen, aber den eigentlichen Zielen nicht förderlichen Merkmalen freihalten. Bei vorhandenem Handlungsspielraum sollten weiterhin auch Gesichtspunkte wie Perzeption oder Interaktivität vertieft bedacht werden, vgl. etwa [2].
Die oben genannte determinierende Verfügbarkeit der Ressourcen gestaltet sich durch Organisationszugehörigkeit, zeitliche Kapazität, Weisungsbefugnis und monetären Aufwand, mit starker Wirkung auf die Durchführungsplanung und -kontrolle. Bei der zuvor genannten Entscheidung für eine Umsetzung in mehreren Kursversionen sollte die Verfügbarkeit möglichst über sämtliche Stufen hinweg geplant werden, um das Risiko der mehrfachen Einarbeitung nicht unnötig in das Projekt hineinzutragen. Der festgestellte Ressourcenbedarf bestimmt neben der Bedeutung des Projekts, gemessen etwa in erwarteter Belegerzahl und mittel- bis langfristiger Reduktion des Kursberatungsaufwands, die rationale Entscheidung für die Managementform, z.B. Matrix- oder Einflussprojektmanagement.

5 Kritische Erfolgsfaktoren des Projekts

In diesem kurzen Abschnitt sollen die oben teilweise über verschiedene Zugänge abgeleiteten kritischen Erfolgsfaktoren für die Entwicklung eines universitären Multimediakurses zusammengetragen werden. Die Form der Darstellung verrät den praktischen Zweck, nämlich als Checkliste für eine konkrete Management- sowie Durchführungsplanung im Projekt.

Kritische Erfolgsfaktoren

- Anwendung von Projektmanagement
- bedeutungsgerechte Managementform
- dezidierte Analyse der Projektziele und Aufstellen des Zielsystems
- anforderungsorientierte Projektleiterwahl
- zielkonforme Produktgestaltung
- Wahl einer handhabbaren (technischen) Plattform
- (retrograde) Zeitplanung, Risikoanalyse hier bei erheblichem Know-how Bedarf oder nicht einschlägig wohlbekannten Ressourcen
- Ressourcensicherung

Im Prinzip stellt diese Liste einen Auszug wesentlicher Aspekte allgemeinen Projektmanagements (siehe auch [4]), angewandt auf die besondere Situation der Multimedia-Kursentwicklung an Hochschulen, dar.

6 Projektfallbeispiel Kurs „Planungs- und Entscheidungstechniken"

Abschließend soll eine mögliche Projektausgestaltung anhand der Erstellung des multimedialen Dateikurses „Planungs- und Entscheidungstechniken"[3] des Lehrstuhls für Operations Research an der FernUniversität in Hagen in Form eines kritischen Erfahrungsberichts aufgezeigt werden. Die tradi-

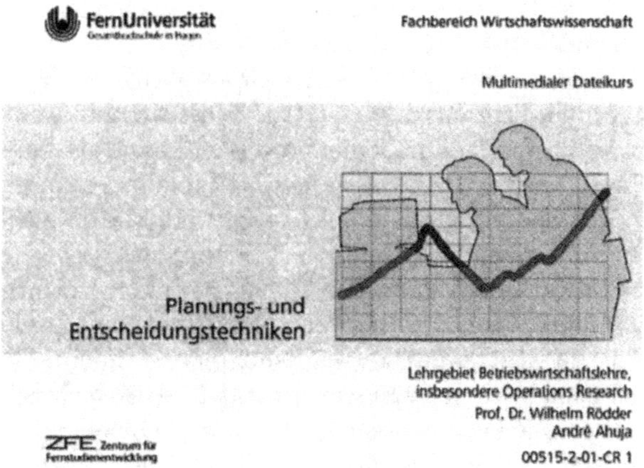

Abbildung1. Titelbild des Kurses „Planungs- und Entscheidungstechniken"

tionell als Printkurs angebotene Pflichtveranstaltung im Hauptstudium des kaufmännischen Diplomstudiengangs hat je Studienjahr rund 3.000 Beleger.

1997 fiel die Entscheidung zur Erstellung eines parallel anzubietenden Multimediakurses. Die wesentlichen Komponenten des Zielsystems waren langfristige Senkung des Betreuungsaufwands, Verbesserung des Wissenstransports (besonders motiviert durch gelegentliche Kritik seitens Studierender und deren Vertretung sowie durch die fortschreitende Umsetzung der Dienstleistungsorientierung), Innovationsorientierung sowie Möglichkeit des externen, evtl. kommerziellen Einsatzes. Nach Benennung des Kursbetreuers zum Projektleiter und Wahl der Form Einflussprojektmanagement führte eine erste Aufwandsanalyse zur Entscheidung für eine Umsetzung in mehreren Stufen, um einen Ersteinsatz im WS 98/99 anstreben zu können. Eine Projektförderung lag nicht vor. Deshalb, und weil die stark textbasierten Kursinhalte eine sequentielle Bearbeitung erfordern, wurde als Entwicklungsumgebung die für den Lehrstuhl bereits lizenzierte Software „Acrobat“gewählt, zu welcher das kostenlose Front-End „Acrobat-Reader“frei erhältlich ist. Auf Grund der seinerzeit nicht unerheblichen Onlinekosten, bei einer erwarteten Datenmenge von mehreren hundert MByte gegenüber Vervielfältigungsstückkosten von ca. 1,16 Euro einer CD-ROM (ab einer Auflage von 1.000 Stück), sollte der Kurs als Hybrid umgesetzt werden, nämlich eine CD mit Internetlink zu Websites mit Ergänzungen und aktuellen Informationen rund um den Kurs. Als Multimediaelemente waren Videos, Selbsttests, Animationen sowie ein Softwarelabor zur linearen Programmierung als Applikation einzusetzen. Gerade die Animation, gleichsam ein elektronischer Tutor, ist in besonderer Weise zur Vermittlung von OR-Verfahren geeignet, sind doch die Algorithmen dezidiert in der Dynamik ihres Ablaufs darstellbar.
Trotz offensichtlich fehlender Kongruenz von Projektbedeutung und Managementformwahl „Einfluss-PM“konnte die erste Version termingerecht eingesetzt werden und stellte den als reinen Dateikurs umgesetzten Lehrtext dar. Obwohl für dieses Produkt kein aktives Marketing betrieben wurde, wählten rund 10% der Beleger die CD-ROM als Kursmedium (zu gleichen Bezugsgebühren wie für den Printkurs) und etwa 16% die Kombination Multimedia- und Printkurs (zu etwa 20% höheren Bezugsgebühren als für einen Einzelkurs). Ein Jahr später wurde die um Videos, Animationen zu den am häufigsten durch Studierende angefragten Themen und Multiple-choice Tests je Kapitel erweiterte Version 2.0 angeboten. Die größten Schwierigkeiten bereiteten hier das fehlende Know-how am Lehrstuhl zur Entwicklung von Animationen sowie die fehlende Erfahrung beim Schreiben eines Drehbuchs für die Videosequenzen, beim Regieführen anlässlich entsprechender Aufnahmetermine sowie der Video-Schnittvorbereitung mit Hilfe eines haushaltsüblichen Videorekorders. Die Schwierigkeiten äußerten sich in einer stark unsicherheitbehafteten Zeit- und Aufwandplanung bzw. einer starken Abhängigkeit der Projektleitung vom Bereich Film des ZFE, einer für Medienentwicklung zuständigen Organisationseinheit der FernUniversität in Hagen.
Abgeschlossen sind mittlerweile die Arbeiten zur dritten und vorerst letzten Version des Kurses, welche ab dem WS 2001/2002 eingesetzt werden wird.

Gemäss einer Entscheidung ausserhalb des Lehrstuhls werden fortan Print- und Multimediakurs zusammen unter einer Kursnummer geführt, und zwar zum Bezugspreis des früheren Einzelkurses. Damit wird die zukünftige jährliche Distribution der Kurs-CD wohl zwischen 2.000 und 3.000 Stück liegen. Version 3.0 erweitert die vorherigen Realisierungen um eine Vielzahl zusätzlicher Animationen sowie ein Softwarelabor, bestehend aus einer professionellen LP-Software und einer Anwendung mit einer Vielzahl von Verfahren zur Lösung des Travelling-Salesman-Problems. Beide Applikationen konnten gebührenfrei für den Einsatz in der universitären Lehre lizenziert werden, allerdings mit Wirkung auf das Teilziel eines möglichen kommerziellen Einsatzes des Produkts. Diese Beobachtung hat allgemeinen Charakter: mehr als nur innerhalb eines Projekts stellen sich gerade bei einer stufenweisen Realisierung (implizite) Zielsystemüberarbeitungen ein. Aufwand-/Nutzenüberlegungen führten bei Version 3.0 zur Entscheidung, keine weiteren Videos zu produzieren. Auf Grund des stellenweisen Ressourcenproblems während der Produktion von Version 2.0 wurden frühzeitig geschätzte Kapazitätsbedarfe an Ressourcen anderer Organisationseinheiten gesichert, so dass man wohl von einer Mischform von Einfluss- und Matrixprojektmanagement sprechen darf. Hierdurch stellten sich verkürzte Reaktionszeiten innerhalb des verteilten Projektteams ein. Allein die Neubesetzung einer studentischen Hilfskraftstelle am Lehrstuhl, welche für Programmiertätigkeiten in das Projekt eingebunden war, mit der Folge einer eingeschränkten Produktivität zusammen mit einer wechselnden zeitlichen Verfügbarkeit, führte kurz vor Realisierungsabschluss zu einer Kürzung des Sachziels um eine Animation zur Sicherung des strikten Zeitziels.

7 Fazit

Der wachsende, nachvollziehbare Bedarf an multimedialer Hochschullehre stellt einen enormen Komplexitätszuwachs in der Aufgabe der Kursentwicklung dar. Derzeit noch fehlende Workflowdefinitionen und mangelnde Anwendung notwendigen Projektmanagements kennzeichnen die momentane Umbruchphase. Für diese Phase geben die Autoren als ersten möglichen rationalen Zugang, insbesondere durch Auswertung eigener Erfahrungen in der Multimedia-Produktion, eine Checkliste kritischer Erfolgsfaktoren für die Projektdurchführungsqualität an. Diese umfasst die Anwendung von Projektmanagement, die Wahl der bedeutungsgerechten Managementform, die dezidierte Analyse der Projektziele und Aufstellen des Zielsystems, die anforderungsorientierte Projektleiterwahl, die zielkonforme Produktgestaltung, die Wahl einer handhabbaren technischen Plattform, die retrograde Zeitplanung sowie die frühzeitige Ressourcensicherung über die ganze, möglicherweise mehrstufige Projektdurchführung. Diese Liste stellt somit eine Projektion von Projektmanagement auf die besondere, oben beschriebene Situation dar. Speziell beim Zielsystem lässt sich feststellen, dass typischerweise die harten

Budget- und vor allem Terminrestriktionen die Ressourcenplanung bzw. die Produktgestaltung in Umfang und Wahl der Multimediaelemente determinieren. Stets wird realistischerweise unterstellt, dass die operativen Projektziele das strategische Ziel der Verbesserung der Lehre selbst verfolgen.

Literatur

1. Ahuja, A. (2001) Managing the development of multimedia lectures, International Conference on Distance Education ICDE.
2. Kerres, M. (1995) Technische Aspekte multimedialer Lehr-Lernmedien, in: Issing, L.J., Klisma, P. (Hrsg): Information und Lernen mit Multimedia, Psychologie-Verl.-Union, Weinheim, S. 25 - 45.
3. Rödder, W., Ahuja, A. (2001) Planungs- und Entscheidungstechniken (PET), Multimedialer Dateikurs 00512, 3., erw. Aufl., FernUniversität in Hagen; Demoversion Online unter http://www.fernuni-hagen.de/BWLOR/lsmm.htm.
4. Schifman, R., Heinrich, G. (2000) Multimedia-Projektmanagement, 2. Aufl., Springer.
5. Seidel, C., Lipsmeier, A. (1989) Computergestütztes Lernen, Verlag für angewandte Psychologie.

Integration von OR-Modellen in die Logistikausbildung: Bearbeitung von Fallbeispielen mit einem Softwarepaket für Studenten

Dieter Feige

Friedrich-Alexander-Universität Erlangen-Nürnberg, Lehrstuhl für Logistik

Summary: Operations Research (OR) ist eine anwendungsorientierte Wissenschaftsdisziplin. OR ist auf die Verbesserung von Managemententscheidungen gerichtet und gehört deshalb in den betriebswirtschaftlichen Studienrichtungen zum Pflichtprogramm. Dennoch haben die akademisch gebildeten Betriebswirte in der Praxis oft große Schwierigkeiten, das Gelernte anzuwenden. Die Fähigkeit, mit modellbasierten Methoden neuartige und nicht vorstrukturierte Realprobleme zu bearbeiten, ist meist nicht ausgeprägt. Diese Fähigkeit muss durch Training erworben werden. Ein Ansatz in dieser Richtung ist die Integration modellbasierter Übungen mit realitätsnahen Fallbeispielen in die Fachausbildung. Für die Bearbeitung dieser Problemstellungen wurde, angelehnt an professionelle Entscheidungsunterstützungssysteme (EUS), ein Lehrsoftwarepaket entwickelt.

Keywords: Ausbildung, Logistik, modellbasierte Entscheidungsunterstützung, EUS-Software für Studenten.

1 Einleitung

In jedem Jahr verlassen junge Absolventen unsere Universität mit dem Ziel, in einem der vielen Bereiche der Logistik eine Tätigkeit aufzunehmen. Sie werden an ihren neuen Wirkungsstätten sehr bald mit schwierigen Aufgaben konfrontiert: Die Unternehmen müssen sich im globalen Wettbewerb behaupten, in europäischen und weltweiten Netzwerken agieren, Kooperationen aufbauen und Prozesse gestalten. Entscheidungen, die in diesen Situationen zu treffen sind, zeichnen sich durch hohe Komplexität aus. Jede Veränderung eines Elements wirkt mehr oder weniger auf das ganze Logistiknetz. [1]

---

[1] Klaus (1993) definiert Logistik als Management von Fließsystemen. Logistik in diesem Sinne betrachtet wirtschaftliche Phänomene als Flüsse in Ketten und Netzen und will deren Strukturen sowie die Prozesse in ihnen gestalten und optimieren.

Von den jungen akademisch gebildeten Logistikern werden dazu Ideen und Problemlösungsansätze erwartet. Aber ihnen scheint plötzlich das erlernte Wissen nutzlos zu sein. Sie können ihre theoretischen Kenntnisse oft nur unzureichend oder überhaupt nicht in den realen Situationen anwenden. Diese Fähigkeit muss erst in der Auseinandersetzung mit den wirklichen Problemen schrittweise erlernt werden. „Der Mensch ist ein lernendes Lebewesen. Wenn er einer Folge von Problemsituationen mit steigendem Schwierigkeitsgrad unterworfen wird und zwar solchen Schwierigkeiten, die dem von ihm bereits erreichten Niveau entsprechen, so steigert sich im Allgemeinen bei ihm die Fähigkeit, mit den Problemen gut fertig zu werden.“ [2]

Wir versuchen, diesen Lernprozess durch eine PC-basierte Übungsveranstaltung bereits im Studium zu fördern, in der praxisorientierte Fallbeispiele mit einem speziell für Studenten entwickelten Programmpaket zu bearbeiten sind. [3]

2 Anforderungen der Praxis

Praktische Entscheidungsprobleme in der Wirtschaft, speziell in der Logistik, sind durch Intransparenz, Heterogenität und Flüchtigkeit der Fragestellung, Schwierigkeit der Bewertung sowie Vernetztheit und Dynamik gekennzeichnet.

Wie tritt eine Problemstellung an den Logistiker heran? Ein Prozess oder ein System ist verbesserungswürdig, Kosten sollen eingespart und Servicegrade verbessert werden. Es gibt möglicherweise einige strategische Überlegungen der Geschäftsleitung über Entwicklungsrichtungen des Unternehmens, aber keine klar formulierte Aufgabe nach dem Schema: „Gegeben ist... , gesucht wird ...“.

Die verfügbaren Daten, in der Regel unvollständig und fehlerhaft, erfordern Ergänzung und Aufbereitung. Qualifizierte Tools, die problembezogene Unterstützung bieten können, fehlen oft. Zudem müssen Entscheidungsvorschläge im Regelfall unter starkem Zeitdruck erarbeitet werden.

Wie ist in solchen Situationen zu handeln? Der Entscheidungsprozess beginnt mit einer Analyse der Situation, ihrer verbalen Beschreibung, der Formulierung von Fragestellungen und möglichen Handlungsrichtungen. Diese Arbeiten erfordern Kreativität und das Vermögen, sich tief in ein Problem hineinzudenken.

Erst nachdem sinnvolle Fragen formuliert wurden, können weitere Schritte folgen:

- Strukturieren und Formalisieren des Entscheidungsproblems
- Abtrennen von quantifizierbaren Teilproblemen

---

[2] Siehe Simon (1966), S.81.

[3] Vgl. Hanssmanns (1995) Plädoyer für ein problembezogenes OR

- Definieren von Bewertungsmethoden
- Wahl geeigneter Modelle zur Problemabbildung
- Ermitteln der Modellparameter
- Finden von Modell-Lösungen
- Ableiten von Schlussfolgerungen für das Realproblem und/oder Formulierung neuer Fragestellungen

Das Entscheiden kann nicht auf die Auswahl einer guten oder der besten Handlungsvariante reduziert werden, sondern es sind vor allem die intellektuellen Anstrengungen der Abbildung des Problems in einem Modell und die Rücktransformation der aus dem Modell gewonnenen Erkenntnisse in das Realproblem, welche den Entscheidungsprozess prägen (Abb. 1). [4]

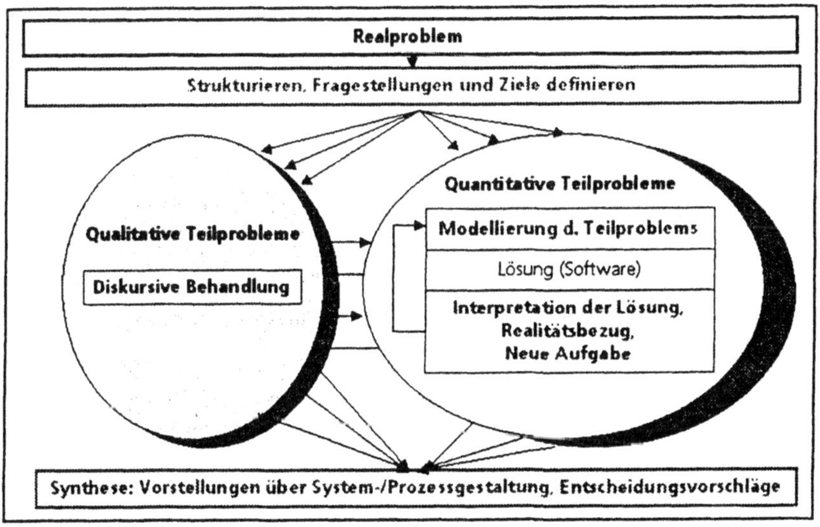

Abbildung 1: Intellektuelle Aufgaben im Entscheidungsprozess

Der Prozess der Entscheidungsfindung ist in der Regel ein rückgekoppelter Lernprozess. Jede neue Erkenntnis aus einem Modellexperiment führt wieder zu neuen Fragestellungen, für die wieder neue Ziele, neue Bedingungen oder neue Modelle erforderlich sind.

---

4 Simon (1966) unterteilt Entscheidungsprozesse in die Phasen Nachrichtensuche (*intelligence*), Planung (*design*) und Auswahl (*choice*). Der Entscheidungsprozess darf nach Simon nicht nur auf die letzte Phase bezogen werden. Alle Phasen sind miteinander verwoben (S.96 ff.).

Schließlich werden aus den Erkenntnissen der Modellexperimente unter kritischer Beachtung der qualitativen Faktoren Handlungsempfehlungen abgeleitet. In vielen Fällen ist das Resultat nicht eine einzige optimale Lösung, sondern besteht aus mehreren Handlungsalternativen, die durch unterschiedliche Chancen und Risiken gekennzeichnet sind und von denen keine als die eindeutig beste Lösung ausgewählt werden kann.

Diesem realen Entscheidungsprozess wollen wir uns in den Übungen annähern. Die Studenten sollen lernen, aus komplexen Problemstellungen die quantifizierbaren Teilaufgaben herauszufiltern und in ein geeignetes Modell zu übertragen. Lösungen sollen unter verschiedenen Kriterien und mit gezielten Änderungen von Modellparametern erzeugt und auf ihre Entscheidungsrelevanz für das Realproblem beurteilt werden. Schließlich sollen die Studenten Entscheidungsvorschläge für das Gesamtproblem formulieren. Dafür sind geeignete Softwaretools erforderlich, mit deren Hilfe modellbasierte Aufgabenstellungen interaktiv bearbeitet werden können.

3 Das Ausbildungskonzept

Die Ausbildung muss auf die künftigen Aufgaben vorbereiten. Dazu gehört die Vermittlung von Kenntnissen ebenso wie auch die Herausbildung von Fähigkeiten und Persönlichkeitseigenschaften. Im Kern handelt es sich um:

- *Wissen* in Gestalt von Kenntnissen über die grundlegenden Modelle und die Eigenschaften von Lösungsverfahren;
- *Können* als Fähigkeit zur eigenständigen Problemstrukturierung und Formulierung von Fragestellungen sowie die Fähigkeit, Kenntnisse auf unterschiedliche und neuartige Aufgaben anzuwenden;
- *Wollen* als Persönlichkeitseigenschaft, die das Ziel besitzt, Gelerntes anzuwenden, Schwierigkeiten als Herausforderung anzunehmen und fehlendes Wissen oder Können sich selbständig anzueignen.

Können und Wollen erwirbt man in der Auseinandersetzung mit Problemen und Schwierigkeiten. Das formale Einprägen von Lehrsätzen schafft dafür zwar wissensmäßige Voraussetzungen, entwickelt jedoch nicht das Problemlösungsverhalten des Studierenden. Dazu soll eine Übungsveranstaltung beitragen, welche die Vorlesung „Computerbasierte Entscheidungsunterstützungssysteme in der Logistik (EUS)“ begleitet. In dieser Übung werden die Studierenden angehalten, sich mit praxisnahen Problemfällen auseinanderzusetzen.

Die Studenten erhalten Fallbeispiele zur selbstständigen Bearbeitung. Anhand von übergebenen Musterlösungen können sie ihre Ergebnisse überprüfen. Dabei ist es wichtig, dass die Fallbeschreibung die Studenten in eine von ihnen als real empfundene Entscheidungssituation versetzt und sie emotional zu Lösungsanstrengungen stimuliert.

Eine typische Fallbeschreibung lautet auszugweise folgendermaßen:

Fallbeschreibung

Nach erfolgreichem Abschluss Ihres Studiums als Dipl.-Kffr. bzw. Dipl.-Kfm. arbeiten Sie bei einer großen Consultingfirma in der dortigen Abt. Logistik. Ihr Vorgesetzter betraut Sie mit folgendem ersten Auftrag:

Die Nordfisch AG ist ein in Hamburg ansässiges dt. Unternehmen, das mit eigener Fangflotte frischen Fisch fängt und täglich über 3 firmeneigene Kühlläger (KL) an 8 kundeneigene Distributionszentren (D) einer großen Kaufhauskette verteilt. ...

Die Verteilung von den Kühllägern KL zu den Distributionszentren D der Kaufhauskette erfolgt in Isoliercontainern; dies impliziert eine Verringerung der Qualität mit steigender Transportzeit.

Es häufen sich bei der Nordfisch AG die Beschwerden, dass in den Filialen der Kaufhauskette, die von den Distributionszentren D_1 und D_6 aus beliefert werden, die Qualität des Fisches nicht den vereinbarten Anforderungen entspricht. ...

Für die Erarbeitung eines Entscheidungsvorschlags sind häufig mehrere Wege möglich und es können auch unterschiedliche akzeptable Lösungsvorschläge ermittelt werden. Eine sinnvolle Vorgehensweise bei obigem Beispiel könnte folgende Schritte umfassen:

- Nachstellen der Ist-Situation und Ursachenforschung (Transportproblem nach Transportkosten, Analyse der Zeitbedarfe).
- Kritische Zeitgrenze ermitteln (Bottleneckproblem, Fahrzeiten).
- Kostenminimierung nach Vorgabe von Zeitbegrenzungen.
- Überprüfen der Effekte von Single-Source-Lösungen.

Von den Studenten wird keine numerische Problemlösung erwartet, die einfach als „richtig" oder „falsch" bewertet werden kann, sondern eine Analyse des Problems, die Entwicklung relevanter Fragestellungen und die Wahl einer sinnvollen Vorgehensweise zur Ermittlung von entscheidungsrelevanten Informationen sowie schließlich die Formulierung von Handlungsempfehlungen. Es sind vor allem die intellektuellen Herausforderungen der Modell- und Verfahrenswahl, der Modellierung von Teilproblemen, die Interpretation und Wertung von Resultaten und das Zusammenfügen von Einzelergebnissen zu einem problemadäquaten Gesamtresultat, welche die Problemlösungskompetenz ausmachen und deshalb herausgebildet werden sollen.

Die dabei zu verwendenden Lösungsverfahren werden durch das Werkzeug – die Software – gestellt. Genaue Algorithmenkenntnisse sind nicht erforderlich, sondern das Lösungsverhalten des Verfahrens muss bekannt sein. Für die Beurteilung der Resultate ist lediglich die Kenntnis nötig, ob es sich um ein exaktes Verfahren handelt oder ob das Verfahren eine Heuristik verwendet, die für das betreffende Problem genügend genaue Resultate liefert.

Der Umfang der Fallstudien muss sich in Grenzen halten, damit sie in verträglicher Zeit bearbeitet werden können. Sie sollen andererseits aber das Gefühl der Nähe zur realen Welt vermitteln und den Drang zur Bewährung in einer als real empfundenen Entscheidungssituation auslösen.

Der Lernerfolg stellt sich erst nach dem wiederholten erfolgreichen Lösen von Aufgaben mit wachsendem Schwierigkeitsgrad ein.

Die oben genannten Anforderungen sind Idealvorstellungen. Im Alltag des Universitätsstudiums sind sie in reiner Form kaum zu verwirklichen. Eine Annäherung an diese Zielvorstellungen ist aber möglich, wenn der etwas höhere Betreuungsaufwand für diese Übungen nicht gescheut wird. Die Bearbeitung einer Fallstudie erfordert gewöhnlich mehrere Stunden intensiver Arbeit. Manche Studierende, die einen „kürzesten Weg“ zum Diplom suchen, empfinden das als unnötige Belastung. Andere Studenten erleben diese Übung zwar als aufwendig und schwierig, aber doch als interessant und gewinnbringend für die berufliche Zukunft.

4 Das EUS-Softwarepaket

Voraussetzung für die Realisierung diese Ausbildungskonzepts sind modellbasierte Softwaretools. Das von uns entwickelte EUS-Lehrprogrammpaket ist vorwiegend auf typische Logistikprobleme ausgerichtet. Es umfasst gegenwärtig die Programme:

LOP: Lineare und gemischt-ganzzahlige Optimierung,[5]

---

[5] In LOP wird zur Lösung von gemischt-ganzzahligen Optimierungsproblemen der von Prof. Suhl /FU Berlin zur Verfügung gestellte Solver MOPS verwendet.

TPP: Transportprobleme (Lineares TPP, Bottleneck TPP, Binäres TPP, kapazitätsbeschränktes TPP, zweiseitig beschränktes TPP),

NWF: Lineare Netzwerkflußprobleme,

TSP: Rundfahrt- und Durchfahrtprobleme, Entfernungsmatrizen aus Netzen,

TPL: Tourenplanung (nach verschiedenen Verfahren).

Die Programme sollen die geistige Auseinandersetzung mit Problemstellungen ermöglichen und zu spielerischen Experimenten anregen. Deshalb wurde auf Interaktion und Visualisierung großer Wert gelegt. Die Benutzeroberfläche ist für alle Programme einheitlich gestaltet und besitzt neben den üblichen Bedienelementen ein Grafik- und ein Textfenster (siehe Abb. 2).

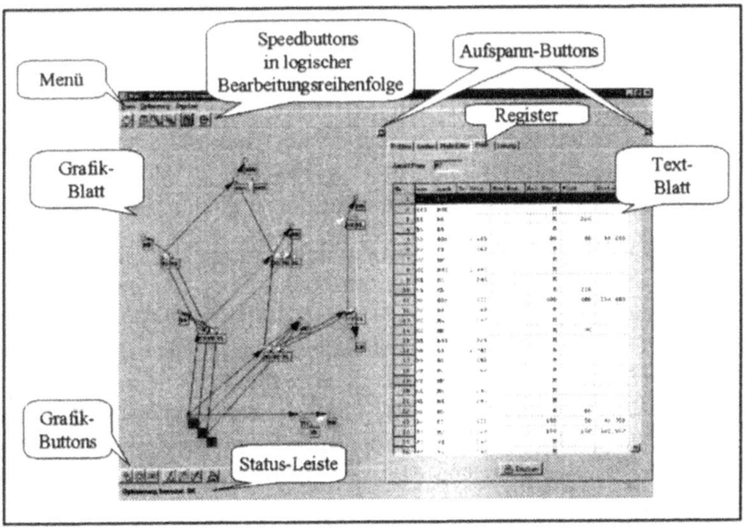

Abbildung 2: Benutzeroberfläche der Programme

Das interessierende Fenster kann über den ganzen Bildschirm aufgespannt werden. Dialogeingaben über die Tastatur und grafische Aktionen mit der Maus sind gleichermaßen möglich und miteinander verknüpft.

Daten und Resultate werden als Textdateien gespeichert und können deshalb mit anderen Programmen (z.B. Excel) bearbeitet werden. Daten für lineare und gemischt-ganzzahlige Optimierungsaufgaben lassen sich auch im MPS-Format [6] importieren und exportieren.

[6] MPS ist die Abkürzung von *Mathematical Programming Standard* und definiert eine Standard-Dateistruktur für mathematische Optimierungsprobleme.

Die studentischen Softwaretools sind weitgehend an professionelle EUS-Werkzeuge angenähert. Sie sind deshalb geeignet, die Studenten auf die aktive Gestaltung von Logistikprozessen unter Nutzung moderner Werkzeuge vorzubereiten. Der semiprofessionelle Charakter dieser Tools erlaubt durchaus auch die Bearbeitung von kleineren Realproblemen.

Zusammenfassung

Die Befähigung der künftigen Logistiker, sich bei praktischen Entscheidungsproblemen modellbasierter Methoden zu bedienen, erfordert nicht nur theoretisches Wissen, sondern auch die Fähigkeit, dieses Wissen sinnvoll anzuwenden. Dazu kann das Üben anhand von praxisnahen Fallbeispielen und deren Lösung mit semiprofessionellen EUS-Softwarewerkzeugen hilfreich sein.

Der Aufwand für die Vorbereitung und Betreuung der Übungen ist nicht gering. Auch von den Studenten wird intensive und eigenständige Arbeit erwartet, wofür Zeit zu investieren ist. Nicht immer findet die geforderte Anstrengung Zustimmung, sondern wird als Last empfunden. Nach Bestehen der Klausuren steigt gewöhnlich der Zustimmungsgrad an und die erbrachte eigene Leistung wird als wertvolle Errungenschaft für die berufliche Zukunft betrachtet.

Es ist vorgesehen, die Fallbeispielsammlung weiter zu verbessern und zu erweitern. Die Lehrsoftwaretools sollen mit neuen Programmen und Funktionen erweitert und mit einem vereinfachten Geographischen Informationssystem (GIS) ergänzt werden. Auf diese Weise wollen wir eine stärkere Annäherung an professionelle EUS-Tools erreichen. Und schließlich denken wir an die Entwicklung einer „Logistik-Toolbox“ für die Unternehmenspraxis. Sie wird professionelle, modellbasierte EUS-Tools für wichtige Problemfälle enthalten und damit das an der Universität Gelernte in der Praxis anwendbar machen.

References

Hanssmann, Friedrich: Vierzig Jahre Operations Research im internationalen Vergleich. Technologie & management, 44. Jg., 1995, H4., S. 175-178.

Klaus, Peter: Die dritte Bedeutung der Logistik. Nürnberger Logistik-Arbeitspapier Nr. 3, Nürnberg, 1993.

Simon, Herbert A., Perspektiven der Automation für Entscheider. Verlag Schnelle Quickborn, 1966.

A Survey of Educational Resources on the Internet

Ralph Scheubrein

University of Hohenheim (510A), D-70593 Stuttgart, Germany,
E-Mail: scheubre@uni-hohenheim.de

Summary:
Today the Internet is a rich source of information on almost every topic. Unfortunately when looking for information on a specific topic the relevant documents are often hard to find because for most documents in the Internet there is not enough standardized classifying metadata available which would provide a possibility to index and query the information in a systematic way. Therefore in this article a survey of Internet resources is given especially about teaching materials for operations research. The selected web sites are classified into information portals, content providers, and application providers.

Keywords: web-based learning, information portal, content provider, application provider

1 Introduction

For many lectures at the university there is the question how and in which way to use the Internet for teaching purposes. While some years ago the "death of the classroom" was predicted most lecturers today use the Internet not as a means to replace the classroom teaching but as an additional medium for communication and education. A typical scenario of using the Internet is providing the class with a course web with selected resources. The students access the course web for preparing the next classroom meeting. Beside administrative information the course web can hold links to content that is used in the classroom teaching. When setting up a course web the lecturer has the possibility to create the resources on his own or to look for suitable resources of high quality on the Internet. This article tries to help in the latter case by giving a survey of resources, which are relevant for teaching operations research. Finding such resources on the Internet with a typical search engine is not so easy: for example when entering the keyword "operations research" in the system Google (http://www.google.com respectively http://www. google.de) about 1,340,000 references are given worldwide and still more than 56,000 links only in Germany. On the other side, a catalog system like Yahoo (http://www.yahoo.com) produces with the same keyword only references to eleven institutes and four organizations. Therefore this survey tries to annotate the most prominent resources available in August 2001. It is of course only a small snapshot due to the immense size and the high dynamic of the Internet. In this survey only resources in English or German language are included. Also only resources, which are free of charge, are presented. This means of course that some interesting, commercial systems like the "Multimedia Educational Technology for Operational Research – MENTOR" (http://www.managementscience.org/mentor/mentor.asp) are not discussed. The resources are classified here in three types. If the main intention of the web site is providing categorized links it is classified as an information portal. In contrast a content provider gives information about a specific topic mainly in the form of hypertext documents. Finally an application provider offers software that can be used for educational purposes.

2 Information Portals

Many international and national organizations for operations research are hosting web sites. These sites typically have some listings of operations research related resources. Also some organizations have recognized the importance of teaching for the promotion of this research area. Consequently they have started initiatives in order to provide the public and their members with teaching materials.

The "International Federation of Operational Research Societies - IFORS" (http://www.ifors.org) has launched the "Educational Resource Initiative" (http://www.umsl.edu/~sauter/ifors) organized by Sauter. The site provides a list of links, which is - at least today - not too long. One project of the initiative is the implementation of the interactive web based "TutORial" organized by Sniedovich, which will be discussed below. Also the online encyclopedia of the Sniedovich's project WORMS will be extended and added to the IFORS web site. It is planned to have both systems completely integrated until summer 2002.

The American "Institute for Operations Research and the Management Sciences – INFORMS" (http://www.informs. org) has collected links in the menu "Education/Students". Moreover this organization has concentrated its education related activities in the initiative INFORM-ED (http://www2.hpu.edu/jkros). Under "readings" and "teaching materials" there are various resources for this topic. Additionally the INFORMS journal "ORMS Today" has a regular column about education and in August 2001 an issue especially dedicated to education. Many articles of this journal can be retrieved online (http://www.lionhrtpub.com/ORMS.shtml). Under "links" this web site also provides short descriptions of some important portals in this area.

The British "Operational Research Society" offers on the web site http://www.orsoc.org.uk teaching materials under "About OR". Two interesting links named "Handling Strategic Problems" and "OR Notes" will be discussed below.

An information portal especially for operations research related software exits at the department of Neumann. This "software library for operations research" (http://www.wior.uni-karlsruhe.de/Bibliothek) classifies numerous commercial and non-commercial systems in the categories "decision support", "fuzzy logic", "genetic algorithms", "integer programming", "linear programming", "mathematics", "modeling languages", "networks", "neural networks", "nonlinear programming", "project planning", "simulation", "stochastic models", and "vehicle routing".

Beside web sites, which are run by organizations, a lot of individuals set up web sites. Their actuality is varying very much depending on the time the individual is currently investing in his system. Probably the most comprehensive and up-to-date web site is "Michael Trick's Operations Research Page" (http://mat.gsia.cmu.edu). From the more than 1000 links stored here, especially the categories "courses" and "computer programs" are interesting for teaching purposes. Under "courses" there are currently 25 complete courses available. The "computer programs" are classified into Java programs, online optimizer and software packages. This web site is surely one of the most interesting starting point for every search of operations research related resources in the Internet.

3 Content Providers

There are various content providers for operations research related knowledge. Apart from learning systems also some operations research related dictionaries are listed here because they provide some useful content in the form of (extended) definitions. One important difference of the systems is, if their content is provided online and can be viewed in the web browser, or if it has to be downloaded and installed on the own computer. For classroom teaching both kinds of systems have specific advantages. An argument for online content providing is that in general no additional software installation is necessary and the content can be updated easily. An argument against online contents is that there is a strong dependence on the provider at the time the content is needed, e.g. the availability of the hosting web site and the speed of the Internet connection are crucial. In addition downloaded software can provide richer multimedia content due to higher data throughput capacity on the local computer compared to an Internet connection.

Since 1995 Sniedovich has developed the "World-Wide-Web for Operations Research and Management Science - WORMS" (http://www.worms.ms.unimelb.edu.au). The content consists of summaries of the newsgroup "sci.op-research", a virtual encyclopedia, a list of dissertations, and - last but not least - a collection of jokes about operations research. It is planned to integrate most of the content in the IFORS web site.

The most comprehensive learning system in German that is free of charge is ORWelt (http://dsor.uni-paderborn.de/de/forschung/wbs/orwelt) developed at the department of Suhl. After the download and installation of the system, there are numerous modules in the areas "Operations Research", "Mathematical Optimization", "Algorithms for Graphs", "Simulation", and "Case Studies". The modules can be examined in the suggested order or can be explored by direct selection from the overview map. The content of each module normally consists of an introduction to the problem, a mathematical formulation, a solution algorithm, and a multiple-choice test.

The system "tutOR" (http://www.tutor.ms.unimelb.edu.au) of the operations research group of the University of Melbourne is focused on the mathematical foundations of operations research. This project was organized by Sniedovich who also directs the project "TutORial" (http://www.ifors.ms.unimelb.edu.au/tutorial) of the IFORS. Both "tutOR" and "tutORial" are interactive, web-based tutorials. They are implemented in HTML/DHTML and JavaScript. Currently their content is almost the same and consists of modules about "linear algebra", "linear programming", "integer programming", "dynamic programming", "simulation", "fun", and "tools". In general the content of each module consists of a description of the problem and an algorithm that iteratively generates a solution.

The British "Operational Research Society" has two interesting resources. Firstly, Pidd has provided the tutorial "Handling Strategic Problems" (http://www.orsoc.org.uk/about/teaching/StrategicProblems/index.htm). This tutorial shows various "soft OR" approaches based on two case studies. This tutorial can be studied online, but also be completely downloaded. Secondly Beasley has compiled his "OR Notes" about "hard OR" (http://mscmga.ms.ic.ac.uk/jeb/or/contents.html). It consists of his extensive lecture notes and some examination papers.

Since 1996 Greenberg offers a "mathematical programming glossary" (http://carbon.cudenver.edu/~hgreenbe/ glossary) and is still extending it. Additionally to the glossary entries there are also some articles to specific topics and some "tours", which link related entries.

On the interface of computer science and mathematics there is the "Free On-line Dictionary of Computing" (http://foldoc.doc.ic.ac.uk/foldoc) of the Imperial College in London. This system offers the possibility of an automatic translation to German, French, Italian, Portuguese, and Spanish – a useful feature even if sometimes amusing new definitions are created (see for example the German translation of "Internet"). This dictionary is mirrored from many web sites and is therefore highly available on the Internet. The text can also be downloaded and examined with a standard text editor.

4 Application Providers

Essential elements of operations research are algorithms and their implementation in software systems. Especially for teaching two types of software are interesting. Firstly, there are programs that visualize an algorithm step by step in order to improve the understanding how the algorithm works. Secondly, one is of course often also interested in the solution to a problem, which can be computed by a software system. Here some providers for both types of software are shown.

At the department of Domschke in the project "Tutorial Environment for Operations Research – TENOR" (http://www.bwl.tu-darmstadt.de/bwl3/forsch/projekte/tenor) there were downloadable programs implemented in German for branch and bound, graph theory, linear optimization, queuing theory and simulation, tabu search and heuristics for the traveling salesman problem.

At the department of Neumann and organized by Trautmann in the system "Produktions und Operations-Management – POM" (http://www.wior.uni-karlsruhe.de/neumann/POM_Master/APS-Matrix/pom.html) there were various modules implemented for supply chain management tasks. Each module consists of an algorithm implemented as a Java applet and an explanation in German. While using the applet is free of charge, the department itself can use the additional feature to test the understanding of its students by so-called "mini-examinations". The student has to answer some questions about the method and enter some data in the applet. In order to prevent the students from just copying the entries from each other, the data of the questions posed is varied randomly and there is no online confirmation if the given answer was correct or not.

In the initiative "Multimedia in der Lehre – MeiLe" (http://www.fh-regensburg.de/meile) Bavaria has supported the development of new media for teaching. Especially for operations research there is the project "Entwicklung multimedialer interaktiver elektronischer Lernprogramme – Emiel" (http://www.fh-augsburg.de/informatik/projekte/_emiel) of Lutz and Meixner. Various modules for teaching and visualizations were developed and can be downloaded. In contrast the modules for the new project will be available online, but only for students of the "Virtuelle Hochschule Bayern" (http://www.vhb.org).

Apart from knowing how an algorithm works of course also often the solution of a specific problem is of interest. In 1994 the Optimization Technology Center of the Northwestern University made the "Network-Enabled Optimization System –

NEOS" (http://www.ece.nwu.edu/OTC) available. The "NEOS Guide" provides case studies showing the mathematical formulation of optimization problems and an extensive glossary with an overview based on a "tree of optimization". The "NEOS Server" makes several optimization engines available online. These engines can handle various model types (e.g. integer/non-integer, linear/non-linear) and different data entry formats (e.g. AMPL, GAMS, MPS, LP). The model, the data, and the commands to the engine can be entered on a web form and the result is returned after a little while. For documentation purposes the result can also be sent via e-mail. As an alternative to using the web browser one can also download a special Java program that handles the data exchange with the server.

5 Conclusion

Today on the Internet a lot of educational resources are available. When designing a course, which makes use of these resources, a lecturer has the task to identify suitable high quality resources and make a pre-selection for his students. Due to the fact that there also exists a huge number of low quality documents, an operations research specific information portal with annotated links would be useful to speed up the search process for lecturers. Ideally, a referee process should take place and comment the quality of the documents listed on this portal. In the long run a special didactic for operations research including the possibility of e-learning using the Internet would be of interest for higher education in this field.

The author apologizes to all authors of valuable Internet resources, which are not mentioned here, and would highly appreciate to get notified of this fact. If you should be interested in the currently compiled list of web links with (my personal) comments please send an e-mail to get the HTML-file for direct use in your web browser.

Appendix: Web Links

Information Portals

"Educational Resource Initiative" of the IFORS http://www.umsl.edu/~sauter/ifors Links to topics like "ORMS Cases", "Materials for Teaching ORMS", "ORMS Tutorials", and "ORMS Software".

INFORM-ED http://www2.hpu.edu/jkros Articles about teaching OR under "readings" and various "teaching materials" of the categories "textbooks", "cases", "software", and "games".
British Operational Research Society http://www.orsoc.org.uk Teaching materials under "About OR".
Software Library for Operations Research http://www.wior.uni-karlsruhe.de/Bibliothek Extensive list of OR-related commercial and non-commercial software.
Michael Trick's Operations Research Page http//mat.gsia.cmu.edu Links to teaching materials can be found under "computer programs" and "courses".
Operations Research and Mathematical Optimization http://www.uni-koeln.de/themen/or Links especially for OR-related software.

Content Providers

World-Wide-Web for Operations Research and Management Science – WORMS http://www.worms.ms.unimelb.edu.au Summaries of the newsgroup "sci.op-research", a virtual encyclopedia, an (outdated) list of Ph.D. thesis, and OR-related jokes.
ORWelt http://dsor.uni-paderborn.de/de/forschung/wbs/orwelt Downloadable extensive German e-learning system with many modules of the categories "Introduction to OR", "Mathematical Optimization", "Graph-based Algorithms", "Simulation", and "Case Studies".
tutOR http://www.tutor.ms.unimelb.edu.au Online learning system on the topics "linear algebra", "linear programming", "integer programming", "dynamic programming", "fun", and "tools". Algorithms are implemented in JavaScript.
tutorial http://www.ifors.ms.unimelb.edu.au/tutorial Successor of tutOR (see above) with currently similar content.
Handling Strategic Problems http://www.orsoc.org.uk/about/teaching/StrategicProblems/index.htm Two case studies from Pidd about "soft OR" which can be viewed online, but also be downloaded.
OR Notes http://mscmga.ms.ic.ac.uk/jeb/or/contents.html Extensive lecture notes from Beasley about "hard OR".

Mathematical Programming Glossary http://carbon.cudenver.edu/~hgreenbe/glossary Explanation of numerous keywords from mathematical programming and some tours which link related entries of the glossary.
Free On-line Dictionary of Computing http://foldoc.doc.ic.ac.uk/foldoc Explanation of numerous keywords from information science and mathematics and the possibility of an automatic translation in several languages.

Application Providers

Tutorial Environment for Operations Research – TENOR http://www.bwl.tu-darmstadt.de/bwl3/forsch/projekte/tenor Downloadable algorithms and visualizations in German for branch and bound, graph theory, linear optimization, queuing theory, simulation, tabu search, and heuristics for the traveling salesman problem.
Produktions- und Operations-Management – POM http://www.wior.uni-karlsruhe.de/neumann/POM_Master/APS-Matrix/pom.html Online Java applets for several methods of production and operations management in the context of supply chain management.
Entwicklung multimedialer interaktiver elektronischer Lernprogramme – Emiel http://www.fh-augsburg.de/informatik/projekte/emiel Many downloadable algorithms and visualizations for example for the traveling salesman problem, the simplex algorithm, branch and bound, simulated annealing, and threshold accepting.
Network-Enabled Optimization System – NEOS Guide http://www-fp.mcs.anl.gov/otc/guide Online learning system about mathematical programming and optimization systems.
Network-Enabled Optimization System – NEOS Server http://www-neos.mcs.anl.gov Online optimization system for various model types (e.g. integer/non-integer, linear/non-linear). Problems can be formulated in several data formats and submitted via web browser or a special downloadable Java tool.

Internet–based Exercises and Mini–exams for Production and Operations Management

Norbert Trautmann

Universität Karlsruhe, Institut für Wirtschaftstheorie und Operations Research, D-76128 Karlsruhe, Germany
e–mail: trautmann@wior.uni-karlsruhe.de

Abstract. The purpose of the internet–based teachware mySCM is that students of economics, informatics and industrial engineering get familiar with quantitative methods for supply chain management. Input–output–relationships of various optimization methods can be detected by sampling input values, parameters, and alternative methods for the same problem. Students can gain extra benefits by passing so–called mini–exams that motivate active learning. mySCM can be used for free, round–the–clock, and any place where access to the Internet is available.

1 Introduction

Several teachware systems for *operations research* are documented in literature, cf. e.g. [3], [6] or [7]. In this paper we present the teachware mySCM for *supply chain management.* The purpose of this internet–based training system is that students get familiar with quantitative methods for supply chain management, especially for production and operations management. The system is an add–on to various lectures given for students of of economics, informatics and industrial engineering. It can be used for free, round–the–clock, and any place. Only access to the Internet and a standard web browser are required. mySCM is available at the website http://www.wior.uni–karlsruhe.de/neumann (section Lehre / Didactics).

mySCM allows students to practise the methods that have been discussed in the classroom and to check their homework. Input–output–relationships of the optimization methods can be detected by sampling input values, parameters, and alternative methods for the same problem. The iterations of the algorithms can be traced step–by–step. For each topic, students can gain extra benefits by passing a so–called mini–exam. During the test, the mySCM algorithms have to be applied.

2 Web–Based Training

The individual modules of mySCM can be accessed via a so–called supply chain planning matrix (cf. figure 1, [5]). This matrix shows how to decompose supply chain management into individual modules with a two–dimensional view of planning. The first dimension is the supply chain section involved, where we differ procurement, production, distribution, and sales. The second dimension is the planning level, differentiated into strategic, tactical, and operational planning.

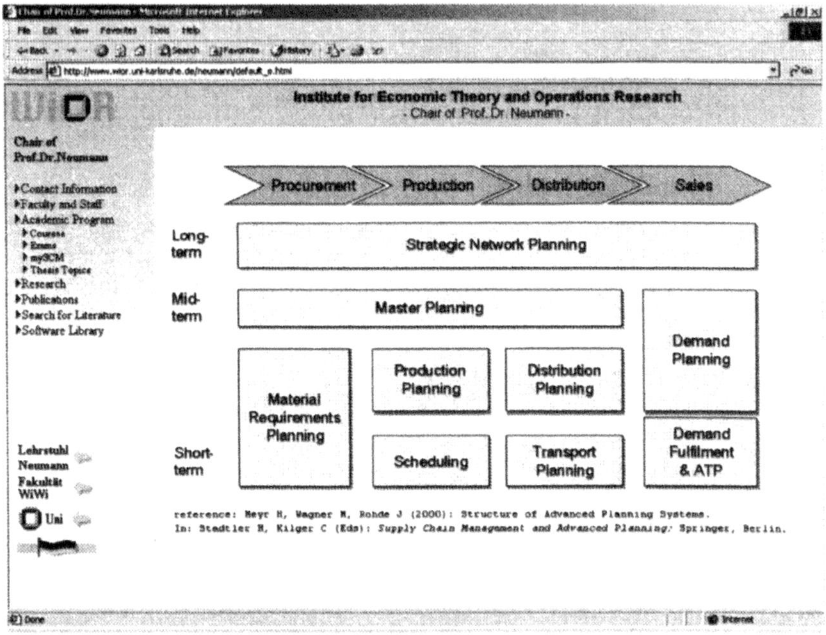

Fig. 1. Supply Chain Planning Matrix

In the current version (August 2001), the following methods are implemented in `mySCM`:

- **Strategic Network Planning:** plant location, warehouse location
- **Demand Planning:** demand forecasting via moving average, exponential smoothing, Holt's method, and Winter's method
- **Material Requirements Planning:** BOM explosion, EOQ, methods of Wagner–Within, Groff, Silver–Meal for deterministic demand, and method of Arris–Harrow–Marschak for uncertain demand
- **Production Planning:** RCPSP for make–to–order–production
- **Scheduling:** priority rule methods for one machine, method of Giffler–Thompson for several machines
- **Demand Fulfillment:** quality control cards

For a detailed description of these methods we refer to [4].

Point–and–click on a module of the matrix leads to a short description of the purpose of the module and the methods available (cf. Figure 2). In what follows, we present the training possibilities at hand of the demand planning module (cf. Figure 3).

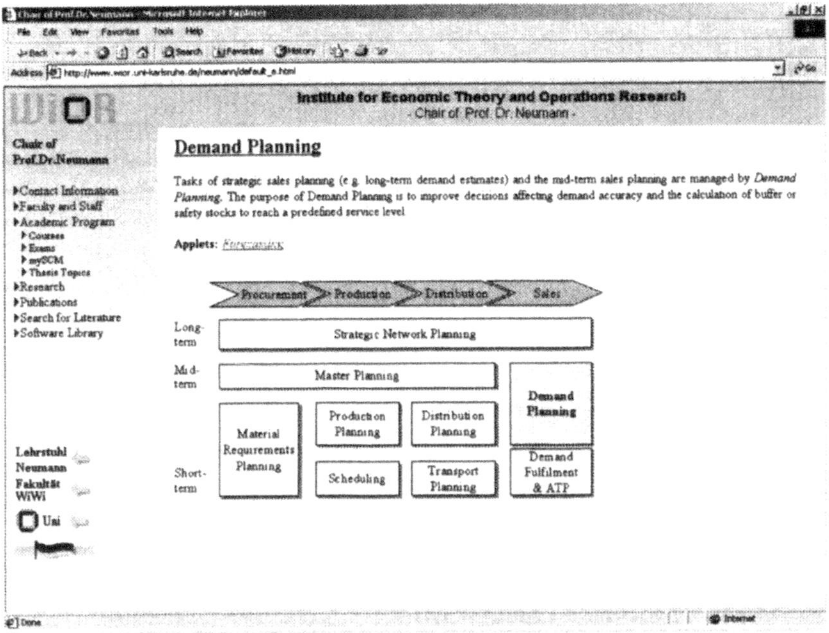

Fig. 2. Supply Chain Planning Module

The structure of the graphical user interface is the same for all methods: The upper half is reserved for the input, and the lower half for the output. The input area consists of input cells and drop down boxes. Additionally a method out of several alternatives as well as method-specific parameters can be chosen. In the output area, the computations of the algorithm are explained step-by-step. Additionally, input and output values are compiled in an easy-to-understand chart. For the demand forecast, the input is some past demand for goods. After choosing a method and parameters, a forecast for the future demand is computed.

By playing with parameters of the chosen method and/or selecting another method, the user detects relationships between input data, output data, and parameters. Depending on the type of past demand — e.g. a seasonal demand or a strongly increasing demand — some methods are more appropriate than others. Even if all methods are discussed during the lecture, for each method an online help in form of a HTML page is offered. Additionally a printed manual is available describing the complete system (cf. [2]).

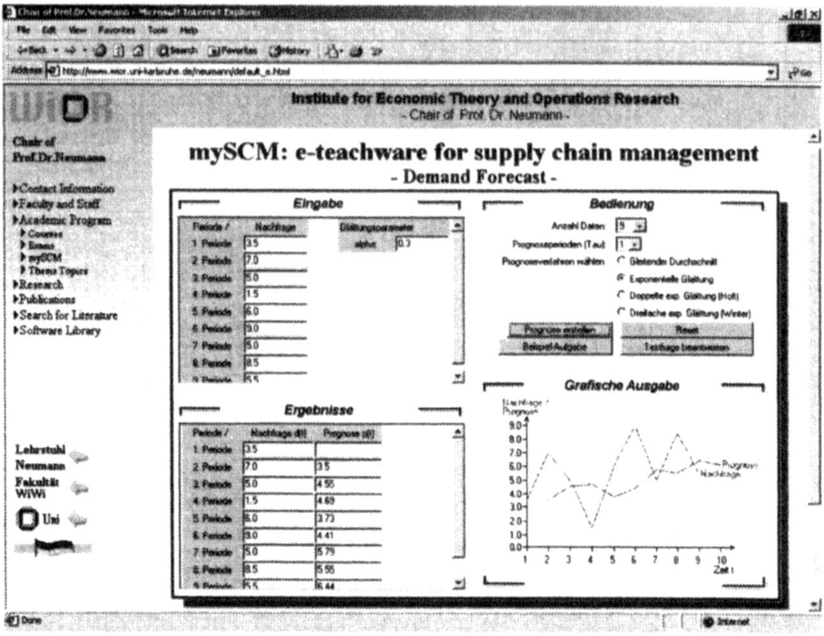

Fig. 3. Graphical user interface: demand forecast

3 Web–Based Mini–Exams

To motivate students to an active learning during the semester, they have the possibility to gain some extra points for the final exam by taking so–called mini–exams for each method of `mySCM`. This exam takes some minutes and has to be taken within two or three weeks after the topic has been discussed in the lecture. To do so, the student has to identify himself with his name and his student ID. If he is registered at the server of the chair, he immediately gets a randomized optimization problem that is not known before (cf. Figure 4). In the demand forecast topic, the server sends the student some sales data of the past twelve months and asks him to choose a method and appropriate parameters to make a good demand forecast. The student now can play with the methods and parameters as described above. There is unlimited time for answering questions, but no way back, i.e. the exam can only be taken once, and one of the multiple–choice answers has to be selected. The results of the mini–exam are published on the bulletin board after the deadline.

Note that for passing the mini–exam, it is necessary to use the applet and to have understood the underlying models and methods. Thus, it is not sufficient to learn facts by heart, but an active learning is motivated.

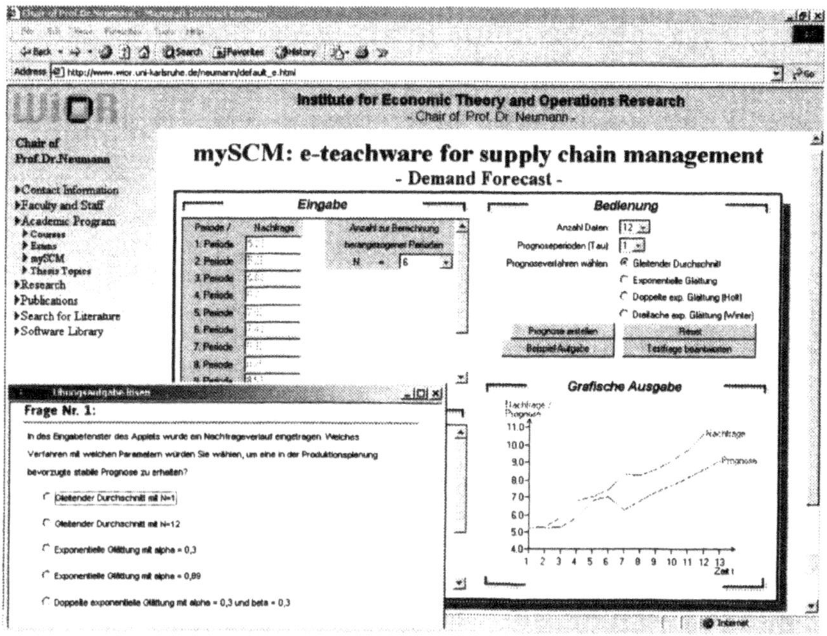

Fig. 4. Mini-exam

4 Use and Feedback

An installation of mySCM is not necessary. The system is completely web-based and coded in JAVA and PERL. Therefore mySCM can be used on Windows PC's, Mac's and Linux workstations. The user interface is the same for all topics and gives a look-and-feel atmosphere. Currently the language used is German. An English version will be available soon. The symbols used are international standard. The system is open for everyone and can be useful in every course about supply chain management and business optimisation.

Looking at the educational approach, the system is not designed to replace a lecture, but as an add-on to a lecture which helps and motivates students to work during the semester. They can check their own computations, and find errors by executing algorithms in a trace mode, that is step-by-step.

The system mySCM has been used in two courses at the University of Karlsruhe: *Production and Operations Management* during two semesters and *Decision Support Systems* during one semester. All together, about 70 students have gone for the mini-exams. Reactions via E-Mail have been entirely positive. Every student got some points, some even all. Every student with more than 50% correct answers in his mini-exams passed the final exam.

5 Conclusions and Outlook

In this paper we presented the internet–based teachware mySCM for quantitative supply chain management. For optimization problems of different supply chain planning levels and sections, students can play with input values, parameters, and alternative methods. Web–based mini–exams offer the possibility of gaining extra points for the final exam by applying the methods to some randomized optimization problems.

In future, it is planned to add further methods, e.g. for master planning and transportation planning. Further, the HTML–based online–help will be extended by some multimedia teaching material. Additionally, some web–based case studies (cf. [1]) will be available that train students in integrated, collaborative supply chain management, i.e., several modules have to be used to solve the cases.

References

1. Bell, P.C., Lanzenauer, C. Haehling von (2000): Teaching objectives: the value of using cases in teaching operational research. Journal of the Operational Research Society **51** (12), 1367–1377
2. Berning, G., Trautmann, N. (1999): Ein Internet–basiertes Lernsystem zum Produktions– und Operations–Management. Report WIOR 562, University of Karlsruhe.
3. Bossert, B., Scheubrein, R. (2001): Stand und Entwicklungstendenzen der Computerunterstützung in der Lehre. In: Fleischmann, B., Lasch, R., Derigs, U., Domschke, W., Rieder, U. (Eds.): Operations Research Proceedings 2000, Springer, Berlin
4. Neumann, K. (1997): Produktions– und Operations–Management. Springer, Berlin
5. Stadtler, H., Kilger, C. (2000): Supply Chain Management and Advanced Planning. Springer, Berlin
6. Suhl, L. (2000): Aktives Lernen mit hypermedialer Lernsoftware. Zeitschrift für Betriebswirtschaft, Ergänzungsheft **3**, 93–106
7. Werners, B. (2001): Internet–basierter Übungsbetrieb in Operations Research. In: Fleischmann, B., Lasch, R., Derigs, U., Domschke, W., Rieder, U. (Eds.): Operations Research Proceedings 2000, Springer, Berlin

Special Section

GOR Diploma Prize

Mathematische Optimierung zur Unterstützung kundenorientierter Disposition im Schienenverkehr

Natalia Kliewer

Decision Support & OR Laboratory, University of Paderborn,
Warburger Str. 100, D-33098 Paderborn, Germany,
Email: kliewer@upb.de, Tel.: 05251/603720

Zusammenfassung Der vorliegende Beitrag befaßt sich mit der Konzeption und Implementierung eines mathematischen Modells, das die Dispositionsaufgabe im fahrplangebundenen Schienenverkehr beschreibt. Im Rahmen der Arbeit werden Methoden der exakten mathematischen Optimierung auf unterschiedlich komplexe Instanzen des Dispositionsproblems angewandt, um zu im Sinne der Kundenorientierung optimalen Lösungen zu gelangen. Das Ergebnis der Optimierung ist keine einzelne Entscheidungsempfehlung, sondern ein Dispositionsplan für einen kompletten Entscheidungszeitraum. Ein weiterer Aspekt der Untersuchung ist die Frage nach der Anwendbarkeit der mathematischen Optimierung bei der Unterstützung des Disponenten im täglichen Bahnbetrieb. Außerdem kann das Optimierungssystem die Entscheidungen im Schulungsbetrieb bewerten sowie die Qualität von Heuristiken beurteilen. Durchgeführte Experimente zur Verifikation des Modells basieren auf realen Fahrplan- und Topologiedaten der Deutschen Bahn AG.

1 Kundenorientierte Disposition

Aus verschiedenen und von der Deutschen Bahn AG oft nicht beeinflußbaren Gründen treten im Bahnnetz Störungen auf. Der entstehende Schaden soll durch Minimierung der Auswirkungen von Störungen begrenzt werden. Dies wird durch die *dispositive Kontrolle* im Schienenverkehr angestrebt.

Die dispositive Kontrolle greift unter anderem dann ein, wenn ein Anschlußkonflikt durch die Anwendung globaler oder lokaler Wartezeitvorschriften nicht gelöst werden kann. Es ist schwierig festzustellen, welche Lösung für eine vorliegende Störung am besten geeignet ist. Noch schwieriger wird es, wenn mehrere Störungen an verschiedenen Stellen im Netz gleichzeitig auftreten. Das ist im täglichen Bahnbetrieb oft der Fall.

Als Reaktion auf Konflikte sollen Entscheidungen so getroffen werden, daß möglichst viele Anschlüsse erhalten bleiben und die verspäteten Züge ihre Verspätungen zumindest teilweise aufholen können. Bei der Konfliktlösung sollten sowohl kostenorientierte Kriterien als auch die Interessen der Passagiere berücksichtigt werden.

Die Kunden(un)zufriedenheit kann auf unterschiedliche Weise gemessen werden. Beispielsweise durch die Anzahl der Verspätungen oder der Pünktlichkeitsgrad der Bahn. Aussagekräftiger ist es, wenn auch die Auswirkungen von Verspätungen auf die betroffenen Reisenden berücksichtigt werden. Eine Aussage wie „40 Passagiere haben eine 20-minütige Reiseverzögerung" besitzt einen höheren Informationsgehalt als die Feststellung „Ein Zug kommt mit 20 Minuten Verspätung an". Aus diesem Grund sollte die Wartezeit der Passagiere bei Dispositionsentscheidungen mitberücksichtigt werden. Bedacht werden sollte, daß eine Passagierwarteminute einen unterschiedlichen Wert haben kann. Das Warten an einem kalten Bahnsteig wird im Vergleich zum Warten in einem komfortablen Zug subjektiv als länger empfunden. Eine Berücksichtigung derartiger Aspekte macht den ohnehin schwierigen Entscheidungsprozeß für die Dispositionsmitarbeiter noch komplexer.

Eines der Ziele der vorliegenden Arbeit ist es, das Dispositionsproblem möglichst genau zu modellieren und die Anwendbarkeit von mathematischen Optimierungsmethoden für die Lösung des Problems zu erforschen. Im Rahmen dieser Arbeit wurde ein Optimierungssystem prototypisch implementiert und auf realen Probleminstanzen getestet. Als Optimierungssoftware wurde die an der Freien Universität Berlin entwickelte Programmbibliothek MOPS (Mathematical OPtimization System) benutzt.

2 Problembeschreibung

Pro Werktag führt die DB AG zur Zeit ca. 38000 Fahrten zwischen 6200 Stationen aus. Somit ist das Netz der Deutschen Bahn eines der dichtesten und am stärksten beanspruchten weltweit. Der Bahnbetrieb wird täglich durch Störungen beeinflußt. Nach einem Sturm können die Bahngleise durch umgestürzte Bäume vorübergehend nicht befahrbar sein. Personalausfall, Schäden am Triebfahrzeug, Wagenstörungen, Weichen- oder Signalstörungen – all das bringt zeitliche (meist Verzögerungen) und/oder räumliche (Umleitungen) Abweichungen vom Fahrplan mit sich.

Im weiteren betrachten wir als Störungen die Zug- und Personalverspätungen jeglichen Ausmaßes oder den kompletten Ausfall von Ressourcen, d. h. von Fahrzeugen, Strecken, Bahnsteigen oder - im Extremfall - Bahnhöfen. Eine Verspätung, die keinen anderen Zug behindert, und durch die kein Reisender seinen Anschlußzug verpaßt, ist zwar eine Abweichung vom Plan, bedarf aber keiner regulierenden Maßnahme.

Einige durch Zugverspätungen verursachte Anschlußprobleme können durch globale Regelsysteme behandelt werden. Es handelt sich dabei um Regelwartezeiten für Anschlußverbindungen in Bahnhöfen, nach denen bei kleinen Verspätungen Züge aufeinander warten, um den umsteigenden Reisenden den Anschluß zu ermöglichen. Auch bei solchen Störungen müssen die Disponenten nicht eingreifen. Die Wartezeitregeln geben ein Zeitfenster vor, in dem ein Anschlußzug auf einen verspäteten Zubringer warten muß. Ein Anschlußkonflikt tritt auf, wenn eine Verspätung dieses Zeitfenster überschreitet. In diesem Fall muß der Disponent entscheiden, ob der Anschlußzug auf den verspäteten Zug warten soll. Wird für das Warten entschieden, bekommt der Abbringer eine Verspätung, die zu Folgekonflikten auf der weiteren Zugfahrt führen kann. Entscheidet der Disponent, daß der Abbringer abfahren darf, ohne zu warten, so verpassen die Passagiere des verspäteten Zuges ihren Anschluß und müssen länger als geplant auf eine passende Verbindung warten.

Im folgenden werden die bei der Modellierung getroffenen Vereinbarungen und Annahmen über Struktur und Verhalten des Systems Bahnbetrieb der DB AG vorgestellt und ein mathematisches Modell der Disposition als gemischt-ganzzahliges lineares Programm formuliert.

Annahmen Es wird angenommen, daß die Züge nach dem Fahrplan fahren, solange keine Störungen auftreten. Die minimalen Fahrzeiten zwischen den Bahnhöfen beziehungsweise die Möglichkeiten des Aufholens von Verspätungen sind vorgegeben. Die Aufenthaltszeiten in den Bahnhöfen stammen aus dem Fahrplan und können bei einer vorhandenen Verspätung verkürzt werden (MAZ - minimale Aufenthaltszeit). Nicht zu vernachlässigen ist die Zeit, die ein Reisender zum Aussteigen, den Übergang von einem Gleis zum anderen und das Einsteigen benötigt. Diese Zeit wird Mindestübergangszeit (MÜZ) genannt.

Als Dispositionskonflikte werden nur die Anschlußkonflikte betrachtet, deren Lösung in einer „Warten oder nicht Warten"-Entscheidung besteht. Diese Konflikte müssen vom Disponenten so gelöst werden, daß die kumulierte Wartezeit der Passagiere minimal ist. Die Informationen für dieses Modell werden dem Fahrplan und aktuellen Zuglaufmeldungen entnommen. Bei den Entscheidungen werden für jeden Anschluß die Anzahl der Übergangsreisenden und für jeden Zughalt die Anzahl der im Zug sitzenden Passagiere berücksichtigt.

Bewertungskriterien Die Qualität der Dispositionsentscheidungen wird hauptsächlich durch Akkumulation der Anzahl der Passagierwarteminuten charakterisiert. Dabei werden die Wartezeit am Gleis und die Wartezeit im Zug unterschiedlich bewertet. Außerdem muß berücksichtigt werden, daß die Unzufriedenheit der Reisenden überproportional mit steigender Wartezeit wächst.

Die Anzahl der „Warten"-Entscheidungen sollte begrenzt werden. Wenn dies nicht geschieht, könnte es dazu kommen, daß kein Zug mehr nach dem Fahrplan fährt. Diese Begrenzung ist besonders wichtig, wenn das Modell für die Bewertung der Disponentenentscheidungen verwendet werden soll. Kann ein Disponent maximal 20 Entscheidungen in einer Stunde treffen, so ist es fair, dem Optimierungssystem ebenfalls „nur" 20 Entscheidungen pro Stunde zu erlauben.

Die Gewichtung dieser Kriterien ist jedoch kein triviales Problem und kann nur von Verantwortlichen des Unternehmens gelöst werden.

3 Grundmodell: Mathematische Formulierung

Variablen Eine geordnete Menge der Ankunfts- und Abfahrtszeiten ($T_{i,j}^{An/Ab}$) der relevanten Züge im Entscheidungszeitraum stellt die Entscheidungsvariablen des Grundmodells dar.

Die binären Variablen $Y_{i,j,k}$ haben folgende Bedeutung: sei Zug k einer der Anschlußzüge des Zuges j im Bahnhof i, so bekommen die $in_{i,k}$ Übergangsreisenden ihren Anschluß genau dann, wenn $Y_{i,j,k} = 1$ ist. Hat $Y_{i,j,k}$ den Wert 0, dann ist der Anschlußzug abgefahren, bevor die Passagiere umsteigen konnten. Die $um_{i,j,k}$ Passagiere haben den Anschlußzug k verpaßt und müssen $takt_{i,k}$ Minuten auf den nächsten Zug in gewünschter Richtung warten. Parameter $müz_{i,j,k}$ bezeichnet die Zeit für den Übergang im Bahnhof i vom Gleis, an dem der Zubringerzug ankommt zu dem Gleis von dem der Abbringerzug abfährt.

Kostenfunktion Die gewichtete Summe der durch Verspätungen und ihre Auswirkungen verursachten Passagierwarteminuten ist zu minimieren. Es gibt zwei Wartezeitkategorien: „Warten im Zug“ (wird gewichtet mit α) und „Warten am Gleis“ (wird gewichtet mit β). Wenn man subjektives Empfinden der wartenden Passagiere berücksichtigt, sollte $\beta \geq \alpha$ gewählt werden.

$$K_1 = \sum_{IK} in_{i,k} * (T_{i,k}^{Ab} - t_{i,k}^{Ab}) * \alpha + \sum_{IJK} um_{i,j,k} * (1 - Y_{i,j,k}) * takt_{i,k} * \beta$$

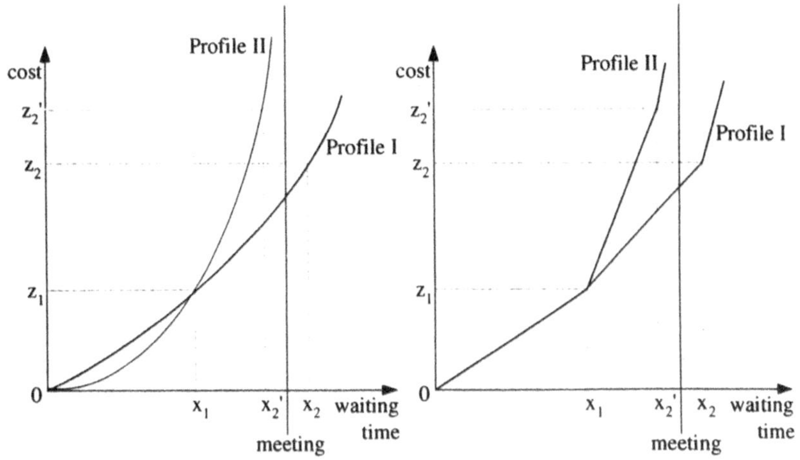

Abbildung1. Kosten des Wartens und ihre lineare Approximation

Das Modell wird um einen nichtlinearen Zielfunktionsverlauf erweitert (Abbildung 1): Die Zielfunktion ist nicht linear in dem Sinne, daß „3 warten

30 Minuten“ schlechter ist, als „30 warten 3 Minuten“. Die Kosten setzen sich nun aus der gewichteten Summe der Passagierwartezeiten aus den $L^1 + L^2$ verschiedenen Wartezeitkategorien zusammen:

$$K_3 = \sum_{IK}(in_{i,k} * \sum_{l^1=1..L^1} z_{l^1} * U_{l^1,i,k}) * \alpha + \sum_{IJK}(um_{i,j,k} * \sum_{l^2=1..L^2} w_{l^2} * V_{l^2,i,j,k}) * \beta$$

Die Variablen $U_{l^1,i,k}$ bzw. $V_{l^2,i,j,k}$ unterliegen den SOS2-Bedingungen (Special Ordered Set vom Typ 2).

Restriktionen Die Ungleichungen 1 beschränken die Ankunfts- und Abfahrtszeiten der Züge durch den gegebenen Fahrplan. Die Gleichungen 2 initialisieren die Ankunfts- oder Abfahrtszeiten aller verspäteten Züge in der jeweils aktuellen Station.

$$T_{i,k}^{An/Ab} \geq t_{i,k}^{An/Ab} \tag{1}$$

$$T_{akt_k,k}^{An/Ab} = \tau_{akt_k}^{An/Ab} \; \forall k \in versp\ddot{a}teteZ\ddot{u}ge \tag{2}$$

Hat ein Zug bei seiner Ankunft in einem Bahnhof eine Verspätung, so könnte diese dadurch vermindert werden, daß die Aufenthaltszeit des Zuges im Bahnhof im Vergleich zu der fahrplanmäßigen Aufenthaltszeit verkürzt wird. Dies ist durch die Ungleichungen 3 garantiert.

Ein Zug der Zuggattung zg kann auf der Strecke s $a_{zg,s}$ Minuten pro 100 Kilometer oder Minuten pro Stunde Soll-Fahrzeit aufholen. Die Fahrzeit zum nächsten Bahnhof wird in den Gleichungen 4 berücksichtigt.

$\forall k \in Z\ddot{u}ge, \forall i \in ZugStationen_k$:

$$T_{i,k}^{Ab} \geq T_{i,k}^{An} + maz_{i,k} \tag{3}$$

$$T_{i2_s,k}^{An} \geq T_{i1_s,k}^{Ab} + (t_{i2_s,k}^{An} - t_{i1_s,k}^{Ab}) * a_{k,s} \tag{4}$$

Anschlußrestriktionen:

$$T_{i,k}^{Ab} - T_{i,j}^{An} - m\ddot{u}z_{i,j,k} \geq M_1 * (Y_{i,j,k} - 1) \; \forall (i,j,k) \in IJK \tag{5}$$

$$T_{i,k}^{Ab} - T_{i,j}^{An} - m\ddot{u}z_{i,j,k} \leq M_2 * Y_{i,j,k} \; \forall (i,j,k) \in IJK \tag{6}$$

Die Wartezeitregelungen besagen folgendes: wenn der Abbringer durch das Warten auf den Zubringer eine Reiseverzögerung bekommen würde, die nicht länger als die vorgegebene Regelwartezeit $rwz_{i,j,k}$ ist, muß er auf diesen verspäteten Zubringer warten. Durch das Setzen des Parameters $rwzOn$ auf 0 können die Regelwartezeit-Restriktionen (7) ausgeschaltet werden.

$$\begin{aligned} rwzOn * (rwz_{i,j,k} - (T_{i,j}^{An} + m\ddot{u}z_{i,j,k} - t_{i,k}^{Ab})) \\ \leq M_3 * Y_{i,j,k} \quad \forall (i,j,k) \in IJK \end{aligned} \tag{7}$$

Modellerweiterungen Auf dem beschriebenen Grundmodell aufbauend werden komplexere Modelle formuliert, die das Dispositionsproblem realitätsnäher beschreiben:

- Berücksichtigung von Prognosedaten – Diese Erweiterung des Grundmodells ermöglicht eine Einbeziehung der Aussagen über die zukünftigen Ereignisse in das Optimierungssystem. Es können Prognosen über die Entwicklung vorhandener Verspätungen oder das Entstehen neuer Verspätungen sein, die von den Dispositionsmitarbeitern, von den Zugführern oder von einer speziell dafür bestimmten Software-Komponente stammen. Solche Prognosen können zusätzlich mit Informationen über ihre Verläßlichkeit versehen werden.
- Streckenbelegungsplanung und Kapazitäten auf Strecken – eine Streckenbelegungsplanung kann für die „wichtigen Strecken", zum Beispiel die mit einer hohen Verkehrsdichte vorgenommen werden. In Abhängigkeit von der Art der verfügbaren Informationen sind zwei Varianten im Modell vorgesehen.
- Begrenzung der Anzahl der Disponentenentscheidungen – Wenn gefordert wird, daß die Anzahl der getroffenen Entscheidungen für das Warten auf einen verspäteten Zubringer begrenzt werden soll, wird das mathematische Modell folgendermaßen erweitert: Man führt eine (weiche) obere Schranke für die Anzahl der von Disponenten geplanten Abfahrtsverzögerungen sowie Strafkosten für die Überschreitung dieser Schranke ein.

4 Ergebnisse

Zahlreiche Tests mit unterschiedlichen Fahrplanausschnitten und Parametereinstellungen zeigen, daß das Optimierungssystem große gemischt-ganzzahlige Optimierungsprobleme in kurzer Zeit lösen kann. Tabelle **??** enthält Ergebnisse einiger charakteristischen Läufe.

Das Optimierungssystem kann unter anderem die Suche nach besseren Wartezeitregeln unterstützen. Die Aufgabe der Wartezeitregelungen der DB AG besteht darin, dem Disponenten einen Teil der Entscheidungen abzunehmen und ihn dadurch zu entlasten. Sie beruhen auf der Annahme, daß es bei kleinen Zubringerverspätungen in meisten Fällen sinnvoller ist, den Abbringer warten zu lassen. Tabelle **??** enthält einige Beispiele dafür, wie die Berücksichtigung der Wartezeitregelungen sich auf die Passagierwartezeiten auswirkt. Für verschiedene Probleminstanzen wurden Testläufe mit und ohne Berücksichtigung der Regelwartezeiten durchgeführt. Bei einer Modellierungszeit unter einer Stunde wurde beim Weglassen der Regelwartezeit-Restriktionen eine Reduzierung der Passagierwartezeit um maximal 5% beobachtet. Ein Vergleich der Parametersätze 1.5.3 und 1.5.4 zeigt eine 15%-tige Kostenreduzierung durch das Weglassen der Regelwartezeit-Restriktionen. Die Anzahl der zu treffenden „Warten"-Entscheidungen wächst stark gegenüber der Vari-

Parametersatz	1.4.2	1.4.3	1.5.1	1.5.2	1.5.3	1.5.4
Zuggattungen	D;EC;IC; ICE;IR; RB;RE	D;EC;IC; ICE;IR; RB;RE	D;EC;IC; ICE;IR; RB;RE	D;EC;IC; ICE;IR; RB;RE	D;EC;IC; ICE;IR; RB;RE	D;EC;IC; ICE;IR; RB;RE
Modell.zeit(h)	1	1	4	4	4	4
Region	F	F	F	F	F	F
RWZ berücks.	ja	nein	ja	nein	ja	nein
Anz. Störungen	14	14	5	5	10	10
Restriktionen	5125	5125	21479	21479	21479	21479
Variablen	4211	4211	17034	17034	17034	17034
Nonzeros	11493	9007	48241	37675	48241	37675
LP Lös.zeit(sec)	7,47	6,87	32,02	29,71	30,16	27,91
IP Lös.zeit(sec)	46,36	27,07	145,5	135,17	134,01	168,34
B&B-Knoten	1263	1027	2087	1878	2976	4143
IP-Iterationen	369	184	178	182	206	177
Wartezeit(min)	165000	161700	131600	131600	137100	116900

Tabelle1. Rolle der Wartezeitregelungen

ante mit Wartezeitregelungen. Andererseits wurde kein höherer Zeitaufwand für die Lösung eines Modells ohne Regelwartezeit-Restriktionen beobachtet.

Diese Ergebnisse sprechen dafür, daß Wartezeitregelungen in der dispositiven Kontrolle sinnvoll sind, solange die Entscheidungen von einem Mitarbeiter ohne Rechnerunterstützung getroffen werden. Wenn die Dispositionsmitarbeiter eine Unterstützung von einem Optimierungssystem erhalten, können die Wartezeitregelungen weggelassen und, für jeden einzelnen Anschluß, Entscheidungen nach Empfehlungen des Optimierers getroffen werden.

5 Bewertung und Einsatzgebiete

Das vorgestellte mathematische Modell ermöglicht die Lösung der Dispositionsaufgabe unter Einbeziehung kundenorientierter Kriterien. Für die Realisierung des Optimierungssystems stehen mehrere alternative Möglichkeiten der Modellierung des Aufholens von Verspätungen und der Streckenbelegungsplanung zur Verfügung. Neben der Konzeption wurde das Modell prototypisch implementiert. Das System ist in der Problemgröße skalierbar: zum einen können unterschiedlich viele Aspekte der Realität bei der Modellierung erfaßt werden, und zum anderen kann die Modellierungstiefe durch die Auswahl des Netzausschnittes, der zu berücksichtigenden Zuggattungen und der Modellierungszeit gesteuert werden.

Die Einbeziehung der Optimierung in ein Disponentenunterstützungssystem scheint vielversprechend zu sein. Ein Dispositionsmitarbeiter wird in diesem Fall durch die Interaktion mit den Optimierungsalgorithmen in seinen Entscheidungen wesentlich unterstützt.

Einsatz in einem Entscheidungsunterstützungssystem für die Disposition Das Optimierungssystem kann als eine der Problemlösungskomponenten in die Gesamtarchitektur für die Unterstützung der Disposition integriert werden. Im realen Betrieb kann die Optimierungskomponente von der Dispatching-Assistent-Komponente als Reaktion auf eine aufgetretene Störung aufgerufen werden. Sie liefert Entscheidungsvorschläge für die Konfliktlösung, die von dem Dispositionsmitarbeiter angenommen oder verworfen werden können.

Einsatz in einem Disponentenschulungssystem Das Optimierungssystem kann zur Bewertung der Entscheidungen von Trainees eingesetzt werden. Eine adäquate Bewertung ist nur dann möglich, wenn die Kostenfunktion des Optimierungsproblems die Kriterien beinhaltet, mit denen auch die Auswirkungen der Entscheidungen im Simulator-Betrieb gemessen wurden. In diesem Fall liefert das Optimierungssystem eine obere Schranke für die Entscheidungsqualität.

Bewertung der Lösungsqualität von Entscheidungsheuristiken Da die im Dispositionsprozeß auftretenden Probleme eine hohe Komplexität aufweisen und die Zeit zur Entscheidungsfindung auf wenige Minuten begrenzt ist, kann eine exakte Optimierung im realen Betrieb nicht immer durchgeführt werden. Gute, aber nicht unbedingt optimale Lösungen können in annehmbarer Zeit von heuristischen Methoden geliefert werden. Bei der Entwicklung solcher Methoden bietet sich der Einsatz des Optimierungssystems zur Bewertung der Qualität der von den Heuristiken berechneten Lösungen an.

Der Prototyp liefert eine Grundlage für viele Experimente. Die bestehenden Wartezeitregelungen der Deutschen Bahn können mit alternativen Regelwartezeitmatrizen verglichen werden, um zu besseren Regelsystemen zu gelangen. Dabei können sowohl globale Regelwartezeit-Systeme als auch lokale Ausnahmen von den Wartezeitregelungen getestet werden. Das Optimierungssystem ermöglicht weiterhin das Testen der neuen Fahrpläne auf Störungssensibilität.

Literatur

1. C. Biederbick, J. Goecke
Simulation im Konfliktmanagement in großen Transportnetzwerken,
Operations Research Proceedings, Springer, 1999
2. U. Suhl
MOPS: A Mathematical OPtimization System,
in: European Journal of Operations Research, 72, S. 312-322, 1994
3. L. Suhl, T. Mellouli
Requirements for, and Design of, an Operations Control System for Railways,
in: N.H.M. Wilson (Ed.): Computer-Aided Transit Scheduling, S. 371-390, Berlin u.a.: Springer, 1999

Special Section

GOR Dissertation Prize

Anlagenbelegungsplanung in der Prozeßindustrie

Norbert Trautmann

Institut WiOR, Universität Karlsruhe, 76128 Karlsruhe, Germany

Zusammenfassung Bei der Anlagenbelegungsplanung in der Prozeßindustrie muß eine im Vergleich zur Situation in der Fertigungsindustrie erheblich schwierigere Optimierungsaufgabe gelöst werden. Restriktionen wie Mindestbestände und begrenzte Kapazitäten von Lagern für Zwischenprodukte, reihenfolgeabhängige Umrüstzeiten von Apparaten, Nachliege- und Verfallszeiten, variable Chargengrößen oder zyklische Materialflüsse bestimmen die Komplexität des Problems. Wir stellen einen Dekompositionsansatz zur Optimierung der Chargenproduktion auf Mehrzweckanlagen vor, der eine Belegungsplanung mit Hilfe von Modellen und Methoden zur ressourcenbeschränkten Projektplanung sowie Standardsoftware zur nichtlinearen Optimierung ermöglicht. Die in einer experimentellen Performanceanalyse nachgewiesene Planungsqualität bei kurzer Rechenzeit erlaubt die Einbindung der Modelle und Methoden in Advanced-Planning-Systeme zum Supply Chain Management wie SAP APO.

1 Einleitung

Seit einiger Zeit werden sogenannte Advanced-Planning-Systeme (APS) von verschiedenen Software-Herstellern angeboten. Diese unterstützen das integrierte und kapazitätsorientierte Supply Chain Management. Die modulare Aufteilung auf Basis der zwei Dimensionen *Supply Chain Sektion* und *Planungshorizont* erklären wir an Hand der in Abbildung 1 gezeigten Architekur des Systems SAP APO (vgl. [8], [10]). Die Supply Chain wird in die Sektionen *Buy* (Beschaffung), *Make* (Produktion), *Store* (Lagerung), *Move* (Transport) und *Sell* (Verkauf) unterteilt. Dem Zeithorizont der Planung entsprechend unterscheiden wir strategische, taktische, operative und sekundengenaue Planung. Hieraus ergeben sich die in Abbildung 1 ersichtlichen Module eines APS:

- *Supply Chain Cockpit:* Benutzerschnittstelle
- *Network Design:* Wahl von Standorten und Produktionstechnologien
- *Demand Planning:* Prognose künftiger Nachfragen
- *Supply Network Planning:* Perioden (Bucket)-orientierte Festlegung von Produktions- und Beschaffungsmengen je Standort sowie Transportmengen zwischen Standorten
- *Procurement Planning:* Planung der Fremdbeschaffung
- *Production Planning:* Unterteilung der Produktionsmengen je Standort und Bucket in Chargen

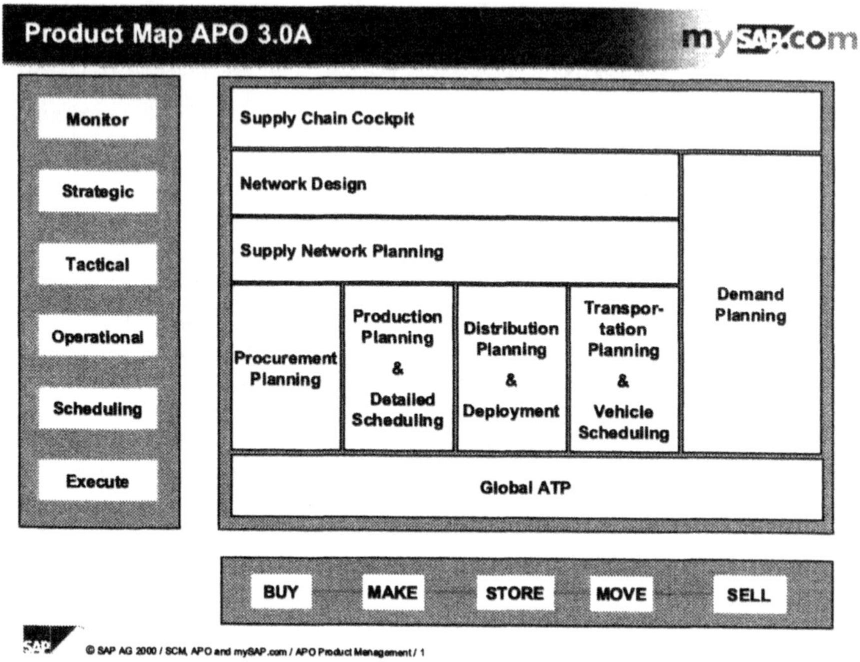

Abbildung 1. Architektur SAP APO

- *Detailed Scheduling:* Zeitliche Zuordnung der Betriebsmittel eines Standorts zur Herstellung der Chargen eines Buckets
- *Distribution Planning & Deployment:* Aufteilung der verfügbaren Produktmengen auf die Bedarfe sowie Planung der Distribution
- *Transportation Planning & Vehicle Scheduling:* Wahl der Transportmittel und -lose sowie Tourenplanung
- *Global ATP:* Globale Verfügbarkeitsprüfung

In [9] werden Modelle und Methoden für die Module Production Planning und Detailed Scheduling vorgestellt, die insbesondere die speziellen Anforderungen der Prozeßindustrie abdecken. Dabei werden im Chargenmodus betriebene Mehrzweckanlagen betrachtet. Die wichtigsten Anforderungen der Prozeßindustrie stellen wir im folgenden an Hand der Fallstudie von [11] (WK–Fall) vor.

Abbildung 2 (vgl. [2]) zeigt die Produktionsanlage des WK–Falls als State–Task–Network. Ein State–Task–Network besteht aus drei Elementen (vgl. [3]): State–Knoten, Task–Knoten und Pfeilen, die den Materialfluß angeben. Ein State entspricht einem Rohstoff, Zwischen- oder Endprodukt, für den bzw. das ein Anfangs-, ein Mindest- sowie ein Höchstbestand vorgegeben ist (Beispiel WK–Fall State 2: Mindestbestand 0ME, Höchstbestand 30ME, Anfangsbestand 20ME). Nicht lagerfähige Zwischenprodukte werden durch die Angabe von ns (non storable) für den entsprechenden State gekennzeich-

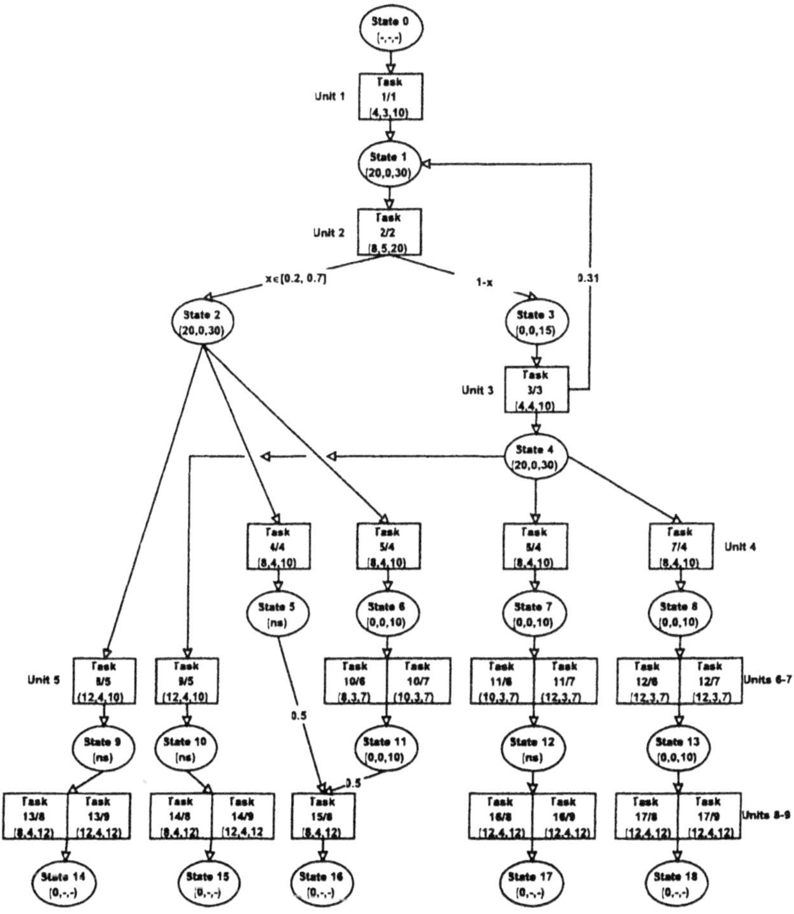

Abbildung2. State–Task–Network WK–Fall

net. Ein Task transformiert einen oder mehrere Eingangsstoffe, d.h. States, in einen oder mehrere Ausgangsstoffe (Beispiel WK–Fall Task 3: Transformation von 100% State 3 in 31% State 1 und 69% State 4, jeweils bezogen auf die Chargengröße). Jeder Transformationsprozeß nimmt einen Apparat in Anspruch. Einige Tasks können auf alternativen Apparaten ausgeführt werden, die sich in Prozeß- und Reinigungszeiten voneinander unterscheiden können (Beispiel WK–Fall Task 10: Ausführung auf Unit 6 mit Prozeßzeit 8ZE, alternativ Ausführung auf Unit 7 mit Prozeßzeit 10ZE). Die mit einer Durchführung des Tasks bearbeitete Materialmenge, d.h. die Chargengröße, kann innerhalb technologisch bedingter Grenzen gewählt werden (Beispiel WK–Fall Task 1: Mindestfüllstand des Apparats 3ME, Fassungsvermögen des Apparats 10ME). Die Prozeßzeit ist dabei unabhängig von der gewählten Chargengröße. Ebenso wie die Chargengröße ist der Anteil der Eingangs- bzw. Ausgangsstoffe bei einigen Tasks flexibel (Beispiel WK-Fall Task 2). Da die Anlage im Chargenmodus betrieben wird, werden die Eingangsstoffe eines Tasks zu Beginn als Ganzes in den benutzten Apparat gefüllt, und die Ausgangsstoffe zum Ende des Prozesses entnommen. Zwischen der Ausführung aufeinanderfolgender Tasks muß der benutzte Apparat gereinigt werden, wenn er zwischen den Tasks leersteht oder wenn der nachfolgende Task ein Produkt höherer Qualität herstellt als der vorhergehende. Im WK-Fall entspricht die Reinigungszeit der halben Prozeßzeit. Die Tasks sind nach aufsteigender Qualität numeriert. Gegeben sind die im Supply Network Planning ermittelten Mengen (Beispiel WK-Fall: $(30, 30, 40, 20, 40)$ für die Produkte $(14, \ldots, 18)$), die im Bucket herzustellen sind. Gesucht ist ein zulässiger Anlagenbelegungsplan, der diese Produktmengen in kürzester Zeit zur Verfügung stellt.

2 Production Planning

Aufgabe des Production Planning ist die Unterteilung der Produktionsmengen eines Buckets für einen Standort in Chargen. Diskrete Variablen sind die Anzahl der Ausführungen der Tasks. Kontinuierliche Variablen sind die Chargengrößen und die Anteile der Eingangs- bzw. Ausgangsprodukte an der Chargengröße der Tasks. Folgende Nebenbedingungen sind zu beachten:

- Die Chargengrößen und die Anteile der Eingangs- bzw. Ausgangsstoffe müssen innerhalb der vorgegebenen Grenzen gewählt werden.
- Die gesamte Menge an Eingangsstoffen eines Tasks wird in Ausgangsstoffe transformiert.
- Ist ein Produkt nicht lagerfähig, so muß für jede Kombination aus herstellendem und verbrauchendem Task die verbrauchte gleich der hergestellten Menge dieses Produkts sein.
- Die nach Abschluß aller Tasks verbleibende Menge zuzüglich des Anfangsbestandes eines Produkts muß die im Bucket nachgefragte Menge abdecken, darf aber nicht die Lagerkapazität überschreiten.

Ziel ist die Minimierung des Arbeitsinhaltes der Apparate im Planungszeitraum, d.h. der Summe der mit der durchschnittlichen Prozeßzeit gewichteten Anzahl der Ausführungen der Tasks.

In [9] wird dieses Problem als nichtlineares gemischt-ganzzahliges Problem formuliert. Mit Hilfe von Standardsoftware zur nichtlinearen Optimierung (Frontline Systems Premium Solver, vgl. [4]) wird auf einem Pentium 800MHz-PC innerhalb von 10 Sekunden eine zulässige Lösung für das Production Planning Problem des WK-Falls ermittelt. Alternativ wird in [6] eine Transformation auf ein lineares gemischt-ganzzahliges Problem vorgeschlagen, das in vergleichbarer Rechenzeit mit Hilfe von Standardsoftware wie z.B. CPLEX gelöst werden kann.

Ergebnis des Production Planning ist eine Menge von Chargen mit fixen Batchgrößen und Materialzu- bzw. -abgängen.

3 Detailed Scheduling

Aufgabe des Detailed Scheduling ist die zeitliche Zuordnung der Betriebsmittel zur Herstellung der im Production Planning ermittelten Chargen. Zeitkontinuierliche Variablen sind die Start- und Endzeitpunkte der Herstellung der Chargen. Binäre Variablen ergeben sich durch die Wahlmöglichkeiten der Betriebsmittel.

In [9] werden eine neue Modellierung von Detailed Scheduling Problemen der Prozeßindustrie sowie ein entsprechendes Lösungsverfahren entwickelt, die auf der ressourcenbeschränkten Projektplanung (vgl. [5]) basieren. Bei der Modellierung als Projekt entspricht die Bearbeitung einer Charge einer Aktivität. Als zu minimierende Zielfunktion wird in [9] eine gewichtete Summe aus Makespan und Terminüberschreitung betrachtet. Ein zulässiger Anlagenbelegungsplan muß folgende Restriktionen erfüllen:

- Zwischen den Start- und Endzeitpunkten der Bearbeitung verschiedener Chargen sind zeitliche Mindest- und Höchstabstände einzuhalten, die sich beispielsweise aus Nachliege- und Verfallszeiten, insbesondere aus nicht lagerfähigen Produkten ergeben.
- Start- und Endzeitpunkt der Bearbeitung der Chargen sind entsprechend den vorgegebenen Kalendern der Betriebsmittel zu wählen. Ein Beispiel für einen Kalender sind Produktionspausen, während derer gewisse Betriebsmittel nicht zur Verfügung stehen. Einige Tasks sind unterbrechbar, d.h., sie dürfen unmittelbar vor einer Pause unterbrochen werden und müssen unmittelbar nach der Pause fortgesetzt werden. Andere Tasks sind nicht unterbrechbar.
- Apparate werden als erneuerbare umzurüstende Ressourcen modelliert. Die zur Umrüstung erforderliche Zeit ist von der Reihenfolge der Tasks abhängig. Zudem ist ein Apparat vor einer Leerzeit zu reinigen, um unerwünschte Folgereaktionen zu vermeiden. Zu einem Zeitpunkt kann auf einem Apparat nur ein Task oder eine Reinigung bzw. Umrüstung stattfinden.
- Für die Lagerung der Rohstoffe, Zwischen- und Endprodukte stehen Lagervorrichtungen wie z.B. Silos oder Tanks zur Verfügung. Eine Lagervorrichtung wird als kumulative Ressource modelliert. Zu Beginn eines Tasks werden die benötigten Eingangsprodukte den Zwischenlagern entnommen, d.h. die Bestände der entsprechenden kumulativen Ressourcen verringert. Die Ausgangsprodukte werden zum Endzeitpunkt des Tasks in die Zwischenlager gefüllt, d.h. die Bestände der kumulativen Ressourcen erhöht. Der Bestand einer kumulativen Ressource ist von oben durch die Lagerkapazität und von unten durch den Sicherheitsbestand des gelagerten Produkts begrenzt.
- Mitarbeiter werden als erneuerbare Ressourcen modelliert. Ein Task beansprucht während der Ausführung entsprechend der benötigten Mitarbeiter Einheiten der erneuerbaren Ressourcen. Zu keinem Zeitpunkt

dürfen mehr Einheiten einer erneuerbaren Ressource beansprucht werden als verfügbar sind.

- Alternative Betriebsmittel werden durch alternative Modi der entsprechenden Aktivitäten des Projekts abgebildet. Die Modi unterscheiden sich also in ihrer Ressourceninanspruchnahme, Dauer und/oder ihren zeitlichen Mindest- und Höchstabständen. Jede Aktivität muß in genau einem ihrer alternativen Modi ausgeführt werden.

In [9] wird ein aus der Literatur bekannter Relaxationsansatz für Projektplanungsprobleme ohne kumulative Ressourcen, umzurüstende Ressourcen und Kalender (vgl. [1], [5]) zu einem Verfahren zur Lösung solcher Detailed Scheduling Probleme erweitert. Dem Verfahren liegt folgendes Generierungsschema zu Grunde (vgl. Abbildung 3):

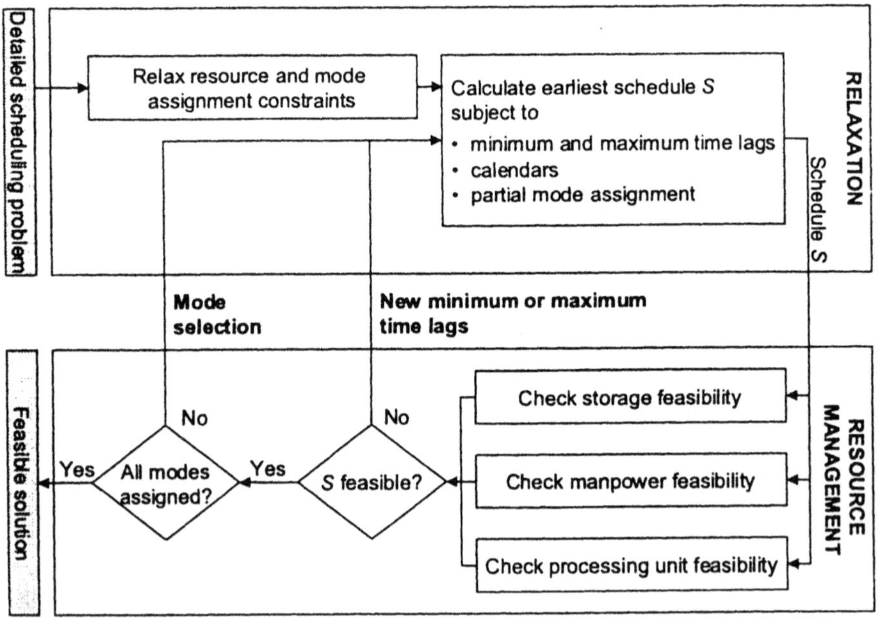

Abbildung3. Relaxationsansatz

1. Zu Beginn des Verfahrens werden die Nebenbedingungen, die das Problem schwer machen, relaxiert. Bei den oben beschriebenen Detailed Scheduling Problemen der Prozeßindustrie sind dies die beschränkte Verfügbarkeit von Lagerkapazität, Eingangsprodukten, Mitarbeitern und Apparaten. Im entsprechenden Projektplanungsproblem werden also die Ressourcenbeschränkungen und Moduswahlrestriktionen relaxiert.
2. Das verbleibende Problem der Zeitplanung, d.h. der Berechnung eines Anlagenbelegungsplans, der jeder Charge den frühestmöglichen Bearbeitungszeitraum zuweist, so daß die zeitlichen Mindest- und Höchstabstände

sowie die Kalenderrestriktionen erfüllt sind, ist mathematisch einfach zu lösen. Ein mögliches Lösungsverfahren (vgl. [9]) basiert auf der Berechnung längster Wege in (Projekt-)Netzwerken mit Hilfe eines Label-Correcting-Verfahrens. Dabei werden zusätzlich Start- und Endzeitpunkte der Aktivitäten entsprechend den Kalenderrestriktionen verzögert.

3. Der so berechnete Anlagenbelegungsplan wird auf Zulässigkeit überprüft. Sind alle Ressourcenbeschränkungen eingehalten und alle Modi gewählt, so ist der Plan zulässig. Andernfalls werden die Möglichkeiten enumeriert, die bestehenden Ressourcenkonflikte durch Einfügen zusätzlicher zeitlicher Mindest- und Höchstabstände zu lösen bzw. einer Aktivität einen Modus zuzuweisen. Anschließend wird mit Schritt 2 fortgefahren, d.h. erneut das Zeitplanungsproblem gelöst.

Um für Probleminstanzen praxisrelevanter Größe innerhalb einer sinnvollen Rechenzeit eine gute zulässige Lösung zu bestimmen, wurde die Implementierung als verkürztes Branch-and-Bound Verfahren durchgeführt, in dem nur ein Teil der Möglichkeiten enumeriert wird, Konflikte aufzulösen bzw. Modi zuzuweisen. Mit Hilfe eines solchen *Filtered-Beam-Search* Verfahrens läßt sich für das Detailed Scheduling Problem des WK-Falls auf einem Pentium 800MHz-PC innerhalb von 10 Sekunden eine gute zulässige Lösung bestimmen.

4 Zusammenfassung und Ausblick

In [9] wird ein Lösungsansatz zur Optimierung der Chargenproduktion auf Mehrzweckanlagen der Prozeßindustrie vorgestellt, der auf einer in Advanced-Planning-Systemen üblichen Dekomposition in die Schritte Production Planning und Detailed Scheduling basiert. Das Production Planning Problem wird als nichtlineares gemischt-ganzzahliges Problem modelliert, das mit Standardsoftware gelöst werden kann. Das Detailed Scheduling Problem wird als ressourcenbeschränktes Projektplanungsproblem modelliert. Zur Lösung wird ein Filtered-Beam-Search Verfahren entwickelt.

Die Leistungsfähigkeit des vorgestellten Ansatzes wird in [9] an Hand einer experimentellen Performance-Analyse mit aus der Literatur bekannten Verfahren zur Anlagenbelegungsplanung (vgl. [2]) verglichen. Mit dem neuen Verfahren werden in einem Bruchteil der von den bekannten Verfahren benötigten Rechenzeit zulässige Lösungen gleicher Qualität erzeugt. Der Grund für die Überlegenheit des neuen Ansatzes ist darin zu sehen, daß bei der Modellierung auf die Verwendung zeitindizierter Variablen verzichtet wurde und damit die Problemgröße unabhängig von der Aufteilung des Zeithorizontes ist.

In [9] werden Erweiterungen des Dekompositionsansatzes um die Berücksichtigung von weiteren Kalendern, allgemeinen Nachliege- und Verfallszeiten, verketteten Anlageneinheiten, planungsabhängigen Prozeßzeiten, zeitlich variierenden Betriebszuständen von Apparaten und Kampagnen sowie die Verwendung des Ansatzes für Ziele wie z.B. Minimierung der Umrüstkosten oder Rescheduling vorgestellt.

Für die Zukunft bietet sich die Übertragung des Ansatzes auf kontinuierlich arbeitende Anlagen an. Die ermittelten Anlagenbelegungspläne können durch eine Rückkopplung zwischen Detailed Scheduling und Production Planning verbessert werden, indem die Größe kritischer Chargen iterativ verändert wird.

Literatur

1. Bartusch, M., Möhring, R., Radermacher, F.J. (1988): Scheduling project networks with resource constraints and time windows. Annals of Operations Research **16**, 201–240
2. Blömer, F., Günther, H.O. (1998): Scheduling of a multi-product batch process in the chemical industry. Computers in Industry **36**, 245–259
3. Kondili, E., Pantelides, C.C., Sargent, R.W.H. (1993): A general algorithm for short-term scheduling of batch operations — I. MILP Formulation. Computers & Chemical Engineering **17**, 211–227
4. Lasdon, L.S., Waren, A.D., Jain, A., Ratner, M. (1978): Design and testing of a generalized reduced gradient code for nonlinear programming. Transactions on Mathematical Software **4**, 34–50
5. Neumann, K., Schwindt, C., Zimmermann, J. (2001): Project Scheduling with Time Windows and Scarce Resources. Lecture Notes in Economics and Mathematical Systems, Vol. 508. Springer, Berlin
6. Schwindt, C. (2001): Solution methods for production planning and detailed scheduling in process industries. Proceedings IEPM 2001, 271–280, Québec City, Canada
7. Schwindt, C., Trautmann, N. (2000): Batch scheduling in process industries. OR Spektrum **22**, 501–524
8. Stadtler, H., Kilger, C. (2000): Supply Chain Management and Advanced Planning. Springer, Berlin
9. Trautmann, N. (2001a): Anlagenbelegungsplanung in der Prozeßindustrie. Gabler, Wiesbaden
10. Trautmann, N. (2001b): Components of Advanced Planning Systems for Process Industries. Proceedings IEPM 2001, 253–262, Québec City, Canada
11. Westenberger, H., Kallrath, J. (1995): Formulation of a job shop problem in process industry. Bayer AG, Leverkusen and BASF AG, Ludwigshafen

List of Authors and Coauthors

Ahuja, André, Dipl.-Kfm. Dipl.-Math., FernUniversität in Hagen, Lehrstuhl für BWL, insb. Operations Research, Profilstr. 8, 58084 Hagen, Germany, phone: +49 23 31 / 9 87 26 39, email: andre.ahuja@fernuni-hagen.de.

Alicke, K., ICON Industrie Consulting GmbH, Hertzstr. 16, 76187 Karlsruhe, Germany.

Bachman, Aleksander, Dr., Wroclaw University of Technology, Institute of Engineering Cybernetics, Janiszewskiego 11/17, 50-372 Wroclaw, Poland, phone: +48 7 13 20 27 45, email: bachman@sprocket.ict.pwr.wroc.pl.

Barber, Federico, Univesidad Politecnica de Valencia, Sistemas Informaticos y computacion, Camino de Vera s/n, 46071 Valencia, Spain, phone: +34 9 63 87 70 00, email: fbarber@dsic.upv.es.

Beekmann, Frank, Dipl.-Math., Gerhard-Mercator-Universität Duisburg, Fachgebiet Wirtschaftsinformatik und Operations Research, Lotharstr. 65, 47057 Duisburg, Germany, phone: +49 2 03 / 3 79 - 34 84, email: beekmann@uni-duisburg.de.

Belov, Gleb, TU Dresden, Numerische Mathematik, Mommsenstr. 13, 01062 Dresden, Germany.

Bessler, Wolfgang, Prof. Dr., Justus-Liebig-Universität Gießen, Professur für Finanzierung und Banken, Licher Str. 74, 35394 Gießen, email: Wolfgang.Bessler@wirtschaft.uni-giessen.de.

Bot, Radu-Ioan, Dipl.-Math., TU Chemnitz, Fakultät für Mathematik , Reichenhainer Str. 41, 09107 Chemnitz, Germany, phone: +49 3 71 / 5 31 44 92, email: bot@mathematik.tu-chemnitz.de.

Breitner, Michael H., PD Dr., TU Clausthal, Fachbereich Mathematik und Informatik, Erzstr. 1, 38678 Clausthal-Zellerfeld, Germany, phone: +49 53 23 / 72 29 59, email: breitner@math.tu-clausthal.de.

Burmester, Tobias, SAP AG, Supply Chain Management, Neurottstr. 16, 69190 Walldorf, Germany, email: tobias.burmester@sap.com.

Caramia, Massimiliano, IAC - Istituto per le Applicazione del Calcolo, Viale del Policlinico 137, 00161 Roma, Italy, email: caramia@iac.rm.cnr.it.

Chrobok, Roland, Dipl.-Phys., Gerhard-Mercator-Universität Duisburg, Physics of Transport and Traffic, Lotharstr. 1, 47048 Duisburg, Germany, phone: +49 2 03 / 3 79 - 31 50, email: chrobok@traffic.uni-duisburg.de.

Derigs, Ulrich , Prof. Dr. Dr., Universität zu Köln, WINFORS, Pohligstr.1, 50969 Köln, Germany, phone: +49 2 21 / 4 70 53 27, email: derigs@winfors.uni-koeln.de.

Doerner, Karl, Universität Wien, Lehrstuhl für Betriebswirtschaftslehre; insb. Produktion und Logistik, Brünner Str. 72, 1210 Wien, Österreich, email: karl.doerner@univie.ac.at.

Eickemeier, Susanne, Dr., Johann Wolfgang Goethe-Universität, Institut für Statisitk und Mathematik, Mertonstr. 17-23, 60054 Frankfurt am Main, Germany, phone: +49 69 / 79 82 36 09, email: eickemeier@wiwi.uni-frankfurt.de.

Engelhardt-Funke, Ophelia, Dipl.-Math., TU-Clausthal, Institut für Mathematik, Erzstraße 1, 38678 Clausthal-Zellerfeld, Germany, phone: +49 53 23 / 72 24 19, email: engelhardt-funke@math.tu-clausthal.de.

Feige, Dieter, Prof. Dr.-Ing., Friedrich-Alexander-Universität Erlangen-Nürnberg, Lehrstuhl für Logistik, Theodorstr. 1, 90489 Nürnberg, Germany, phone: +49 9 11 / 5 88 79 - 16, email: feige_d@avk.fraunhofer.de.

Felden, Carsten, Dipl.-Kfm., Gerhard-Mercator-Universität Duisburg, Fachgebiet Wirtschaftsinformatik und Operations Research, Lotharstraße 65, 47057 Duisburg, Germany, phone: +49 2 03 / 3 79 - 26 30, email: felden@uni-duisburg.de.

Ferris, Michael C., Prof. Dr., University of Wisconsin, Computer Sciences Department, 1210 West Dayton Street, 53706 Madison, Wisconsin, USA, phone: +1 60 82 62 42 81, email: ferris@cs.wisc.edu.

Fleßa, Steffen, Prof. Dr., Evang. Fachhochschule Nürnberg, Fachbereich Pflegemanagement, Bärenschanzstr. 4, 90429 Nürnberg, Germany, phone: +49 9 11 / 27 25 38 86, email: Steffen.Flessa@evfh-nuernberg.de.

Florek, Stefan, Dipl.-Ing., RWE Trading GmbH, Business System Development - Knowledge Management, Huyssenallee 2, 45128 Essen, Germany, phone: +49 2 01 / 1 21 79 17, email: stefan.florek@rwetrading.com.

Fröhling, Magnus, Dipl.-Wi.-Ing., Universität Karlsruhe (TH), Institut für Industriebetriebslehre und Industrielle Produktion, Hertzstr. 16, 76187 Karlsruhe, Germany, phone: +49 7 21 / 6 08 44 00, email: magnus.froehling@wiwi.uni-karlsruhe.de.

Gawiejnowicz, Stanislaw, Dr., Adam Mickiewicz University, Faculty of Mathematics and Computer Science, Umultowska 87, 61-614 Poznan, Poland, phone: +48 61 82 73 - 147, email: stgawiej@amu.edu.pl.

Gelbrich, Katja, Dipl.-Kffr., Technische Fakultät Dresden, Lehrstuhl für Marketing, Mommsenstr. 13, 01062 Dresden, Germany, phone: +49 3 51 / 46 33 60 21, email: gelbrich@rcs.urz.tu-dresden.de.

Geldermann, Jutta, Dr., Universität Karlsruhe (TH), Institut für Industriebetriebslehre und industrielle Produktion , Hertzstr. 16, 76187 Karlsruhe, Germany, phone: +49 7 21 / 6 08 45 83, email: jutta.geldermann@wiwi.uni-karlsruhe.de.

Giret, Adriana, Univesidad Politecnica de Valencia, Sistemas Informaticos y computacion, Camino de Vera s/n, 46071 Valencia, Spain, phone: +34 9 63 87 70 00, email: agiret@dsic.upv.es.

Gold, Hermann, Dr., Infineon Technologies AG, CFE R PMT, Wernerwerkstr. 2, 93049 Regensburg, Germany, phone: +49 9 41 / 2 02 - 14 66, email: hermann.gold@infineon.com.

Goldstein, Gary A., International Resources Group Ltd., 1211 Connecticut Ave. NW, 20036 Washington, DC, USA, phone: +01 6 31 / 7 25 - 79 29, email: ggoldstein@irgltd.com.

Grothmann , Ralph, Universität Bremen, FB 7- Wiwi, Hochschulring 4, 28359 Bremen, Germany, email: agroth@uni-bremen.de.

Gündra, Hartmut, Geomer GmbH, Geoinformatik & Ressourcenmanagement, Redtenbacherstr. 5, 69126 Heidelberg, Germany.

Gutjahr, Walter J., Universität Wien, Institut für Statistik und Decision Support Systeme, Universitätsstr. 5/9, 1010 Wien, Österreich, email: walter.gutjahr@univie.ac.at.

Hamacher, Horst W., Prof. Dr. , Universität Kaiserslautern, Fachbereich Mathematik, Postfach 3049, 67653 Kaiserslautern, Germany, phone: +49 6 31 / 2 05 - 22 67 , email: hamacher@mathematik.uni-kl.de .

Hanne, Thomas, Dr., Fraunhofer Institut Techno- und Wirtschaftsmathematik, Gottlieb-Daimler-Str. 49, 67663 Kaiserslautern, Germany, phone: +49 6 31 / 2 05 44 70, email: hanne@itwm.fhg.de.

Hartl, Richard F., Prof., Universität Wien, Lehrstuhl für Betriebswirtschaftslehre; insb. Produktion und Logistik, Brünner Str. 72, 1210 Wien, Österreich, email: richard.hartl@univie.ac.at.

Hilbert, Andreas, Dr., Universität Augsburg, Lehrstuhl für Mathematische Methoden der Wirtschaftswissenschaften, 86135 Augsburg, Germany, email: andreas.hilbert@wiso.uni-augsburg.de.

Hildenbrandt, Regina, PD Dr., TU Ilmenau, Institut für Mathematik, Fachgebiet Operations Research und Stochastik, Postfach 10 05 65, 98684 Ilmenau, Germany, phone: +49 36 77 / 69 32 62, email: Regina.Hildenbrandt@Mathematik.TU-Ilmenau.DE.

Hubschneider, Hans, Dr.-Ing., PTV Planung Transport Verkehr AG, Stumpfstr. 1, 76131 Karlsruhe, Germany, phone: +49 7 21 / 9 65 15 05, email: Hans.Hubschneider@ptv.de.

Hüllermeier, Eyke, Dr., Universität Dortmund, FB Informatik; Automaten- und Schaltwerktheorie, 44221 Dortmund, Germany, phone: +49 2 31 / 7 55 63 72, email: eyke@ls1.cs.uni-dortmund.de.

Inderfurth, Karl, Prof. Dr., Universität Magdeburg, Fakultät für Wirtschaftswissenschaft; Produktion und Logistik, Postfach 4120, 39016 Magdeburg, Germany, phone: +49 3 91 / 6 71 87 98, email: inderfurth@ww.uni-magdeburg.de.

Italiano, Giuseppe, Università di Roma, Dip. Informatica - Sistemi e Produzione, via di Tor Vergata 110, 00133 Roma, Italy, email: italiano@disp.uniroma2.it.

Janiak, Adam, Prof. Dr., Wroclaw University of Technology, Institute of Engineering Cybernetics, Janiszewskiego 11/17, 50-372 Wroclaw, Poland, phone: +48 7 13 20 27 45, email: janiak@ict.pwr.wroc.pl.

Kalcsics, Jörg, Dipl.-Math., Fraunhofer Institut für Techno- und Wirtschaftsmathematik, Optimierung, Gottlieb-Daimler-Str. 49, 67663 Kaiserslautern, Germany, phone: +49 6 31 / 2 05 - 44 79, email: kalcsics@itwm.fhg.de.

Kempf, Alexander, Prof. Dr., Universität zu Köln, ABWL und Finanzierungslehre, Albertus-Magnus-Platz, 50923 Köln, Germany, email: kempf@wiso.uni-koeln.de.

Kleine, Andreas, Dr., Universität des Saarlandes, Rechts- und Wirtschaftswissenschaftliche Fakultät, Fachgebiet Unternehmensforschung, Postfach 151150, 66041 Saarbrücken, Germany, phone: +49 6 81 / 3 02 - 21 37, email: ak@wiwi.uni-sb.de.

Kliewer, Natalia, Universität Paderborn, Decision Support & OR Laboratory, Warburger Str. 100, 33098 Paderborn, Germany, phone: +49 52 51 / 60 37 20, email: kliewer@upb.de.

Kolonko, Michael, Prof. Dr., TU-Clausthal, Institut für Mathematik, Erzstraße 1, 38678 Clausthal-Zellerfeld, Germany, phone: +49 53 23 / 72 24 10, email: kolonko@math.tu-clausthal.de.

Kreuzberg, Klaus, Universität zu Köln, Seminar für ABWL und Finanzierungslehre, Albertus-Magnus-Platz, 50923 Köln, Germany, phone: +49 2 21 / 4 70 69 67, email: kreuzberg@wiso.uni-koeln.de.

Kurc, Wieslaw, Adam Mickiewicz University, Faculty of Mathematics and Computer Science, Matejki 48/49, 60-769 Poznan, Poland, email: wkurc@amu.edu.pl.

Liebchen, Christian, TU Berlin, Combinatorial Optimization and Graph Algorithms, Straße des 17. Juni 136, 10623 Berlin, Germany, phone: +49 30 / 31 42 57 39, email: liebchen@math.TU-Berlin.DE.

Marek, Marcin, M. Sc., Wroclaw University of Technology, Institute of Engineering Cybernetics, Janiszewskiego 11/17, 50-372 Wroclaw, Poland, phone: +48 7 13 20 27 45, email: jola@wroc.kn.pl.

Melo, Teresa, Dr., Fraunhofer Institut für Techno- und Wirtschaftsmathematik, Optimierung, Gottlieb-Daimler-Str. 49, 67663 Kaiserslautern, Germany, email: melo@itwm.fhg.de.

Memmel, Christoph, Universität zu Köln, Seminar für ABWL und Finanzierungslehre, Albertus-Magnus-Platz, 50923 Köln, Germany, phone: +49 2 21 / 4 70 69 68, email: memmel@wiso.uni-koeln.de.

Meyer, Matthias, Dr., University of Munich, Empirical Research and Quantitative Business Administration, Kaulbachstr. 45, 80539 München, Germany, email: matthias.meyer@bwl.uni-muenchen.de.

Mues, Christopher, Universität zu Köln, Zentrum für Angewandte Informatik, Weyertal 80, 50931 Köln, Germany, phone +49 2 21 / 4 70 60 50, email: Christopher.Mues@zpr.uni-koeln.de.

Müller, Stefan, Prof. Dr., Technische Fakultät Dresden, Lehrstuhl für Marketing, Mommsenstr. 13, 01062 Dresden, Germany, phone: +49 3 51 / 46 33 71 48, email: mls@rcs.urz.tu-dresden.de.

Neuneier, Ralph, Siemens AG, Otto Hahn Ring 6, 81730 Muenchen, Germany.

Nickel, Nils-Holger, Dipl.-Wirt.-Inf., Universität zu Köln, WINFORS, Pohligstr. 1, 50969 Köln, Germany, phone: +49 2 21 / 4 70 53 33, email: nickel@winfors.uni-koeln.de.

Nickel, Stefan, Dr., Fraunhofer Institut für Techno- und Wirtschaftsmathematik, Optimierung, Gottlieb-Daimler-Str. 49, 67663 Kaiserslautern, Germany, phone: +49 6 31 / 2 05 - 45 58, email: nickel@itwm.fhg.de.

Nikulin, Yury, Belarusian State university, Mechanics-mathematics department, Logoyskaya street, 220050 Minsk, phone: +17 2 09 50 47, email: nkln@mmf.bsu.unibel.by.

Nunge, Sandrine, Dr., DFIU - Deutsch-Französisches Institut für Umweltforschung, Hertzstr. 16, 76187 Karlsruhe, Germany, phone: +49 7 21 / 6 08 44 59, email: sandrine.nunge@wiwi.uni-karlsruhe.de.

Opfer, Heiko, Dipl.-Ök., Justus-Liebig-Universität Gießen, Professur für Finanzierung und Banken, Licher Str. 74, 35396 Gießen, email: Heiko.Opfer@wirtschaft.uni-giessen.de.

Oriolo, Gianpaolo, Università di Roma Tor Vergata, Dip. Informatica-Sistemi e Produzione, via di Tor Vergata 110, 00133 Roma, Italy, email: oriolo@disp.uniroma2.it.

Oswald, Marcus, Dipl.-Math., Ruprecht-Karls-Universität Heidelberg, Institut für Informatik, Im Neuenheimer Feld 368, 69120 Heidelberg, Germany, phone: +49 62 21 / 54 57 48, email: Marcus.Oswald@Informatik.Uni-Heidelberg.de.

Pacifici, Andrea, Università di Roma, Dip. Informatica - Sistemi e Produzione, via del Politecnico 1, 00133 Roma, Italy, phone: +39 67 25 97 36 0, email: pacifici@disp.uniroma2.it.

Pankowska, Lidia, Adam Mickiewicz University, Faculty of Mathematics and Computer Science, Matejki 48/49, 60-769 Poznan, Poland, email: lpankow@amu.edu.pl.

Paul, Henning, Dr., Universität Augsburg, Lehrstuhl für Mathematische Methoden der Wirtschaftswissenschaften, 86135 Augsburg, Germany, email: henning.paul@wiso.uni-augsburg.de.

Peeters, Leon, Rotterdam School of Management, Decision and Information Sciences, P.O. Box 1738, 3000 DR Rotterdam, The Netherlands.

Perugia, Alessandro, Centro Ricerche ENEA Casaccia, via Anguillarese 301, 00060 S. Maria di Galeria - Roma, Italy, phone: +39 63 04 83 69 5, email: perugia@casaccia.enea.it.

Pickl, Stefan W., Dr., Universität zu Köln, Mathematisches Institut, ZAIK, Weyertal 80, 50931 Köln, Germany, phone: +49 2 21 / 4 70 60 24, email: pickl@zpr.uni-koeln.de.

Pottmeier, Andreas, Dipl.-Phys., Gerhard-Mercator-Universität Duisburg, Physics of Transport and Traffic, Lotharstr. 1, 47048 Duisburg, Germany, phone: +49 2 03 / 3 79 - 44 35, email: pottmeier@traffic.uni-duisburg.de.

Reinelt, Gerhard, Ruprecht-Karls-Universität Heidelberg, Institut für Informatik, Im Neuenheimer Feld 368, 69120 Heidelberg, Germany, phone: +49 62 21 / 54 57 48, email: Gerhard.Reinelt@Informatik.Uni-Heidelberg.de.

Remme, Uwe, Universität Stuttgart, Institut für Energiewirtschaft und Rationelle Energieanwendung, Pfaffenwaldring 31, 70550 Stuttgart, Germany, phone: +49 7 11 / 7 80 61 53, email: ur@ier.uni-stuttgart.de.

Rentz, Otto, Prof. Dr., Universität Karlsruhe (TH), Institut für Industriebetriebslehre und Industrielle Produktion, Hertzstr. 16, 76187 Karlsruhe, Germany, phone: +49 7 21 / 6 08 44 60, email: otto.rentz@wiwi.uni-karlsruhe.de.

Reucher, Elmar, Dipl.-Math., FernUniversität in Hagen, LS für BWL, insb. Operatons Research, Profilstr. 8, 58084 Hagen, Germany, phone: +49 23 31 / 9 87 - 25 15, email: elmar.reucher@fernuni-hagen.de.

Rödder, Wilhelm, Prof. Dr., FernUniversität in Hagen, Lehrstuhl für BWL, insb. Operations Research, Profilstr. 8, 58084 Hagen, Germany, phone: +49 23 31 / 9 87 26 35, email: wilhelm.roedder@fernuni-hagen.de.

Rudolph, Andreas, Dr., IBM Deutschland, Wissenschaftliches Zentrum Heidelberg, Vangerowerstr. 18, 69115 Heidelberg, Germany, phone: +49 62 21 / 59 - 44 56, email: arudolph@de.ibm.com.

Salido, Miguel A., Univesidad Politecnica de Valencia, Sistemas Informaticos y computacion, Camino de Vera s/n, 46071 Valencia, Spain, phone: +34 6 07 87 05 80, email: msalido@dsic.upv.es.

Scheffran, Jürgen, Dr., PIK-Potsdam-Institut für Klimafolgenforschung, Dept. Global Change and Social System, Postfach 601203, 14412 Potsdam, Germany, phone: +49 3 31 / 2 88 25 28, email: scheffran@pik-potsdam.de.

Scheithauer, Guntram, Dr., TU Dresden, Numerische Mathematik, Mommsenstr. 13, 01062 Dresden, Germany, phone: +49 3 51 / 46 33 20 02, email: scheit@math.tu-dresden.de.

Schellmann, Ulrich, Universität Stuttgart, Institut für Energiewirtschaft und Rationelle Energieanwendung, Pfaffenwaldring 31, 70550 Stuttgart, Germany, email: ls@ier.uni-stuttgart.de.

Scheubrein, Ralph, Dr., Universität Hohenheim, Lehrstuhl 510A, Schloß Osthof-Nord, 70593 Stuttgart, Germany, email: scheubre@uni-hohenheim.de.

Schlenzig, Christoph, Dr.-Ing. Universität Stuttgart, Institut für Energiewirtschaft und Rationelle Energieanwendung, Pfaffenwaldring 31, 70550 Stuttgart, Germany, phone: +49 7 11 / 6 85 75 74, email: cs@ier.uni-stuttgart.de.

Schön, Cornelia, Universität Karlsruhe (TH), Institut für Wirtschaftstheorie und OR, 76128 Karlsruhe, Germany, phone: +49 7 21 / 6 08 61 94, email: schoen@wior.uni-karlsruhe.de.

Schreckenberg, Michael, Prof. Dr., Gerhard-Mercator-Universität Duisburg, Physics of Transport and Traffic, Lotharstr. 1, 47048 Duisburg, Germany, phone: +49 2 03 / 3 79 - 35 52, email: schreckenberg@uni-duisburg.de.

Schultmann, Frank, Dr., Universität Karlsruhe (TH), Institut für Industriebetriebslehre und Industrielle Produktion, Hertzstr. 16, 76187 Karlsruhe, Germany, phone: +49 7 21 / 6 08 44 58, email: frank.schultmann@wiwi.uni-karlsruhe.de.

Schwaiger, Manfred, Prof. Dr., University of Munich, Empirical Research and Quantitative Business Administration, Kaulbachstr. 45, 80539 München, Germany, email: schwaiger@bwl.uni-muenchen.de.

Schwindt, Christoph, Dr., Universität Karlsruhe (TH), Institut für Wirtschaftstheorie und Operations Research, Collegium am Schloss IV, 76128 Karlsruhe, Germany, phone: +49 7 21 / 6 08 32 26, email: schwindt@wior.uni-karlsruhe.de.

Seebregts, Ad J., Energy Research Centre of the Netherlands (ECN), Policy Studies unit, P.O. Box 1, 1755 ZG Petten, The Netherlands, phone: +31 2 24 56 40 90, email: seebregts@ecn.nl.

Smekens, Koen, Energy Research Centre of the Netherlands (ECN), Policy Studies unit, P.O. Box 1, 1755 ZG Petten, The Netherlands.

Spreiter, Quirin, FernUniversität in Hagen, Lehrstuhl für BWL, insb. Operations Research, Profilstr. 8, 58084 Hagen, Germany, email: quirin.spreiter@gmx.de.

Steland, Ansgar, Dr., Ruhr-Universität Bochum, Fakultät für Mathematik, Universitätsstr. 150, 44780 Bochum, Germany, email: ansgar.steland@ruhr-uni-bochum.de.

Strauß, Christine, Prof. Dr., Universität Wien, Institut für Betriebswirtschaftslehre; Lehrstuhl für Produktion und Logistik, Brünner Str. 72, 1210 Wien, Österreich, phone: +43 1 42 77 / 3 80 92, email: christine.strauss@univie.ac.at.

Stummer, Christian, Dr., Universität Wien, Institut für Betriebswirtschaftslehre, Brünner Str. 72, A-1210 Wien, Österreich, phone: +43 1 42 77 / 3 81 46, email: christian.stummer@univie.ac.at.

Sucky, Eric, Johann Wolfgang Goethe-Universität Frankfurt am Main, Seminar für Logistik & Verkehr, Mertonstr. 17, 60054 Frankfurt am Main, Germany, phone: +49 69 / 79 82 32 63, email: esucky@wiwi.uni-frankfurt.de.

Tenfelde-Podehl, Dagmar, Dipl.-Math. techn., Universität Kaiserslautern, Fachbereich Mathematik, Postfach 3049, 67653 Kaiserslautern, Germany, phone: +49 6 31 / 2 05 - 38 78, email: tenfelde@mathematik.uni-kl.de.

Theiler, Ursula, Dr., Risk Training, Carl-Zeiss-Str. 11, 83052 Bruckmühl, Germany, phone: +49 80 62 / 80 55 44, email: theiler@risk-training.de.

Trautmann, Norbert, Dr., Universität Karlsruhe (TH), Institut für Wirtschaftstheorie und Operations Research, Collegium am Schloss IV, 76128 Karlsruhe, Germany, phone: +49 7 21 / 6 08 42 81, email: trautmann@wior.uni-karlsruhe.de.

Voelker, Meta, University of Wisconsin, Computer Sciences Department, 1210 West Dayton Street, 53706 Madison, WI, USA, phone: +1 60 82 62 43 94, email: voelker@cs.wisc.edu.

Wahle, Joachim, Dipl.-Phys., , TraffGo GmbH, Grabenstr. 132, 47047 Duisburg, Germany, phone: +49 2 03 / 3 79 - 35 28, email: Wahle@traffgo.com.

Wajs, Wieslaw, University of Mining and Metallurgy, Department of Automatics, Mickiewicz Avenue 30, 30-059 Kraków, Poland.

Wanka, Gert, Prof. Dr., TU Chemnitz, Fakultät für Mathematik, Reichenhainer Str. 41, 09107 Chemnitz, Germany, phone: +49 3 71 / 5 31 - 85 60, email: g.wanka@mathematik.tu-chemnitz.de.

Weber, Gerhard W., PD Dr., TU Chemnitz, Fakultät für Mathematik, Reichenhainer Str. 41, 09107 Chemnitz, Germany, phone: +49 3 71 / 5 31 26 69, +49 2 21 / 4 70 60 47, email: weber@mathematik.tu-chemnitz.de, gwweber@zpr.uni-koeln.de.

Woeginger, Gerhard J., Prof., University of Twente, Department of Mathematics, P.O. Box 217, 7500 AE Enschede, The Netherlands, phone: +31 53 / 4 89 34 62, email: g.j.woeginger@math.utwente.nl.

Zimmermann, Hans-Georg, Dr., Siemens AG, CT IC 4, Otto Hahn Ring 6, 81730 Muenchen, Germany, phone: +49 89 / 6 36 - 4 44 95, email: Georg.Zimmermann@mchp.siemens.de.